T0180404

Lecture Notes in Computer Science 12916

More information about this subseries at http://www.springer.com/series/7412

Elisa H. Barney Smith ·
Umapada Pal (Eds.)

Document Analysis and Recognition – ICDAR 2021 Workshops

Lausanne, Switzerland, September 5–10, 2021
Proceedings, Part I

Springer

Editors
Elisa H. Barney Smith
Boise State University
Boise, ID, USA

Umapada Pal 🆔
Indian Statistical Institute
Kolkata, India

ISSN 0302-9743   ISSN 1611-3349 (electronic)
Lecture Notes in Computer Science
ISBN 978-3-030-86197-1   ISBN 978-3-030-86198-8   (eBook)
https://doi.org/10.1007/978-3-030-86198-8

LNCS Sublibrary: SL6 – Image Processing, Computer Vision, Pattern Recognition, and Graphics

This Springer imprint is published by the registered company Springer Nature Switzerland AG
The registered company address is: Gewerbestrasse 11, 6330 Cham, Switzerland

Foreword

Our warmest welcome to the proceedings of ICDAR 2021, the 16th IAPR International Conference on Document Analysis and Recognition, which was held in Switzerland for the first time. Organizing an international conference of significant size during the COVID-19 pandemic, with the goal of welcoming at least some of the participants physically, is similar to navigating a rowboat across the ocean during a storm. Fortunately, we were able to work together with partners who have shown a tremendous amount of flexibility and patience including, in particular, our local partners, namely the Beaulieu convention center in Lausanne, EPFL, and Lausanne Tourisme, and also the international ICDAR advisory board and IAPR-TC 10/11 leadership teams who have supported us not only with excellent advice but also financially, encouraging us to setup a hybrid format for the conference.

We were not a hundred percent sure if we would see each other in Lausanne but we remained confident, together with almost half of the attendees who registered for on-site participation. We relied on the hybridization support of a motivated team from the Lule University of Technology during the pre-conference, and professional support from Imavox during the main conference, to ensure a smooth connection between the physical and the virtual world. Indeed, our welcome is extended especially to all our colleagues who were not able to travel to Switzerland this year. We hope you had an exciting virtual conference week, and look forward to seeing you in person again at another event of the active DAR community.

With ICDAR 2021, we stepped into the shoes of a longstanding conference series, which is the premier international event for scientists and practitioners involved in document analysis and recognition, a field of growing importance in the current age of digital transitions. The conference is endorsed by IAPR-TC 10/11 and celebrates its 30th anniversary this year with the 16th edition. The very first ICDAR conference was held in St. Malo, France in 1991, followed by Tsukuba, Japan (1993), Montreal, Canada (1995), Ulm, Germany (1997), Bangalore, India (1999), Seattle, USA (2001), Edinburgh, UK (2003), Seoul, South Korea (2005), Curitiba, Brazil (2007), Barcelona, Spain (2009), Beijing, China (2011), Washington DC, USA (2013), Nancy, France (2015), Kyoto, Japan (2017), and Syndey, Australia in 2019.

The attentive reader may have remarked that this list of cities includes several venues for the Olympic Games. This year the conference was hosted in Lausanne, which is the headquarters of the International Olympic Committee. Not unlike the athletes who were recently competing in Tokyo, Japan, the researchers profited from a healthy spirit of competition, aimed at advancing our knowledge on how a machine can understand written communication. Indeed, following the tradition from previous years, 13 scientific competitions were held in conjunction with ICDAR 2021 including, for the first time, three so-called "long-term" competitions, addressing wider challenges that may continue over the next few years.

Other highlights of the conference included the keynote talks given by Masaki Nakagawa, recipient of the IAPR/ICDAR Outstanding Achievements Award, and Mickaël Coustaty, recipient of the IAPR/ICDAR Young Investigator Award, as well as our distinguished keynote speakers Prem Natarajan, vice president at Amazon, who gave a talk on "OCR: A Journey through Advances in the Science, Engineering, and Productization of AI/ML", and Beta Megyesi, professor of computational linguistics at Uppsala University, who elaborated on "Cracking Ciphers with 'AI-in-the-loop': Transcription and Decryption in a Cross-Disciplinary Field".

A total of 340 publications were submitted to the main conference, which was held at the Beaulieu convention center during September 8–10, 2021. Based on the reviews, our Program Committee chairs accepted 40 papers for oral presentation and 142 papers for poster presentation. In addition, nine articles accepted for the ICDAR-IJDAR journal track were presented orally at the conference and a workshop was integrated in a poster session. Furthermore, 12 workshops, 2 tutorials, and the doctoral consortium were held during the pre-conference at EPFL during September 5–7, 2021, focusing on specific aspects of document analysis and recognition, such as graphics recognition, camera-based document analysis, and historical documents.

The conference would not have been possible without hundreds of hours of work done by volunteers in the organizing committee. First of all we would like to express our deepest gratitude to our Program Committee chairs, Joseph Lladós, Dan Lopresti, and Seiichi Uchida, who oversaw a comprehensive reviewing process and designed the intriguing technical program of the main conference. We are also very grateful for all the hours invested by the members of the Program Committee to deliver high-quality peer reviews. Furthermore, we would like to highlight the excellent contribution by our publication chairs, Liangrui Peng, Fouad Slimane, and Oussama Zayene, who nego-tiated a great online visibility of the conference proceedings with Springer and ensured flawless camera-ready versions of all publications. Many thanks also to our chairs and organizers of the workshops, competitions, tutorials, and the doctoral consortium for setting up such an inspiring environment around the main conference. Finally, we are thankful for the support we have received from the sponsorship chairs, from our valued sponsors, and from our local organization chairs, which together enabled us to put in the extra effort required for a hybrid conference setup.

Our main motivation for organizing ICDAR 2021 was to give practitioners in the DAR community a chance to showcase their research, both at this conference and its satellite events. Thank you to all the authors for submitting and presenting your out-standing work. We sincerely hope that you enjoyed the conference and the exchange with your colleagues, be it on-site or online.

September 2021

Andreas Fischer
Rolf Ingold
Marcus Liwicki

Preface

Our heartiest welcome to the proceedings of the ICDAR 2021 Workshops, which were organized under the 16th International Conference on Document Analysis and Recognition (ICDAR) held in Lausanne, Switzerland during September 5–10, 2021.

We are delighted that this conference was able to include 13 workshops. The workshops were held in Lausanne during September 5–7, 2021. Some were held in a hybrid live/online format and others were held entirely online, with space at the main conference for in-person participants to attend. The workshops received over 100 papers on diverse document analysis topics, and these volumes collect the edited papers from 12 of the workshops.

We sincerely thank the ICDAR general chairs for trusting us with the responsibility for the workshops, and for assisting us with the complicated logistics in order to include remote participants. We also want to thank the workshop organizers for their involvement in this event of primary importance in our field. Finally, we thank the workshop presenters and authors without whom the workshops would not exist.

September 2021

Elisa H. Barney Smith
Umapada Pal

Organization

Organizing Committee

General Chairs

Andreas Fischer	University of Applied Sciences and Arts Western Switzerland
Rolf Ingold	University of Fribourg, Switzerland
Marcus Liwicki	Luleå University of Technology, Sweden

Program Committee Chairs

Josep Lladós	Computer Vision Center, Spain
Daniel Lopresti	Lehigh University, USA
Seiichi Uchida	Kyushu University, Japan

Workshop Chairs

Elisa H. Barney Smith	Boise State University, USA
Umapada Pal	Indian Statistical Institute, India

Competition Chairs

Harold Mouchère	University of Nantes, France
Foteini Simistira	Luleå University of Technology, Sweden

Tutorial Chairs

Véronique Eglin	Institut National des Sciences Appliquées, France
Alicia Fornés	Computer Vision Center, Spain

Doctoral Consortium Chairs

Jean-Christophe Burie	La Rochelle University, France
Nibal Nayef	MyScript, France

Publication Chairs

Liangrui Peng   Tsinghua University, China
Fouad Slimane   University of Fribourg, Switzerland
Oussama Zayene   University of Applied Sciences and Arts Western
                         Switzerland, Switzerland

Sponsorship Chairs

David Doermann   University at Buffalo, USA
Koichi Kise   Osaka Prefecture University, Japan
Jean-Marc Ogier   University of La Rochelle, France

Local Organization Chairs

Jean Hennebert   University of Applied Sciences and Arts Western
                         Switzerland, Switzerland
Anna Scius-Bertrand   University of Applied Sciences and Arts Western
                         Switzerland, Switzerland
Sabine Süsstrunk   École Polytechnique Fédérale de Lausanne,
                         Switzerland

Industrial Liaison

Aurélie Lemaitre   University of Rennes, France

Social Media Manager

Linda Studer   University of Fribourg, Switzerland

Workshops Organizers

W01-Graphics Recognition (GREC)

Jean-Christophe Burie   La Rochelle University, France
Richard Zanibbi   Rochester Institute of Technology, USA
Motoi Iwata   Osaka Prefecture University, Japan
Pau Riba   Universitat Autnoma de Barcelona, Spain

W02-Camera-based Document Analysis and Recognition (CBDAR)

Sheraz Ahmed   DFKI, Kaiserslautern, Germany
Muhammad Muzzamil   La Rochelle University, France
  Luqman

W03-Arabic and Derived Script Analysis and Recognition (ASAR)

Adel M. Alimi	University of Sfax, Tunisia
Bidyut Barań Chaudhur	Indian Statistical Institute, Kolkata, India
Fadoua Drira	University of Sfax, Tunisia
Tarek M. Hamdani	University of Monastir, Tunisia
Amir Hussain	Edinburgh Napier University, UK
Imran Razzak	Deakin University, Australia

W04-Computational Document Forensics (IWCDF)

Nicolas Sidère	La Rochelle University, France
Imran Ahmed Siddiqi	Bahria University, Pakistan
Jean-Marc Ogier	La Rochelle University, France
Chawki Djeddi	Larbi Tebessi University, Algeria
Haikal El Abed	Technische Universitaet Braunschweig, Germany
Xunfeng Lin	Deakin University, Australia

W05-Machine Learning (WML)

Umapada Pal	Indian Statistical Institute, Kolkata, India
Yi Yang	University of Technology Sydney, Australia
Xiao-Jun Wu	Jiangnan University, China
Faisal Shafait	National University of Sciences and Technology, Pakistan
Jianwen Jin	South China University of Technology, China
Miguel A. Ferrer	University of Las Palmas de Gran Canaria, Spain

W06-Open Services and Tools for Document Analysis (OST)

Fouad Slimane	University of Fribourg, Switzerland
Oussama Zayene	University of Applied Sciences and Arts Western Switzerland, Switzerland
Lars Vögtlin	University of Fribourg, Switzerland
Paul Märgner	University of Fribourg, Switzerland
Ridha Ejbali	National School of Engineers Gabes, Tunisia

W07-Industrial Applications of Document Analysis and Recognition (WIADAR)

Elisa H. Barney Smith	Boise State University, USA
Vincent Poulain d'Andecy	Yooz, France
Hiroshi Tanaka	Fujitsu, Japan

W08-Computational Paleography (IWCP)

Isabelle Marthot-Santaniello	University of Basel, Switzerland
Hussein Mohammed	University of Hamburg, Germany

W09-Document Images and Language (DIL)

Andreas Dengel	DFKI and University of Kaiserslautern, Germany
Cheng-Lin Liu	Institute of Automation of Chinese Academy of Sciences, China
David Doermann	University of Buffalo, USA
Errui Ding	Baidu Inc., China
Hua Wu	Baidu Inc., China
Jingtuo Liu	Baidu Inc., China

W10-Graph Representation Learning for Scanned Document Analysis (GLESDO)

Rim Hantach	Engie, France
Rafika Boutalbi	Trinov, France, and University of Stuttgart, Germany
Philippe Calvez	Engie, France
Balsam Ajib	Trinov, France
Thibault Defourneau	Trinov, France

Contents – Part I

ICDAR 2021 Workshop on Camera-Based Document Analysis and Recognition (CBDAR)

ICDAR 2021 Workshop on Arabic and Derived Script Analysis and Recognition (ASAR)

Contents – Part II

ICDAR 2021 Workshop on Open Services and Tools for Document Analysis (OST)

ICDAR 2021 Workshop on Industrial Applications of Document Analysis and Recognition (WIADAR)

ICDAR 2021 Workshop on Computational Paleography (IWCP)

ICDAR 2021 Workshop on Document Images and Language (DIL)

ICDAR 2021 Workshop on Graphics Recognition (GREC)

GREC 2021 Preface

Graphics recognition is the subfield of document recognition dealing with graphic entities such as tables, charts, illustrations in figures, and notations (e.g., for music and mathematics). Graphics often help describe complex ideas much more effectively than text alone. As a result, recognizing graphics is useful for understanding the information content in documents, the intentions of document authors, and for identifying the domain of discourse in a document. Since the 1980s, researchers in the graphics recognition community have addressed the analysis and interpretation of graphical documents (e.g., electrical circuit diagrams, engineering drawings, etc.), handwritten and printed graphical elements/cues (e.g., logos, stamps, annotations, etc.), graphics-based information retrieval (comics, music scores, etc.) and sketches, to name just a few of the challenging topics in this area.

The GREC workshops provide an excellent opportunity for researchers and practitioners at all levels of experience to meet and share new ideas and knowledge about graphics recognition methods. The workshops enjoy strong participation from researchers in both industry and academia.

The aim of this workshop is to maintain a very high level of interaction and creative discussions between participants, maintaining a *workshop* spirit, and not being tempted by a 'mini-conference' model.

The 14th edition of the International Workshop on Graphic Recognition (GREC 2021) built on the success of the thirteen previous editions held at Penn State University (USA, 1995), Nancy (France, 1997), Jaipur (India, 1999), Kingston (Canada, 2001), Barcelona (Spain, 2003), Hong Kong (China, 2005), Curitiba (Brazil, 2007), La Rochelle (France, 2009), Seoul (South Korea, 2011), Lehigh (USA, 2013), Nancy (France, 2015), Kyoto (Japan, 2017), and Sydney (Australia, 2019).

Traditionally, for each paper session at GREC, an invited presentation describes the state of the art and open questions for the session's topic, which is followed by short presentations of each paper. Each session concludes with a panel discussion moderated by the invited speaker, in which the authors and attendees discuss the papers presented along with the larger issues identified in the invited presentation.

For this 14th edition of GREC, the authors had the opportunity to submit short or long papers depending on the maturity of their research. From 14 submissions, we selected 12 papers from authors in eight different countries, comprising nine long papers and three short papers. Each submission was reviewed by at least two expert reviewers, with the majority (10 papers) reviewed by three expert reviewers. We would like to take this opportunity to thank the Program Committee members for their meticulous reviewing efforts.

A hybrid mode was setup to welcome both on-site and online participants. The workshop consisted of three sessions dealing with the issues of "Handwritten Graphics", "Typeset Graphics", and "Handwritten Text and Annotations". Timothy Hospedales, from the Institute of Perception, Action and Behaviour of the University of

Edinburgh, UK, offered an interesting keynote talk focused on the topic of sketch analysis, a recent issue addressed by many researchers of our community.

September 2021

Jean-Christophe Burie
Richard Zanibbi
Motoi Iwata
Riba Fierre

Organization

General Chair

Jean-Christophe Burie — La Rochelle University, France

Program Committee Chairs

Richard Zanibbi — Rochester Institute of Technology, USA
Motoi Iwata — Osaka Prefecture University, Japan
Pau Riba — Universitat Autònoma de Barcelona, Spain

Steering Committee

Alicia Fornés — Universitat Autònoma de Barcelona, Spain
Bart Lamiroy — Université de Lorraine, France
Rafael Lins — Federal University of Pernambuco, Brazil
Josep Lladós — Universitat Autònoma de Barcelona, Spain
Jean-Marc Ogier — Université de la Rochelle, France

Program Committee

Sébastien Adam — University of Rouen Normandy, France
Eric Anquetil — INSA, France
Samit Biswas — Indian Institute of Engineering Science and Technology, India
Jorge Calvo-Zaragoza — University of Alicante, Spain
Bertrand Coüasnon — INSA, France
Mickaël Coustaty — La Rochelle Université, France
Kenny Davila — Universidad Tecnológica Centroamericana, Honduras
Sounak Dey — Universitat Autònoma de Barcelona, Spain
Alicia Fornés — Universitat Autònoma de Barcelona, Spain
Ichiro Fujinaga — McGill University, USA
Alexander Gribov — Environmental Systems Research Institute, USA
Nina S. T. Hirata — University of São Paulo, Brasil
Bart Lamiroy — University of Reims Champagne-Ardenne, France
Christoph Langenhan — Technical University of Munich, Germany
Rafael Lins — Federal University of Pernambuco, Brasil

Josep Llados	Universitat Autònoma de Barcelona, Spain
Alexander Pacha	Vienna University of Technology, Austria
Umapada Pal	Indian Statistical Institute, India
Oriol Ramos-Terrades	Universitat Autònoma de Barcelona, Spain
Romain Raveaux	University of Tours, France
Christophe Rigaud	La Rochelle Université, France
K. C. Santosh	University of South Dakota, USA

Relation-Based Representation for Handwritten Mathematical Expression Recognition

Thanh-Nghia Truong[1] ⓘ, Huy Quang Ung[1] ⓘ, Hung Tuan Nguyen[2] ⓘ,
Cuong Tuan Nguyen[1](✉) ⓘ, and Masaki Nakagawa[1] ⓘ

[1] Department of Computer and Information Sciences, Tokyo University of Agriculture and
Technology, 2-24-16 Naka-cho, Koganei-shi, Tokyo 184-8588, Japan
fx4102@go.tuat.ac.jp, nakagawa@cc.tuat.ac.jp
[2] Institute of Global Innovation Research, Tokyo University of Agriculture and Technology,
2-24-16 Naka-cho, Koganei-shi, Tokyo 184-8588, Japan

Abstract. This paper proposes relation-based sequence representation that
enhances offline handwritten mathematical expressions (HMEs) recognition.
Commonly, a LaTeX-based sequence represents the 2D structure of an HME as
a 1D sequence. Consequently, the LaTeX-based sequence becomes longer, and
HME recognition systems have difficulty in extracting its 2D structure. We propose a new representation for HMEs according to the relations of symbols, which
shortens the LaTeX-based representation. We use an offline end-to-end HME
recognition system that adopts weakly supervised learning to evaluate the proposed representation. Recognition experiments indicate that the proposed relation-
based representation helps the HME recognition system achieve higher performance than the LaTeX-based representation. In fact, the HME recognition system
achieves recognition rates of 53.35%, 52.14%, and 53.13% on the dataset of the
Competition on Recognition of Online Handwritten Mathematical Expressions
(CROHME) 2014, 2016, and 2019, respectively. These results are more than 2
percentage points higher than the LaTeX-based system.

Keywords: Relation-based representation · Offline recognition · Handwritten
mathematical expressions · End-to-end recognition

1 Introduction

Mathematical expressions play an essential role in scientific documents since they are
indispensable for describing problems, theories, and solutions in math, physics, and many
other fields. Due to the rapid emergence of pen-based or touch-based input devices such
as digital pens, tablets, and smartphones, people have begun to use handwriting interfaces
as an input method. Although the input method is natural and convenient, it is useful if
and only if handwritten mathematical expressions (HMEs) are correctly recognized.

There are two approaches to recognize handwriting based on the type of patterns.
One approach uses the sequences of pen-tip, or finger-top coordinates, collected from
modern electronic devices, termed as online patterns. A sequence of coordinates from

© Springer Nature Switzerland AG 2021
E. H. Barney Smith and U. Pal (Eds.): ICDAR 2021 Workshops, LNCS 12916, pp. 7–19, 2021.
https://doi.org/10.1007/978-3-030-86198-8_1

pen/finger-down to pen/finger-up is called a stroke. Online input allows strokes to be separated from time sequence information, but it is troubled by writing order variation and stroke duplication. Another approach processes handwritten images captured from scanner or camera termed as offline patterns. Offline patterns are also easily converted from online patterns by discarding temporal information. Generally, offline recognition has a problem of segmentation, but it is free from various stroke orders or duplicated strokes. It is also free from the order of symbols written to form an HME.

In this paper, we focus on offline HME recognition since it can be used to recognize HMEs written on paper and those written on tablets widely spreading into education. HME recognition has been studied for decades from the early top-down methods [1] and bottom-up methods [2] to grammar-based [3–5], tree-based [6], and graph-based methods [7] until Deep neural network (DNN) was introduced in recent years [8–10]. Generally, three subtasks are involved in both online and offline HME recognition [11, 12] that are symbol segmentation, symbol recognition and structural analysis. If these three subtasks are executed independently, however, the total task of HME recognition is limited because the isolated subtasks are separately optimized.

The DNN based approach has recently been successfully used to parse the structures directly from HME training samples. It deals with HME recognition as an input-to-sequence problem [13], where an input can be either an HME image (OffHME) or an online HME (OnHME) pattern. This approach is flexible and achieves good performance because it uses shared global context to recognize HMEs. It is also liberated from manual tuning of each component and combination. However, the approach requires a large number of training data to improve the generalization of the DNN-based models. Moreover, DNNs also have difficulty extracting a 2D structure of an HME from a 1D LaTeX sequence when 1D LaTeX sequences are used for ground truth [6]. This problem is revealed in low recognition rates, as shown for the CROHME 2014, 2016, and 2019 testing sets, i.e., the expression recognition rates are around 50% [11, 12, 14]. Thus, there is a demand for appropriate 2D structural representations of HMEs for DNNs.

In this work, we propose relation-based representation instead of the LaTeX-based representation to improve the expression recognition rate of DNN-based OffHME recognizers. We conduct experiments on the CROHME datasets [11, 12, 14] to demonstrate the improvement of the state-of-the-art offline HME recognition systems [8, 10], by training them with our proposed relation-based representation. We present the following contributions:

(1) We propose relation-based representation rather than the LaTeX-based representation for HMEs so that DNNs straightforwardly extract the 2D structure of HME.
(2) We evaluate the proposed representation with an end-to-end OffHME recognition [10]. The OffHME recognition model trained with the LaTeX-based representation achieves 51.34%, 49.16%, and 50.21% on the CROHME 2014, 2016, and 2019 testing sets, respectively. The relation-based representation helps the recognition model achieve expression recognition rates of 53.35%, 52.14%, and 53.13% on the CROHME testing sets.

The rest of the paper is organized as follows. Section 2 provides an overview of the state-of-the-art. Section 3 presents our method proposed in this paper. Section 4 describes our dataset and experiments. Section 4 highlights the results of the experiments. Section 5 draws the conclusion and future works.

2 Related Works

In this section, we review the sequence transcription-based end-to-end approach for OffHME recognition and its problems. Our proposed relation-based representation combined with the OffHME recognition system proposed in [9] tries to address these problems.

2.1 Sequence Transcription Approach for HME Recognition.

In recent years, many researchers use Convolutional Neural Network (CNN) and Long Short-Term Memory (LSTM) to solve computer vision and related tasks. CNN and LSTM showed their strength in solving automatic segmentation, detection, and recognition [15–17].

The encoder-decoder framework has been successfully applied to HME recognition [8–10]. This approach jointly handled symbol segmentation, symbol recognition, and classification of spatial relations for HME recognition while general grammar-based approach attempted to solve the three sub-tasks separately and its grammar had to be carefully designed to avoid the lack of generality and extensibility. Many encoder-decoder systems [8–10, 13] show their strength in solving HME recognition.

An end-to-end method is composed of many jointly optimized components so that it can achieve a global optimization rather than a local optimization. An end-to-end HME recognition method consists of two main components, an encoder and a decoder, as shown in Fig. 1, which predicts an output, e.g., a label sequence corresponding to a given input image. The encoder, an LSTM or a BLSTM network (for online data) or CNN (for offline data), accepts an input pattern and encodes information of the input to high-level features. The decoder, which is an LSTM network, or a recurrent neural network (RNN) with gated recurrent units (GRU), constructs a label sequence using the encoded high-level features.

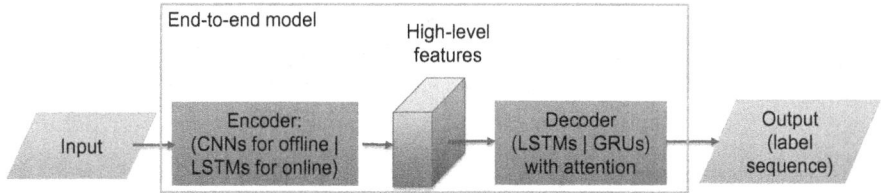

Fig. 1. General network structure of end-to-end methods.

2.2   Representation of MEs in Sequence Transcription Model

Latex-Based Sequence Transcription. Generally, an end-to-end OffHME recognition method has two main problems in extracting the 2D structure of an HME. First, it does not efficiently learn the structures of HMEs as the end-to-end HME recognition model sometimes outputs sequences that disobey the LaTeX grammar [18, 19]. For example, the decoder might generate an ungrammatical sequence like **x^{2** and **x^{2}}** for the expression **x^{2}**. This problem suggests that a 1D LaTeX sequence is not effective in representing the 2D structures of mathematical expressions. In fact, LaTeX needs many symbols (e.g., **{,}**, _ and ^) to represent the mathematical scope of its sub-expression that makes it long and complex. Secondly, a LaTeX sequence has some ambiguities in representing a 2D structure that causes the inconsistency of context. LaTeX uses special symbols (e.g., _ and ^) to represent spatial relations among sub-expressions. The _ symbol represents both the subscript and below relations (e.g., **a_{i}** and **\sum_{i}i**). Similarly, the ^ symbol represents both the superscript and upper relations (e.g., **a^{i}** and **\sum_{i= 0}^{N}i**).

Tree-Based Sequence Transcription. Tree-based representation [19] could deal with these ambiguities since it uses relation attributes. However, the method is inefficient because it produces a long sequence of the subtree for complex HMEs.

3   Our Approach

In this section, we propose a relation-based representation for HMEs. Then, we show an OffHME recognition system to evaluate the representation.

3.1   Relation-Based Representation

As we mentioned above, a LaTeX sequence represents the 2D structure of an HME as a 1D sequence. In more detail, it uses the pair of symbols **{** and **}** to define a mathematical scope for each sub-expression, as shown in Fig. 2. Therefore, a sub-expression contains not only its string but also the relation symbol and the two symbols **{** and **}**. We reduce

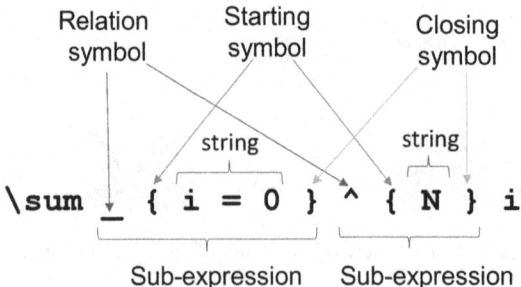

Fig. 2. Example of expression with its 2D structure.

Table 1. Examples of LaTeX-based representations and corresponding relation-based representations.

#	LaTeX-based representations	Relation-based representations	No. of reduced symbols
1	`a ^ {2}`	`a \sup 2 \end`	1
2	`a _ {2}`	`a \sub 2 \end`	1
3	`\sqrt {2 x}`	`\sqrt 2 x \end`	1
4	`\lim _ {x \rightarrow 0} x ^ {2}`	`\lim \below x \rightarrow 0 \end x \sup 2 \end`	2
5	`\int _ {2} ^ {3} x d x`	`\int \below 2 \end \above 3 \end x d x`	2
6	`\sum _ { i = 1} ^ { \infty} a _ {i}`	`\sum \below i = 1 \end \above \infty \end a \sub i \end`	3
7	`\frac {a} {b}`	`\frac a \end b \end`	2

the number of symbols to represent each sub-expression by reusing the relation symbol as its starting symbol and using the single symbol **\end** as its closing symbol for its scope. In this way, we can reduce one symbol for each sub-expression.

To solve the problem of ambiguity in the relation of the LaTeX-based representation, we use **\sub** and **\below** for the subscript and below relations, respectively, instead of _. Moreover, we use **\sup** and **\above** for the superscript and above relations, respectively, instead of ^. The number of symbols does not change, but we can remove the ambiguity.

Table 1 shows some examples of LaTeX-based representations and the corresponding relation-based representations. For every basic expression that has only one sub-expression, such as _{} or ^{}, we convert it to relation-based representation by changing the relation symbol and using **\end** as the closing symbol for its scope. For every complex expression that has multiple sub-expressions such as a sum expression **\sum_{}^{}** or an integral expression **\int_{}^{}**, we separate its sub-expression _{}^{} into two basic sub-expressions _{} and ^{} and then convert them as basic expressions. For every fraction expression **\frac{}{}**, especially, we dispense with the **\above** and **\below** relation symbols assuming the first sub-expression is above and the second sub-expression is below a fraction bar.

For the relation-based representation, we do not use {,}, _ and ^ but now use **\end**, **\sub**, **\below**, **\sup**, and **\above**. Therefore, the number of symbol classes in the dictionary increases by one. On the other hand, we can reduce one symbol for every sub-expression inside an expression. Eventually, we can reduce a large number of symbols for complex HMEs. It should be noted that Latex-based and relation-based representations can be converted to each other.

3.2   End-to-End HME Recognition

The recent state-of-the-art offline HME recognition model, DenseWAP [20], is proposed to use DenseNet for the encoder of the Watch, attend and parse (WAP) model [8]. We integrate a symbol classifier into the DenseWAP model, which is named as Rec-wsl [10]. The Rec-wsl system consists of three principal components: a convolutional neural network-based encoder, a gated recurrent unit-based decoder with an attention mechanism, and an added symbol classifier, as shown in Fig. 3. In the training process, Rec-wsl uses both the decoder and the symbol classifier for training the model. Note that the branch of the symbol classifier is trained by weakly supervised learning. In the evaluation process, Rec-wsl uses only the decoder to generate the output sequence.

In its forward step, the encoder extracts high-level features from an HME image and feeds them to the decoder using the attention mechanism to generate an output, as shown by the blue arrows in Fig. 3. The encoded high-level features are also transferred to the symbol classifier to determine whether the symbols exist in the HME image. In the backward step of the training process, the backpropagation of gradients is computed from both the decoder and symbol classifier outputs. Then, it is used to update the entire Rec-wsl model, as shown by the red arrows in Fig. 2. Thus, Rec-wsl is optimized through two loss functions, an expression-level loss function of the decoder and a symbol-level loss function of the symbol classifier.

The decoder is trained by minimizing the cross-entropy loss (CE) of a predicted relation-based sequence from an input HME. Each word y_t is predicted word-by-word to generate a complete output sequence $(y_1, \ldots, y_T)$ for the input HME, and the expression-level loss for each batch of HMEs is obtained as shown in Eq. (1).

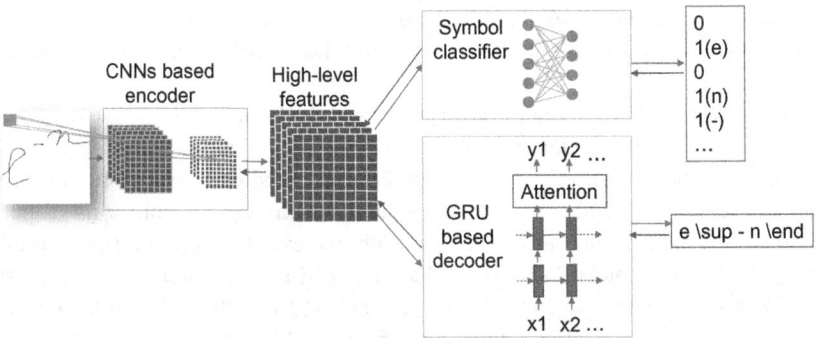

Fig. 3. Network structure of Rec-wsl.

$$
\begin{aligned}
loss_{expression} &= \sum_{i=1}^{batchsize} \sum_{t=1}^{T} -log\, p_{reg}(y_t | HME_i) \\
&= \sum_{i=1}^{batchsize} \sum_{t=1}^{T} -log\, p(y_t | g_{1:t-1}; y_{1:t-1}; \varphi_i)
\end{aligned} \tag{1}
$$

where

- t: t^{th} predicted time step.
- $g_{1:t-1}$: LaTeX ground-truth symbols from the first to t^{th} symbol.
- p_{reg}: symbol prediction probability.
- $y_{1:t-1}$: previously predicted symbols at t^{th} time step.
- φ_i: context of the ith HME in the batch of HMEs.

The symbol classifier predicts the probability of occurrence of every symbol in the dictionary while minimizing the binary cross-entropy loss (BCE). For each class c, there are two possibilities, i.e., occurring or not in an input image. For each batch of HMEs, the symbol-level loss is obtained as shown in Eq. (2).

$$
\begin{aligned}
loss_{symbol} &= \sum_{i=1}^{batchsize} \sum_{c=1}^{C} BCE_{c,i} \\
&= \sum_{i=1}^{batchsize} \sum_{c=1}^{C} -\left(t_{c,i} \log y_{c,i} + \left(1 - t_{c,i}\right) \log \left(1 - y_{c,i}\right)\right)
\end{aligned}
\tag{2}
$$

where

- $t_{c,i}$: binary target of $\{0, 1\}$ for the ith HME, with 0 indicating that the class c does not occur in the image and 1 indicating the opposite.
- $y_{c,i}$: occurrence probability of class c for the ith HME.
- C: total number of symbol classes in the dictionary.

Rec-wsl uses the combined loss of the symbol-level BCE loss ($loss_{symbol}$) and the expression-level CE loss ($loss_{expression}$), as shown in Eq. (3).

$$
loss_{combined} = loss_{expression} + \alpha \, loss_{symbol}
\tag{3}
$$

where α is the weight of $loss_{symbol}$. When α is near zero, the model primarily uses the expression-level loss. In our experiments, we choose α to be 0.5, based on the best value in the validation experiment on the CROHME 2014 validation set.

4 Experiments

In this section, we first briefly describe the datasets used for evaluation and then show the results of the experiments as well as the in-depth analysis of the results. We show the performance of the proposed representation by comparing it with the benchmark trained with LaTeX-based representation [10]. Then, we compare our OffHME recognition system with the other systems on the CROHME 2014, 2016, and 2019 testing sets.

4.1 Datasets

Data samples and evaluation metrics were introduced in CROHME [11, 12, 14]. Revisions of the datasets and metrics from CROHME 2011 to CROHME 2019 were made for more precise evaluation. For testing on the CROHME 2014 and 2016 testing sets, we use the CROHME 2014 training set to train the proposed model and use the CROHME 2013 testing set for validation. For testing on the CROHME 2019 testing set, we use the training set and validation set provided by the contest.

In our experiments, we use the CROHME tool LgEval [12] to evaluate our proposed model. We mainly use the expression recognition rate (ExpRate) for the metrics, which shows the percentage of correctly predicted HMEs over all HMEs.

4.2 Ablation Experiments of Relation-Based Method

In this section, we evaluate the proposed method using the CROHME validation tool on the CROHME 2014, 2016, and 2019 testing sets.

Table 2 shows the comparison of the OffHME recognition systems, DenseWAP and Rec-wsl, trained with the LaTeX-based representation and proposed relation-based representation (DenseWAP trained with the LaTeX representation is termed LaTeX-based DenseWAP while that with the relation-based representation is Relation-based DenseWAP and similarly for Rec-wsl). In comparison with the LaTeX-based representation, the relation-based representation substantially improved the ExpRate values of both HME recognition systems (DenseWAP and Rec-wsl) by more than 2 points compared with the LaTeX-based representation on the CROHME testing sets.

4.3 Detailed Experiment Results on CROHME Datasets

Table 3, Table 4, and Table 5 compare the ExpRate of the proposed method with other offline HME recognition systems on the CROHME 2014, 2016, and 2019 testing sets, respectively. These tables consist of correct ExpRate (Correct), ExpRate with a single error excepted (≤ 1), that with up to 2 errors excepted (≤ 2), and that with up to 3 errors excepted (≤ 3).

Table 2. Comparison of ExpRate (%) on different CROHME testing sets.

System	Dataset		
	CROHME 2014	CROHME 2016	CROHME 2019
LaTeX-based DenseWAP	48.28	44.90	47.29
Relation-based DenseWAP	51.42	46.99	49.71
LaTeX-based Rec-wsl	51.34	49.16	50.21
Relation-based Rec-wsl	**53.35**	**52.14**	**53.13**

Table 3. Comparison of ExpRate (%) on the CROHME 2014 testing set.

System	Criterion			
	Correct	≤1	≤2	≤3
MyScript [11]	62.68	72.31	75.15	76.88
Valencia [11]	37.22	44.22	47.26	50.20
Nantes [11]	26.06	33.87	38.54	39.96
Tokyo [11]	25.66	33.16	35.90	37.32
WAP [8]	46.55	61.16	65.21	66.13
TAP [23]	46.86	61.87	65.82	66.63
PAL-v2 [9]	48.88	64.50	69.78	73.83
LaTeX-based DenseMSA [19]	43.0 ± 1.0	57.8 ± 1.4	61.9 ± 1.8	63.2 ± 1.7
Tree-based DenseMSA [19]	49.1 ± 0.9	64.2 ± 0.9	67.8 ± 1.0	68.6 ± 1.6
LaTeX-based Rec-wsl	51.34	64.50	69.78	73.12
Relation-based Rec-wsl	**53.35**	**65.21**	**70.28**	**72.82**

Table 4. Comparison of ExpRate (%) on the CROHME 2016 testing set.

System	Criterion			
	Correct	≤1	≤2	≤3
MyScript [12]	67.65	75.59	79.86	–
Wiris[12]	49.61	60.42	64.69	–
Tokyo[12]	43.94	50.91	53.70	–
Sao Paolo[12]	33.39	43.50	49.17	–
Nantes[12]	13.34	21.02	28.33	–
WAP [8]	44.55	57.10	61.55	62.34
TAP [18]	41.3	–	–	–
PAL-v2 [9]	49.61	64.08	70.27	73.50
LaTeX-based DenseMSA [19]	40.1 ± 0.8	54.3 ± 1.0	57.8 ± 0.9	59.2 ± 0.8
Tree-based DenseMSA [19]	48.5 ± 0.9	62.3 ± 0.9	65.3 ± 0.7	65.9 ± 0.6
LaTeX-based Rec-wsl	49.16	57.63	60.33	61.63
Relation-based Rec-wsl	**52.14**	**63.21**	**69.40**	**72.54**

The MyScript system was built on the principle that segmentation, recognition, and interpretation have to be handled concurrently at the same level to produce the best candidate. They used a large number of extra training samples of HME patterns. The Valencia system parsed expressions using two-dimensional stochastic context-free grammars. The WAP system [8] used an end-to-end network to learn directly LaTeX sequences from

OffHME patterns. The TAP system [13] used an end-to-end network to learn directly LaTeX sequences from OnHME patterns. PAL-v2 [21] used paired adversarial learning to learn semantic invariant features. LaTeX-based DenseMSA [20] was the updated WAP model that used multiple DenseNet-based encoders with multiple scales to deal with different symbol sizes. Tree-based DenseMSA [19] was the system that used the DenseMSA model to learn tree representation instead of the LaTeX-based representation. LaTeX-based DenseMSA and Tree-based DenseMSA used multiple HME recognition models and averaged the recognition rate of the models. LaTeX-based Rec-wsl [10] was the benchmark to compare with our system, where they used a LaTex-based recognition system.

Table 5. Comparison of ExpRate (%) on the CROHME 2019 testing set.

System	Criterion			
	Correct	≤ 1	≤ 2	≤ 3
TAP [18]	41.7			
LaTeX-based DenseMSA [19]	41.7 ± 0.9	55.5 ± 0.9	59.3 ± 0.5	60.7 ± 0.6
Tree-based DenseMSA [19]	51.4 ± 1.3	66.1 ± 1.4	69.1 ± 1.2	69.8 ± 1.1
LaTeX-based Rec-wsl	50.21	62.14	66.56	68.81
Relation-based Rec-wsl	53.13	63.89	68.47	70.89

For the comparison of ExpRate on the CROHME 2014 testing set, our proposed relation-based Rec-wsl system achieved 2.01 points higher ExpRate with 53.35% compared to 51.34% of the benchmark than LaTeX-based Rec-wsl. Relation-based Rec-wsl system also achieved better ExpRate with 1, 2, and 3 errors excepted than LaTeX-based Rec-wsl.

As shown in Table 4, the relation-based Rec-wsl accounts for the ExpRate of 52.14% on the CROHME 2016 testing set. This result is competitive to the other participant systems in the contest without extra samples and state-of-the-art systems. Note that MyScript used a large number of extra training samples of HME patterns, as mentioned above. The Wiris system won the CROHME 2016 competition using a pre-trained language model based on more than 590,000 formulas from the Wikipedia formula corpus [22].

For ExpRate on the CROHME 2019 testing set in Table 5, our improvement relation-based Rec-wsl outperforms the state-of-the-art methods for OffHME (Tree-based DenseMSA) and that for OnHME (TAP) [18]. However, ExpRate of our model is still limited compared with other systems in the CROHME 2019 contest. Note that these systems used extra training HME patterns to train their HME recognition model.

4.4   In-depth Analysis

In this section, we make an in-depth analysis of our system to understand better and explore the directions for improving recognition rate in the future.

Table 6. Comparison of total number of symbols and maximum sequence length in the CROHME datasets.

Dataset	Metric			
	Total number of symbols		Max sequence length	
	LaTeX-based	Relation-based(*)	LaTeX-based	Relation-based(*)
CROHME 2014 training set	136,873	116,894 (85.40%)	96	76 (79.17%)
CROHME 2014 testing set	15,897	13,600 (85.55%)	204	157 (76.96%)
CROHME 2016 testing set	20,505	17,280 (84.27%)	108	88 (81.48%)
CROHME 2019 training set	19,947	17,056 (85.51%)	88	73 (82.95%)
CROHME 2019 testing set	157,030	133,906 (85.27%)	96	76 (79.17%)

(*) Values of the "Relation-based" column are in the form of "number of symbols (per-centage in comparison with the LaTeX based representation)"

Table 6 lists the total number of symbols and the number of symbols of the longest sequence in the CROHME datasets. We also show the percentage of the reduction in the number of symbols when using the proposed relation-based representation instead of the LaTeX based representation. Values of the "Relation-based" column are in the form of "number of symbols (percentage in comparison with the LaTeX based representation)" As shown in Table 6, we reduced the number of symbols by 15% to 20% when the relation-based representation is adopted instead of the LaTeX-based representation. We expect that the shorter sequences are more accessible for the decoder to predict. The relation-based Rec-wsl achieves higher performance than the LaTeX-based Rec-wsl.

However, the proposed system still outputs misrecognized candidates that might not follow the mathematical grammar. Note that we do not use any mathematical rules in the parsing stage of the HME recognition system. There is room to improve the end-to-end HME recognition system by integrating mathematical rules.

5   Conclusion

In this work, we proposed relation-based representation for HMEs and used it instead of the LaTeX-based representation. We used an end-to-end OffHME recognition model [10] to evaluate the proposed representation. Our recognition system trained with the relation-based representation has achieved the expression recognition rates of 53.35%, 52.14, and 53.13% from 51.34%, 49.16%, and 50.21% by the same system trained with the LaTeX-based representation on the CROHME 2014, 2016, and 2019 testing sets, respectively. The results are also competitive to the state-of-the-art methods that do not use extra data. The proposed representation helps the HME recognition model learn better the 2D structures of HMEs when generating the candidates. In our future work, we plan to improve the HME recognition system using math grammatical constraints.

Acknowledgement. This work is being partially supported by the Grant-in-Aid for Scientific Research (A) 19H01117 and that for Early-Career Scientists 21K17761.

References

1. Anderson, R.H.: Syntax-directed recognition of hand-printed two-dimensional mathematics. In: Interactive Systems for Experimental Applied Mathematics - Proceedings of the Association for Computing Machinery Inc. Symposium, pp. 436–459. ACM (1967)
2. Chang, S.K.: A method for the structural analysis of two-dimensional mathematical expressions. Inf. Sci. **2**, 253–272 (1970)
3. Álvaro, F., Sánchez, J.A., Benedí, J.M.: Recognition of on-line handwritten mathematical expressions using 2D stochastic context-free grammars and hidden Markov models. Pattern Recognit. Lett. **35**, 58–67 (2014)
4. Belaid, A., Haton, J.P.: A syntactic approach for handwritten mathematical formula recognition. IEEE Trans. Pattern Anal. Mach. Intell. PAMI-**6**, 105–111 (1984)
5. Garain, U., Chaudhuri, B.B.: Recognition of online handwritten mathematical expressions. IEEE Trans. Syst. Man, Cybern. Part B. **34**, 2366–2376 (2004)
6. Zhang, T., Mouchère, H., Viard-Gaudin, C.: A tree-BLSTM-based recognition system for online handwritten mathematical expressions. Neural Comput. Appl. **32**(9), 4689–4708 (2018). https://doi.org/10.1007/s00521-018-3817-2
7. Zhang, T., Mouchere, H., Viard-Gaudin, C.: Online handwritten mathematical expressions recognition by merging multiple 1D interpretations. In: Proceedings of 15th International Conference on Frontiers in Handwriting Recognition, pp. 187–192 (2016)
8. Zhang, J., et al.: Watch, attend and parse: an end-to-end neural network based approach to handwritten mathematical expression recognition. Pattern Recognit. **71**, 196–206 (2017)
9. Wu, J.W., Yin, F., Zhang, Y.M., Zhang, X.Y., Liu, C.L.: Image-to-markup generation via paired adversarial learning. In: Proceedings of Joint European Conference on Machine Learning and Knowledge Discovery in Databases, pp. 18–34 (2018)
10. Truong, T., Nguyen, C.T., Phan, K.M., Nakagawa, M.: Improvement of end-to-end offline handwritten mathematical expression recognition by weakly supervised learning. In: Proceedings of 17th International Conference on Frontiers in Handwriting Recognition, pp. 181–186 (2020)
11. Mouchere, H., Viard-Gaudin, C., Zanibbi, R., Garain, U.: ICFHR 2014 competition on recognition of on-line handwritten mathematical expressions. In: Proceedings of 14th International Conference on Frontiers in Handwriting Recognition, pp. 791–796 (2014)
12. Mouchère, H., Viard-gaudin, C., Zanibbi, R., Garain, U.: ICFHR 2016 competition on recognition of online handwritten mathematical expressions. In: Proceedings of 15th International Conference on Frontiers in Handwriting Recognition, pp. 607–612 (2016)
13. Zhang, J., Du, J., Dai, L.: Track, attend, and parse (TAP): an end-to-end framework for online handwritten mathematical expression recognition. IEEE Trans. Multimed. **21**, 221–233 (2019)
14. Mahdavi, M., Zanibbi, R., Mouchère, H.: ICDAR 2019 CROHME + TFD: competition on recognition of handwritten mathematical expressions and typeset formula detection. In: Proceedings of 15th International Conference on Document Analysis and Recognition, pp. 1533–1538 (2019)

15. Shelhamer, E., Long, J., Darrell, T.: Fully convolutional networks for semantic segmentation. IEEE Trans. Pattern Anal. Mach. Intell. **39**, 640–651 (2017)
16. Liu, W., Anguelov, D., Erhan, D., Szegedy, C., Reed, S., Fu, C.-Y., Berg, A.C.: SSD: single shot multibox detector. In: Leibe, B., Matas, J., Sebe, N., Welling, M. (eds.) ECCV 2016. LNCS, vol. 9905, pp. 21–37. Springer, Cham (2016). https://doi.org/10.1007/978-3-319-464 48-0_2
17. Sundermeyer, M., Schlüter, R., Ney, H.: LSTM neural networks for language modeling. In: Proceedings of 13th Annual Conference of the International Speech Communication Association, pp. 194–197 (2012)
18. Zhang, J., Du, J., Yang, Y., Song, Y.-Z., Dai, L.: SRD: a tree structure based decoder for online handwritten mathematical expression recognition. IEEE Trans. Multimed. **1**, 1–10 (2020)
19. Zhang, J., Du, J., Yang, Y., Si, Y.S., Lirong, W.: A tree-structured decoder for image-to-markup generation. In: Proceedings of 37th International Conference on Machine Learning, pp. 11076–11085 (2020)
20. Zhang, J., Du, J., Dai, L.: Multi-scale attention with dense encoder for handwritten mathematical expression recognition. In: Proceedings of 24th International Conference on Pattern Recognition, pp. 2245–2250 (2018)
21. Wu, J.-W., Yin, F., Zhang, Y.-M., Zhang, X.-Y., Liu, C.-L.: Handwritten mathematical expression recognition via paired adversarial learning. Int. J. Comput. Vision **128**(10–11), 2386–2401 (2020). https://doi.org/10.1007/s11263-020-01291-5
22. Zanibbi, R., Aizawa, A., Kohlhase, M., Ounis, I., Topic, G., Davila, K.: NTCIR-12 MathIR task overview. In: Proceedings of 12th NTCIR Conference, pp. 299–308 (2016)
23. Zhang, J., Du, J., Dai, L.: A GRU-based encoder-decoder approach with attention for online handwritten mathematical expression recognition. In: Proceedings of 14th International Conference on Document Analysis and Recognition, pp. 902–907 (2017)

A Public Ground-Truth Dataset for Handwritten Circuit Diagram Images

Felix Thoma[1,2], Johannes Bayer[1,2(✉)], Yakun Li[1], and Andreas Dengel[1,2]

[1] Smart Data and Knowledge Services Department,
DFKI GmbH, Kaiserslautern, Germany
`{felix.thoma,johannes.bayer,yakun.li,andreas.dengel}@dfki.de`
[2] Computer Science Department, TU Kaiserslautern, Kaiserslautern, Germany

Abstract. The development of digitization methods for line drawings – especially in the area of electrical engineering – relies on the availability of publicly available training and evaluation data. This paper presents such an image set along with annotations. The dataset consists of 1152 images of 144 circuits by 12 drafters and 48539 annotations. Each of these images depicts an electrical circuit diagram taken by consumer grade cameras under varying lighting conditions and perspectives. A variety of different pencil types and surface materials has been used. For each image, all individual electrical components are annotated with bounding boxes and one out of 45 class labels. In order to simplify a graph extraction process, different helper symbols like junction points and crossovers are introduced, while texts are annotated as well. The geometric and taxonomic problems arising from this task as well as the classes themselves and statistics of their appearances are stated. The performance of a standard Faster RCNN on the dataset is provided as an object detection baseline.

Keywords: Circuit diagram · Ground truth · Line drawing

1 Introduction

The sketch of a circuit diagram is a common and important step in the digital system design process. Even though the sketch gives the designer a lot of freedom to revise, it is still time consuming to transform the sketch into a formal digital representation for existing tools to simulate the designed circuit. There are mainly two approaches which have been applied to address the transformation from sketch into a useful format to be used by existing simulation tools. The first approach allows the user to draw the diagram in an on-line fashion on a screen and tries to recognize the sketch by interpreting pen strokes, like shown in [2,7] and [3]. The second off-line approach assumes that the diagrams are drawn with a pen on paper from which an image is captured afterwards. This approach relies on machine learning and computer vision algorithms, like shown in [4,8,9] and [1], to process the images. It is important to collect large and high quality training data for the machine learning and computer vision

© Springer Nature Switzerland AG 2021
E. H. Barney Smith and U. Pal (Eds.): ICDAR 2021 Workshops, LNCS 12916, pp. 20–27, 2021.
https://doi.org/10.1007/978-3-030-86198-8_2

approach, especially when deep learning algorithms are applied. However, most of this literature [1, 4, 8, 9] exposes only little of the data it relies on publicly, and to the best of the knowledge of this paper's authors, there is no publicly available comparable dataset.

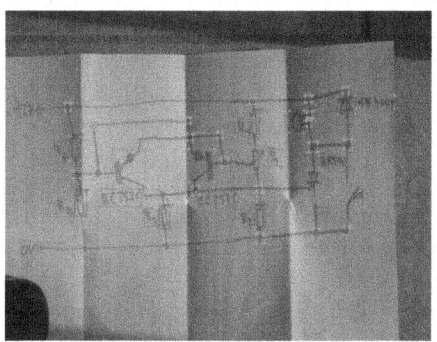 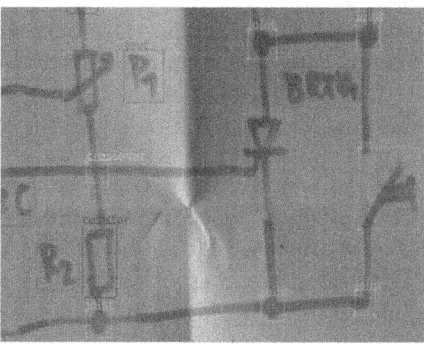

(a) Full Image   (b) Detail showing Junction, Crossover and Text Annotations

Fig. 1. Dataset sample with annotations

The dataset described in this paper mainly addresses this issue and is publicly available[1] under the MIT Licence[2]. Previous datasets described in literature will be summarized in Sect. 2. Afterwards, characteristics of the hand-drawn circuit diagrams images contained in the dataset will be stated in Sect. 3. In the end, Sect. 6 will present basic recognition results using a Faster-RCNN [10] on the presented dataset.

2 Related Work

Recent literature on off-line approaches mentioned in Sect. 1 mainly focus on the algorithms to recognize sketches but the datasets they rely on are usually not publicly accessible. While [1] presented a comprehensive system to recognize hand drawn sketches and transform the recognition results to Verilog code, there was no description of the concrete number of images and the variety of symbols. In [9], the use of Fourier Descriptors as feature representation of hand-drawn circuit components and subsequent SVM classification is proposed. As test data, only 10 hand drawn circuit sketches have been used, but only one circuit sketch was shown as an example. The authors of [4] gave information of the number of nodes and components included in their hand drawn sketches but without revealing the exact number of sketches. [8] mentioned that an evaluation on 100 hand drawn circuit diagrams. However, the sample quantity is relatively small.

---

[1] https://osf.io/ju9ck/.
[2] https://opensource.org/licenses/MIT.

3   Images

Every image of the dataset contains a hand-drawn electrical circuit (see Fig. 1). Every of the 12 drafters was instructed to draw 12 circuits, each of them twice. The drafters (with varying engineering ba) were guided to produce reasonable results. They where supplied with publicly available circuits to reproduce or could create something on their own. For these two versions, the drawers were allowed to alter the circuit geometrically. Each of the drafts was photographed four times, resulting in an overall image count of 1152. Taking multiple images of the same drawing was based on the following considerations:

- The dataset can conveniently be used for style transfer learning
- Since different images of the same drawing contain the same annotation items, they can be used for automatic verification of the labeling process
- The cost of creating a single sample is reduced
- Different real-world captured images are a realistic and rich replacement for artificial data augmentation
- Different samples of the same drawing must be carefully annotated to ensure that the object detection task is well-defined (no ambiguities)
- Enforcing diverse capturing conditions aim to support camera feeds (Fig. 2)

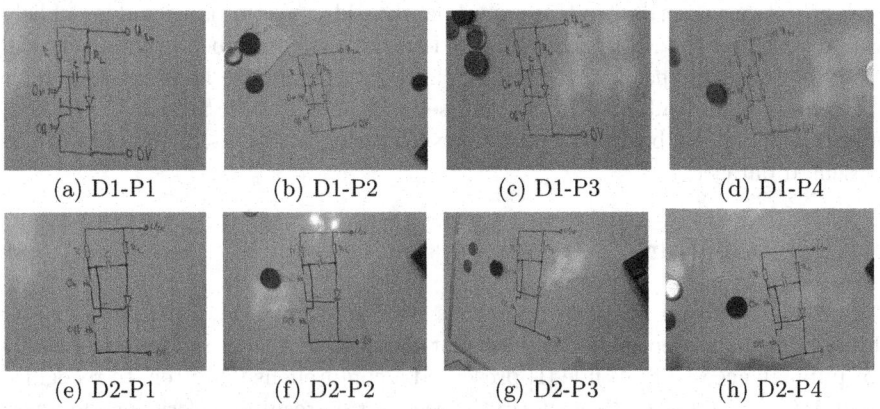

| (a) D1-P1 | (b) D1-P2 | (c) D1-P3 | (d) D1-P4 |

| (e) D2-P1 | (f) D2-P2 | (g) D2-P3 | (h) D2-P4 |

Fig. 2. From each circuit 8 images samples are created: four images of two drawings. The two drawings could differ not only in background and pencil type, but also in their layout.

Semantically, identical circuits can be drawn using different syntactic representations. For example, a voltage source can be represented by a single DC symbol or instead a VCC and a GND symbol can be used. Terminal symbols with textual voltage level descriptions can serve as a third option.

3.1 Drawing Surfaces and Instruments

The depicted surface types include: plain paper of various colors, ruled paper, squared paper, coordinate paper, glas, whiteboard, cardboard and aluminum foil. Pens, pencils, permanent markers and whiteboard markers of different colors have been utilized for drawing. A few drawings contain multiple colors or text that has been highlighted by a dedicated color. Buckling, kinking, bending, spots, transillumination, and paper cracks can be found as distortions on the samples. Some samples have been treated by corrective means while others have been created using a ruler as a drawing aid rather than free-hand sketching.

3.2 Capturing

All images have been captured by consumer cameras and no artificial image distortion has been applied. Some images have been cropped, and company logos have been obliterated. The images have been taken from different positions and orientations. That way, naturally observable distortions like (motion) blur and light reflexes of surfaces as well as ink are captured in the dataset. Despite these disturbances no major occlusions are present. More precisely, circuit symbols and their connections are recognizable by humans and the circuits are oriented right side up. Often, circuit drawings are surrounded by common (office environment) objects. The sample image sizes range between 864 by 4160 and 396 by 4032 and are stored as JPEG.

4 Annotations

Bounding box annotations stored in the PascalVOC [5] format have been chosen for symbols and texts as a trade-off between effort in ground truth generation and completeness in capturing the available information from the images. The downsides of this decision are the missing of rotation information of the symbols as well as the lack of wire layout capturing. The latter will be addressed by the auxiliary classes below. The annotations have been created manually using labelIMG [12].

4.1 Auxiliary Classes

In order to address the lack of wire-line annotation and to simplify the extraction of a circuit graph, auxiliary classes are introduced:

- **junction** represents all connections or kink-points of wire edges. Since wire corners and junctions share the same semantic from the electrical graph point of view, they share the same label.
- **crossover** represents the optical crossing of a pair of lines that is not physically connected.
- **terminal** denotes an 'open' wire ending, i.e. circuit input or output lines.

- **text** encloses a line of text which usually denotes physical quantities or the name of a terminal or a component.

   Some symbols (like potentiometers and capacitors) have outgoing lines or corners (like transistors or switches) which are considered part of their definition. Hence, these line (parts) are also included in the bounding boxes, and junction annotations within them are neglected. Likewise, text annotations related to symbols are only omitted if they are part of the symbol definition (like an 'M' in the motor symbol). If they are however explaining component parts (like integrated circuit pins) or the component itself (like integrated circuit type labels), they are annotated, even if they are located inside the symbol.

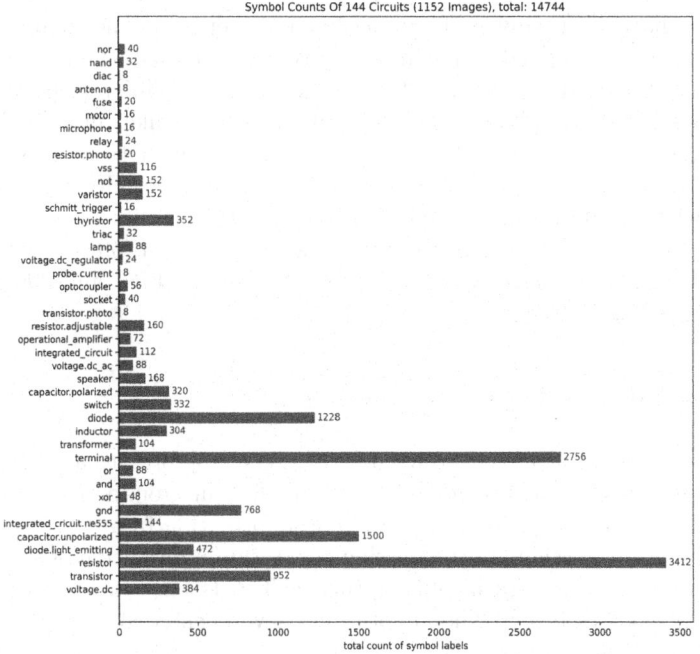

Fig. 3. Total count of label instances in the full dataset. Junctions, crossover and text are not considered due to their dominance in the dataset.

4.2 Symbol Classes

The dataset contains both IEEE/ANSI as well as IEC/DIN symbols. Individual circuit drawings are allowed to contain both standards, and a mix of analogue and digital symbols as well. Some drafters also used non standard-conform symbols. For example, a speaker that is explicitly drawn with a coil to show its type of construction.

The classes have been chosen to represent *electrically* connectable components. For example, if there is a full-wave rectifier depicted as four diodes, each of the diodes is annotated individually. Likewise, a relay label is used as an annotation rather than a switch and a inductor (or the two inductors of a transformer), since the components are coupled *inductively*. However, if the individual parts are separated so that other components are in between, an individual labeling is used as a makeshift. To simplify annotation, some of the electrical taxonomy is neglected. For example, all bipolar transistors, Mosfets and IGBTs are labeled as *transistors*.

4.3 Geometry

Generally, the bounding box annotations have been chosen to completely capture the symbol of interest, while allowing for a margin that includes the surrounding space. Often, there is an overlap between tightly placed symbols. The task of fully capturing all parts of the symbol was not well-defined in the context of curvy straight lines connected by rounded edge corners.

5 Statistics

The dataset has a total amount of 48 539 annotations among 45 classes. The junction class occurs 18 221 times, the text class 14 263 times and the crossover class 1 311 times. All other electrical symbol counts are listed in detail in Fig. 3. The frequency of symbol occurrences per circuit are displayed in Fig. 4.

6 Baseline Performance

A suggested division of the dataset into subsets for training (125 circuits), validation (6 circuits) and testing (12 circuits, from the second drafter), is provided so that images of every circuit (and therefore drawing) are exclusively assigned to one of the sets only.

In order to establish a baseline performance on the dataset, a Faster RCNN [10] with a ResNet152 [6] backbone was applied using the Torchvision [11] implementation, a learning rate of 0.007 and a batch size of 8, yielding an mAP score of 52%.

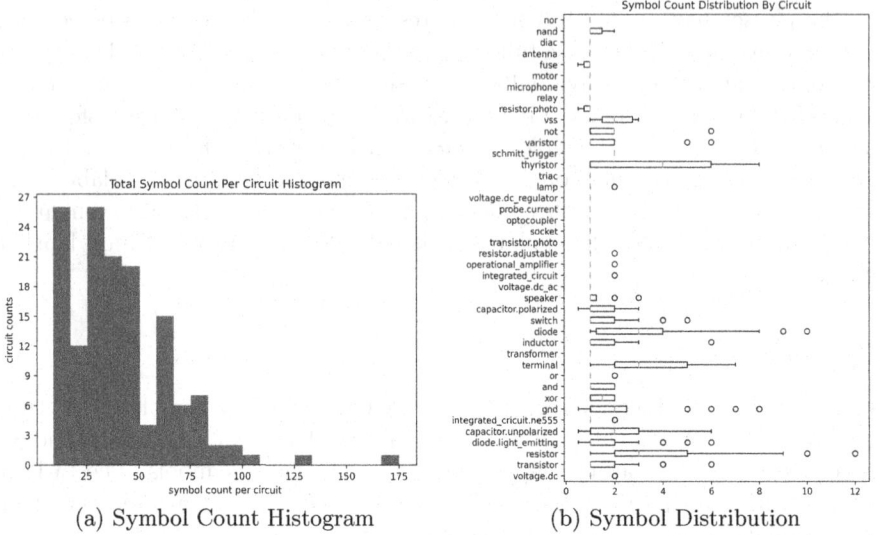

(a) Symbol Count Histogram   (b) Symbol Distribution

Fig. 4. a) Total symbol count per circuit. The histogram has 20 evenly spaced bins. b) Distribution of symbol count per circuit that the symbol is part of. Junctions, crossover and text are not considered due to their dominance in the dataset.

Acknowledgement. The authors coardially thank Thilo Pütz, Anshu Garg, Marcus Hoffmann, Michael Kussel, Shahroz Malik, Syed Rahman, Mina Karami Zadeh, Muhammad Nabeel Asim and all other drafters who contributed to the dataset. This research was funded by the German Bundesministerium für Bildung und Forschung (Project SensAI, grant no. 01IW20007).

References

1. Abdel-Majeed, M., et al.: Sketic: a machine learning-based digital circuit recognition platform. Turkish J. Electr. Eng. Comput. Sci. **28**, 2030–2045 (2020)
2. Alvarado, C., Davis, R.: Sketchread: a multi-domain sketch recognition engine. In: 2004 Proceedings of the 17th Annual ACM Symposium on User Interface Software and Technology, UIST 2004, pp. 23–32, New York, NY, USA. Association for Computing Machinery. https://doi.org/10.1145/1029632.1029637. ISBN 1581139578
3. Alvarado, C., et al.: LogiSketch: a free-sketch digital circuit design and simulation SystemLogiSketch. In: Hammond, T., Valentine, S., Adler, A., Payton, M. (eds.) The Impact of Pen and Touch Technology on Education. HIS, pp. 83–90. Springer, Cham (2015). https://doi.org/10.1007/978-3-319-15594-4_8 ISBN 1581139578
4. Edwards, B., Chandran, V.: Machine recognition of hand-drawn circuit diagrams. In: 2000 IEEE International Conference on Acoustics, Speech, and Signal Processing, Proceedings (Cat. No. 00CH37100), vol. 6, pp. 3618–3621 (2000)
5. Everingham, M., et al.: The pascal visual object classes (VOC) challenge. Int. J. Comput. Vis. **88**(2), 303–338 (2010). https://doi.org/10.1007/s11263-009-0275-4
6. He, K., et al.: Deep residual learning for image recognition. CoRR, abs/1512.03385, (2015). http://arxiv.org/abs/1512.03385

7. Liwicki, M., Knipping, L.: Recognizing and simulating sketched logic circuits. In: Khosla, R., Howlett, R.J., Jain, L.C. (eds.) KES 2005, Part III. LNCS (LNAI), vol. 3683, pp. 588–594. Springer, Heidelberg (2005). https://doi.org/10.1007/11553939_84

8. Moetesum, M., et al.: Segmentation and recognition of electronic components in hand-drawn circuit diagrams. EAI Endorsed Trans. Scalable Inf. Syst. **5**, e12 (2018)

9. Patare, M.D., Joshi, M.: Hand-drawn digital logic circuit component recognition using SVM. Int. J. Comput. Appl. **143**, 24–28 (2016)

10. Ren, S., et al.: Faster R-CNN: towards real-time object detection with region proposal networks. CoRR, abs/1506.01497 (2015). http://arxiv.org/abs/1506.01497

11. torchvision. https://github.com/pytorch/vision

12. Tzutalin. Labelimg. https://github.com/tzutalin/labelImg

A Self-supervised Inverse Graphics Approach for Sketch Parametrization

Albert Suso[✉], Pau Riba, Oriol Ramos Terrades, and Josep Lladós

Computer Vision Center and Computer Science Department, Universitat Autònoma de Barcelona, Catalunya, Spain
albert.suso@e-campus.uab.cat, {priba,oriolrt,josep}@cvc.uab.cat

Abstract. The study of neural generative models of handwritten text and human sketches is a hot topic in the computer vision field. The landmark SketchRNN provided a breakthrough by sequentially generating sketches as a sequence of waypoints, and more recent articles have managed to generate fully vector sketches by coding the strokes as Bézier curves. However, the previous attempts with this approach need them all a ground truth consisting in the sequence of points that make up each stroke, which seriously limits the datasets the model is able to train in. In this work, we present a *self-supervised* end-to-end inverse graphics approach that learns to embed each image to its best fit of Bézier curves. The self-supervised nature of the training process allows us to train the model in a wider range of datasets, but also to perform better after-training predictions by applying an overfitting process on the input binary image. We report qualitative an quantitative evaluations on the *MNIST* and the *Quick, Draw!* datasets.

Keywords: Inverse graphics · Sketch parametrization · Bézier curve · Chamfer distance · Symbol recognition

1 Introduction

The pervasive nature of touchscreen devices has motivated the emergence of the sketch modality for a broad range of tasks. Examples of the novel tasks dealing with sketch nature are: sketch classification [35], sketch-based image retrieval [9], sketch-to-image synthesis [6] or sketch-guided object detection [33]. In the abstract nature of sketches lays the main difficulty of these representations. Sketches have been used over centuries or millennia to transmit information even before written records. By means of few strokes, humans have been able to create understandable abstractions of the real world. Human beings, are able to decipher this condensed information although the absence of visual cues such as color and texture that are present, for instance, in natural images. In addition, the variability on this modality is extremely huge and can range from amateur to expert sketches or even from realistic to stylistic or artistic representations. All in all, makes the sketch a natural, yet hard, modality for humans to interact with these devices.

E. H. Barney Smith and U. Pal (Eds.): ICDAR 2021 Workshops, LNCS 12916, pp. 28–42, 2021.
https://doi.org/10.1007/978-3-030-86198-8_3

Human sketches consist of a set of strokes that are traditionally represented as raster images or as sequences of 2D points. However, approaches dealing with such representations must deal with important drawbacks. On the one hand, treating sketches as a digitized sequence of 2D points leads to dense and noisy representations. On the other hand, raster images suffer from non-smooth drawings that limits its representational capacity. From a Graphics Recognition perspective, raster-to-vector conversion has been the standard procedure for representing line drawings in a compact way, preserving the original shape invariant to geometric transformations. While vectorial representations have been adopted as a standard in domains like architecture or engineering [10], in sketching it is not so frequent. Recently, BézierSketch [8], proposed to overcome the drawbacks of sequences of 2D points for sketch representations by generating a more compact and smooth representation by means of bézier curves. In such case, the large sequence of points is transformed to few parametrized curves. However, such approach cannot handle raster representations.

In the present work, we propose an inverse graphics approach able to generate an approximation of a sketch image by means of few strokes, *i.e.* bézier curves. The proposed methodology, is able to obtain the desired representation in a self-supervised manner. Moreover, for such cases where a high fidelity representation is required, we propose to employ a single image overfitting technique. In addition, we present an evaluation on two well known datasets on hand drawn sketches and handwritten digits.

To summarize, the main contributions of this work are:

- A novel self-supervised inverse graphics approach for mapping raster binary images to a set of parametrized Bézier curves.
- A new probabilistic Chamfer loss that leads to a significantly better training than the original Chamfer distance.
- An overfitting mechanism to obtain high quality representations.

The rest of this paper is organized as follows. Section 2 reviews the relevant works of the state of the art. Section 3 presents the proposed methodology to encode raster images into a set of Bézier curves. Section 4 conducts an extensive evaluation on two datasets, namely on hand drawn sketches and handwritten symbols. Finally, Sect. 5 draws the conclusions and the future work.

2 Related Work

2.1 Inverse Graphics and Parametrized Curves

Bézier curves and Splines are powerful tools in Computer Graphics, that have been widely used in interactive and surface design [3,29]. Traditional optimization algorithms that fit data by Bézier curves and Splines require expensive *per-sample* alternating optimization [21,26] or iterative inference in expensive generative models [18,27], which make them unsuitable for large scale or online applications. Recent works have taken the approach of learning a neural network that maps strokes to Bézier curves in a single shot [8], solving the limitations of the previous attempts. However, all these methods require as input the sequence of 2D points that made up each stroke, which makes them unsuitable to work on image data.

Inverse graphics is the line of work that aims to estimate 3D scene parameters from raster images without supervision, by predicting the input of a render program that can reconstruct the image [17,28]. A specialized case of inverse graphics is to estimate parameters of 2D curves, and recent models have been proposed to tackle this problem. An RNN based agent named SPIRAL [23] learned to draw characters in terms of brush Bézier curves, but it had the drawback of being extremely costly due to its reliance in Policy Gradient [31] and a black-box renderer. Another recent RNN based model [8] learned to draw sketches in a self-supervised manner by fitting Bézier curves to strokes. It uses a white-box Bézier curve renderer, but it can only handle vector representations of input sketches.

In contrast to the state of the art approaches to curve fitting and 2D inverse graphics, our model is designed to handle raster sketch and handwritten data, learns in a self-supervised manner, and makes single-shot predictions.

2.2 Transformers and Parallel Decoding

Transformers were introduced by Vaswani *et al.* [34] as an attention-based building block for machine translation tasks. Attention mechanisms [1] aggregate information from an entire input sequence, and the Transformer architecture introduced self-attention layers that update each element by aggregating information from the whole sequence. Transformers perform global computations and have perfect memory, which makes them better than RNNs on long sequences. In the last years, Transformers have been replacing RNNs in many problems in natural language processing, speech processing and computer vision [4,15,22,24,32].

Transformers were first used as auto-regressive models, able to generate one output token at a time similarly to the previous sequence-to-sequence models [30]. However, the time complexity motivated the development of parallel sequence generation in the domains of machine translation [11], word representation learning [15], speech recognition [5] and object detection [4]. We also combine transformers and parallel decoding for their suitable trade-off between computational cost and ability to perform global computations required for sketch embedding.

3   Methodology

3.1   Mathematical Background

A main step when training inverse graphics models is rendering the estimated curve in a canvas. As explained in Sect. 2.1, this rendering is done through parametric curves like Bézier or Splines. In this section, we introduce the main mathematical concepts required to understand the proposed architecture, described in Sect. 3.3.

Bézier curves are parametric curves defined by a sequence of M control points $\mathcal{CP} = (\mathcal{P}_m)_{m=0}^{M-1}$. A Bézier curve B with M control points $\mathcal{CP}$ is formally defined as:

$$B(t; \mathcal{CP}) = \sum_{m=0}^{M-1} \mathcal{B}_{m,M-1}(t) \cdot \mathcal{P}_m, \tag{1}$$

where $t \in [0,1]$ is the parameter of the curve and $\mathcal{B}_{m,M-1}(t) = \binom{M-1}{m} t^m (1 - t)^{M-1-m}$. The curve starts at $\mathcal{P}_0$ and ends at $\mathcal{P}_{M-1}$, whereas the control points $(\mathcal{P}_1, \ldots, \mathcal{P}_{M-2})$ control its trajectory as shown in Fig. 1.

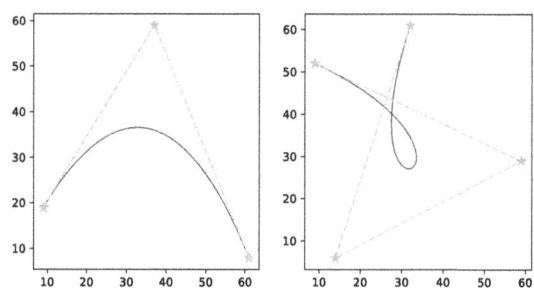

Fig. 1. Examples of Bézier curves with 3 and 4 control points, respectively.

Probabilistic Bézier curves are an extension of Bézier curves so that control points are replaced by a set of distributions driving their position [14]. Their definition is similar to usual Bézier curves: given the M control points $\mathcal{CP} = (P_i)_{i=1}^{M-1}$ of a Bézier curve its *probabilistic* version defines M normal distributions centered in each control point P_i and covariance matrix Σ_i. Therefore, each point $B(t; \mathcal{CP})$ of the probabilistic Bézier curve follows a normal distribution:

$$B(t; \mathcal{CP}) \sim N(\mu(t), \Sigma(t)) \tag{2}$$

with

$$\mu(t) = \sum_{i=0}^{M-1} \mathcal{B}_{i,M-1}(t) P_i \quad \text{and} \quad \Sigma(t) = \sum_{i=0}^{M-1} \mathcal{B}_{i,M-1}(t)^2 \Sigma_i \tag{3}$$

Basically, the probabilistic Bézier curve is a stochastic process with a 2D gaussian density function $f_t(x, y; \mathcal{CP})$ for $t \in [0,1]$. Using this probabilistic construction, we can compute a differentiable *probability map*, shown in Fig. 2, given by:

$$\text{pmap}(x, y; \mathcal{CP}) = \max_{t \in [0,1]} f_t(x, y; \mathcal{CP}). \tag{4}$$

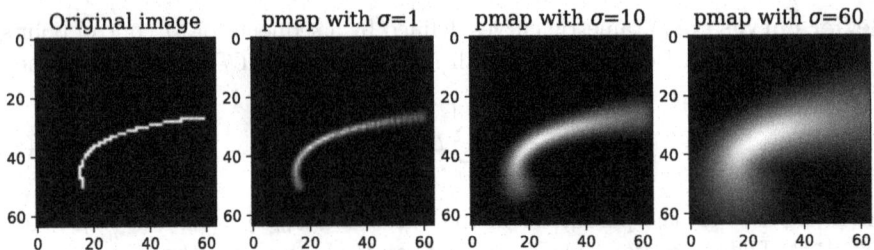

Fig. 2. Image of a Bézier curve $\mathcal{B}$ and three probability maps generated with the control points P_i of $\mathcal{B}$ as the means μ_i of the probabilistic control points, and for different values of σ in the covariance matrix $\Sigma = \sigma I$, shared by all the probabilistic control points.

Chamfer distance is a measure of dissimilarity between two finite sets of points S_1, S_2 [7,20] and it is given by:

$$d_{CD}(S_1, S_2) = \frac{1}{|S_1|} \sum_{p_i \in S_1} \min_{p_j \in S_2} ||p_i - p_j||_2 + \frac{1}{|S_2|} \sum_{p_j \in S_2} \min_{p_i \in S_1} ||p_j - p_i||_2 \quad (5)$$

Traditionally, the Chamfer distance has been used in computer vision with binary images to compare shape contours, line-based drawings and sketches. In this case, the Chamfer distance is computed as:

$$d_{CD}(I_1, I_2) = \frac{\langle I_1, \mathrm{dmap}(I_2) \rangle}{||I_1||_1} + \frac{\langle \mathrm{dmap}(I_1), I_2 \rangle}{||I_2||_1} \quad (6)$$

where $\mathrm{dmap}(\cdot)$ is a distance transform operator applied to binary images. Thus, each pixel of the resulted image after applying the $\mathrm{dmap}(I)$ operator has value equal to the euclidean distance between this pixel and the nearest non-zero pixel in I, see Fig. 3.

3.2   Problem Formulation

Let $\mathcal{X}$ be a hand-drawn symbol dataset, containing binary symbol images. Let us define $\mathcal{Y}_{N,M}$ as the space of all possible combinations of N Bézier curves of M control points. Note that $N, M \in \mathbb{N}$ are fixed and application dependent.

Then, given a binary image $x_i \in \mathcal{X}$ as input, the proposed model $\mathbf{f}(\cdot)$ has the objective to yield an approximation $\hat{y}_i \in \mathcal{Y}$ of the image x_i. Thus, given any renderer $\mathbf{r}(\cdot)$ and a measure of similarity between two images $\mathbf{s}(\cdot, \cdot)$ the problem is formally defined as:

$$\theta = \arg\max_{\theta} \mathbf{s}(\mathbf{r}(\hat{y}_i), x_i) = \arg\max_{\theta} \mathbf{s}(\mathbf{r}(\mathbf{f}(x_i; \theta)), x_i) \quad (7)$$

where θ are the learnable parameters of the model $\mathbf{f}(\cdot)$. Note, that standard renderers $\mathbf{r}(\cdot)$ are not differentiable and, therefore, this cannot be directly optimized by backpropagation.

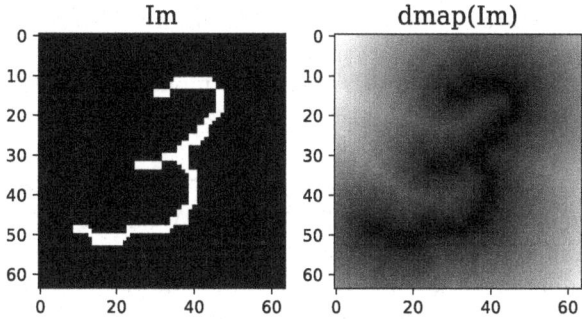

Fig. 3. Example of an image and its distance map.

The proposed framework is divided in two main components. The model $\mathbf{f}(\cdot)$ which transforms the input image to the corresponding set of Bézier curves. The second component is the loss function that makes possible a self-supervised learning process that provides the embedding with the desired properties.

3.3 Model Architecture

The proposed model learns a function $\mathbf{f}(\cdot)$ that given a binary image $x_i \in \mathcal{X}$ as an input, yields to its best fit approximation $\hat{y}_i \in \mathcal{Y}$ in terms of a set of N Bézier curves of M control points.

The full model can be seen in Fig. 4. It has three main components: a convolutional (CNN) backbone $\psi(\cdot)$ that extracts image features, an encoder-decoder transformer, and a final fully-connected layer that outputs the $N \times M$ control points corresponding to the desired N Bézier curves.

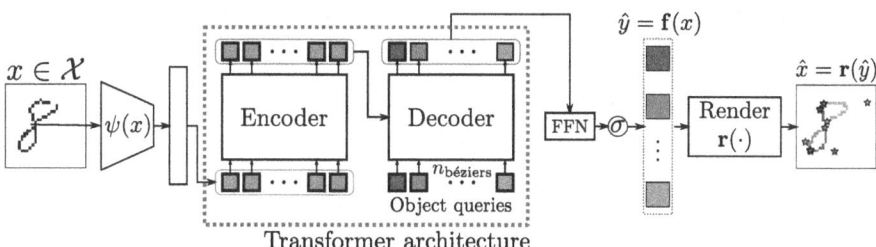

Fig. 4. Overview of the proposed model. In the output example, the model has predicted 3 Bézier curves of 3 control points. Each control point appears in the image drawn as a star.

CNN Backbone. We use a 6-block ResNet [13] that transforms the initial image $x \in \{0,1\}^{1 \times 64 \times 64}$ to a feature map $f \in \mathbb{R}^{512 \times 2 \times 2}$ of whom we extract 4 feature vectors of size 512.

Transformer Encoder. We use a 6-layer transformer encoder following the standard implementation described in [34] except that it does not have positional encoding. The transformer encoder receives as input the sequence of 4 feature vectors generated by the CNN backbone.

Transformer Decoder. The decoder consists of a 6-layer transformer decoder that maps N input embeddings of size 512 into N output embeddings of size 512. The two differences with the original work [34] are that we do not use positional encoding and that we are able to decode the N objects in parallel instead of sequentially in an autoregressive manner. The N input embeddings, namely object queries, are learnt following the work of Carion *et al.* [4].

Linear Layer. The final prediction is computed by a simple fully-connected layer that outputs the $N \times M$ control points corresponding to the desired N Bézier curves. Thus, each of the N curves is represented by M control points of 2 real numbers. The control points are then passed through a sigmoid activation and finally, they are multiplied by the size of the image. In this way, the proposed model is independent of the input image size.

3.4 Objective Functions

This section describe the two proposed loss functions, both allowing a self-supervised training of the proposed model. The first loss is simply an implementation of the Chamfer distance predicted by the model, and the second loss is a probabilistic approximation of the Chamfer distance. Note that the proposed losses are independent and the training process is regulated by one of them.

Chamfer Loss. Let $x \in \mathcal{X}$ be the input image and S be the set of points of the sketches contained in x. Let S_p be the sequence of points sampled from the N Bézier curves defined by the predicted control points $\hat{y} = \mathbf{f}(x)$. We apply Eq. 1 to $\hat{y}$, where t takes values between 0 and 1 with a step of $\frac{1}{\tau}$, to obtain $S_p = \{p_i\}_{i=1}^{\tau N}$ with its points p_i being differentiable with respect to $\hat{y}$. Then, the proposed Chamfer loss is just the Chamfer distance:

$$\mathcal{L}_{\mathrm{CD}} = d_{\mathrm{CD}}(S, S_p) = \frac{1}{|S_p|} \sum_{p_j \in S_p} \min_{p_i \in S} ||p_j - p_i||_2 + \frac{1}{|S|} \sum_{p_i \in S} \min_{p_j \in S_p} ||p_i - p_j||_2 \quad (8)$$

Probabilistic Chamfer Loss. Let $x \in \mathcal{X}$ be the input image of size K. Let $\hat{y} = \{\mathcal{CP}_n\}_{n=1}^{N}$ be the set of the Bézier curves predicted by the model and $\mathcal{CP}_n = (P_{nm})_{m=0}^{M-1}$ be the set of control points of the n-th predicted Bézier curve, as defined before in Sect. 2.1. Let S_p be the sequence of points of $\hat{y}$ computed as for the first loss. Then our probabilistic version of the Chamfer distance is given by

$$\mathcal{L}_{\mathrm{PCD}} = \frac{\langle \hat{x}, \mathrm{dmap}(x) \rangle}{||\hat{x}||_1} + \frac{1}{|S|} \sum_{p_i \in S} \min_{p_j \in S_p} ||p_i - p_j||_2 \quad (9)$$

where

$$\hat{x}[k,l] = \max_{\mathcal{CP} \in \hat{y}} \text{pmap}(k,l; \mathcal{CP}) \quad \forall k,l = 0, \ldots, K-1. \tag{10}$$

We highlight that $\hat{x}$ is an image of the same size that the input image x that is differentiable with respect to the probabilistic Bézier curves parameters ($\hat{y}$ and Σ). Also, $p_j \in S_p$ are differentiable with respect the predicted control points $\hat{y}$.

4   Experimental Evaluation

4.1   Dataset and Implementation Details

MNIST Dataset [19]: The well known *MNIST* dataset contains 60000 28×28 images of handwritten thick digits. Our inverse graphics framework is only able to produce curves without width, and therefore we have applied a preprocessing step mainly consisting in a skeletonization of the digits. We have also placed the 28×28 digits in a random position into 64×64 images.

Quick, Draw! Dataset [12]: *Quick, Draw!* is a large sketch dataset collected thanks to an online drawing game in which players had to draw an object from a given category within a time-limit. It consists in 50 million drawings across 345 categories, but we have only trained the model in the categories *apple*, *axe*, *banana* and *baseball bat*. Each draw comes as a sequence of coordinates with different scales, we have resized all the sequences to fit in a 64×64 image and we draw them by joining the consecutive coordinates with lines.

Implementation Details: We train our model with the Adam [16] optimizer with an initial learning rate of $5 \cdot 10^{-5}$ that is decreased by a factor of $10^{-0.5}$ if the loss is not reduced during 8 epochs. We use a batch size of 64. The training takes 120 epochs into *MNIST*, and 200 epochs into *Quick, Draw!*. The whole framework was implemented with PyTorch [25] deep learning tool. The code is available at https://github.com/AlbertSuso/InverseGraphicsSketchParametrization_grecPaper.

4.2   Model Evaluation

In this section, we provide a quantitative and qualitative evaluation of our model. In addition, we carefully compare the two proposed losses as well as the parameters that define the number of Bézier curves per image and the number of control points per curve.

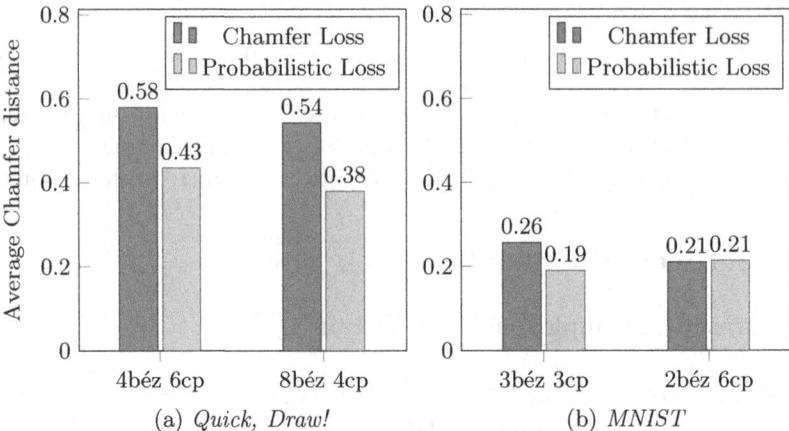

(a) *Quick, Draw!*   (b) *MNIST*

Fig. 5. Chamfer distance validation metric obtained with models that predict different number of Bézier curves per image and number of control points per curve, trained using the *Chamfer Loss* and the *Probabilistic Chamfer Loss*. The models have been trained over the classes *apple*, *axe*, *banana* and *baseball bat* of the *Quick, Draw!* dataset (a) and over the *MNIST* dataset (b).

Figure 5 shows the average Chamfer distance that different models have obtained over the validation set. The different models differ on the number of Bézier curves they predict for each input image, on the number of control points per curve and on the loss function they have trained on. As we can see, the *Probabilistic Chamfer Loss* significantly outperforms the original *Chamfer Loss* in nearly all the experiments in both datasets. Our hypothesis is that this happens because, as shown in Fig. 6, the probability map provides information in a neighborhood of the Bézier curve, and this helps to smooth the loss surface. Looking at the results it is clear that, as expected due to its greater complexity, the *Quick, Draw!* dataset has proven to be more challenging than the *MNIST*.

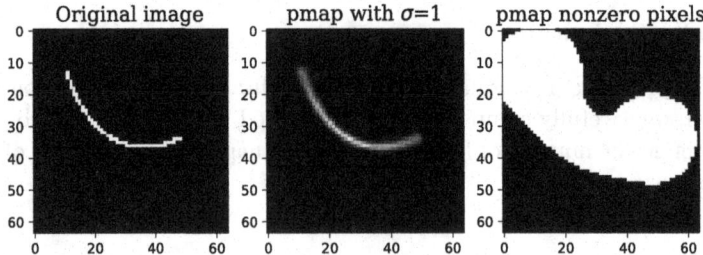

Fig. 6. Image of a Bézier curve, its probability map with $\Sigma = \sigma I$ and a binarization of the probability map showing its nonzero pixels.

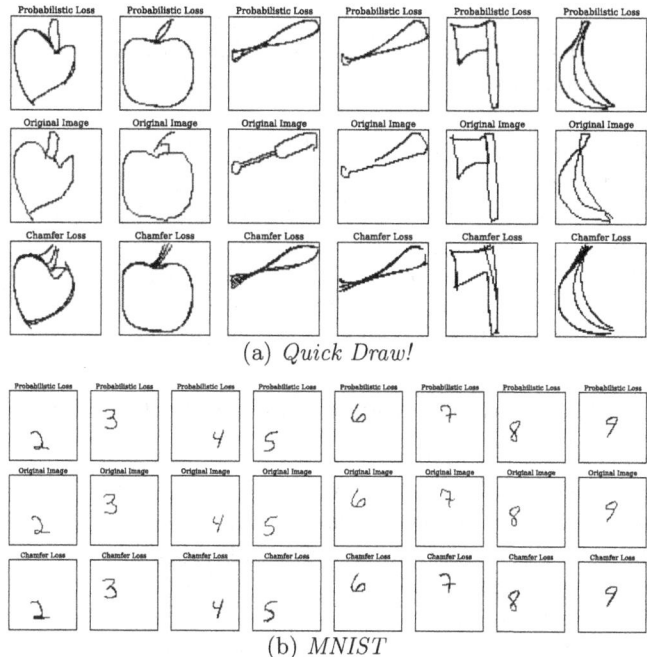

(a) *Quick Draw!*

(b) *MNIST*

Fig. 7. Images of (a) *Quick, Draw!* and (b) *MNIST* dataset (second row) and its predicted images by a model trained with the *Probabilistic Chamfer Loss* (first row) and by a model trained with the original *Chamfer Loss* (third row). The model trained on *MNIST* predicts 3 Bézier curves of 3 control points each (per image), whereas the model trained on *Quick, Draw!* predicts 8 Bézier curves of 4 control points each.

Figure 7 show a qualitative evaluation of the model and both losses over the *Quick, Draw!* and the *MNIST* datasets. As can be seen, while the results obtained with the *Chamfer Loss* are not bad, the results obtained with the *Probabilistic Chamfer Loss* are far better from the human point of view. It is also interesting the fact that, specially in the *Quick, Draw!* dataset, the predicted images are smoother than the original ones and sometimes even add details that enhance the visualization from the human point of view.

We highlight that, although the model outputs a fixed number of Bézier curves, it is able to adapt to simpler input images. When the input image is simple and the required number of curves is lower than N, the model produces repeated output curves and therefore implicitly decreases the number of predicted Bézier curves.

4.3   Zero-Shot Evaluation

Despite presenting a self-supervised approach which does not require any class information, the set of images used for training might effect the final result. Therefore, we present a study on the generalization properties of our model to verify that it does, indeed, learn a generic model able to obtain the proper Bézier representation. Thus, we propose a zero-shot evaluation where images belonging to unseen classes have been used. This experimental setting has been applied to both datasets.

The results in Fig. 10 (see before overfitting bars in blue) show that the model generalizes reasonably well to unseen classes. Note that despite the model has never seen these classes before, it is able to obtain a fairly good approximation. Note that some of the selected classes do require a higher number of curves for a perfect fit and, therefore, the model is not able to obtain an accurate solution. However, we conclude that our model does not overfit to any specific category but still cannot deal with highly detailed sketches such as "sweater".

4.4 Towards High Fidelity Approximations via Overfitting

The self-supervised nature of the training process allows us to obtain highly accurate predictions by applying an overfitting process over the image being considered. Basically, the process consists in performing between 50 and 250 training steps over the same input image to obtain the set of curves whose Chamfer distance with respect to the original image is the lowest. Figure 8 shows an example of this overfitting process. There, we can observe that increasing the number of steps, our model is able to recover fine-grain details that where missed by a single inference pass.

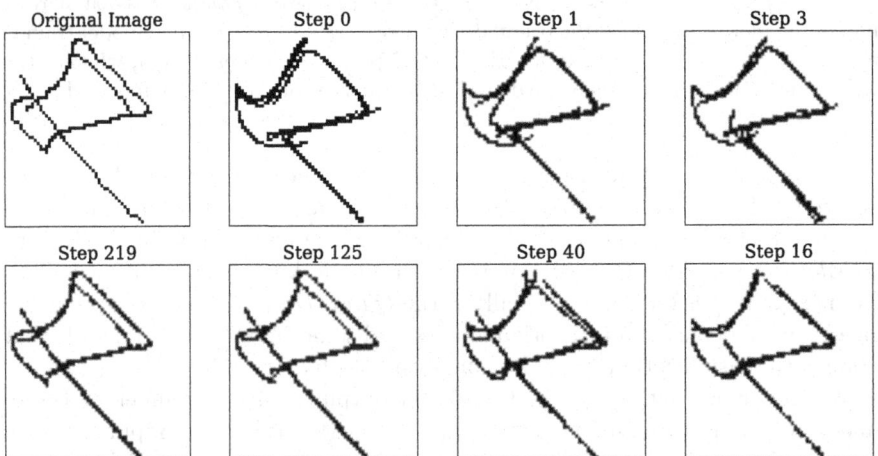

Fig. 8. Example of an overfitting process. The top-left image is the original image, the *Step 0 image* is the one predicted without overfitting and the other images have been obtained during the process and are arranged clockwise.

The overfitting technique, while not being the principal objective of the paper, provides new possibilities with the objective of obtaining a high resolution and high-fidelity representation of sketches in a vectorized format. Figure 9

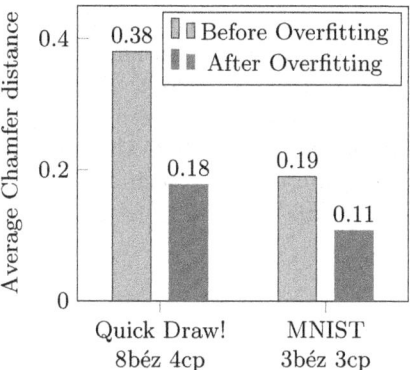

Fig. 9. Comparison on the average Chamfer distance between our best models for the *MNIST* and *Quick, Draw!* datasets before and after overfitting.

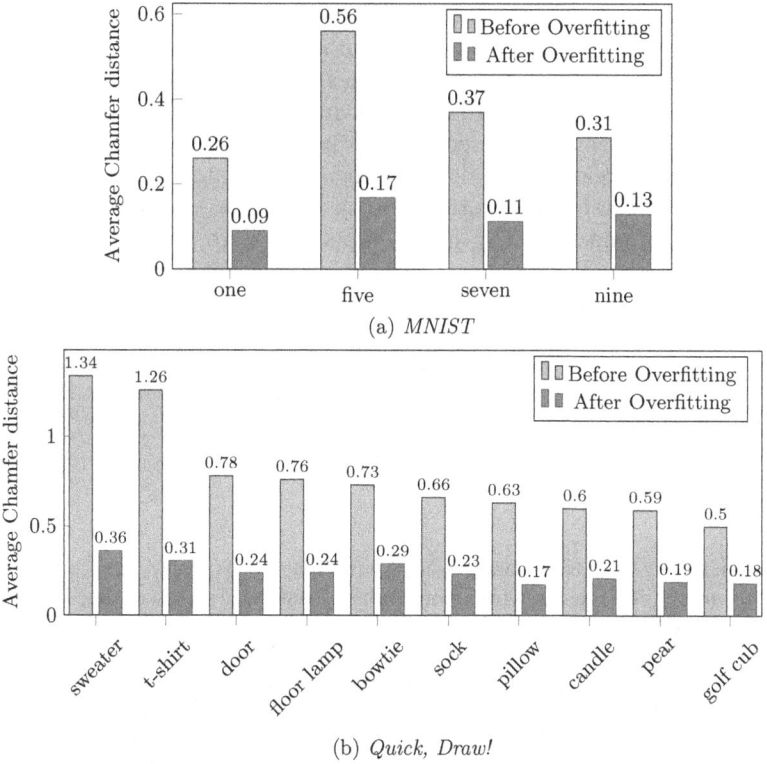

(a) *MNIST*

(b) *Quick, Draw!*

Fig. 10. Chamfer distance comparison for the the (a) *MNIST* and (b) *Quick, Draw!* datasets in a zero-shot setting. The model has been trained over a training set consisting of the classes *two*, *three*, *four*, *six* and *eight* for the *MNIST* and; *apple*, *axe*, *banana* and *baseball bat* for the *Quick, Draw!*. In addition, The model obtained a chamfer evaluation metric of 0.25 and 0.38 over the training classes before the overfitting for each dataset respectively.

shows the improvement over the trained model before and after the training steps for both datasets. In particular, we considered the best model from Fig. 5.

Figure 10 provides a zero-shot comparison of the same process. After few training iterations have been considered, there is a huge improvement over the average Chamfer distance for each class. Note that even hard examples that, in the ideal case would require more curves, such as *sweater*, are able to decrease their distance by a huge margin.

5   Conclusion

In this work we have proposed a study on an inverse graphics problem for hand-written digit or sketch images. Even though previous works have studied such abstract representations for problems such as classification, retrieval, or generation, the online nature of these images have been often neglected. A relevant exception is the work of Bhunia *et al.* [2] that exploits this feature to self supervise an sketch encoder. However, their proposed vectorization generate long and noisy point-wise data. In comparison, our system is able to generate compact representations in terms of Bézier curves. Our model has demonstrated its potential for generating approximations of a given sketch using a fix number of Bézier curves without requiring any ground-truth information. In addition, if a high accuracy approximation is required, an overfitting strategy has been adopted.

The research developed in this work opens some interesting research lines. In comparison to the before-mentioned work [2], the obtained model is able to develop a deeper understanding of the input image to generate the required set of Bézier curves. Therefore, we will explore the potential of this work for pretraining a feature extractor in problems such as classification and retrieval.

Acknowledgment. This work has been partially supported by the Spanish projects RTI2018-095645-B-C21 and FCT-19-15244, the Catalan project 2017-SGR-1783, and the CERCA Program/Generalitat de Catalunya.

References

1. Bahdanau, D., Cho, K.H., Bengio, Y.: Neural machine translation by jointly learning to align and translate. In: International Conference on Learning Representations, ICLR (2015)
2. Bhunia, A.K., Chowdhury, P.N., Yang, Y., Hospedales, T., Xiang, T., Song, Y.Z.: Vectorization and rasterization: Self-supervised learning for sketch and handwriting. In: CVPR (2021)
3. de Boor, C.: A Practical Guide to Spline, vol. 27, January 1978. https://doi.org/10.2307/2006241
4. Carion, N., Massa, F., Synnaeve, G., Usunier, N., Kirillov, A., Zagoruyko, S.: End-to-end object detection with transformers. In: Vedaldi, A., Bischof, H., Brox, T., Frahm, J.-M. (eds.) ECCV 2020, Part I. LNCS, vol. 12346, pp. 213–229. Springer, Cham (2020). https://doi.org/10.1007/978-3-030-58452-8_13

5. Chan, W., Saharia, C., Hinton, G., Norouzi, M., Jaitly, N.: Imputer: sequence modelling via imputation and dynamic programming. In: International Conference on Machine Learning, ICML, pp. 1403–1413 (2020)
6. Chen, W., Hays, J.: SketchyGAN: towards diverse and realistic sketch to image synthesis. In: CVPR (2018)
7. Dantanarayana, L., Dissanayake, G., Ranasinge, R.: C-log: a chamfer distance based algorithm for localisation in occupancy grid-maps. CAAI Trans. Intell. Technol. 1(3), 272–284 (2016)
8. Das, A., Yang, Y., Hospedales, T., Xiang, T., Song, Y.-Z.: BézierSketch: a generative model for scalable vector sketches. In: Vedaldi, A., Bischof, H., Brox, T., Frahm, J.-M. (eds.) ECCV 2020, Part XXVI. LNCS, vol. 12371, pp. 632–647. Springer, Cham (2020). https://doi.org/10.1007/978-3-030-58574-7_38
9. Dey, S., Riba, P., Dutta, A., Lladós, J., Song, Y.Z.: Doodle to search: practical zero-shot sketch-based image retrieval. In: CVPR, pp. 2179–2188 (2019)
10. Egiazarian, V., et al.: Deep vectorization of technical drawings. In: Vedaldi, A., Bischof, H., Brox, T., Frahm, J.-M. (eds.) ECCV 2020, Part XIII. LNCS, vol. 12358, pp. 582–598. Springer, Cham (2020). https://doi.org/10.1007/978-3-030-58601-0_35
11. Gu, J., Bradbury, J., Xiong, C., Li, V.O., Socher, R.: Non-autoregressive neural machine translation. In: International Conference on Learning Representations, ICLR (2018)
12. Ha, D., Eck, D.: A neural representation of sketch drawings. In: International Conference on Learning Representations, ICLR (2018)
13. He, K., Zhang, X., Ren, S., Sun, J.: Deep residual learning for image recognition. In: CVPR (2016)
14. Hug, R., Hübner, W., Arens, M.: Introducing probabilistic bézier curves for n-step sequence prediction. In: AAAI Conf. Artif. Intell., vol. 34, issue 06, pp. 10162–10169 (2020). https://doi.org/10.1609/aaai.v34i06.6576
15. Kenton, J.D., Ming-Wei, C., Toutanova, L.K.: BERT: pre-training of deep bidirectional transformers for language understanding. In: NAACL-HLT, pp. 4171–4186 (2019)
16. Kingma, D.P., Ba, J.: Adam: a method for stochastic optimization. In: International Conference on Learning Representations, ICLR (2015)
17. Kulkarni, T.D., Whitney, W.F., Kohli, P., Tenenbaum, J.: Deep convolutional inverse graphics network. In: International Conference on – Neural Information Processing Systems (2015)
18. Lake, B.M., Salakhutdinov, R., Tenenbaum, J.B.: Human-level concept learning through probabilistic program induction. Science 350(6266), 1332–1338 (2015)
19. Lecun, Y., Bottou, L., Bengio, Y., Haffner, P.: Gradient-based learning applied to document recognition. Proc. IEEE 86(11), 2278–2324 (1998). https://doi.org/10.1109/5.726791
20. Liu, M.Y., Tuzel, O., Veeraraghavan, A., Chellappa, R.: Fast directional chamfer matching, pp. 1696–1703 (2010). https://doi.org/10.1109/CVPR.2010.5539837
21. Liu, Y., Wang, W.: A revisit to least squares orthogonal distance fitting of parametric curves and surfaces. In: Chen, F., Jüttler, B. (eds.) GMP 2008. LNCS, vol. 4975, pp. 384–397. Springer, Heidelberg (2008). https://doi.org/10.1007/978-3-540-79246-8_29
22. Lüscher, C., et al.: RWTH ASR systems for librispeech: hybrid vs attention. In: Proceedings of the Interspeech, pp. 231–235 (2019)
23. Mellor, J.F., et al.: Unsupervised doodling and painting with improved spiral. arXiv preprint arXiv:1910.01007 (2019)

24. Parmar, N., et al.: Image transformer. In: International Conference on Machine Learning, ICML, pp. 4055–4064 (2018)
25. Paszke, A., et al.: Pytorch: An imperative style, high-performance deep learning library. arXiv preprint arXiv:1912.01703 (2019)
26. Plass, M., Stone, M.: Curve-fitting with piecewise parametric cubics. In: Proceedings of the annual conference on Computer Graphics and Interactive Techniques, pp. 229–239 (1983)
27. Revow, M., Williams, C., Hinton, G.: Using generative models for handwritten digit recognition. IEEE PAMI **18**(6), 592–606 (1996). https://doi.org/10.1109/34.506410
28. Romaszko, L., Williams, C.K., Moreno, P., Kohli, P.: Vision-as-inverse-graphics: obtaining a rich 3D explanation of a scene from a single image. In: IEEE International Conference on Computer Vision, pp. 851–859 (2017)
29. Salomon, D.: Curves and Surfaces for Computer Graphics. Springer-Verlag, New York (2005). https://doi.org/10.1007/0-387-28452-4
30. Sutskever, I., Vinyals, O., Le, Q.V.: Sequence to sequence learning with neural networks. In: ICONIP, pp. 3104–3112 (2014)
31. Sutton, R.S., McAllester, D.A., Singh, S.P., Mansour, Y., et al.: Policy gradient methods for reinforcement learning with function approximation. Adv. Neural Inf. Process. Syst. **99**, 1057–1063 (1999)
32. Synnaeve, G., et al.: End-to-end ASR: from supervised to semi-supervised learning with modern architectures. arXiv preprint arXiv:1911.08460 (2019)
33. Tripathi, A., Dani, R.R., Mishra, A., Chakraborty, A.: Sketch-guided object localization in natural images. In: Vedaldi, A., Bischof, H., Brox, T., Frahm, J.-M. (eds.) ECCV 2020, Part VI. LNCS, vol. 12351, pp. 532–547. Springer, Cham (2020). https://doi.org/10.1007/978-3-030-58539-6_32
34. Vaswani, A., et al.: Attention is all you need. In: Advances in Neural Information Processing Systems (2017)
35. Zhang, H., Liu, S., Zhang, C., Ren, W., Wang, R., Cao, X.: SketchNet: sketch classification with web images. In: CVPR (2016)

Border Detection for Seamless Connection of Historical Cadastral Maps

Ladislav Lenc[1,2(✉)], Martin Prantl[1,2], Jiří Martínek[1,2], and Pavel Král[1,2]

[1] Department of Computer Science and Engineering, Faculty of Applied Sciences,
University of West Bohemia, Plzeň, Czech Republic
{llenc,perry,jimar,pkral}@kiv.zcu.cz
[2] NTIS - New Technologies for the Information Society, Faculty of Applied Sciences,
University of West Bohemia, Plzeň, Czech Republic

Abstract. This paper presents a set of methods for detection of important features in historical cadastral maps. The goal is to allow a seamless connection of the maps based on such features. The connection is very important so that the maps can be presented online and utilized easily. To the best of our knowledge, this is the first attempt to solve this task fully automatically. Compared to the manual annotation which is very time-consuming we can significantly reduce the costs and provide comparable or even better results.

We concentrate on the detection of cadastre borders and important points lying on them. Neighboring map sheets are connected according to the common border. However, the shape of the border may differ in some subtleties. The differences are caused by the fact that the maps are hand-drawn. We thus aim at detecting a representative set of corresponding points on both sheets that are used for transformation of the maps so that they can be neatly connected. Moreover, the border lines are important for masking the outside of the cadastre area.

The tasks are solved using a combination of fully convolutional networks and conservative computer vision techniques. The presented approaches are evaluated on a newly created dataset containing manually annotated ground-truths. The dataset is freely available for research purposes which is another contribution of this work.

Keywords: Historical document images · Cadastral maps · Fully convolutional networks · FCN · Computer vision

1 Introduction

In the recent years, there has been a growing interest in the field of historical document and map processing. Such materials already exist in digital form (scanned digital images) and the efforts of current research lie in the automatic processing of such documents (e.g. information retrieval, OCR and full-text search but also

© Springer Nature Switzerland AG 2021
E. H. Barney Smith and U. Pal (Eds.): ICDAR 2021 Workshops, LNCS 12916, pp. 43–58, 2021.
https://doi.org/10.1007/978-3-030-86198-8_4

computer vision tasks such as noise reduction, localization and segmentation of regions of interest and many others). All such tasks are very important for making the documents easily accessible and usable.

This paper deals with historical cadastral Austro-Hungarian maps from the 19th century. The maps are available as isolated map sheets covering particular cadastre areas. To be able to fully utilize such materials it is very important to create a continuous seamless map that can be easily presented online.

The map sheets are arranged in a rectangular grid [1]. Each cell of the grid can contain one or more map sheets (if the grid lies on the border of two or more cadastre areas). A preliminary step is to place the map sheets into corresponding cells according to textual content contained in the sheets. In this work, we concentrate on merging of sheets lying on the same position in the grid. The sheets have to be connected according to the common border line present in both sheets. However, the maps are hand-drawn and there might be tiny differences in the position and shape of corresponding border lines. We thus have to transform the sheets so that they can be neatly put together. Two map sheets belonging to the same cell in the grid are shown in Fig. 1.

Our goal is twofold. First, we have to find a set of corresponding points in the two adjacent sheets that allow the transformation. The second goal is to find the border line which allows us to mask the regions outside the cadastre area.

The presented methods are developed as a part of a robust system for a seamless connection of map sheets which will require a minimum manual interaction of the user. This is the first attempt to solve this task fully automatically. We utilize machine learning approaches based on fully convolutional neural networks (FCN) complemented by analytic computer vision approaches. It is important to note that there is a potential of using the same system for processing of cadastre maps from neighboring countries that were also part of the Austro-Hungarian empire and have the same form (i.e. Slovakia, Hungary, Austria and parts of Ukraine, Serbia or Croatia).

An important contribution is the creation of an annotated dataset which is used for training of neural networks as well as for evaluation of the methods. This dataset is freely available for research purposes.

2   Related Work

To the best of our knowledge there are no studies utilizing deep learning for exactly the same task. We thus present relevant papers utilizing neural networks for map analysis and processing with a particular focus on segmentation methods based on neural networks.

A complete survey of historical cadastral maps digitization process is provided by Ignjatić et al. in [2]. The authors try to pave the way for employing deep neural networks in this field, highlighting potential challenges in map digitization.

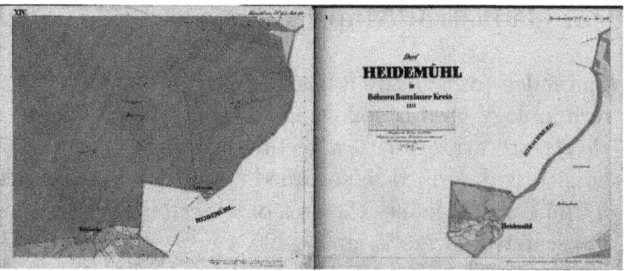

Fig. 1. Example of map sheets belonging to the same cell; The map areas are complements to each other; The outer area of one sheet is the inner area of the other one.

Convolutional neural networks are used for the analysis of aerial maps by Timilsina et al. in [3]. The objective of this work is to identify tree coverage of cadastre parcels.

Illegal building detection from satellite images is investigated by Ostankovich and Afanasyev in [4]. Their approach integrates various computer vision techniques with the *GoogLeNet* model and obtains reasonable and acceptable results that are verified against cadastral maps.

A method for cadastre border detection utilizing FCNs was presented by Xia et al. in [5]. They were able to effectively extract cadastral boundaries, especially when a large proportion of cadastral boundaries is visible. Another example of cadastre border extraction has been explored by Fetai et al. [6].

Nyandwi et al. [7] present detection and extraction method of visible cadastral boundaries from very high-resolution satellite images. They compared their approach with human analysis and they were able to obtain good results on rural parcels. On the other hand, the methods for the urban area extractions have significant room for improvement.

Kestur et al. [8] presented an approach for road detection from images acquired by sensors carried out by an unmanned aerial vehicle (UAV). They obtained promising results with a novel U-shaped FCN (UFCN) model that is able to take inputs of arbitrary size.

A combination of neural networks and mathematical morphology for historical map segmentation was proposed in [9]. The approach first utilizes CNN based network for edge detection. Follows morphological processing which helps to close the shapes and extract final object contours.

To sum up the related work, we can say that variants of CNN, especially the FCN architecture are suitable for our task of detection and segmentation of the important map features.

3   Historical Cadastral Maps

This section is intended to facilitate a basic understanding of the maps we work with and the terminology used in following sections.

The map sheets are arranged to a rectangular grid. The position (cell) of the sheet in the grid is described in so called "nomenclature". Each map sheet contains a map frame which defines the area of the cell that the sheet covers. The resolution of the scanned sheets is circa 8400×6850 pixels and the dimension of the map frame is usually around 7750×6200 pixels. The area inside the map frame can be divided into the map itself and the blank area outside the map. The map area is delimited by cadastre border which is marked as a black line with map symbols (dots, triangles, crosses etc.) determining the type of the border.

Our main concern is finding map sheets from neighboring cadastre areas that belong to the same cell. Such sheets are the most problematic ones for the creation of the seamless map. We focus on detection of landmarks ("Grenzsteine" from German) that are marked as black/red dots lying on the cadastre border line. These points are important because their positions are physically marked in the terrain and they are present on both map sheets covering a particular cell.

Another type of important points are significant direction changes on the border line. We will denote such points as "break points". These points are important mainly for map sheets containing no landmark positions. Both types of points can be used for transformation and connection of the map sheets.

The outside of the border line is usually accompanied by a colored (mostly red) area denoted as "edge-line". The border line, landmarks and edge-line are illustrated in Fig. 2.

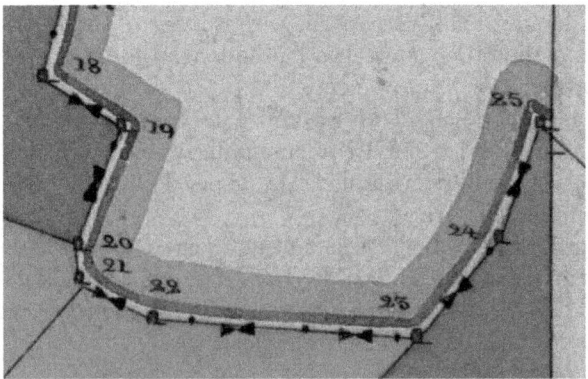

Fig. 2. Detail of a cadastre border with landmarks (red/black symbols) and red edge-line on the outside of the border (Color figure online)

4   Proposed Detection Methods

The scheme of the overall map sheet processing pipeline is depicted in Fig. 3. We assume that the input map sheet is already cropped according to the bounding box given by the map frame. Our task is to find a set of points that can be used for connection with the corresponding (complementing) map sheet. We utilize four separate algorithms, namely *Landmark Detector, Border Line Detector, Break Point Detector* and *Edge-line Detector*, to extract all necessary information. We first search for explicitly drawn landmarks and also identify the border line. If the amount of landmark points is low or we find no landmarks at all, the border line is further processed by the *Break Point Detector* that seeks significant direction changes in the border line and such points are used together with the landmark points to create a set of *Map Border Key-points* that are used for map transformation and connection. The *Edge-line Detector* is useful for removing false positive landmark points. Moreover, the relative edge-line position to the border line is crucial for determining the inner and outer area of the map. According to the border line and edge-line we can construct a map area mask which is subsequently utilized for map area masking.

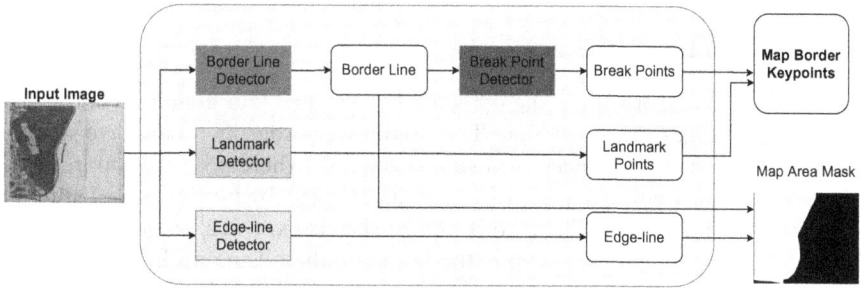

Fig. 3. Overall scheme of the map sheet processing pipeline

4.1   Landmark Detector

The landmark detection algorithm utilizes an FCN trained to predict landmark positions. We use the architecture proposed by Wick and Puppe [10]. In our preliminary experiments we compared this architecture with U-Net [11], which is one of the first and most popular FCN architectures, and obtained comparable results. Moreover, U-Net has ten times more parameters and is computationally much more demanding.

Because of the large size of the input images and the need of preserving relatively small details, we chose a patch-based approach instead of processing the whole (resized) map sheet at once. The training is performed on rectangular image crops that were extracted randomly from annotated training examples. We only ensure that each extracted patch contains at least one landmark if there are any in the map sheet. See Sect. 5.1 for examples of annotated ground-truths.

In the prediction process, we divide the map sheet into a set of overlapping patches and predict each of them by the network. The prediction result is a composition of individual patches predictions. It is a black and white image where white points represent the predicted landmarks.

We then apply a post-processing step based on the connected components analysis. After that we perform a reduction of false positives. We reduce components with size smaller than a specified area threshold (50 px). The centres of remaining components comprise the resulting set of landmark positions.

The further filtration of the resulting points is done according to the detected edge-line (see Sect. 4.4). We keep only the points that are sufficiently close (defined by distance threshold 20 px) to the edge-line.

4.2   Border Line Detector

The border line detector uses the same neural network architecture as the landmark detector. The training is also performed using patches randomly extracted from the training images. We take only patches coinciding with the border line. The trained network is used for patch-wise prediction of a map sheet. The final prediction result contains the predicted border line mask.

4.3   Break Point Detector

The input of this algorithm is the detected border line. Our goal is to find significant direction changes on the line. The predicted border line may have varying width and it is also disconnected in some cases. We therefore apply morphological closing with a circular structuring element (size 10 pixels) in order to fill small holes in the border. Then we compute the skeleton [12] to get a one-pixel wide representation. The skeleton is further simplified using an implementation of *Douglas-Peucker* algorithm [13]. This way we obtain a line (or several line segments) represented by a series of points.

The final step is the reduction of non-significant angles. We inspect all triplets of consecutive points (P_1, P_2 and P_3) and calculate difference in direction of vectors (P_1, P_2) and (P_2, P_3). If this direction difference α is lower than specified angle threshold (we use a value of 10 degrees), we discard point P_2. Figure 4 illustrates the process of direction difference computation.

4.4   Edge-Line Detector

We use traditional image-processing techniques (edge detection, flood fill, morphology etc.) for this task. We divide the pipeline into the following major steps and describe each of them separately.

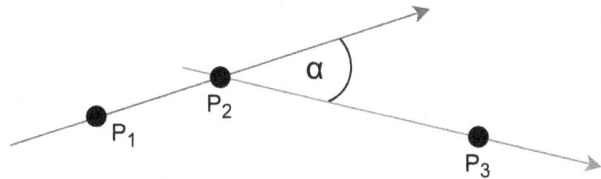

Fig. 4. Illustration of direction change at three consecutive border-line points

Edge Detection. The goal of this step is to create a binary edge map. To partially suppress noise in the input, median filter (with kernel size 9) is used. Edges are then detected by Sobel operator [14] for each RGB channel separately. A single-channel version is created by taking maximal difference among the channels. To create the binary edge map, we use Otsu's thresholding [15].

Components Detection. In this step, we identify areas (components) with similar colors. We first transform the RGB color space into HSV that is more suitable for this task.

Then we detect initial components by iterating the image pixel by pixel. For every pixel P of the image that is not already part of any component or is not an edge pixel, we start the flood filling algorithm [16]. The newly processed pixel Q is added to the current component only if it satisfies the following conditions:

1. Q is not an edge pixel - its value in edge map corresponds to the background;
2. Differences between pixels P and Q in H and S channels are lower than a specified threshold (pixels are considered to have a similar color).

The result of the initial components detection can be seen in the left part of Fig. 5. It can be observed that there are many small, noise-like components. We first filter "holes" formed by edge detection by merging them with the neighboring component (if there is just one).

Second, we iteratively search for small components that have only one unique surrounding neighbor. This process utilizes a max-heap implementation. For the remaining components, we apply morphological opening which removes fine details and jagged edges. Thus we improve stability and performance. The result of the filtering can be seen in the right part of Fig. 5.

Components Filtering. We further filter the remaining components. We focus on long and narrow components that should represent edge-line candidates. To obtain the width of the components, we use signed distance field (SDF) [17]. The length of the component is obtained from its skeleton [12]. Components with sufficient length and width (according to the specific threshold) are marked as edge-line candidates.

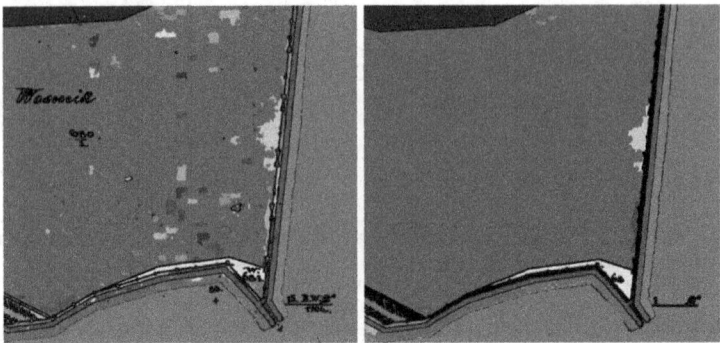

Fig. 5. Component detection example. Each component has a different color, the edges are black. Left: Initial components; Right: After filtering

Nonetheless, after the previous step, we still have many false positives (e.g. roads, rivers). To remove them, we exploit the knowledge that there is a border line with map symbols (mostly "dots") in the immediate vicinity of the edge-line as shown in the middle part of Fig. 6.

The "dots" are detected using the V channel of the HSV color space.

We apply morphological opening to eliminate thin lines and texts that can cause false detections. Then we apply *Harris corner detector* [18] followed by another morphological opening for noise removal. The result is depicted in the right part of Fig. 6.

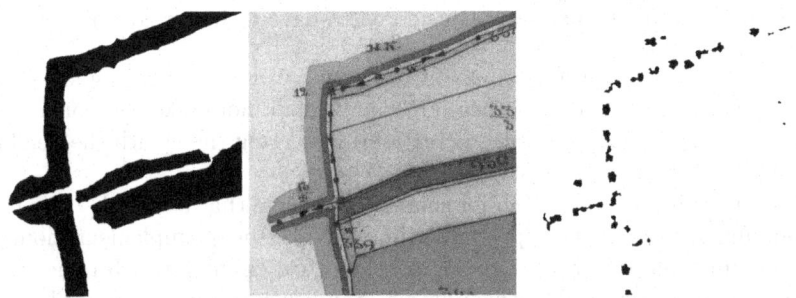

Fig. 6. Left: all detected edge-line candidates; Middle: original image; Right: "dots" on the border line

To filter out false positives, we use following conditions:

1. Majority of "dots" is placed around one side of the candidate;
2. The distance between neighboring "dots" is in a specific range based on the input images. In our experiments it was 20–100 px;

3. The distance between each "dot" and skeleton of the edge-line is in a specific range (10–40 px).

The final detected edge-lines together with associated "dots" are presented in Fig. 7.

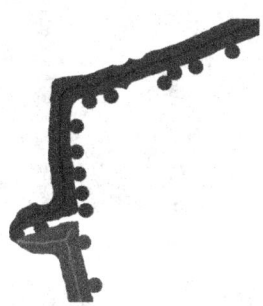

Fig. 7. The final detected edge-lines and their corresponding "dots"; The central line is a vectorized representation of the edge-line.

5 Experiments

5.1 Dataset

In order to obtain the training data for proper FCN training, we have created a dataset containing 100 map sheets. Annotated ground-truths also allow us to evaluate the presented methods.

The dataset is split into training (70 images), validation (10 images) and testing (20 images) parts. For each of the map sheets we have manually annotated a set of ground-truth images for landmarks, border line, break points and edge-line. The ground-truths for break points and edge-line were created only for the testing part because they are not used for network training.

All ground-truth types are stored as binary images with black background and the objects of interest are marked with white color. Landmarks are marked as white circles with diameter corresponding approximately to the real size of the drawn landmarks.

Border lines are drawn as a white line, trying to have a similar width as the original border line. Break point ground-truths were annotated in the same way as landmarks – white dots represent significant direction changes in the border line. Edge-line ground-truths contain white mask for the red edge-line. The latter two types of ground-truths are used only for evaluation purposes.

All types of ground truths are shown in Fig. 8. Note that the crop contains just one explicitly drawn landmark symbol. However, the break-point ground-truth contains 4 points in border line direction changes.

The dataset is freely available for research and education purposes at[1].

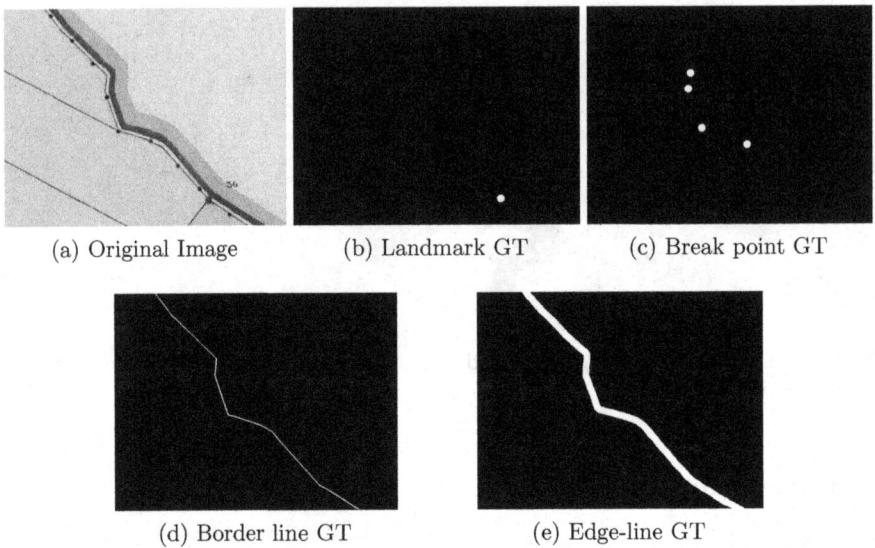

(a) Original Image   (b) Landmark GT   (c) Break point GT

(d) Border line GT   (e) Edge-line GT

Fig. 8. Example of a map sheet fragment with corresponding ground-truths

5.2   Evaluation Criteria

For the evaluation of landmark and break point detectors, we present the standard precision (P), recall (R) and F1 score $(F1)$. The edge-line is represented as a set of consecutive points and thus we use the same evaluation criteria. We first count true positives (TP), false positives (FP) and false negatives (FN) as follows.

For each ground-truth point, we identify the closest predicted point. If their distance is smaller than a certain distance threshold T the point is predicted correctly (TP). Otherwise, it is counted as either FP (present only in prediction) or FN (present only in ground-truth).

Visualization of the evaluation process is depicted in Fig. 9. It shows a fragment of a processed map sheet with marked TP, FP and FN points. Green circles denote TP points, yellow circles are used for FN and yellow crosses for FP points. The colored circles show the distance threshold T used for computation of the scores.

---

[1] https://corpora.kiv.zcu.cz/map_border/.

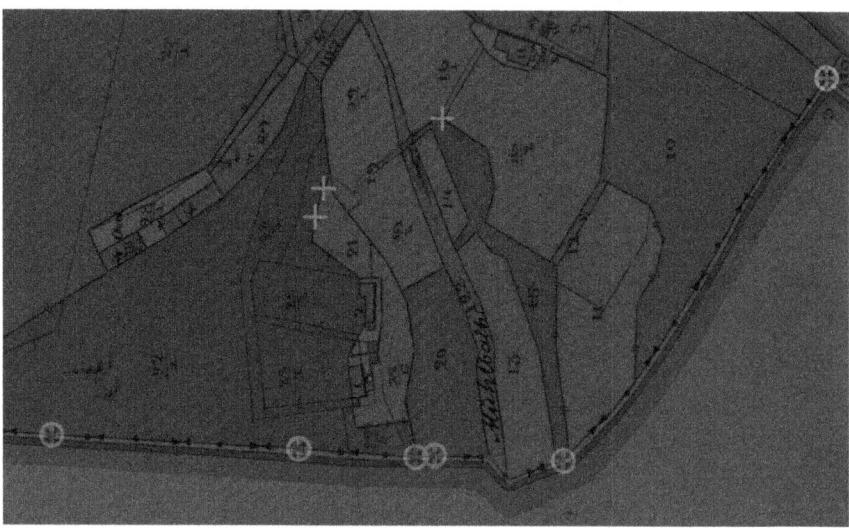

Fig. 9. Visualized TP (green circles), FP (yellow crosses) and FN (yellow circle) points (Color figure online)

5.3   Landmark Detection

The crucial part of the landmark detector is the network trained for landmark prediction. We performed an experiment with different network configurations in order to find suitable parameters. We utilized several sizes of the training patches. The patch size determines the context the network takes into consideration and can thus influence the overall result. We also must consider that larger patches are computationally more demanding.

We also varied the number of patches extracted from one training map sheet. The main reason was to find a minimum amount of patches that provides enough variability to learn the features.

We use the cross-entropy loss for network training and as an optimizer we utilize Adam [19] with initial learning rate set to 0.001. We apply early-stopping based on validation loss to prevent overfitting as well. Results for different patch sizes and different numbers of training patches extracted from one image are presented in Table 1. The best performing configuration is typeset in bold.

We report the scores for two values of distance threshold T. Smaller threshold causes lower scores, however, it ensures that the predicted points are really close to the ground-truth ones. The ground-truth images contain 33 break points in average. The presented recalls are thus sufficient for our goal (5–10 detected landmarks are usually enough for the transformation and connection).

When using $T = 5$, the average distance of the corresponding points is 1.3 px. Using $T = 2$ we reduce this value to 0.8 px. We have performed a comparative experiment with landmarks annotated by two human annotators. On three randomly selected map sheets we obtained average distance of 1.5 px between the annotations. We thus can state that the presented approach performs better than a human annotator and the resulting set of detected landmarks is usable for map sheets connection.

Table 1. Evaluation of the landmark detection algorithm with different sizes and numbers of the training patches extracted from one map sheet

Patch size ($w \times h$)		Threshold 5			Threshold 2		
		P	R	F1	P	R	F1
10 patches	320 × 240	37.6	34.7	36.1	22.2	19.8	20.9
	640 × 480	65.0	56.8	60.6	31.5	27.9	29.6
	960 × 720	85.5	76.2	80.6	62.2	49.7	55.2
	1280 × 960	86.5	81.3	83.8	66.0	57.0	61.2
25 patches	320 × 240	53.9	37.9	44.5	33.3	21.0	25.8
	640 × 480	74.6	36.4	48.9	54.5	25.1	34.4
	960 × 720	84.5	80.6	82.5	61.7	53.4	57.2
	1280 × 960	82.2	31.7	45.7	63.9	21.1	31.7
50 patches	320 × 240	79.3	76.2	77.7	54.8	48.7	51.6
	640 × 480	85.2	84.7	84.9	62.7	58.3	60.4
	960 × 720	86.0	71.7	78.2	65.2	50.6	57.0
	1280 × 960	84.5	85.3	84.9	61.2	57.6	59.4
75 patches	320 × 240	74.2	90.5	81.5	56.0	64.9	60.1
	640 × 480	78.9	84.7	81.7	58.6	59.7	59.2
	960 × 720	**88.3**	**82.0**	**85.1**	**67.3**	**58.2**	**62.4**
	1280 × 960	88.4	38.7	53.9	71.7	28.0	40.3
100 patches	320 × 240	78.7	89.2	83.6	59.0	59.3	59.1
	640 × 480	76.5	37.7	50.5	53.9	24.6	33.7
	960 × 720	86.5	80.0	83.2	59.6	50.9	54.9
	1280 × 960	87.1	81.6	84.2	65.2	56.1	60.3

5.4 Border Line and Break Points Detection

In this section, we experiment with neural network configurations in the same way as for the landmark detection. We do not evaluate the predicted border line. Instead of that, the evaluation is performed for the detected break points using the same criteria as in the case of landmarks. Also the settings and hyperparameters of the network are the same.

The results are summarized in Table 2. We report the scores for two values of distance threshold, namely $T = 5$ and $T = 2$. We can observe that the scores are significantly lower compared to the landmark detection. However, the recall exceeding 60 % is still sufficient for finding enough points usable for the map sheet connection. It is also important to note that the break point detector is applied only on map sheets containing insufficient amount of landmarks which happens only in a minority of cases.

5.5   Edge-Line Detection

To validate the results of the proposed solution, we have compared our detected edge-lines against the ground-truth data.

Table 2. Evaluation of the break point detection algorithm with different sizes and numbers of training patches

Patch size ($w \times h$)		Threshold 5			Threshold 2		
		P	R	F1	P	R	F1
10 patches	320×240	13.6	33.6	19.3	4.0	9.4	5.6
	640×480	10.8	34.7	16.5	3.1	7.2	4.4
	960×720	14.9	38.3	21.5	3.6	9.2	5.2
	1280×960	25.1	58.7	35.2	7.1	18.1	10.2
25 patches	320×240	6.4	42.0	11.1	0.9	6.2	1.6
	640×480	19.6	50.8	28.3	6.6	16.5	9.4
	960×720	20.7	46.6	28.7	7.3	14.5	9.7
	1280×960	34.0	45.9	39.1	7.6	17.0	10.5
50 patches	320×240	14.2	54.6	22.6	4.7	16.9	7.4
	640×480	26.0	63.7	36.9	7.5	18.3	10.7
	960×720	28.3	52.8	36.8	9.5	17.5	12.3
	1280×960	32.3	50.2	39.3	8.6	19.3	12.0
75 patches	320×240	15.9	50.6	24.1	6.4	20.5	9.8
	640×480	20.3	49.2	28.8	6.4	15.8	9.2
	960×720	35.1	46.5	40.0	9.7	17.6	12.5
	1280×960	37.5	52.6	43.8	6.3	12.8	8.5
100 patches	320×240	14.5	57.1	23.2	5.1	17.1	7.9
	640×480	30.7	53.9	39.2	4.1	9.0	5.7
	960×720	32.6	58.5	41.9	**12.9**	**24.9**	**17.0**
	1280×960	**33.9**	**65.5**	**44.7**	8.9	18.4	12.0

To overcome problems with the detected edge-lines that may contain noise, we have vectorized the edge-lines in both datasets. The lines are represented as sets of equally distant points. Therefore, instead of pixel-based comparisons, we use vector-based calculations that are more robust. The downside is that the points from ground-truth and the results may not be aligned. For comparison, we have used the solution with a threshold defined in Sect. 5.2. We have tested three different threshold values T selected to be near the median distance between points in the ground-truth dataset (median distance is 27.3 px). We present the results of the edge-line detection in Table 3.

Table 3. Evaluation of edge-line detection algorithm

Threshold	P	R	F1
20	73.6	75.4	74.5
30	80.1	81.5	81.2
40	81.5	86.4	83.9

This table shows clearly, that the best results are obtained for the threshold value of 40. This configuration gives very high F1 score (F1 = 83.9%). This quality of detection is sufficient for our needs. We do not require detected edge-line to be pixel accurate. Our goal is to have a correct orientation against the border line.

5.6 Final Results

In this final experiment, we used the above described methods on the whole task of the seamless map sheet connection and visualize the results.

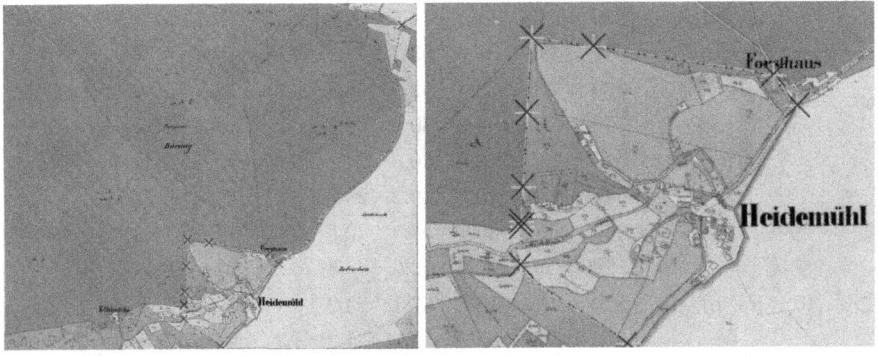

Fig. 10. Example of connected maps (left) with the detail of the connection in the complicated areas (right)

Figure 10 shows one example of the resulting connection of two neighboring map sheets. The corresponding detected break points on these sheets are marked by the different blue cross symbols. Based on the manual analysis of the small sample containing 100 of resulting images, we can claim that the proposed approaches are sufficient to be integrated into the final system.

6 Conclusions and Future Work

We have presented a set of methods for the detection of important features in historical cadastral maps. The methods represent the fundamental part of a system for seamless connection of the individual map sheets. We have focused on finding important points lying on cadastre border usable for seamless connection of neighboring map sheets. Another task is the detection of the cadastre border which allows masking of the area outside the map. The presented methods are built upon FCN networks in combination with traditional computer vision techniques. Only the task of edge-line detection is solved using solely traditional computer vision techniques. The reason is a relatively more complicated annotation of ground-truths for this task and also the sufficient results obtained by the presented solution.

The methods are evaluated on a newly created dataset containing ground-truths for all four solved tasks. We have compared several configurations of the algorithms and searched for the best performing ones. We have further visually demonstrated that the best implemented methods have sufficient accuracy to be integrated into the final system.

We have utilised a standard FCN architecture and there is surely a room for further improvements using different architectures. We can also concentrate on strategies of networks training, e.g. balance between positive and negative training samples etc.

Acknowledgement. This work has been partly supported by Grant No. SGS-2019-018 Processing of heterogeneous data and its specialized applications.

References

1. Timár, G., Molnár, G., Székely, B., Biszak, S., Varga, J., Jankó, A.: Digitized maps of the Habsburg Empire - the map sheets of the second military survey and their georeferenced version (2006)
2. Ignjatić, J., Nikolić, B., Rikalović, A.: Deep learning for historical cadastral maps digitization: overview, challenges and potential (2018)
3. Timilsina, S., Sharma, S., Aryal, J.: Mapping urban trees within cadastral parcels using an object-based convolutional neural network. ISPRS Ann. Photogram. Remote Sens. Spatial Inf. Sci. **4**, 111–117 (2019)
4. Ostankovich, V., Afanasyev, I.: Illegal buildings detection from satellite images using googlenet and cadastral map. In: 2018 International Conference on Intelligent Systems (IS), pp. 616–623. IEEE (2018)
5. Xia, X., Persello, C., Koeva, M.: Deep fully convolutional networks for cadastral boundary detection from UAV images. Remote Sens. **11**(14), 1725 (2019)
6. Fetai, B., Oštir, K., Kosmatin Fras, M., Lisec, A.: Extraction of visible boundaries for cadastral mapping based on UAV imagery. Remote Sens. **11**(13), 1510 (2019)
7. Nyandwi, E., Koeva, M., Kohli, D., Bennett, R.: Comparing human versus machine-driven cadastral boundary feature extraction. Remote Sens. **11**(14), 1662 (2019)
8. Kestur, R., et al.: UFCN: a fully convolutional neural network for road extraction in RGB imagery acquired by remote sensing from an unmanned aerial vehicle. J. Appl. Remote Sens. **12**(1), 016020 (2018)
9. Chen, Y., Carlinet, E., Chazalon, J., Mallet, C., Duménieu, B., Perret, J.: Combining deep learning and mathematical morphology for historical map segmentation, arXiv preprint arXiv:2101.02144 (2021)

10. Wick, C., Puppe, F.: Fully convolutional neural networks for page segmentation of historical document images. In: 13th IAPR International Workshop on Document Analysis Systems (DAS), pp. 287–292. IEEE (2018)

11. Ronneberger, O., Fischer, P., Brox, T.: U-Net: convolutional networks for biomedical image segmentation. In: Navab, N., Hornegger, J., Wells, W.M., Frangi, A.F. (eds.) MICCAI 2015. LNCS, vol. 9351, pp. 234–241. Springer, Cham (2015). https://doi.org/10.1007/978-3-319-24574-4_28

12. Zhang, T.Y., Suen, C.Y.: A fast parallel algorithm for thinning digital patterns. Commun. ACM **27**(3), 236–239 (1984). https://doi.org/10.1145/357994.358023

13. Douglas, D.H., Peucker, T.K.: Algorithms for the reduction of the number of points required to represent a digitized line or its caricature. Cartographica: Int. J. Geogr. Inf. Geovisualization **10**(2), 112–122 (1973)

14. Kanopoulos, N., Vasanthavada, N., Baker, R.L.: Design of an image edge detection filter using the sobel operator. IEEE J. Solid-State Circuits **23**(2), 358–367 (1988)

15. Otsu, N.: A threshold selection method from gray-level histograms. IEEE Trans. Syst. Man Cybern. **9**(1), 62–66 (1979)

16. Burtsev, S., Kuzmin, Y.: An efficient flood-filling algorithm. Comput. Graphics **17**(5), 549–561 (1993). http://www.sciencedirect.com/science/article/pii/0097849 39390006U

17. Felzenszwalb, P.F., Huttenlocher, D.P.: Distance transforms of sampled functions. Theory Comput. **8**(19), 415–428 (2012). http://www.theoryofcomputing.org/articles/v008a019

18. Harris, C., Stephens, M.: A combined corner and edge detector. In: Proceedings of the 4th Alvey Vision Conference, pp. 147–151 (1988)

19. Kingma, D.P., Ba, J.: Adam: a method for stochastic optimization, arXiv preprint arXiv:1412.6980 (2014)

Data Augmentation for End-to-End Optical Music Recognition

Juan C. López-Gutiérrez[1], Jose J. Valero-Mas[2], Francisco J. Castellanos[2], and Jorge Calvo-Zaragoza[2(✉)]

[1] Alicante, Spain
[2] Department of Software and Computing Systems,
University of Alicante, Alicante, Spain
{jjvalero,fcastellanos,jcalvo}@dlsi.ua.es

Abstract. Optical Music Recognition (OMR) is the research area that studies how to transcribe the content from music documents into a structured digital format. Within this field, techniques based on Deep Learning represent the current state of the art. Nevertheless, their use is constrained by the large amount of labeled data required, which constitutes a relevant issue when dealing with historical manuscripts. This drawback can be palliated by means of Data Augmentation (DA), which encompasses a series of strategies to increase data without the need of manual labeling new images. This work studies the applicability of specific DA techniques in the context of end-to-end staff-level OMR methods. More precisely, considering two corpora of historical music manuscripts, we applied different types of distortions to the music scores and assessed their contribution in an end-to-end system. Our results show that some transformations are much more appropriate than others, leading up to a 34.5% of relative improvement with respect to scenario without DA.

Keywords: Optical music recognition · Data augmentation · Deep learning

1 Introduction

Music sources, as one of the cornerstones of human culture, has historically been preserved and transmitted through written documents called music scores [22]. In this sense, there exist archives with a vast amount of compositions which only exist in physical formats or, in some cases, as scanned images [18]. Note that retrieving digital versions of such pieces is of remarkable interest, not only in terms of cultural heritage preservation, but also for tasks such as dissemination, indexing or musicological studies, among many others [11].

For a long time, this digitization process has been carried out in a manual fashion, resulting in tedious and prone-to-error tasks with limited scalability.

J. C. López-Gutiérrez—Independent Researcher.
This work was supported by the Generalitat Valenciana through grant APOSTD/2020/256, grant ACIF/2019/042, and project GV/2020/030.

E. H. Barney Smith and U. Pal (Eds.): ICDAR 2021 Workshops, LNCS 12916, pp. 59–73, 2021.
https://doi.org/10.1007/978-3-030-86198-8_5

Fortunately, the outstanding advances in technology during the last decades have enabled the emergence of the automated version of this transcription process, namely Optical Music Recognition (OMR).

OMR represents the field that studies how to computationally read music notation in written documents to store the content in a structured format [4]. While proposals in this field have typically relied on traditional computer vision strategies, adequately designed for a particular notation and engraving mechanism [19], recent developments in machine learning, and more especially in the so-called Deep Learning paradigm [14], have led to a considerable renewal of learning-based approaches which allow more general formulations.

Within this paradigm, end-to-end frameworks operating at the staff level stand for the current state of the art in OMR [2,6]. These schemes map the series of symbols within a single staff that appear on an image onto a sequence of music symbols. In addition to their effectiveness, it highlights their ability to be trained without an explicit alignment between the image and the position of the music symbols.

Nevertheless, in the case of historical documents, it is usual to find a lack of labeled data, which could limit the applicability of state-of-the-art methods. In addition, the great amount of manuscripts scattered all over the world awaiting to be digitized hinders their labeling, necessary to reliably apply modern Deep learning strategies. One of the most common strategies for addressing this deficiency is the so-called *Data Augmentation* (DA) paradigm [21]. DA is a family of procedures which creates new artificial samples by performing controlled distortions on the initial data. In general, this solution is capable of providing more robust and proficient models than those trained with only the initial unaltered data. The reader may check the work by Journet et al.[12] for a thorough revision of such processes applied to image data.

While DA techniques are quite common in the context of learning-based systems when involving images, the existing procedures are typically devised for general-purpose tasks. In this regard, we consider that certain distortions may be more adequate and realistic than others in the OMR context. For instance, while flipping an image is a common DA process, in the case at issue, this type of distortion is not realistic. However, contrast changes or simulating bulges do represent alterations typically found when scanning ancient music books. Moreover, focusing on the neural end-to-end models introduced before, their input is a staff image that has been previously cropped from a full page. Therefore, some augmentation processes, such as rotation or padding, should be carried out in the context of the original location of the page and not just applied to the segmented sample. It must be highlighted that, while there exist some precedents to our work as the one by Baró et al. [3] in which DA was studied in the context of handwritten music scores, such DA processes have not yet been evaluated on the particular neural end-to-end architectures considered in our case.

For all above, this work proposes a collection of DA processes specifically adapted for neural staff-level end-to-end OMR systems. Additionally, instead of considering the DA and training stages independently, we perform these two processes in a joint manner by directly distorting the images used at each iteration of the training procedure of the model. The results obtained when considering several corpora of historical documents prove the validity of our proposal, reaching relative improvement figures of up to 34.5% when compared to the non-augmented scenario.

The rest of the work is structured as follows: Sect. 2 presents the proposal of the work; Sect. 3 introduces the experimental set-up used for assessing our methodology; Sect. 4 presents and discusses the results obtained; finally, Sect. 5 concludes the work and indicates some avenues for further research.

2 Methodology

Our proposal builds upon neural end-to-end OMR systems that work at the staff level; i.e., given an image of a single staff, our goal is retrieving the series of symbols that appear therein. It is important to note that the meaning of musical symbols relies on two geometrical pieces of information: shape and height (vertical position of the symbol in the staff), which typically indicate duration and pitch, respectively. In this regard, and following previous work in this field [1, 6], each possible combination of shape and height is considered as a unique symbol.

Concerning this, our recognition approach can be defined as a *sequence labeling* task [9], where each input must be associated to a sequence of symbols from a predefined alphabet, but without the need of aligning with respect to the image.

Formally, let $\mathcal{T} = \{(x_i, \mathbf{z}_i) : x_i \in \mathcal{X}, \mathbf{z}_i \in \mathcal{Z}\}_{i=1}^{|\mathcal{T}|}$ represent a set of data, where sample x_i is drawn from the image space $\mathcal{X}$, and $\mathbf{z}_i = (z_{i1}, z_{i2}, \ldots, z_{iN_i})$ corresponds to its transcript in terms of music-notation symbols. Note that $\mathcal{Z} = \Sigma^*$ where Σ represents the music-symbol vocabulary. The recognition task can be formalized as learning the underlying function $g : \mathcal{X} \to \mathcal{Z}$.

The rest of the section introduces the neural approach used for approximating function $g(\cdot)$, as well as the different DA techniques to be considered.

2.1 Neural End-to-End Recognition Framework

We consider a Convolutional Recurrent Neural Network (CRNN) to approximate $g(\cdot)$, given its good results in end-to-end OMR approaches [2, 6, 20].

CRNN constitutes a particular neural architecture formed by an initial block of *convolutional* layers, which aim at learning the adequate set of features for the task, followed by another group of *recurrent* stages, which model the temporal dependencies of the elements from the initial feature-learning block [20].

To attain a proper end-to-end scheme, the CRNN is trained using the Connectionist Temporal Classification (CTC) algorithm [10], which allows optimizing the weights of the neural network using unsegmented sequential data. In our case, this means that, for a given staff image $x_i \in \mathcal{X}$, we only have its associated sequence of characters $\mathbf{z}_i \in \mathcal{Z}$ as its expected output, without any correspondence at pixel level or similar input-output alignment. Due to its particular training procedure, CTC requires the inclusion of an additional *"blank"* symbol within the Σ vocabulary, i.e., $\Sigma' = \Sigma \cup \{blank\}$.

The output of the CRNN can be seen as a *posteriorgram*; i.e., the probability of each $\sigma \in \Sigma'$ to be located in each frame of the input image. Most commonly, the actual sequential prediction can be obtained out of this posteriorgram by using a *greedy* approach, which keeps the most probable symbol per step, merges repeated symbols in consecutive frames, and eventually removes the *blank* tokens.

2.2 Data Augmentation Procedures

Once the neural end-to-end recognition scheme has been introduced, we shall now present the particular DA framework proposed in this work. In contrast to other existing augmentation processes that perform an initial processing of the entire corpus to derive a new (larger) set of data, our proposal is directly embedded in the network pipeline for performing these transformations in an online fashion.

Concerning the actual transformations considered in this work, all the alterations are based on possible distortions that a given score image may depict, especially those found in historical documents. The following sections provide a concise description for each of the transformations considered, as well as a graphical example. For the sake of compactness, let us consider Fig. 1 as the reference (undistorted) staff image for the sections below.

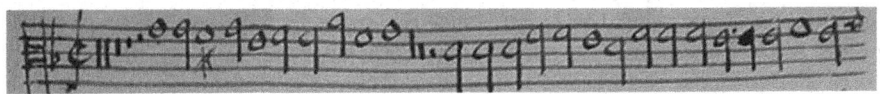

Fig. 1. Reference staff image for the different data transformation examples.

Contrast Variation. This first augmentation process tries to alter the images to simulate different capture devices. For instance, depending on the type of lens, the images obtained may have a different color distribution. This is why a contrast enhancement may have been applied to the image by means of image quality improvement techniques as Contrast Limited Adaptive Histogram Equalization algorithm [17]. In this sense, we consider the use of this algorithm by applying it with a probability of 50% for any given image for each iteration of the training process, and remaining unaltered otherwise. The tile size parameter is set to 8×8. Figure 2 shows an example of the resulting altered image.

Erosion and Dilation Operations. The thickness of the symbols in historical scores varies greatly because they are written with a quill. To emulate this effect, the staff symbols can be made thicker and narrower through erosion and dilation image processing operations, respectively.

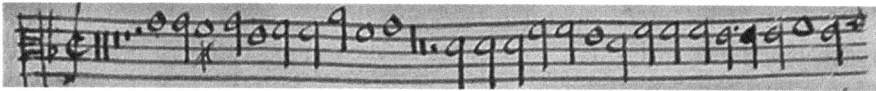

Fig. 2. Resulting staff sample after altering Fig. 1 with the 'Contrast' augmentation.

To achieve such effect, at each training iteration, we randomly decide whether to apply an erosion, a dilation or leave the staff undistorted. In case an erosion or a dilation must be applied, the size of the kernel is determined randomly and uniformly between 2 and 4.

Figures 3a and 3b show the result of applying the erosion and dilation processes for the example staff image, respectively.

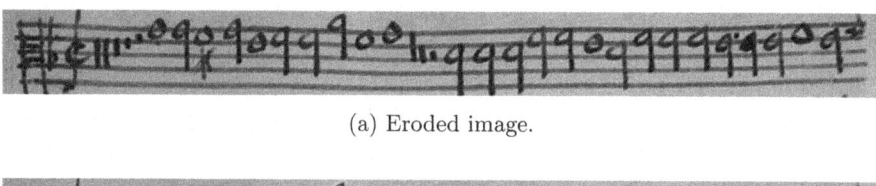

(a) Eroded image.

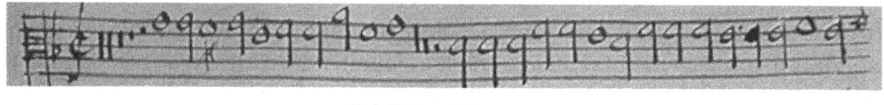

(b) Dilated image.

Fig. 3. Resulting staff sample after altering Fig. 1 with the 'Erosion' and 'Dilation' augmentations.

Margin Alteration. The process of localizing each individual staff on a full-page image is highly arbitrary, regardless of whether it is done manually or by computational means. Therefore, both training and test samples will be presented with varying degrees of margins with respect to the staff itself. As we shall observe in the results, this has a clear impact on the accuracy of the model.

In order to increase the robustness with respect to this phenomenon, we randomly modify the position of the bounding box containing a single staff. This modification is performed based on sampling four values drawn from a normal distribution with mean $\mu = 0$ and variance $\sigma^2 = 10$, i.e. $\mathcal{N}(0, 10)$, each of which is added to each of the four values defining the two corners of a bounding box.

In image processing, this alteration is applied by adding pad to the images with a constant value. This is not something negative but in our case we would be losing possible valuable information, since we do not have only the image of the staff but also the image of the complete page where it is located. To preserve this information, the modification described above is applied before cropping the staff.

Figure 4a shows an example of distortion of this region using the usual image-processing strategy, while Fig. 4b shows how the context can be preserved with our own process.

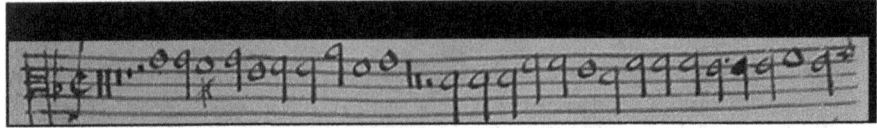

(a) Generic augmentation (not used in this work).

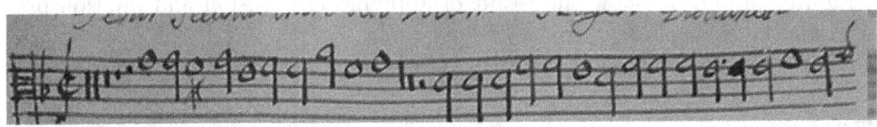

(b) Custom augmentation for our task (used in this work).

Fig. 4. Resulting staff sample after altering Fig. 1 with the 'Margin' augmentation.

Rotation Transformation. Another typical condition in score images is an imperfect alignment with respect to the horizontal axis. This might be caused by the process of capturing the image or by the nature of the engraving mechanism in the physical source.

To mimic this effect, at each training iteration, a variable-angle rotation is applied to the samples. This means uniformly choosing a random value in the range $[-3°, 3°]$ to serve as the angle of the rotation. Note that no rotation is also possible.

It is worth highlighting that the rotation is applied to the full page image before cropping the staff. Likewise the previous augmentation case, the contextual information of the staff location is, therefore, preserved. This produces images as that in Fig. 5b, whereas a generic rotation would produce what is depicted in Fig. 5a.

Wavy Pattern. Another possible phenomenon is the emergence of wavy patterns caused by either the process of engraving or, more commonly, the scanning, especially when book formats are involved. To emulate this effect, we consider the use of the fish-eye transformation already implemented in the OpenCV library. For each sample in each training iteration, we apply it with a probability of 50%.

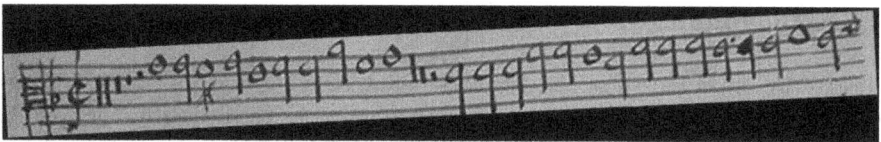

(a) Generic augmentation (not used in this work).

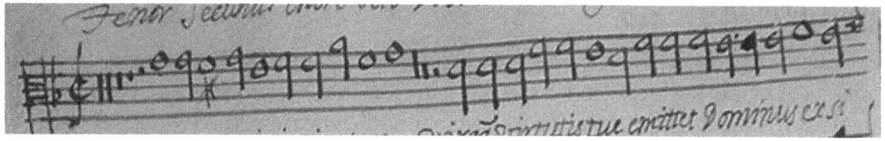

(b) Custom augmentation for our task (used in this work).

Fig. 5. Resulting staff sample after altering Fig. 1 with the 'Rotation' augmentation.

We manually tuned the parameters of the distortion for the case at issue. Following the notation in the OpenCV implementation, vector D of distorted coefficients is set to $D = [0, -0.25, 1, 0]$, whereas camera matrix K is given by

$$K = \begin{pmatrix} w/3 & 0 & w/2 \\ 0 & -w/2 & 0 \\ 0 & 0 & 1 \end{pmatrix}$$

where w denotes the width of the image.

Figure 6 shows an example of this process applied to the reference staff considered. Note that the gaps introduced by this distortion are filled with the average pixel values of the sample.

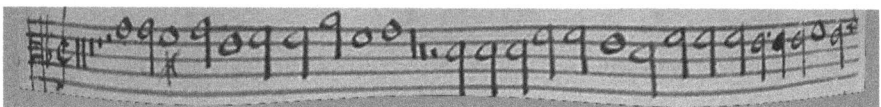

Fig. 6. Resulting staff sample after altering Fig. 1 with the 'Wavy' augmentation.

All. In addition to the mentioned augmentation procedures described above, we also study the case in which all these are applied sequentially for each training iteration. It must be remarked that the order of the operations considered is equal to that in which they have been presented in this paper, although this sequence could have been applied in a different order. It is also worth noting that all these operations have a non-null probability of applying no changes to the image to be augmented, according to the criteria detailed for each procedure.

3   Experiments

This section introduces the corpora considered for assessing the validity of the proposed methodology as well as the evaluation metrics used. Finally, the particular neural end-to-end recognition model contemplated in the work is described.

3.1   Corpora

We consider two corpora of music scores depicting Mensural notation, i.e., the music engraving system used for most part of the XVI and XVII centuries in the Western music tradition. These two sets are now described:

- CAPITAN [5]: Handwritten manuscript of ninety-six pages dated from the 17th century of *missa* (sacred music). An example of a particular page from this corpus is depicted in Fig. 7a.
- SEILS [16]: Collection of one hundred and fifty-five typeset pages corresponding to an anthology of Italian madrigals of the 16th century, particularly that to Symbolically Encoded Il Lauro Secco. Figure 7b shows a page of the collection.

   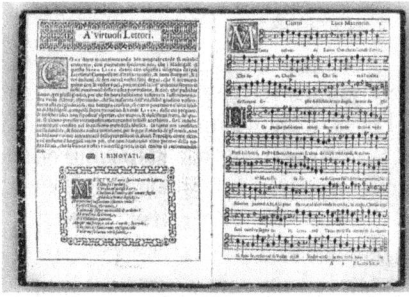

(a) CAPITAN corpus.   (b) SEILS corpus.

Fig. 7. Page examples of the two corpora used in the experiments.

Given that our proposal considers a neural end-to-end model which works at the staff level, it is necessary to first process the complete pages of the introduced corpora to segment those image excerpts. While both collections include the manually-annotated bounding boxes for segmenting those regions of interest, in order to increase the cases of study, in this work we also consider the possibility of automatically extracting those regions. In this regard, we shall study two different scenarios which differ on the nature of these bounding boxes:

- **Annotated**: Case in which the different staves are extracted using the bounding box annotations of the original corpora.

– **Predicted**: In this scenario, the different bounding boxes are estimated using the approach by Castellanos et al. [7] based on the so-called Selectional Auto-Encoders.

It must be noted that both scenarios provide the same number of staves, thus not differing the number of experimental samples depending on the modality. In this respect, the details of each corpus are provided in Table 1 in terms of the engraving mechanism, number of pages, staves, and vocabulary size.

Table 1. Details of the corpora in terms of the number of pages, staves, and the cardinality of the vocabulary.

Corpus	Engraving	Pages	Staves	Vocabulary
CAPITAN	Handwritten	97	737	320
SEILS	Typeset	150	1,278	182

Finally, we consider a 5-fold cross-validation (5-CV) set-up with three data partitions —train, validation, and test— corresponding to 60%, 20%, and 20% of the whole set of staves, respectively. Since we are considering two different scenarios for extracting these staves, in our experiments we shall assess the performance of the recognition framework when trained with either the annotated or the predicted bounding boxes, considering that test data always comprises the predicted boxes.

3.2 Metrics

Regarding the performance evaluation of the recognition framework, we considered the Symbol Error Rate (SER) as in other neural end-to-end OMR works [2,6]. This metric may be defined as the average number of editing operations (insertions, deletions, or substitutions) necessary to match the sequence predicted by the model with that of the ground truth, normalized by the length of the latter. Mathematically, it may be defined as:

$$\text{SER}\,(\%) = \frac{\sum_{i=1}^{|\mathcal{S}|} \text{ED}\,(\mathbf{z}_i,\ \mathbf{z}_i')}{\sum_{i=1}^{|\mathcal{S}|} |\mathbf{z}_i|} \tag{1}$$

where $\text{ED}\,(\cdot,\cdot)$ stands for the string Edit distance [15], $\mathcal{S}$ a set of test data, and $\mathbf{z}_i$ and $\mathbf{z}_i'$ the target and estimated sequences, respectively.

3.3 CRNN Configuration

Due to its reported good results when tackling these corpora [1,7], this work replicates the neural configuration from Calvo et al. [6] as the base neural configuration for the experiments. Table 2 thoroughly describes this architecture.

Table 2. CRNN configuration considered. Notation: $\text{Conv}(f, w_c \times h_c)$ stands for a convolution layer of f filters of size $w_c \times h_c$ pixels, BatchNorm performs the normalization of the batch, $\text{LeakyReLU}(\alpha)$ represents a leaky rectified linear unit activation with negative slope value of α, $\text{MaxPool2D}(w_p \times h_p)$ stands for the max-pooling operator of dimensions $w_p \times h_p$ pixels, $\text{BLSTM}(n)$ denotes a bidirectional long short-term memory unit with n neurons, and $\text{Dropout}(d)$ performs the dropout operation with d probability.

Layer 1	Layer 2	Layer 3	Layer 4	Layer 5	Layer 6
Conv(64, 5 × 5)	Conv(64, 5 × 5)	Conv(128, 3 × 3)	Conv(128, 3 × 3)	BLSTM(256)	BLSTM(256)
BatchNorm	BatchNorm	BatchNorm	BatchNorm	Dropout(0.50)	Dropout(0.50)
LeakyReLU(0.20)	LeakyReLU(0.20)	LeakyReLU(0.20)	LeakyReLU(0.20)		
MaxPool(2 × 2)	MaxPool(2 × 1)	MaxPool(2 × 1)	MaxPool(2 × 1)		

Finally, this architecture was trained using the backpropagation method provided by CTC with the ADAM optimizer [13], a fixed learning rate of 10^{-3}, and a batch size of 16 samples. We set a maximum of $1,000$ epochs, keeping the network weights of the best validation result. All images were scaled at the input of the model to a height of 64 pixels, forcing the maintenance of the aspect ratio for its width, and converted to grayscale for simplification reasons. It must be remarked that another color space may be used.

4   Results

This section presents and discusses the results obtained. Since the experiments have been performed in a cross-validation scheme, the figures reported constitute the average values obtained for each of the cases considered. These results are the ones obtained with the test data partition since, as commented, the validation one was used for optimizing the weights of the neural network.

Table 3 shows the obtained error rates for the introduced DA processes applied to the two corpora considered. Note that the influence of whether the bounding boxes are annotated or extracted is also assessed.

Regarding the CAPITAN corpus, for the case of the annotated bounding boxes, the base recognition model obtains an error rate of 14.34%. Once the different DA processes are introduced, this error rate decreases, being the overall improvements in the range of 1% to 4%, depending on the precise method considered. Among all of them, the overall minimum is achieved by the *Rotation* process with a rate of 9.34%, which supposes a 34.5% of relative improvement with respect to the base case. Note that applying all the augmentation processes also remarkably improves the base result, achieving an error rate of 10.88%, i.e., a relative improvement of roughly a 24%.

When considering the predicted bounding boxes for the CAPITAN set, a similar trend to that of the annotated boxes is observed. In this case, though, the base recognition error is set to 12.86%, which is almost a 2% better than the previous scenario. It must be noted that, while the augmentation processes generally suppose an improvement in the recognition rate, the use of the *Erosion*

Table 3. Results obtained in terms of the Symbol Error Rate (%) for the different DA processes and corpora considered. Note that the results have been divided into two row sections according to the source of the bounding boxes—annotated or predicted. The *All* and *No aug.* rows represent applying all and none augmentation processes, respectively. Highlighted values in bold represent the best results achieved for each corpus and bounding box source.

Augmentation procedure	Symbol error rate (%)	
	CAPITAN	SEILS
Annotated bounding boxes		
Contrast	13.14	4.97
Erosion & dilation	11.89	5.50
Margin	10.57	**2.96**
Rotation	**9.39**	3.07
Wavy	10.63	3.76
All	10.88	3.43
No aug.	14.34	5.24
Predicted bounding boxes		
Contrast	11.31	3.95
Erosion & dilation	12.92	4.40
Margin	10.02	**2.84**
Rotation	**9.52**	3.17
Wavy	10.01	3.40
All	12.11	3.78
No aug.	12.86	4.01

& dilation method slightly increases the base error rate. Nevertheless, as in the previous scenario, the *Rotation* process is the one achieving the best overall performance with a figure of 9.52%, which supposes a relative improvement of almost a 26% with respect to the base case. While the improvement observed when jointly considering all augmentation processes is not as remarkable as in the previous case, it must be mentioned that the result does outperform the base case.

Moving on to the SEILS corpus, it is quite noticeable the lower error rates achieved with respect to the CAPITAN set. This difference in performance, which ranges between 6% and 9%, is due to the typeset nature of this set, since the printed engraving of this corpus is considerably more uniform and, therefore, easier than the handwritten one of CAPITAN.

Focusing on the case of annotated bounding boxes, an error rate of 5.24% is achieved when no augmentation process is applied. As in the CAPITAN case, the inclusion of the different augmentation processes generally results in a performance improvement, with the sole exception of the *Erosion & dilation* case in which the error rate increases. In this scenario, the *Margin* process achieves the best overall recognition rate with a 2.96% of error, which supposes a 43.5% of relative improvement with respect to the base case. Also note that, when jointly considering all augmentation processes, the error rate decreases to 3.43%, i.e., a relative boosting compared to the base case of 35.5%.

With regards to the predicted bounding boxes, it can be again observed that all the methods reduce the base error rate except for the *Erosion & dilation* case, whose application supposes a decrease in the recognition performance of almost 0.40%. The *Margin* method, conversely, achieves the minimum error rate with a value of 2.84%, which supposes roughly a 29% of relative improvement. Note that, when considering all augmentation processes, the base error rate lowers to 3.78% which, despite not being the sharpest reduction, represents almost a 6% of relative improvement.

In general terms, it is observed that the *Margin* and *Rotation* processes are the ones which achieve the best overall recognition rates. Indeed, according to the experiments, the *Rotation* is the best DA process on CAPITAN, whereas *Margin* is that on SEILS. Since CAPITAN is a handwritten manuscript, the variability in the staff skew within each page is greater than in the printed cases, so that we can assume that the *Rotation* DA process reinforces this aspect by adding more variability in the training data, and, therefore, obtaining a more robust learning model. Oppositely, due to the printed engraving of SEILS, the staff-region skew is less noticeable, making that a greater rotation variability in the training data is not so profitable, since the test data does not present this characteristic. Note, however, that these two DA procedures are the best cases for both corpora, regardless the engraving, with the exception of the *Wavy* process, which also retrieves competitive results, especially in the case of using predicted image staves.

On the contrary, attending to the presented figures, the *Erosion & dilation* distortion may be deemed as the least competitive since, in most cases, its use implies an increase in the overall error rate. Nevertheless, the joint use of all the augmentation processes consistently improves the results with respect to the case of not considering any of these methods, thus stating their relevance in the context of neural-based recognition schemes.

Finally, it must be noted that the use of predicted boxes outperforms the manually-annotated scenario, most likely because of the fact that the test staves considered are extracted using the automated approach. If we focus on the error rate figures, we only observe improvements close to 1.5% for CAPITAN and 1% for SEILS. While this may be considered as marginal reductions, note the reduced improvement margin that these corpora depict, especially SEILS. So that, although the predicted staves from CAPITAN improves the results from 14.34% to 12.86% in the base scenarios, and those from SEILS from 5.24% to 4.01%, we may observe that the relative improvement is close to 10.3% and 23.5%, respectively, thus being a substantial reduction factor with respect to those cases in which the annotated bounding boxes are used.

5   Conclusions

Music scores have historically represented the main vehicle for music preservation. Nevertheless, physical formats are unavoidably associated to gradual degradation when they are not preserved under strict conditions. The digitization of these documents would enable to easily preserve, index or even disseminate this valuable heritage.

Optical Music Recognition (OMR) stands for the family of computational methods meant to read music notation from scanned documents and transcribe their content into a digital structured format. This field is deemed as one of the key processes in music heritage preservation since, in contrast to the prone-to-error manual annotation campaigns, such techniques are inherently scalable.

Currently, neural end-to-end models are considered the state-of-the-art in OMR tasks. Nevertheless, these models generally require of a large amount of data for achieving competitive performance rates. This fact constrains its application to historical music documents given the scarce existence of annotated data.

In this regard, data augmentation processes, which represent those methods that create artificial samples by performing controlled distortions on the initial data for increasing its variability and providing robust recognition models, have been typically considered in learning-based systems for image-related tasks. However, note that these processes are meant for general recognition and/or classification tasks, hence being of remarkable relevance a proper study of the most adequate transformation methods for OMR.

This work performs a comparative study of different augmentation processes for neural end-to-end OMR methods in the context of historical documents. Results obtained suggest that two transformations are the most adequate for this type of data, leading up to a 34.5% of relative improvement with respect to the non-augmented scenario. More precisely, these methods are the margin alteration, which represents the case in which the staff is not centered in the image, and the rotation distortion, which simulates a certain degree of slanting of the staff.

In light of this results, for future work, we consider checking the validity of these results by extending it to other historical music corpora. Besides, simulating other distortions as, for instance, slanting or ink bleeding may also report a boost in performance as they constitute common problems observed in this type of documents. In addition, as an alternative to directly distorting existing data, we aim at exploring the use of Generative Adversarial Networks for artificially creating data samples [8]. Finally, we also consider studying the performance of other learning-based schemes suitable for this task as *Sequence-to-Sequence* or *Transformer* models.

References

1. Alfaro-Contreras, M., Valero-Mas, J.J.: Exploiting the two-dimensional nature of agnostic music notation for neural optical music recognition. Appl. Sci. **11**(8), 3621 (2021)
2. Baró, A., Badal, C., Fornés, A.: Handwritten historical music recognition by sequence-to-sequence with attention mechanism. In: 17th International Conference on Frontiers in Handwriting Recognition, ICFHR 2020, Dortmund, Germany, 8–10 September 2020, pp. 205–210. IEEE (2020)
3. Baró, A., Riba, P., Calvo-Zaragoza, J., Fornés, A.: From optical music recognition to handwritten music recognition: a baseline. Pattern Recogn. Lett. **123**, 1–8 (2019)
4. Calvo-Zaragoza, J., Jan, H., Jr., Pacha, A.: Understanding optical music recognition. ACM Comput. Surv. (CSUR) **53**(4), 1–35 (2020)
5. Calvo-Zaragoza, J., Toselli, A.H., Vidal, E.: Handwritten music recognition for mensural notation: formulation, data and baseline results. In: 2017 14th IAPR International Conference on Document Analysis and Recognition (ICDAR), vol. 1, pp. 1081–1086. IEEE (2017)
6. Calvo-Zaragoza, J., Toselli, A.H., Vidal, E.: Handwritten music recognition for mensural notation with convolutional recurrent neural networks. Pattern Recogn. Lett. **128**, 115–121 (2019)
7. Castellanos, F.J., Calvo-Zaragoza, J., Inesta, J.M.: A neural approach for full-page optical music recognition of mensural documents. In: Proceedings of the 21th International Society for Music Information Retrieval Conference, ISMIR, pp. 23–27 (2020)
8. Creswell, A., White, T., Dumoulin, V., Arulkumaran, K., Sengupta, B., Bharath, A.A.: Generative adversarial networks: an overview. IEEE Signal Process. Mag. **35**(1), 53–65 (2018)
9. Graves, A.: Supervised sequence labelling. In: Graves, A. (ed.) Supervised Sequence Labelling with Recurrent Neural Networks, pp. 5–13. Springer, Heidelberg (2012). https://doi.org/10.1007/978-3-642-24797-2_2
10. Graves, A., Fernández, S., Gomez, F., Schmidhuber, J.: Connectionist temporal classification: labelling unsegmented sequence data with recurrent neural networks. In: Proceedings of the 23rd International Conference on Machine Learning, ICML 2006, pp. 369–376. ACM, New York (2006)
11. Jan, H., Jr., Kolárová, M., Pacha, A., Calvo-Zaragoza, J.: How current optical music recognition systems are becoming useful for digital libraries. In: Proceedings of the 5th International Conference on Digital Libraries for Musicology, pp. 57–61 (2018)
12. Journet, N., Visani, M., Mansencal, B., Van-Cuong, K., Billy, A.: Doccreator: a new software for creating synthetic ground-truthed document images. J. Imaging **3**(4), 62 (2017)
13. Kingma, D.P., Ba, J.: Adam: a method for stochastic optimization. In: 3rd International Conference on Learning Representations, San Diego, USA (2015)
14. LeCun, Y., Bengio, Y., Hinton, G.: Deep learning. Nature **521**(7553), 436–444 (2015)
15. Levenshtein, V.I.: Binary codes capable of correcting deletions, insertions, and reversals. Soviet Physics Doklady **10**(8), 707–710 (1966)
16. Parada-Cabaleiro, E., Batliner, A., Schuller, B.W.: A diplomatic edition of il lauro secco: ground truth for OMR of white mensural notation. In: ISMIR, pp. 557–564 (2019)

17. Pizer, S.M., et al.: Adaptive histogram equalization and its variations. Comput. Vis. Graph. Image Process. **39**(3), 355–368 (1987)
18. Pugin, L.: The challenge of data in digital musicology. Front. Digital Humanit. **2**, 4 (2015)
19. Rebelo, A., Capela, G., Cardoso, J.S.: Optical recognition of music symbols. Int. J. Doc. Anal. Recognit. (IJDAR) **13**(1), 19–31 (2010)
20. Shi, B., Bai, X., Yao, C.: An end-to-end trainable neural network for image-based sequence recognition and its application to scene text recognition. IEEE Trans. Pattern Anal. Mach. Intell. **39**(11), 2298–2304 (2017)
21. Shorten, C., Khoshgoftaar, T.M.: A survey on image data augmentation for deep learning. J. Big Data **6**(1), 60 (2019)
22. Treitler, L.: The early history of music writing in the west. J. Am. Musicol. Soc. **35**(2), 237–279 (1982)

Graph-Based Object Detection Enhancement for Symbolic Engineering Drawings

Syed Mizanur Rahman[1,2], Johannes Bayer[1,2(✉)], and Andreas Dengel[1,2]

[1] Smart Data & Knowledge Services Department,
DFKI GmbH, Kaiserslautern, Germany
{mizanur_rahman.syed,johannes.bayer,andreas.dengel}@dfki.de
[2] Computer Science Department, TU Kaiserslautern, Kaiserslautern, Germany

Abstract. The identification of graphic symbols and interconnections is a primary task in the digitization of symbolic engineering diagram images like circuit diagrams. Recent approaches propose the use of Convolutional Neural Networks to the identification of symbols in engineering diagrams. Although recall and precision from CNN based object recognition algorithms are high, false negatives result in some input symbols being missed or misclassified. The missed symbols induce errors in the circuit level features of the extracted circuit, which can be identified using graph level analysis. In this work, a custom annotated printed circuit image set, which is made publicly available in conjunction with the source code of the experiments of this paper, is used to fine-tune a Faster RCNN network to recognise component symbols and blob detection to identify inter-connections between symbols to generate a graph representation of the extracted circuit components. The graph structure is then analysed using graph convolutional neural networks and node degree comparison to identify graph anomalies potentially resulting from false negatives from the object recognition module. Anomaly predictions are then used to identify image regions with potential missed symbols, which are subject to image transforms and re-input to the Faster RCNN, which results in a significant improvement in component recall, which increases to **91%** on the test set. The general tools used by the analysis pipeline can also be applied to other Engineering Diagrams with the availability of similar datasets.

Keywords: Graph convolutional network · Circuit diagram · Graph refinement

1 Introduction

Graph-based symbolic engineering drawings (like circuit diagrams or piping and instrumentation diagrams) use graphical symbols and line segments to represent the components of technical facilities or devices as well as their interconnections.

© Springer Nature Switzerland AG 2021
E. H. Barney Smith and U. Pal (Eds.): ICDAR 2021 Workshops, LNCS 12916, pp. 74–90, 2021.
https://doi.org/10.1007/978-3-030-86198-8_6

In addition, these images can contain texts to provide further information about individual components. The digitization of such images implies the extraction of these information - components, interconnections and texts to extract a complete graph description of the digitized source document. An early attempt for such an extraction is described in [20]. Symbols connected through lines or multiple segments are a feature of many engineering diagrams such as mechanical engineering diagrams and Piping and Instrumentation Diagrams (P&IDs).

In this paper, The GraphFix framework[1] is proposed as an approach to digitize circuit diagrams, but is envisioned to be applicable to other similar document types with the help of suitable datasets. GraphFix is a multi-stage information extraction framework to identify different types of information in an engineering document. First of all, a Faster RCNN [15] is trained to identify component symbols, which form the node of the extracted graph. Based on that, blob detection is used to predict the connections (wiring) between the components, which make up the graph's edges.

The resulting graph structure is amenable to the application of graph refinement and error detection algorithms. There are two main types of errors in the component proposal list: false positives - component proposals in regions of the diagram, which do not have any identifiable symbol and false negatives - which can either result from a symbol in the input image being misclassified or not recognised as a symbol region, the latter type are referred to as Unmarked False Negatives (UFNs) in this work and result in graph anomalies, which can be detected with the help of Graph Convolutional Networks (GCNs) or with node degree comparison.

An attempt is also made to refine symbol labels in the component list generated by the Faster RCNN, using graph level features and symbol characteristics such as position and size. Some methods achieve up to **60%** recall@1, but the refinement does not help achieve any improvement in the overall recall of the framework when these predictions are combined with the Faster RCNN results. However, using graph anomalies, UFNs can be detected, which results in an improvement in recall@1 of up to **2–4%**.

This paper is further structured as follows: Sect. 2 describes related work in digitization of engineering diagrams (EDs) and circuit diagrams in particular. This section also briefly introduces topics on graph refinement and other concepts touched upon in this work. Section 3 provides information on the printed circuit ED dataset used to train and test GraphFix as well as the different data augmentation techniques used to improve the object recognition module's performance. The different processing steps required by GraphFix are explained under Methodology 4 and a review of the overall performance and of the different refinement and error detection techniques is presented in the Results Sect. 5. In conclusion, Sect. 6 summarises the salient contributions of this work as well limitations of the framework, which can be addressed in future research.

---

[1] https://github.com/msyed-unikl/GraphFix.

2   Related Work

Digitization of document types such as electric circuits, floor plans and P&IDs share many common features. [12] lists the identification of symbols, interconnections and text and some of main challenges in the digitization of engineering diagrams such as circuit diagrams. Circuit diagrams can utilize different standards and conventions to represent symbols for circuit components, resulting in difficulty in producing a well defined dataset, which captures variations in symbol style, pose and scale [12]. Initial attempts to identify symbols in circuit components such as [2] extract graphical primitives such as lines and circles from images and apply rule based templates to identify component symbols. [11] proposes a system to identify symbols in hand drawn engineering diagrams based on subgraph isomorphism by representing symbols and drawings as relational graphs using which the system could also learn to identify new symbols.

More recent approaches such as [13] extract graphical features from hand drawn circuits and inputs them to a neural network for symbol recognition. Other notable attempts utilizing machine learning based approaches as opposed to rule matching include [1], which proposes a probabilistic-SVM classifier using Histogram of Oriented Gradients (HOG) and Radon Transform features and [4], which used geometric analysis to analyse vertical, horizontal and circuit space features to identify symbols in electric and electronic symbols. Direct comparisons of the effectiveness of the different methods is not possible as these systems are trained on different (and often private) datasets with varying sets of symbols despite the availability of a standard dataset for electrical circuits [18].

Alternate approaches to symbol detection using Convolutional Neural Networks (CNNs) were proposed by [6] and more recent attempts in this direction include [17] that combines deep learning based Faster RCNN [15] and semantic segmentation with other statistical methods for such as morphology and component filtration for the vectorization of floor plans. [14] employs a VGG-19 based Fully Convolutional Neural Network to identify symbols in P&ID diagrams. [22] uses a Region based Convolutional Network (RCNN) to generate region proposals for 'symbols' and 'dummy' regions in P&ID images. [21] also proposes the use of RCNNs for the identification of symbols in P&ID images. GraphFix follows a similar approach by training a Faster RCNN module to train electric component symbols in the custom printed circuit ED dataset.

Multiple computer vision based approaches have also been applied task of identifying connections between symbols in EDs. [4] applies a Depth First Search by considering darkened pixels as nodes and introducing edges between nodes for adjacent pixels. [14] applies Hough Lines Transform to identify pipelines between symbols in P&IDs. GraphFix uses blob detection to identify connections between identified component symbols. Blobs are regions in an image, which differ from their surroundings in terms of image features such as the pixel colour [19]. Blob detection can be carried out with a number of algorithms such as Laplacian of Gaussian (LoG) or Difference of Gaussian (DoG) [19].

Symbol extraction as well as connection identification from ED images are associated with potential errors such as misclassification (for symbol extraction only), false negatives and false positives. Some works have attempted to use graph or network level features to identify errors in the extraction process. The use of graph based rules to modify graph structures extracted from P&ID diagrams is proposed in [3]. [14] creates a forest structure with the extracted components and uses properties of P&ID diagrams to detect false positive pipelines identified by the Hough Transform algorithm.

[5] applies morphological operations to the shape-graph space of a tree of connected components extracted from maps to filter out components not corresponding to expected layers. In [16], graph features are used to detect node labels for regions in floorplan images by converting floorplans to Region Adjacency Graphs and using Zernike moments as node attributes. Graph Neural Networks and Edge Networks proposed in [16] achieve up to **100%** accuracy on ILIPso dataset [8] in predicting node labels. Experiments attempting to predict node labels (component symbols) using graph or regional features in this work (Graph Refinement) showed limited success and have much lower accuracy compared to the Faster RCNN module. The GraphFix pipeline achieves its improvement over the Faster RCNN's recall by identifying regions in the circuit image with symbols that have been missed by the Faster RCNN module and this is done by identifying anomalies in the graph extracted from the circuit images using node degree comparison and Graph Convolutional Networks (GCNs).

GCNs provide a semi supervised graph based approach to predict node labels [9]. The Fourier basis on a graph is defined as the eigenvalues of the graph Laplasian, which is defined as $D - A$ — where D is the diagonal degree matrix (diagonal elements are the degrees of the nodes and other elements are 0) and A is the adjacency matrix for the Graph [10].

Graph convolutions can be defined on the Fourier basis, but such an approach is prohibitively expensive for large graphs as the calculation of eigenvector matrix is $O(N^2)$ where N is the number of nodes in the graph. Localised spectral filters to make spectral convolutions computationally feasible were proposed in [7]. [7] also put forward the use of a truncated expansion of Chebyshev polynomials as an approximation of the eigendecomposition of the Laplasian. [9] further develops this model to propose a GCN model which can be represented by

$$Z = \tilde{D}^{-\frac{1}{2}} \tilde{A} \tilde{D}^{-\frac{1}{2}} X \Theta$$

In this equation from [9], $\tilde{D} = \Sigma_{ij}\tilde{A}_{ij}$. Where $\tilde{A} = A + I_N$. A represents the adjacency matrix for the graph and $X \in \mathcal{R}^{N \times C}$ is the input signal (with C real input parameters for N nodes). $\Theta \in \mathcal{R}^{C \times F}$, where F is the number of filter maps. The GCN layer is 1-localized convolution and multiple GCN layers can also be stacked, which approximates higher level localization convolutions [9].

3 Printed Circuit ED Dataset

Computer Aided Design (CAD) software used to generate circuit diagrams can use different symbols and standards to represent symbols. A dataset used to train an object detection module to identify various types of symbols for component types should consist of diagrams from various sources, to capture component symbols differing in style, pose, and other visual features and characteristics. Hand drawn circuits are not included in the ground truth to avoid extreme heterogeneity[2].

The ground truth needed to run training, validation and testing for the Faster RCNN module and the graph algorithms is generated by scraping an online source[3] for circuit diagram images, which are converted to a standard image format. Component symbols in these images are then manually annotated[4] to mark symbols in the image with the corresponding component labels. Certain conventions and rules are necessary in order to maintain consistency in the use of labels for components across multiple diagram standards. Some of the important labelling conventions are:

- Circuit components are labeled such that there are no overlapping components.
- Wires are not annotated. However, wire crossings and overlaps are labelled as symbols
- Simple variations in symbol style, line colour and spatial orientation are tolerated and grouped under the same label. However, when a different symbol type is used constantly with a qualified component symbol (such as a 'ground' and 'digital ground'), separate label categories are created
- If a composite component (consisting of multiple smaller components) is repeatedly encountered, then the composite symbol is treated as a label category. For example, a rectifier bridge is labelled as one rectifier bridge as opposed to four diodes

Ground truth graphs for the electric circuits, component regions or bounding boxes from the manually annotated symbol list are input to blob detection to identify interconnections. The resulting graphs carrying symbols and connecting edges then comprise the ground truth for graph based methods.

In total, the ground truth consists of **218** annotated printed circuit EDs. These are divided into **182** images for training, **18** diagrams for validation at the time of training and **18** images are set aside for testing the framework. **85** symbol categories have been identified and the dataset has **8697** annotations. For each image in the training set, **8** variants are generated by randomly applying image transforms such as scaling, horizontal flipping, colour inversion and Gaussian noise (Fig. 1).

4   Methodology

GraphFix proposes a modular extraction of information from diagram images and their refinement. After image pre-processing (such as format conversion), the Faster RCNN object recognition module identifies component symbols in

---

[2] https://github.com/msyed-unikl/printerd-circuit-ED-dataset.

[3] discovercircuits.com.

[4] LabelImg - https://github.com/tzutalin/labelImg.

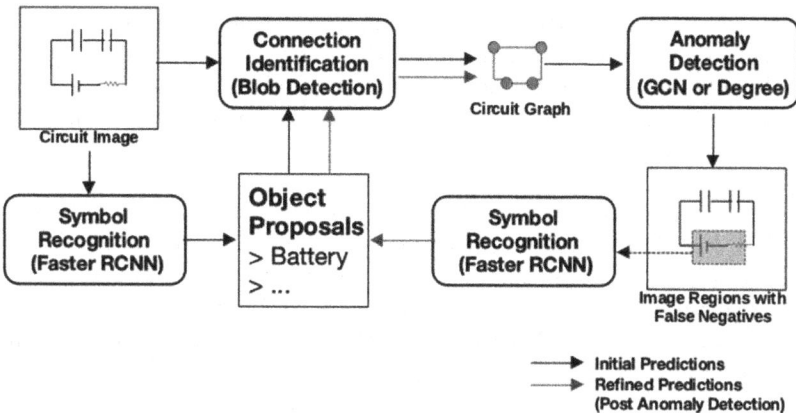

Fig. 1. Circuit digitization workflow proposed by GraphFix.

the image and their locations. This information is then used to identify connections between components using blob detection. The graph generated with the extracted symbols and connections can be subject to graph refinement, where graph anomalies are detected and used to identify component symbols missed by the object recognition module. Finally, select regions of the diagram are subject to image augmentation (cropping and scaling) and input again to the Faster RCNN module to identify these missing symbols.

4.1 Symbol Recognition

The classification head of a pre-trained (COCO Dataset) Faster RCNN module with a ResNet 50 backbone[5] is replaced with a new head to classify component symbols and the model is retrained (or fine-tuned) on the training dataset.

Object proposals from the Faster RCNN consist of bounding box, label and confidence scores. Multiple proposals with different confidence scores can be generated for the same component symbol. GraphFix merges overlapping bounding boxes to create a 'prediction cluster' with labels sorted by the confidence score. The list of prediction clusters are then compared with components in the ground truth to generate precision and recall metrics for the symbol detection task.

With grid search, training parameters such as the optimiser algorithm, training batch size and learning rate can be adjusted to maximise recall@1 on the validation set (see Table 1).

4.2 Connection Identification

After component symbols and their locations are identified, connections between components can be extracted using blob detection and some simple and intuitive

---

[5] https://pytorch.org/vision/stable/models.html#object-detection-instance-segmentation-and-person-keypoint-detection.

computer vision related operations. First, all component regions are blanked out by merging them into the background. After this, the image is converted to grayscale, colours are inverted and a threshold is applied to remove noise. A blob detection algorithm is then applied to identify unique blobs and a bounding box is identified for each blob.

The corners of these bounding boxes are checked for proximity to two or more symbol components and blobs in proximity of two or three symbols are shortlisted. In case the blob bounding box is in proximity to two symbols, then a connection between the symbols (a graph edge between the symbol nodes) is identified. If three symbols are in proximity to the blob bounding box, two of the three symbols are shortlisted such that the area of the rectangle between the two symbol 'touch points' is the maximum (Fig. 2).

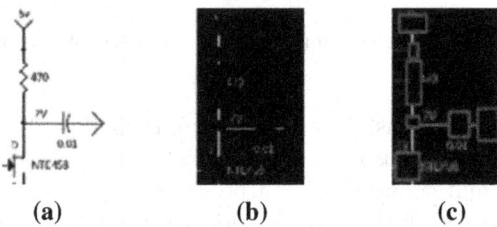

(a)   (b)   (c)

Fig. 2. a. Section of circuit for connection identification b. After symbol removal, thresholding and colour inversion c. Identification of blobs with vertices in vicinity of component symbols (discovercircuits.com).

There are some drawbacks to this technique such as:

- Component symbols must be erased before blob detection is applied. This is easily achieved with the ground truth as symbols are identified manually. However, for testing, the Faster RCNN object proposals can completely miss symbols (UFNs). This results in missing connections linked to these symbols and also, 'overshot' connections, which is noticed when symbols such as wire crossings or junctions are missed. In the latter case, the symbol is misidentified as a part of a wire connection in the image and incorrect connections can be added to the graph.
- The proposed method also misses complicated wire representations which loop around a component symbol.

Despite these drawbacks, the method identifies connections with a high precision and recall. However, these parameters cannot be quantitatively measured with the current printed circuit ED dataset as it does not list connections between symbols and hence the algorithms's output is suitable for manual assessment. Blob detection can also be used to identify text annotations and on circuit images in the printed circuit ED dataset, it performs on par with the EAST deep learning based text recognition algorithm [23]. However, the application of OCR systems to image regions with text results in inaccurate character and symbol recognition and thus not considered for graph refinement.

4.3   Anomaly Detection

Most errors seen in the output of the block detection based connection identification stem from symbols missed by the Faster RCNN module. Identifying
these Unmarked False Negatives (UFNs) can lead to the correction of a number of such errors. These false negatives do not have any corresponding node in
the extracted electric graph as no object proposals corresponding to the ground
truth object exist in the Faster RCNN output (in contrast there are also other
false negatives in the ground truth, which have overlapping object predictions
with incorrect labels). When blob detection is carried out on output from the
Faster RCNN, connections between UFNs and nodes connected to UFNs in the
ground truth will be missed or misidentified by the blob detection. These nodes
connected to the UFNs in the ground truth are present in the Faster RCNN
output (ignoring the possibility of two UFNs connected to one another) and
are termed anomalies. Two methods - node degree comparisons and GCNs are
applied to the task of identifying anomalies (Fig. 3).

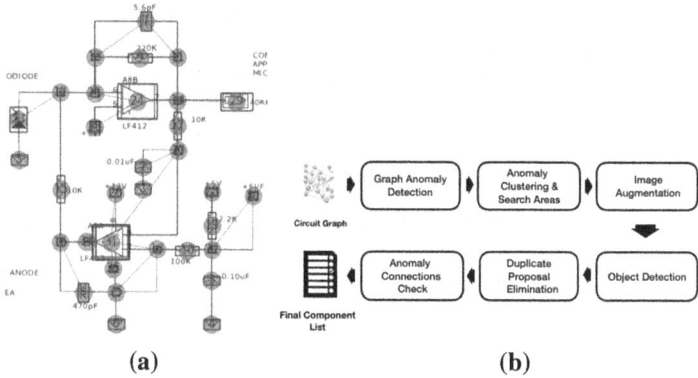

(a)   (b)

Fig. 3. a. Sample graph generated from symbol and connection identification steps
(discovercircuits.com). b. Unmarked False Negative (UFN) Identification Process with
Image Augmentation.

4.4   Training Set for Anomaly Detection

To train models to identify anomalies, training set images are input to the Faster
RCNN model and blob detection to generate graphs for the circuits. The anomalies in these graphs can be identified by comparing them with the ground truth
and identifying nodes connected to UFNs. For Faster RCNN trained on the basic
dataset up to **5%** of nodes in ground truth are UFNs and **8%** of nodes in the
Anomaly Detection training set are anomalies. This results in under-sampling
of anomaly nodes, which has to be handled while training models.

4.5   Anomaly Detection Techniques

Node Degree Comparison. Using the Anomaly Detection training set, two discreet probability distributions of the observed degree for each symbol type can be calculated. The first distribution is for instances where the symbol is encountered as an anomaly and the second distribution is for symbol instances that do not occur as anomalies.

These two distributions for a symbol type are represented as $P(D_n = d_n | A_n = True)$ and $P(D_n = d_n | A_n = False)$, which are the probabilities that the degree D of the node n is an observed natural number d_n given A_n is **True** indicating the node n is anomalous or **False**, which denotes that node n is not an anomaly.

These distributions are then used to predict if node occurrences in the test set graphs are anomalies.

A_n is True if

$$P(D_n = d_n | A_n = True) > P(D_n = d_n | A_n = False)$$

Using Bayes Theorem, an alternate can also be suggested where instead of using $P(D_n = d_n | A_n = True)$ and $P(D_n = d_n | A_n = False)$, $P(A_n = True | D_n = d_n)$ and $P(A_n = False | D_n = d_n)$ can be estimated with prior probabilities $P(A_n = True)$ and $P(A_n = False)$ and the condition becomes:

A_n is True if

$$P(D_n = d_n | A_n = True) \cdot P(A_n = True) > P(D_n = d_n | A_n = False) \cdot P(A_n = False)$$

In practice, $P(A_n = True)$ is usually very small (less than **0.1**) and therefore anomaly detection with Bayesian priors have an extremely low recall.

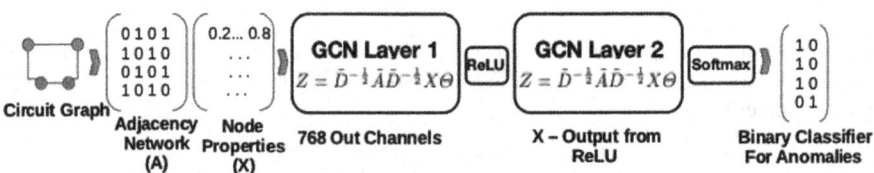

Fig. 4. A two layer GCN network to detect anomalies in circuits. In a three layer GCN, the output of the second layer (768 out channels) is input to a second ReLU activation and then input to a third GCN layer followed by a softmax function

GCNs: The anomaly training dataset can also be used to train a GCN based network to identify anomalies in test set graph outputs from the Faster RCNN and blob detection. In addition to anomalies in the Faster RCNN output in the training set, nodes are artificially dropped in the training set graphs, to generate additional anomalies. All neighbours previously connected to a dropped node can be marked as anomalies. Multiple variants of each training set circuit graph are generated and nodes are dropped randomly. The data is then used to train GCN networks to identify anomaly nodes. Multiple architectures are trained by varying the training parameters such as number of GCN layers, probability of a node being dropped and the number of variants for each electric graph to maximise recall for anomaly detection (Fig. 4).

The performance of the two anomaly detection methods is presented in the Results section.

4.6   Detection of False Negatives

After identification of anomalies in the Faster RCNN output in electric graphs, GraphFix proceeds to use this information to identify the UFNs causing the anomalies by using the following observations:

– Anomalies are connected to the UFN in the ground truth, so a UFN is expected to be in the proximity of anomalies in most cases
– There should be a connection between an anomaly and new object proposals for UFNs. This is useful in dropping false positive UFN proposals
– A UFN can be connected to one or more anomalies in the ground truth - this assumption allows for the clustering of anomalies in close proximity

Some further assumptions are made to cluster anomalies and to identify suitable search areas in which UFNs may potentially be located:

– Distances between components in the circuit are distributed normally with mean $avg\ distance$ and standard deviation $std\ dev$ - then the distance between the anomaly and a UFN can be represented in these terms
– $avg\ distance$ and $std\ dev$ can also be used to identify anomaly clusters - since anomalies in a cluster should be connected to the same UFN. Graph-Fix uses a heuristics based method to cluster anomalies by first defining $D_c = \alpha_c * avg\ distance + \beta_c * std\ dev$. With α_c and β_c set based on experimentation. Each node is assigned to a unique cluster initially. Two clusters are aggregated if the minimum distance between nodes of the two clusters is below D_c. The previous step is repeated till no more clusters can be aggregated. If an anomaly cluster has more than one nodes, the UFN is expected to be located 'near the center' of these anomaly nodes

An anomaly cluster with a single node can result from a UFN positioned either above or below or to the left or right of the anomaly. Hence a square search area with side $S_c = 2 * D_p$ where $D_p = \alpha_p * avg\ distance + \beta_p * std\ dev$ with α_p and β_p values set using trial and error is assigned to such single node clusters. Clusters with two anomalies result in a rectangular search area which is centered at the centroid of the two anomaly components and both the length and breadth of the rectangle are set to S_c.

For clusters with more than two anomalies, the search area is chosen as the minimum rectangle that covers all the anomalies in the cluster - assuming that the anomalies are resulting from one UFN surrounded by the anomalies.

After finalizing search areas, two alternate methods are tested to identify UFNs within the selected regions:

- By default, Faster RCNN outputs proposals with a confidence score greater than **0.05**, the threshold for Faster RCNN object proposals is lowered to **0.01** and the object proposals with confidence scores between **0.05** and **0.01** are stored in a separate list. This list can be searched for object proposals in the search areas
- Search areas can be cropped and subject to image transform such as stretching. The transformed image is input to the Faster RCNN module to identify object proposals. The new proposals have to be converted back to the coordinates of the original circuit image

Experiments on the printed circuit ED dataset show that image augmentation yields better results than lowering the Faster RCNN confidence score and was selected for further processing. Proposals identified using image augmentation are compared with the original output of the Faster RCNN to identify and delete repeats corresponding to components already identified. Finally, blob detection is carried out to identify connections between anomaly nodes and the new object proposals. Any new object proposals not connected to anomalies are dropped as false positives to arrive at the final list of extracted components.

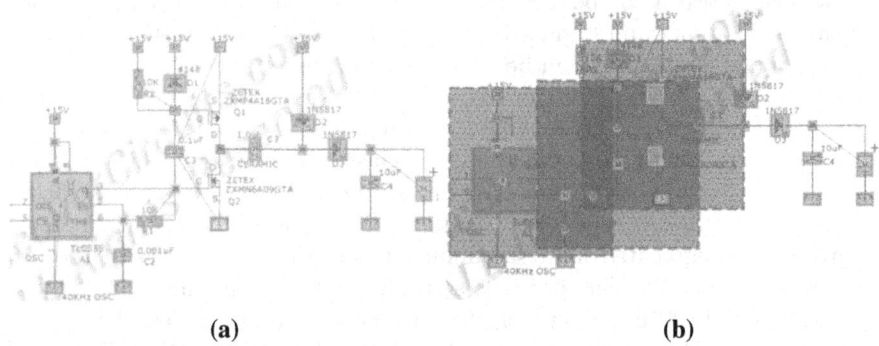

(a)   (b)

Fig. 5. a. Correctly identified anomalies (red borders with red fill) and false negative anomalies (red borders with yellow fill) b. Search areas (green dashed borders) and detected Unmarked False Negatives UFNs (green fill) on a sample circuit (discovercircuits.com). (Color figure online)

4.7   Graph Refinement

Representing electric circuits as graphs with components as nodes and interconnections as edges allows the possibility to apply graph refinement techniques to the extracted information. Graph refinement can be applied for a number of tasks in graphs such as (i) Error Detection (identify wrong node labels or erroneous connections) (ii) Identification of missing edges

For graphs generated from engineering diagrams, the detection of incorrectly labeled nodes is of special interest, as this can potentially improve symbol recall. Two techniques (symbol location and size based prediction, and GCNs) to identify node labels using graph and image features have been presented here. The methods are trained on ground truth graph data and tested on the graphs generated using Faster RCNN output on test set diagrams.

Symbol Location and Size Based Prediction. Circuit diagrams utilise certain conventions such as the placement of voltage sources at the top and representing ground connections at the bottom of diagrams. In addition, the dimensions of a symbol bounding box are a good indicator of the label type. This information is normalised for components in images and is used to train a neural network to predict the symbol labels. The trained neural network is then used to predict node labels based on the same information from the Faster RCNN output on the test set.

Graph Convolutional Networks. Both the node embeddings and symbol location and size based label prediction methods do not make use of all graph features available, therefore, a 2 layer GCN is trained to predict the label of a node taking the graph adjacency matrix and node data as input. Two versions of the GCN are tested, in one version, symbol position and size data is included as node data. In a second version, Faster RCNN prediction scores for the labels are also appended to the node data.

5   Results

Extraction results for the GraphFix framework are presented under four sections - results from fine-tuning the Faster RCNN module with different approaches, anomaly detection and the improvement in extraction after anomaly detection an finally, the results from other graph refinement techniques. The precision@1 and recall@1 metrics are used to quantify the performance on symbol detection and anomaly detection tasks.

5.1   Fine-Tuning of Faster RCNN

The Faster RCNN object detection module provides a good baseline for symbol extraction task. Data augmentation and parameter optimization also help improve the performance even further and a recall@1 of **89.47%** is achieved with a combination of these techniques (see Table 1).

Table 1. Results of faster RCNN training on test set

Model	Dataset	Precision	Recall	F-Measure	Max Recall
Faster RCNN	Basic	83.82%	83.16%	83.49%	85.44%
Faster RCNN	Augmented	86.91%	87.37%	87.14%	**89.47%**

5.2 Anomaly Detection

To compare node degree and GCN based anomaly detection, Recall@1 and Precision@1 metrics are calculated by comparing the list of anomaly nodes in the anomaly dataset with the list of anomalies predicted for the two methods for the test set (see Table 2).

These metrics vary from one GCN training to another, therefore the results for the model with the best recall in **3** training attempts are reproduced below. Additionally, the number of variants for a model is decided by increasing the number till no further improvement in recall is observed. In contrast, these metrics remain constant for the node degree comparison model unless the underlying Faster RCNN model is retrained (Table 3).

Table 2. Anomaly detection using node degree based methods

Anomaly detection method	Precision@1	Recall@1	F-Measure
Node degree	30.60%	58.95%	40.29%

Table 3. Best results for anomaly detection using GCNs

GCN	Node Drop %	Variants	Precision@1	Recall@1	F-Measure
2 layer GCN	20.00%	8	53.49%	27.38%	36.22%
2 layer GCN	25.00%	8	52.00%	30.95%	38.80%
2 layer GCN	33.33%	6	49.02%	30.86%	37.88%
2 layer GCN	50.00%	6	41.82%	27.38%	33.09%
3 layer GCN	25.00%	10	45.90%	33.33%	38.62%
3 layer GCN	30.00%	8	**46.03%**	**34.52%**	**39.46%**
3 layer GCN	33.33%	6	42.42%	33.33%	37.33%
3 layer GCN	40.00%	4	42.42%	33.33%	37.33%

GCN based anomaly detection offers higher precision than the node degree comparison, but the node degree method delivers higher recall. Additionally, some observations can be made about the low F-measure for both models:

- Errors from previous processing steps i.e. Faster RCNN and blob detection for connection detection result in errors both in the training and test datasets for anomaly detection

– The node degree comparison model outputs many false positives, which correspond to issues resulting in the graph structure due to problems other than UFNs such as false positives in Faster RCNN output, incorrect output from blob detection as well as use of infrequent circuit symbols or wire configurations

– Improvements in GCN precision and recall with data augmentation (node drops in variants) leads to the hypothesis that the method can achieve better results with additional training data

Since anomaly detection is carried out to detect UFNs, node degree comparison is chosen for further experiments because of its higher recall. This is because, more search areas increase the chances of detecting UFNs.

5.3   Improving Faster RCNN Recall Using Anomaly Detection

Using anomaly predictions from node degree comparison, a fixed image transformation (scaling by a factor of **1.15**) is applied to search areas to identify new object proposals. New proposals overlapping with existing components or without connections to anomalies are dropped and precision@1 and recall@1 metrics for the expanded component list are calculated (see Fig. 5).

The anomaly detection method is applied to multiple Faster RCNN models to observe the consistency in the improvement of results (see Table 4).

Table 4. Improved symbol recall with UFN & anomaly detection

Model	Dataset	Faster RCNN results			Results after anomaly detection		
		Precision	Recall	F-Meas	Precision	Recall	F-Meas
Faster RCNN I	Basic	83.82%	83.16%	83.49%	75.09%	86.84%	80.54%
Faster RCNN II	Augmented	85.79%	87.37%	86.57%	79.67%	89.34%	84.23%
Faster RCNN III	Augmented	85.60%	89.21%	87.37%	80.93%	**91.05%**	85.69%

Anomaly detection based identification of UFNs results in a consistent increase improvement in the recall metrics. A slight drawback is the drop in precision, which is largely a result of rectangular corners and segments of wires being identified as wire connections after image augmentation. Simple domain specific rules can eliminate many of these wrong proposals. However, this is not implemented to keep the proposed framework general.

5.4   Graph Refinement

For graph refinement techniques, two parameters are tested - recall@1 for the refinement method itself and the recall of a combined model which adds confidence scores of the refinement method and the Faster RCNN model trained on the basic dataset (see Table 5).

Table 5. Recall for graph based label prediction methods. Recall for Faster RCNN model - **83.55%**

Refinement method	Recall	Recall combined with Faster RCNN
Neural network with component position and size	58.95%	80.39%
2 layer GCN	41.05%	81.47%
2 layer GCN with Faster RCNN confidence scores	80.06%	81.84%

6   Conclusion

GraphFix successfully demonstrates the use of graph level features to improve symbol and connection identification in circuit diagrams images. The graph level abstraction of the proposed enhancement techniques allows for the framework to be applied to other symbolic engineering document types such as P&IDs, subject to availability of suitable datasets, providing further scope for research in this topic, additionally, the framework potentially supports symbol extraction approaches other than Faster RCNN, which provide symbol class and position information as well as other connection identification methods.

Although GraphFix achieves a high recall on the tested dataset, multiple improvements such as identifying connection leads for components, improving the OCR recognition and identifying improvements to the graph refinement stage can improve the accuracy of the digitization and increase the practical utility of the extraction. Additionally, graph level features can also be extracted from available electronic versions of engineering diagrams (such as circuit netlists), providing for a more ground truth for the graph based refinement models.

References

1. Agarwal, S., Agrawal, M., Chaudhury, S.: Recognizing electronic circuits to enrich web documents for electronic simulation. In: Lamiroy, B., Dueire Lins, R. (eds.) GREC 2015. LNCS, vol. 9657, pp. 60–74. Springer, Cham (2017). https://doi.org/10.1007/978-3-319-52159-6_5
2. Bailey, D., Norman, A., Moretti, G., North, P.: Electronic schematic recognition. Massey University, Wellington, New Zealand (1995)
3. Bayer, J., Sinha, A.: Graph-based manipulation rules for piping and instrumentation diagrams (2020)
4. De, P., Mandal, S., Bhowmick, P.: Recognition of electrical symbols in document images using morphology and geometric analysis. In: 2011 International Conference on Image Information Processing, pp. 1–6 (2011)
5. Drapeau, J., Géraud, T., Coustaty, M., Chazalon, J., Burie, J.-C., Eglin, V., Bres, S.: Extraction of ancient map contents using trees of connected components. In: Fornés, A., Lamiroy, B. (eds.) GREC 2017. LNCS, vol. 11009, pp. 115–130. Springer, Cham (2018). https://doi.org/10.1007/978-3-030-02284-6_9

6. Fu, L., Kara, L.B.: From engineering diagrams to engineering models: visual recognition and applications. Comput. Aided Des. **43**(3), 278–292 (2011)

7. Hammond, D.K., Vandergheynst, P., Gribonval, R.: Wavelets on graphs via spectral graph theory. Appl. Comput. Harmon. Anal. **30**(2), 129–150 (2011)

8. Héroux, P., Le Bodic, P., Adam, S.: Datasets for the evaluation of substitution-tolerant subgraph isomorphism. In: Lamiroy, B., Ogier, J.-M. (eds.) GREC 2013. LNCS, vol. 8746, pp. 240–251. Springer, Heidelberg (2014). https://doi.org/10.1007/978-3-662-44854-0_19

9. Kipf, T.N., Welling, M.: Semi-supervised classification with graph convolutional networks. arXiv preprint arXiv:1609.02907 (2016)

10. Maier, A.: Graph deep learning — part 1. https://towardsdatascience.com/graph-deep-learning-part-1-e9652e5c4681

11. Messmer, B.T., Bunke, H.: Automatic learning and recognition of graphical symbols in engineering drawings. In: Kasturi, R., Tombre, K. (eds.) GREC 1995. LNCS, vol. 1072, pp. 123–134. Springer, Heidelberg (1996). https://doi.org/10.1007/3-540-61226-2_11

12. Moreno-García, C.F., Elyan, E., Jayne, C.: New trends on digitisation of complex engineering drawings. Neural Comput. Appl. **31**(6), 1695–1712 (2018). https://doi.org/10.1007/s00521-018-3583-1

13. Rabbani, M., Khoshkangini, R., Nagendraswamy, H., Conti, M.: Hand drawn optical circuit recognition. Procedia Comput. Sci. **84**, 41–48 (2016). Proceeding of the Seventh International Conference on Intelligent Human Computer Interaction (IHCI 2015). http://www.sciencedirect.com/science/article/pii/S1877050916300783

14. Rahul, R., Paliwal, S., Sharma, M., Vig, L.: Automatic information extraction from piping and instrumentation diagrams. arXiv preprint arXiv:1901.11383 (2019)

15. Ren, S., He, K., Girshick, R., Sun, J.: Faster R-CNN: towards real-time object detection with region proposal networks. IEEE Trans. Pattern Anal. Mach. Intell. **39**(6), 1137–1149 (2016)

16. Renton, G., Héroux, P., Gaüzère, B., Adam, S.: Graph neural network for symbol detection on document images. In: 2019 International Conference on Document Analysis and Recognition Workshops (ICDARW), vol. 1, pp. 62–67. IEEE (2019)

17. Surikov, I.Y., Nakhatovich, M.A., Belyaev, S.Y., Savchuk, D.A.: Floor plan recognition and vectorization using combination UNet, faster-RCNN, statistical component analysis and ramer-douglas-peucker. In: Chaubey, N., Parikh, S., Amin, K. (eds.) COMS2 2020. CCIS, vol. 1235, pp. 16–28. Springer, Singapore (2020). https://doi.org/10.1007/978-981-15-6648-6_2

18. Valveny, E., Delalandre, M., Raveaux, R., Lamiroy, B.: Report on the symbol recognition and spotting contest. In: Kwon, Y.-B., Ogier, J.-M. (eds.) GREC 2011. LNCS, vol. 7423, pp. 198–207. Springer, Heidelberg (2013). https://doi.org/10.1007/978-3-642-36824-0_19

19. Xiong, X., Choi, B.J.: Comparative analysis of detection algorithms for corner and blob features in image processing. Int. J. Fuzzy Logic Intell. Syst. **13**(4), 284–290 (2013)

20. Yu, B.: Automatic understanding of symbol-connected diagrams. In: Proceedings of 3rd International Conference on Document Analysis and Recognition, vol. 2, pp. 803–806. IEEE (1995)

21. Yu, E.S., Cha, J.M., Lee, T., Kim, J., Mun, D.: Features recognition from piping and instrumentation diagrams in image format using a deep learning network. Energies **12**(23), 4425 (2019)

22. Yun, D.Y., Seo, S.K., Zahid, U., Lee, C.J.: Deep neural network for automatic image recognition of engineering diagrams. Appl. Sci. **10**(11), 4005 (2020)
23. Zhou, X., et al.: East: an efficient and accurate scene text detector. In: Proceedings of the IEEE Conference on Computer Vision and Pattern Recognition, pp. 5551–5560 (2017)

ScanSSD-XYc: Faster Detection for Math Formulas

Abhisek Dey[iD] and Richard Zanibbi[✉][iD]

Rochester Institute of Technology, Rochester, NY 14623, USA
{ad4529,rxzvcs}@rit.edu

Abstract. Detecting formulas offset from text and embedded within textlines is a key step for OCR in scientific documents. The Scanning Single Shot Detector (ScanSSD) detects formulas using visual features, applying a convolutional neural network within windows in a large document image (600 dpi). Detections are pooled to produce page-level bounding boxes. The system works well, but is rather slow. In this work we accelerate ScanSSD in multiple ways: (1) input and output routines have been replaced by matrix operations, (2) the detection window stride (offset) can now be adjusted separately for training and testing, with fewer windows used in testing, and (3) merging with non-maximal suppression (NMS) in windows and pages has been replaced by merging overlapping detections using XY-cutting at the page level. Our fastest model processes 3 pages per second on a Linux system with a GTX 1080Ti GPU, Intel i7-7700K CPU, and 32 GB of RAM.

Keywords: Math formula detection · Single-Shot Detector (SSD)

1   Introduction

ScanSSD [4] by Mali *et al.* detects math formula regions as bounding boxes around formulas embedded in text or offset using a Convolutional Neural Network (CNN). It was based on the original SSD [2] which is a single pass network with a VGG-16 [8] backbone to locate and classify regions. Unlike SSD which was made to detect objects in the wild, a number of modifications allowed ScanSSD to obtain strong performance for math formula detection.

Figure 1 shows an overview of the detection process in ScanSSD. To mitigate the issue of low recall on large PDF images and non-square aspect ratios, ScanSSD uses a sliding window detection strategy. A 1200×1200 window was used to stride across the PDF images generated at 600dpi to generate sub-images resized to 512×512 as the input to the network. A stride of 10% or 120 pixels across and down was used as the striding factor. Detections from these windows are passed through a second stage, comprised of a page-level pixel-based voting algorithm to determine the page-level predictions. Pixels vote based on the number of detection boxes they intersect, after which pixel scores are binarized, with formula detections defined by connected components in the binarized voting grid. Initial detections are then cropped around connected components in the original page image that they contain or intersect.

© Springer Nature Switzerland AG 2021
E. H. Barney Smith and U. Pal (Eds.): ICDAR 2021 Workshops, LNCS 12916, pp. 91–96, 2021.
https://doi.org/10.1007/978-3-030-86198-8_7

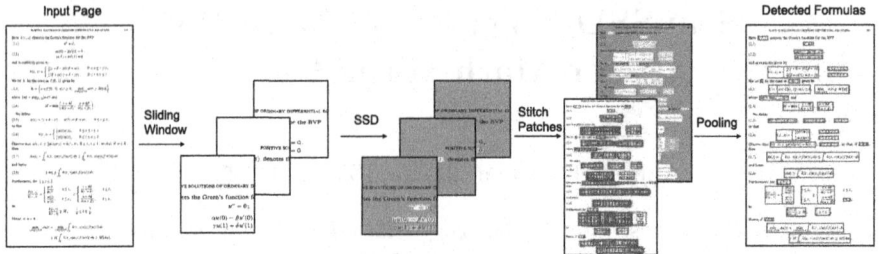

Fig. 1. Sliding window-based formula detection in ScanSSD.

ScanSSD obtains state-of-the-art results with a 79.6 F-score (IoU 0.5) on the TFD-IDCDAR2019v2 dataset, but is quite slow. In this work we have made changes to ScanSSD that allow us to retain comparable accuracy, while decreasing execution times by more than 300 times. Details are provided below.

2   ScanSSD-XYc

ScanSSD-XYc streamlines detection and eliminates redundant operations in ScanSSD, leading to decreases in execution and training times. Substantial changes were made across the pre-processing, windowing and pooling stages, as discussed below.

Pre-processing. A major bottleneck in ScanSSD was the pre-processing stage, where all windows were generated as separate files. Furthermore, inefficient loading and parallelization resulted in the GPU waiting for batches. To address this, the I/O framework was completely revised. Windows for a page are now generated using a single tensor operation applied to the page image. Ground-truth regions that wholly or partially fall within windows are also identified by another tensor operation. This decreased execution times by 28% over ScanSSD, before incorporating the additional modifications discussed below.

In the original ScanSSD windowing algorithm, many windows used in training contain no ground-truth regions. SSD detects objects at multiple scales, using a grid of initial detection regions at each scale. This leads to thousands of SSD candidate predictions in a single window. Even where target formulas are present in a window, the vast majority of candidate detections are true negatives. To reduce this imbalance for negative examples, in addition to hard negative mining windows without ground-truth regions are ignored in training. Page images are also padded at their edges to be an even multiple of 1200 pixels high and wide, matching the area covered by the fixed-size sliding window.

Windowing. ScanSSD uses a fixed 10% stride for training and inference; we instead use a smaller 5% stride in the training stage to see more examples

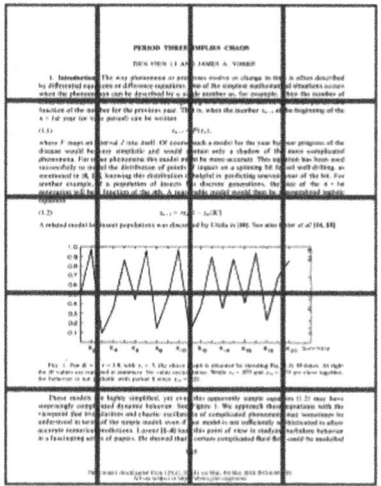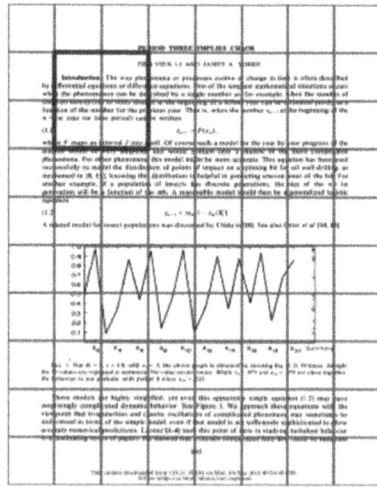

Fig. 2. Left: blue lines show regions for 100% strides of a 1200 × 1200 window, with each sub-region seen once by the network. **Right:** red lines illustrate regions at left split into four by 50% strides of the same 1200 × 1200 window (shown in blue). Except for cells around the border, each sub-region indicated by the red lines is seen four times. (Color figure online)

of the same target formula alongside other page contents, and then run testing/inference using large strides (e.g., 75%) for faster detection. A stride of 60 pixels (5%) was used for training, and differing strides from 10% to 100% were used for execution. As shown in Fig. 2, smaller strides enable the network to see sub-regions multiple times, and improve network convergence. The page edge regions are seen only once, but predominantly consist of white margins.

XY-Cutting to Merge Window-Level Predictions. ScanSSD filters predicted regions twice: once at the window-level and again at the page level. Window-level predictions undergo non-maximal suppression (NMS), and are then stitched together at the page level. Window-level regions vote pixel-wise, producing a vote map over the image which is binarized. Remaining connected components are treated as the final detections. A post-processing step then fits detections tightly around the connected components in the original page image that they intersect. This is an expensive process, illustrated in Fig. 1.

As shown in Fig. 3, ScanSSD-XYc simply filters window-level detections with less than 50% confidence and then pools detections at page level (Fig. 3 left). XY-cutting [1,5] segments documents into axis-aligned rectangular regions using pixel projections. We use XY-cutting as a partitioning algorithm over detected formula boxes, recursively segmenting the page until regions contain only one set of overlapping boxes. Finally, overlapping boxes are merged to predict the final formula regions (Fig. 3 right). Although the worst case time complexity (all overlapping boxes) with n boxes is $O(n^2)$ for both NMS and XY-Cuts [7], for

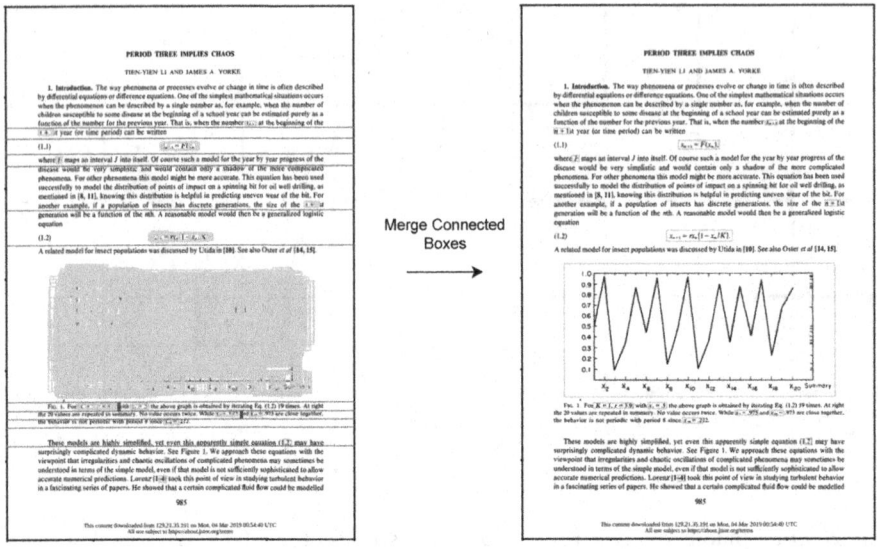

Fig. 3. Left: XY-cutting of Pooled Window-Level Predictions. Orange horizontal lines show splits along the y-axis, and blue vertical lines show splits in the x-direction. Cutting is performed only on detection boxes, and stops when all detection boxes in a region overlap. **Right:** Final page-level predictions after merging overlapping boxes.

XY-cut the worst case is rare. XY cuts recursively groups boxes based on spatial positions, and successive splits may be checked independently of one other. In ScanSSD-XYc, cropping is used only for computing evaluation metrics, as blank space at detection borders may be ignored in our applications, but unfairly penalizes detections when using IoU detection metrics.

3   Results

We present preliminary results for ScanSSD-XYc on the TFD-ICDAR2019v2 [3] dataset, which fixes some missing annotations in TFD-ICDAR2019. There are 446 training page images, and 236 testing images taken from 48 PDFs for math papers. Page images are document scans (600 dpi). Experiments were run using a Core i7-7700K CPU with a GTX 1080Ti GPU and 32 GB of RAM.

Table 1 compares ScanSSD-XYc with systems that participated in the ICDAR 2019 CROHME competition [3] and ScanSSD. ScanSSD-XYc performs comparably to the original ScanSSD at 72.4 F-score (IoU of 0.5). RIT 1 was a default implementation of YoloV3, and RIT 2 was an earlier version of SSD [6]. To diagnose the lower recall scores observed for ScanSSD-XYc, we studied the effect of stride size at test-time on accuracy and speed. Table 2 breaks down results by stride size: a smaller stride produces more windows for a page image.

While 100% strides had the smallest number of windows and fastest execution time (3.1 pages/second), a 75% stride gave us the highest accuracy, while still

Table 1. Comparison of preliminary results from ScanSSD-XYc (using 75% stride)

Model	IoU 0.5			IoU 0.75		
	Precision	Recall	F-score	Precision	Recall	F-score
TFD-ICDAR2019 (Original)						
ScanSSD	85.1	75.9	80.2	77.4	69.0	73.0
RIT 2	83.1	67.0	75.4	75.3	62.5	68.3
RIT 1	74.4	68.5	71.3	63.2	58.2	60.6
Mitchiking	36.9	27.0	31.2	19.1	13.9	16.1
TFD-ICDAR2019v2						
ScanSSD	84.8	74.9	79.6	78.1	69.0	73.3
ScanSSD-XYc	84.8	63.1	72.4	77.2	57.4	65.9

Table 2. ScanSSD-XYc results on TFD-ICDAR2019v2 for different strides

Stride (%)	IoU 0.5			IoU 0.75			Secs/Page
	Precision	Recall	F-score	Precision	Recall	F-score	
10	41.6	63.1	50.2	36.4	55.3	43.9	28.4
25	79.1	62.6	69.9	71.5	56.6	63.1	4.8
50	83.0	**63.4**	71.9	75.3	**57.6**	65.2	1.3
75	**84.8**	63.1	**72.4**	**77.2**	57.4	**65.9**	0.6
100	80.7	61.4	69.7	71.0	54.0	61.3	**0.3**

processing 1.56 pages/second. ScanSSD-XYc using the smallest stride (10%) is over three times (300%) faster than ScanSSD using the same stride, which takes 90 s/page.

As seen in Table 2, the 75% stride produces the best balance between speed and accuracy, roughly matching the precision of the original ScanSSD, with some reduction in recall (10.8% for IoU of 0.5, 11.6% for IoU of 0.75). Opportunities to increase recall are described in the next Section. In Table 2 we see little change in recall for different strides, while precision varies. Furthermore, the presence of figures in the evaluation set impacted precision, as these were detected as math (i.e., false positives). The training set contains 9 of 36 documents with at least one figure, while the test set contains 4 of 10 documents with at least one figure. Non-text regions such as figures and tables are often detected wholly or in part by ScanSSD-XYc (see Fig. 3), and we plan to address this in future work.

We expected the smallest stride to perform best, but smaller strides produce more partial predictions, which are fit less tightly around targets at page level, and result in more false positives.

4   Conclusion

We have presented an accelerated version of the Scanning Single Shot Detector in this paper. Combining window and page level operations, and eliminating NMS and post-processing led to much shorter execution times with some loss in recall. The entire system was refactored to facilitate faster, scalable tensorized I/O operations. Our long-term goal is to improve the usability of SSD-based detection for use in formula indexing for retrieval applications. We hope to improve detection effectiveness by exploring causes of low recall for small formulas, and issues with over-merging across and between text lines.

We think that setting a relative threshold in the page level XY-cuts based on the sizes of the boxes and detection confidences can mitigate the issue of low recall due to over-merging. We will also attempt to use additional training (beyond two epochs) and different weight initializations and other tuning parameters to better optimize the detector parameters.

Acknowledgements. We thank the anonymous reviewers for their helpful comments, along with Alex Keller, Ayush Shah, and Jian Wu for assistance with the design of ScanSSD-XYc. This material is based upon work supported by grants from the Alfred P. Sloan Foundation (G-2017-9827) and the National Science Foundation (IIS-1717997 (MathSeer) and 2019897 (MMLI)).

References

1. Ha, J., Haralick, R., Phillips, I.: Recursive X-Y cut using bounding boxes of connected components. In: Proceedings of 3rd International Conference on Document Analysis and Recognition, vol. 2, pp. 952–955 (1995). https://doi.org/10.1109/ICDAR.1995.602059
2. Liu, W., et al.: SSD: single shot MultiBox detector. In: Leibe, B., Matas, J., Sebe, N., Welling, M. (eds.) ECCV 2016. LNCS, vol. 9905, pp. 21–37. Springer, Cham (2016). https://doi.org/10.1007/978-3-319-46448-0_2
3. Mahdavi, M., Zanibbi, R., Mouchere, H., Viard-Gaudin, C., Garain, U.: ICDAR 2019 CROHME + TFD: competition on recognition of handwritten mathematical expressions and typeset formula detection. In: Proceedings of the International Conference on Document Analysis and Recognition, ICDAR, pp. 1533–1538 (2019). https://doi.org/10.1109/ICDAR.2019.00247
4. Mali, P., Kukkadapu, P., Mahdavi, M., Zanibbi, R.: ScanSSD: scanning single shot detector for mathematical formulas in PDF document images. arXiv (2020)
5. Nagy, G., Seth, S.: Hierarchical representation of optically scanned documents. In: Proceedings - International Conference on Pattern Recognition, pp. 347–349 (1984)
6. Redmon, J., Farhadi, A.: YOLOv3: an incremental improvement (2018). http://arxiv.org/abs/1804.02767
7. Samet, H.: Foundations of Multidimensional and Metric Data Structures. Series on Computer Graphics and Geometric Modeling. Morgan Kaufmann, San Francisco (2005). (XY-cut described in Section 2.1.2)
8. Simonyan, K., Zisserman, A.: Very deep convolutional networks for large-scale image recognition. In: International Conference on Learning Representations (2015)

Famous Companies Use More Letters in Logo: A Large-Scale Analysis of Text Area in Logo

Shintaro Nishi(✉), Takeaki Kadota, and Seiichi Uchida(iD)

Kyushu University, Fukuoka, Japan
shintato.nishi@kyushu-u.ac.jp

Abstract. This paper analyzes a large number of logo images from the LLD-logo dataset, by recent deep learning-based techniques, to understand not only design trends of logo images and but also the correlation to their owner company. Especially, we focus on three correlations between logo images and their text areas, between the text areas and the number of followers on Twitter, and between the logo images and the number of followers. Various findings include the weak positive correlation between the text area ratio and the number of followers of the company. In addition, deep regression and deep ranking methods can catch correlations between the logo images and the number of followers.

Keywords: Logo image analysis · DeepCluster · RankNet

1 Introduction

Logo is a graphic design for the public identification of a company or an organization. Therefore, each logo is carefully created by a professional designer, while considering various aspects of the company. In other words, each logo represents the policy, history, philosophy, commercial strategy, etc., of the company, along with its visual publicity aim.

Logo can be classified roughly into three types: logotype, logo symbol, and their mixed[1]. Figure 1 shows several examples of each type. *Logotype* is a logo comprised of only letters and often shows a company name or its initial letters. *Logo symbol* is a logo comprised of some abstract mark or icon or pictogram. *Mixed* is a logo comprised of a mixture of logotype and logo symbol.

The general purpose of this paper is to analyze the logo images from various aspects to understand the relationship among their visual appearance, text area, and the popularity of the company. For this purpose, we utilize various state-of-the-art techniques of text detection and image clustering, and a large logo dataset, called LLD-logo [16]. To the authors' best knowledge, logo image

---

[1] There are several variations in the logo type classification. For example, in [1], three types are called "text-based logo", "iconic or symbolic logo", and "mixed logo", respectively.

© Springer Nature Switzerland AG 2021
E. H. Barney Smith and U. Pal (Eds.): ICDAR 2021 Workshops, LNCS 12916, pp. 97–111, 2021.
https://doi.org/10.1007/978-3-030-86198-8_8

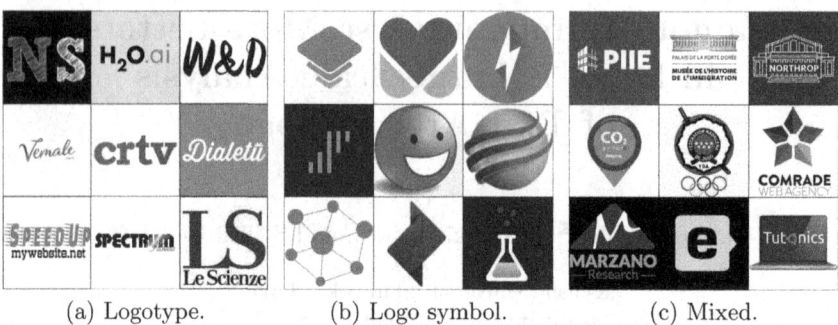

(a) Logotype.   (b) Logo symbol.   (c) Mixed.

Fig. 1. Logo image examples.

analysis has not been attempted on a large scale or objective way using recent machine-learning-based image analysis techniques. Through the analysis, it is possible to reveal not only the trends of logo design but also the relationship between a company and its logo. The analysis results are helpful to design new logos that are appropriate to the company's intention.

The first analysis is how texts are used in logos. Specifically, we detect the text area in each logo image; if no text, the logo image is a logo symbol. If text only, it is a logotype. Otherwise, it is a mixed logo. Using 122,920 logo images from [16] and a recent text detection technique [4], it is possible to understand the ratio of the three logo classes. Moreover, it is possible to quantify the text area ratio and the text location in mixed logos.

The second and more important analysis is the relationship between the text area ratio and the popularity of the company. The logo images in the LLD-logo dataset are collected from Twitter. Therefore it is possible to know *the number of followers*, which is a good measure of the company's popularity. Since most logotypes are company names, the analysis will give a hint to answering whether famous companies tend to appeal their names in their logo or not. Since this relationship is very subtle, we try to catch it by a coarse view using DeepCluster [7].

The last analysis is the relationship between the whole logo image (including both text areas and symbols) and the company's popularity by using regression analysis and ranking analysis. If it is possible to realize a regression function and/or a ranking function with reasonably high accuracy, they are very useful as references of better logo design that fits the company's popularity. The main contributions of this paper are summarized as follows:

- This paper provides the first large-scale objective analysis of the relationship among logo images, texts in logo, and company popularity. The analysis results will give various hints for the logo design strategy.
- The robust estimation using DeepCluster revealed the positive correlation between the popularity of a company and the text area ratio in its logo.
- Regression and ranking analyses showed the possibility of estimating the absolute popularity (i.e., the number of followers) or relative popularity from logo

images; this result suggests we can evaluate the goodness of the logo design by the learned regression and ranking functions.

- In addition to the above results, we derive several reliable statistics, such as the ratio among three logo types (logotypes, logo symbols, and mixed logos) and the text area ratio and text location in logo images.

2   Related Work

2.1   Logo Design Analysis

Logo design has been studied from several aspects, especially, marketing research. According to Aîdr et al. [1], logo is classified into three types, "a text defined a logo", "iconic or symbolic logo", and "A mixed logo." In this paper, we call them logotype, logo symbol, and mixed logo, respectively. A brief survey by Zao [23] discusses the recent diversity of logos. Sundar et al. [19] analyzed the effect of the vertical position of a logo on the package. Luffarelli et al. [12] analyzed how the symmetry and asymmetry in logo design give different impressions by using hundreds of crowd-workers; they conclude that asymmetric logos give more *arousing* impressions.

Recently, Luffarelli et al. [11] report an inspiring result that the *descriptiveness* of logos positively affects the brand evaluation. Here, the descriptiveness means that the logo describes explicitly what the company is doing by texts and illustrations. This report, which is also based on subjective analysis using crowd-workers, gives insights into the importance of texts in logo design.

Computer science has recently started to deal with logo design, stimulated by public datasets of logo images. Earlier datasets, such as UMD-Logo-Database[2] were small. In contrast, nowadays, the dataset size becomes much larger. WebLogo-2M [18] is the earliest large dataset prepared for the logo detection task. The dataset contains 194 different logos captured in 1,867,177 web images. Then, Sage et al. [16] have published the LLD-logo dataset, along with the LLD-icons dataset. The former contains 122,920 logo images collected from Twitter and the latter 486,377 favicon images (32×32 pixels) via web-crawling. They used the dataset for logo synthesis. Recently, Logo-2k+ dataset is released by Wang et al. [21], which contains 2,341 different logos in 167,140 images. The highlight of this dataset is that the logo images collected via web-crawling are classified into 10 company classes (Food, Clothes, Institution, Accessories, etc.). In [21], the logo images are applied to a logo classification task.

Research to quantify logo designs by some objective methodologies is still not so common. One exceptional trial is Karamatsu et al. [9], where the favicons from LLD-icons are analyzed by a top-rank learning method for understanding the trends of favicon designs in each company type. This paper also attempts to quantify the logo design by analyzing the relationships among logo images, text areas in them, and their popularity (given as the number of followers on Twitter).

---

[2] Not available now. According to [15], it contained 123 logo images.

2.2   Clustering and Ranking with Deep Representation

We will use DeepCluster [7] for clustering the LLD-logo images. Recently, the combination of clustering techniques with deep representation becomes popular [2,14]. DeepCluster is a so-called self-supervised technique for clustering, i.e., an unsupervised learning task, and has already been extended to tackle new tasks. For example, Zhan et al. [22] developed an online deep clustering method to avoid the alternative training procedure in DeepCluster. Tang et al. [20] apply deep clustering to unsupervised domain adaptation tasks.

In this paper, we combine DeepCluster with the learning-to-rank technique. Specifically, we will combine RankNet [6] (originally with multi-layer perceptron) with the convolutional neural network trained as DeepCluster. The idea of RankNet is also applied to many applications, such as image attractiveness evaluation [13]. The recent development from RankNet is well-summarized [8].

3   Logo Image Dataset—LLD-logo [16]

In this paper, we use the LLD-logo dataset by Sage et al. [16]. LLD-logo contains 122,920 logo images collected from Twitter profile images using its API called Tweepy. Several careful filtering operations have been made on the collection to exclude facial images, harmful images and illustrations, etc. The size of the individual logo images in the dataset is 63×63–400×400.

The logo images in the LLD-logo dataset have two advantages over the logo image datasets that contain camera-captured images, such as Logo-2K [21]. First, logo images from LLD-logo are so-called born-digital and thus have no disturbance from uneven light, geometric distortion, low resolution, blur, etc.

The second advantage of LLD-logo is meta-data, including *the number of followers*, which is a very good index to understand the popularity of the individual companies[3]. Figure 2 shows the distribution of the number of followers of the companies in the LLD-logo dataset. Note that the bins of this histogram are logarithmic; this is a common way to analyze the popularity, especially the Twitter follower distributions [3,5,10,17]. This distribution shows that the companies with $10^2 \sim 10^4$ followers are the majority and there are several exceptional companies with less than 10 followers or more than 10^7 followers[4].

4   Analysis 1: How Much Are Texts Used in Logo?

4.1   Text Detection in Logo Image

As shown in Fig. 1, logo images often include texts. Especially, a logotype is just a text (including a single letter); that is, the text fills 100% of the logo area.

---

[3] The number of followers is provided in the file LLD-logo.hdf5, which is provided with LLD-logo image data. Precisely, the resource of meta_data/twitter/user_objects in this hdf5 file contains the followers_count data, which corresponds to the number of followers.

[4] For the logarithmic plot, we exclude the companies with zero follower. The number of such companies is around 1,000 and thus no serious effect in our discussion.

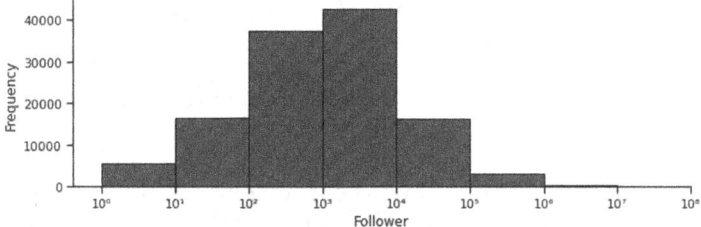

Fig. 2. Distribution of followers of the companies listed in LLD-logo dataset.

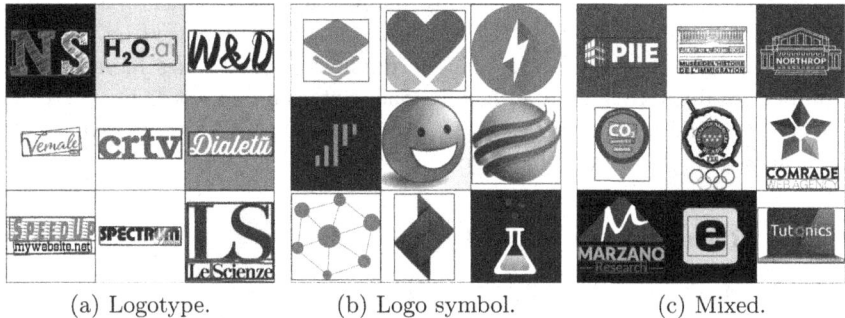

(a) Logotype.   (b) Logo symbol.   (c) Mixed.

Fig. 3. Text detection from logo images by CRAFT [4]. The red boxes are the detected text areas, whereas the blue boxes are the bounding box of the whole logo image.

In contrast, a logo symbol does not include any text. A mixed log will have a different text area ratio. In other words, by observing the text area ratio in the individual logo images, we can understand not only the frequency of the three types, but also the ratio trend in the mixed type.

Recent scene text detection techniques allow us to extract the text area from a logo image, even when the texts are heavily decorated, rotated, and overlaid on a complex background. We use CRAFT [4], which is one of the state-of-the-art scene text detectors. Figure 3 shows several text detection results by CRAFT. They prove that CRAFT can detect the texts in logos very accurately, even with the above difficulties. We, therefore, will use the text bounding boxes detected by CRAFT in the following experiments.

Note that the detection results by CRAFT are not completely perfect. Especially, in logo designs, the border between texts and illustrations is very ambiguous. For example, the logo of "MARZANO", the calligraphic 'M' in the blue triangle is missed. The logo of "Tutonics", 'o' is not detected as a letter due to its illustration-like design. As indicated by those examples, it is impossible to expect to have perfect extraction results from the logo images. Even so, CRAFT still extracts truly text parts quite accurately in most cases.

4.2   Text Area Ratio and the Number of Text Boxes

In the following analysis, we often use two simple metrics: text area ratio and the number of text boxes. The text area ratio is the ratio of the text area in the whole logo area. The text area is specified by all the bounding boxes (red boxes in Fig. 3) in the logo image. The whole logo area, which is shown as the blue bounding box in Fig. 3, is defined by the bounding box of the design element on the logo. Therefore, the whole logo area is often smaller than the image size. For example, the whole logo area of the logo of "NS" is almost the same as the text bounding box, and thus its text area ratio is about 100%[5].

The number of text boxes is simply defined as the number of text boxes detected by CRAFT. Since CRAFT can separate a multi-line text into the lines and then give the boxes at each line, the number of text boxes increases with the number of lines. If there is sufficient space between two words in a text line, CRAFT gives different text boxes for the words. If not, CRAFT will give a single box, as shown in the "SpeedUp" logo.

4.3   The Ratio of Three Logo Types

As the first analysis, we classify each logo image into one of three logo types using the text area ratio. This classification is straightforward: if the text area ratio is 0%, it will be a logo symbol. If the text area ratio is more than 90%, we treat it as a logotype. Otherwise, it is a mixed logo. Note that we set 90% as the class boundary between logotype and mixed (instead of 100%) because the bounding boxes of the whole logo and the text area are not always completely the same, even for logotypes.

The classification result of all 122,920 logos in the large logo set shows that the ratio of three types are: 4%(logotype) : 26%(logo symbol) : 70% (mixed). The fact that mixed logos are clear majority indicates that most companies show their name with some symbol in their logo. Surprisingly, logotypes are very minority. Logo symbols are much more than logotypes but less than half of mixed logos.

4.4   Distribution of the Text Area Ratio

Figure 4 (a) shows the histogram of the text area ratios, where logo symbols are excluded. This suggests that the text areas often occupy only 10–30% of the whole logo area. Figure 4 (b) shows the two-dimensional histogram to understand the relationship between the text area ratio and the number of text boxes. The histogram says that the most frequent case is a single text box with 10–30% occupancy. Logos with multiple text boxes exist but not so many. It should also be noted that the text area ratio and the number of text boxes are not positively correlated; even if the ratio increases, the text boxes do not.

---

[5] As we will see later, the text area ratio sometimes exceeds 100%. There are several reasons behind it; one major reason is the ambiguity of the whole logo's bounding box and the individual text areas' bounding boxes. If the latter slightly becomes the former, the ratio exceeds 100%. Some elaborated post-processing might reduce those cases; however, they do not significantly affect our median-based analysis.

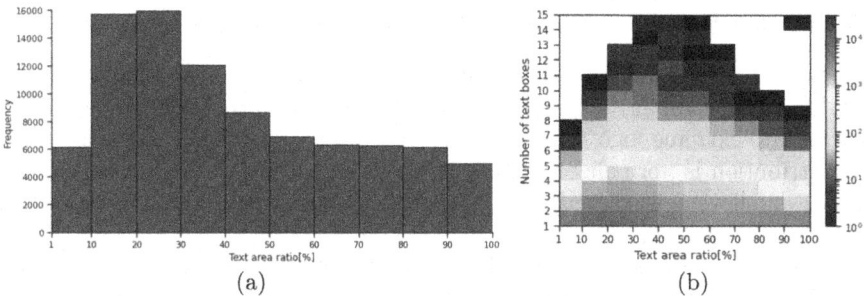

Fig. 4. (a) Distribution of the text area ratios. (b)Two-dimensional histogram showing the relationship between the text area ratio and the number of text boxes.

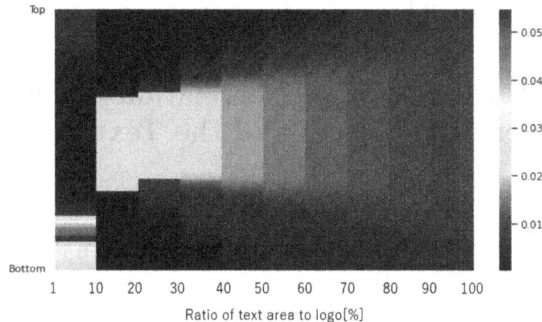

Fig. 5. Distribution of text regions in the vertical direction for different text area ratios. The frequency is normalized at each vertical bin (i.e., each text area ratio).

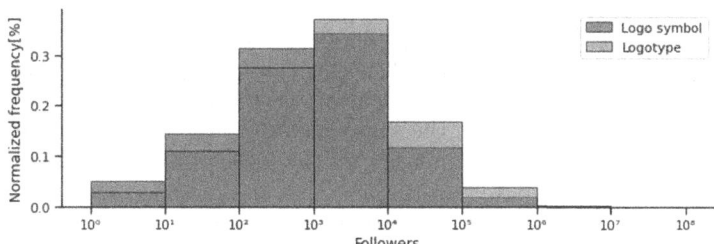

Fig. 6. Distributions of Twitter followers for logotype and logo symbol.

4.5 The Location of the Text Area

Figure 5 visualizes the vertical location of the text area for different text area ratios. Since the logo height is normalized for this visualization, the top and bottom of the histogram correspond to the top and bottom of logo images, respectively. This visualization reveals that when a small text area (<10%) is presented on a logo image, it is mostly located at the bottom of the logo. However, when the text area occupies more than 10% of the whole logo area, it is mostly located around the center of the logo.

4.6 Does a Famous Company Show the Name on Its Logo or Not?

As noted in Sect. 3, the LLD-logo dataset includes the number of followers for each logo (i.e., company). Figure 6 shows the distributions of the follower numbers for two extreme logo types, i.e., logotype and logo symbol. For each type, the distribution is normalized so that its total becomes 1. As noted before, the number of followers is treated in a logarithmic manner according to the standard way.

These distributions prove that the average number of followers is larger when the logo is logotypes than logo symbols. One might have supposed that popular companies need not show their name (because they are always famous enough) and use logo symbols more. However, among the logotype users, the famous companies are not minority [6]. The relationship between the number of followers and the text area ratio will be analyzed in more detail in Sect. 5.5.

5 Analysis 2: Cluster-Wise Correlation Analysis Between the Number of Followers and the Text Area Ratio

5.1 Logo Image Clustering by DeepCluster [7]

DeepCluster is a clustering technique with a representation learning function. Its idea is rather simple based on the typical pseudo-labeling technique. Given a set of image samples, DeepCluster first outputs their representations (i.e., feature vectors) via convolutional layers. Second, the k-means clustering is performed to the feature vectors, after applying PCA for dimensionality reduction. Third, the cluster ID of each pseudo-label is assigned to each sample as its pseudo-label. Fourth, the convolutional layers are re-trained along with the attached classification layers so that the whole neural network can classify the samples according to their pseudo-labels. By repeating those steps, we have k clusters for the set. Since DeepSets relies on a self-trained discriminability, we do not need to try various similarity metrics for the suitable clustering result. Moreover, we can expect that it can capture the similarity among logos in rather an abstract level (than the bitmap level), based on the fact that it shows accurate performance at the classification, detection, and segmentation task on Pascal VOC [7].

We apply DeepCluster to 40,000 logo images randomly selected from LLD-logo. The number of clusters is fixed at $k = 128$. All the images are rescaled to be 224×224 pixels. Figure 7 (a) shows the size s_i of each cluster $i \in [1, k]$. The average cluster size $\bar{s}$ is around 300. The #128-th cluster is a huge miscellaneous cluster and contains 2,948 images (i.e., about 6% of all the images). This cluster seems to contain all the "outsiders" around the edge of the logo image distribution. Therefore, this miscellaneous cluster will not show any clear trend within it and is excluded from the later analysis.

---

[6] Recall that each distribution is normalized. As noted before, logotypes are just 4%, whereas logo symbols are 26%. Thus, as the absolute numbers, the famous companies use more logo symbols than logotypes.

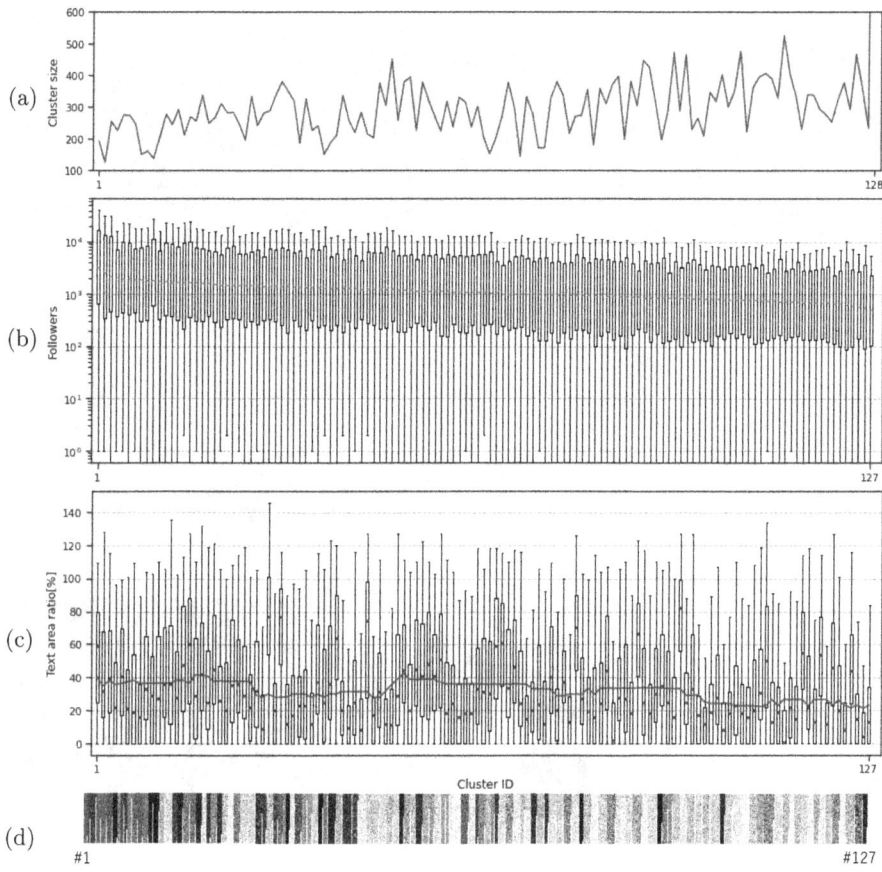

Fig. 7. (a) Cluster size s_i, (b) the box plot of the numbers of followers, and (c) the box plot of text area ratios, for each cluster i. In (a)-(c), the horizontal axis is the cluster ID i and sorted by the descending order of f_i (orange), which is the median value of the number of followers. The blue curve in (c) shows a smoothed transition of t_i (red), which is the median values of text area ratios. (d) Overview of all clusters. From left to right, the logo images of each cluster are shown as very tiny images. (Color figure online)

5.2 The Number of Followers at Each Cluster

Figure 7 (b) shows the box plot showing the number of followers of the logos that belong to each cluster. Let $x_{i,j}$ denote the jth logo image of the ith cluster ($j \in [1, s_i]$) and $f_{i,j}$ is the number of its followers. In this figure (also (a) and (c)), the clusters are sorted in the descending order of the median of the number of followers; this means that the condition $f_i \geq f_{i-1}$ always holds for all i, where $f_i = \mathrm{Med}_j f_{i,j}$. Note that the outlier dots are not plotted in the box plot for better visibility.

Fig. 8. Logo examples of several representative clusters. For each cluster, the tiny thumbnail images of all the logo images in the cluster is shown above the examples, for grasping the general color trend of the cluster. Those thumbnail images are also shown in Fig. 7(d). Only in the large miscellaneous cluster #128, not all images are shown.

Since many logo images have around $f_{i,j} = 10^2 \sim 10^4$ followers as already shown by Fig. 2, the medians $\{f_i\}$ (indicated by orange) are also around 10^3; however, they are still different. For example, the median $f_1 \sim 3,000$, whereas $f_{127} \sim 500$.

5.3   The Text Area Ratio of Each Cluster

Figure 7 (c) is the box plot showing the text area ratios of the logos that belong to each cluster. Let $t_{i,j}$ denote the text area ratio of the jth logo in the ith cluster and $t_i = \mathrm{Med}_j t_{i,j}$. The median t_i is depicted as a red dot in (c).

Some clusters have larger t_i (i.e., more logotypes), whereas some have smaller t_i (i.e., more logo symbols). This means that DeepCluster forms clusters while considering the text areas in logo images. At the same time, there is no cluster whose box height is almost zero; this means that no logotype-only cluster and no logo symbol-only cluster.

Note again that the clusters are arranged in the descending order of f_i and not t_i in Fig. 7. Therefore, no monotonicity condition holds for t_i. However, in Sect. 5.5, we will see that there is a weak trend that t_i tends to be larger than $t_{i'}$ for $i < i'$. This suggests the followers f_i and texts t_i are weakly correlating.

5.4   Logos in Several Clusters

Figure 8 shows the logo image examples from several clusters. For each cluster, six representative examples are shown. A tiny thumbnail showing all the logos in the cluster is also shown above the examples, for capturing the trend (especially color) of the cluster.

Clusters #1, #2, ..., #6 are the six clusters with the highest f_i, and Clusters #125, #126 and #127 are the lowest. From those examples, we can observe that the DeepCluster composes its clusters by extracting various features from logo images.

Especially, the top six clusters have very consistent color and shape styles within the cluster. The thumbnail images of those clusters prove this color homogeneity within each cluster. Except for #4, logos with a red background and a white large text are common among them. Although we do not give any prior knowledge about text areas to DeepCluster, it could find "text-like" elements as features for the clustering.

It is interesting to observe that red logos indeed occupy most of the top clusters. Since the clusters' order is determined by the median f_i, we cannot say the red logos *always* correlate the companies with high popularity. However, at least the cluster-wise average (by means of median) shows this trend. Figure 7(d) supports this trend. In contrast, the logo images from Clusters #125–127 (having the lowest f_i) often show finer structures and pictorial elements (rather than texts) in their design.

Cluster #128 is the miscellaneous cluster with an extremely large number of logos (2,948). As expected from its size, it contains logos with various appearances. Since DeepCluster is a classifier-based clustering method, it may realize

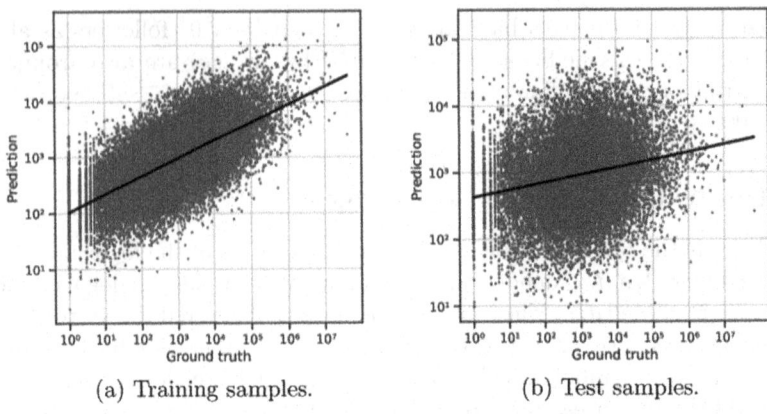

(a) Training samples.　　　　　　(b) Test samples.

Fig. 9. Prediction result of the number of followers from logo images.

such a "rejection" class. This can be confirmed from its colorful thumbnail image. We excluded this cluster from the trend analysis because it seems to have no clear trend.

Clusters #96 and #113 are the clusters with the maximum and minimum f_i, respectively. The former contains logotypes; the latter contains logo symbols and mixed logos where texts are often surrounded by a large background element.

5.5 Cluster-Wise Correlation Analysis Between the Number of Followers and the Text Area Ratio

The red bar in Fig. 7 (c) shows the median of text area ratios of logos in each cluster. The red dots seem to fluctuate from Cluster #1 (leftmost) to Cluster #127 (rightmost); however, careful observation will find the trend that the text area ratio decreases along with the cluster ID. The blue curve is a smoothed transition of the median values (red dots) of text area ratios. For the smoothing, the median filter with a width of 13 is applied in the Cluster-ID direction, to catch the trend of the curve robustly.

Considering that cluster ID is relative to the number of followers, the slope of the blue line catches the weak trend that *the number of followers correlates positively with the text area ratio.* In other words, the more followers a company has, the more texts its logo contains. Of course, this correlation does not always hold—however, this cluster-wise median-based robust estimation reveals this weak trend. It should be emphasized that this conclusion coincides with our previous analysis in Sect. 4.6 with Fig. 6.

6   Analysis 3: Estimation of the Number of Followers from Logo Image

6.1   Regression-Based Estimation

In this section, we conduct a challenging task of estimating the number of followers from a logo image by regression analysis. This task is challenging because the number of followers depends on so many factors, including the company's product, the market trends, the size of the company, etc. However, we tackle this task because of two positive evidences. First, the recent deep regression models show their powerful ability to deal with extremely nonlinear regression tasks. Second, as we observed already, we could catch the weak correlation between the text area ratio and the number of followers, by the cluster-wise analysis in Sect. 5.5.

The deep regression is used to estimate the number of followers for the input log image. All logo images are re-scaled to 224×224 pixels. The network structure is DenseNet-169, and the standard MSE loss function is used. For training, validating, and testing, 30,000, 10,000, and 20,000 logo images randomly selected from LLD-logo are used, respectively. The standard early stopping rule (no improvement on the validation accuracy for 10 epochs) is used to terminate the training process. During training and testing, the number of followers is treated in a logarithmic manner due to the same reason of Sect. 3.

Figures 9 (a) and (b) show the scattered plots of the ground-truth (GT) and prediction (PR) for the training samples and test samples, respectively. The black line is the linear regression result between GT and PR. Surprisingly, the linear regression result shows that GT and PR are clearly correlating; this means that the prediction of the number of followers is possible to some extend. (At least, it is not impossible.) The Pearson correlation coefficients between GT and PR for the training and test sets are about 0.64 and 0.23, respectively. The test correlation is not very strong but it is enough to confirm the existence of the correlation. The p-values are almost zero and satisfy p for both sets, and thus we can also confirm that GT and PR are actually correlated.

6.2   Ranking-Based Estimation

As another formulation of the estimation task, we conduct a learning-to-rank experiment with the idea of RankNet [6]. The idea of RankNet is to determine a ranking function $r(x)$ that satisfies the condition $r(x_1) > r(x_2)$ when x_1 should be ranked higher than x_2. In our case, x represents a logo image and if the logo image x_1 more followers than x_2, the condition should be satisfied. If we have an accurate ranking function $r(x)$, it can predict a *relative* popularity, such as $r(x_1) > r(x_2)$ or $r(x_1) < r(x_2)$, for a given pair of images $\{x_1, x_2\}$.

RankNet realizes a nonlinear ranking function by its neural network framework. The original RankNet [6] is implemented by a multi-layer perception, and we substitute it with DenseNet-169 to deal with a stronger nonlinearity and image inputs. The same training, validating, and testing sets as the regression experiment are used. From those sets, 30,000, 10,000, and 20,000 "pairs" are randomly created, and used for training, validating, and testing RankNet with DenseNet-169. The cross-entropy loss was used to train RankNet. Again, the standard early stopping rule (no improvement on the validation accuracy for 10 epochs) is used for the termination.

Unlike the regression task, the performance of the (bipartite) learning-to-rank task is evaluated the successful ranking rate for the test pairs $\{x_1, x_2\}$. If x_1 should be higher ranked than x_2 and the learned RankNet correctly evaluates $r(x_1) > r(x_2)$, this is a successful case. The chance rate is 50%.

The successful ranking rates for the 30,000 training pairs and 10,000 test pairs are 60.13% and 57.19%, respectively. This is still surprisingly high accuracy because RankNet determines the superiority or inferiority of two given logo images with about 60% accuracy (i.e., not 50%). This result also proves that the logo images themselves contain some factor that correlates with the number of followers.

7   Conclusion

This paper analyzed logo images from various viewpoints. Especially, we focus on three correlations between logo images and their text areas, between the text areas and the number of followers on Twitter, and between the logo images and the number of followers. The first correlation analysis with a state-of-the-art text detector [4] revealed that the ratio of logotypes, logo symbols, and mixed symbols is 4%, 26%, and 70%, respectively. In addition, the ratio and the vertical location of the text areas are quantified. The second correlation analysis with DeepCluster [7] revealed the weak positive correlation between the text area ratio and the number of followers. The third correlation analysis revealed that deep regression and deep ranking methods could catch some hints for the popularity (i.e., the number of followers) and the relative popularity, respectively, just from logo images.

As summarized above, the recent deep learning-based technologies, as well as large public logo image datasets, help to analyze complex visual designs, i.e., logo images, in an objective, large-scale, and reproducible manner. Further research attempts, such as the application of explainableAI techniques to deep regression and deep ranking, company-type-wise analysis, font style estimation, etc., will help to understand the experts' knowledge on logo design.

Acknowledgment. This work was supported by JSPS KAKENHI Grant Number JP17H06100.

References

1. Adîr, G., Adîr, V., Pascu, N.E.: Logo design and the corporate identity. Procedia. Soc. Behav. Sci. **51**, 650–654 (2012)

2. Aljalbout, E., Golkov, V., Siddiqui, Y., Strobel, M., Cremers, D.: Clustering with deep learning: taxonomy and new methods. arXiv preprint arXiv:1801.07648 (2018)
3. Ardon, S., et al.: Spatio-temporal analysis of topic popularity in twitter. arXiv preprint arXiv:1111.2904 (2011)
4. Baek, Y., Lee, B., Han, D., Yun, S., Lee, H.: Character region awareness for text detection. In: CVPR (2019)
5. Bakshy, E., Hofman, J.M., Mason, W.A., Watts, D.J.: Everyone's an influencer: quantifying influence on twitter. In: ACM WSDM (2011)
6. Burges, C., et al.: Learning to rank using gradient descent. In: ICML (2005)
7. Caron, M., Bojanowski, P., Joulin, A., Douze, M.: Deep clustering for unsupervised learning of visual features. In: Ferrari, V., Hebert, M., Sminchisescu, C., Weiss, Y. (eds.) Computer Vision – ECCV 2018. LNCS, vol. 11218, pp. 139–156. Springer, Cham (2018). https://doi.org/10.1007/978-3-030-01264-9_9
8. Guo, J., et al.: A deep look into neural ranking models for information retrieval. Inf. Proc. Manag. **57**(6), 102067 (2020)
9. Karamatsu, T., Suehiro, D., Uchida, S.: Logo design analysis by ranking. In: ICDAR (2019)
10. Lerman, K., Ghosh, R., Surachawala, T.: Social contagion: an empirical study of information spread on digg and twitter follower graphs. arXiv preprint arXiv:1202.3162 (2012)
11. Luffarelli, J., Mukesh, M., Mahmood, A.: Let the logo do the talking: the influence of logo descriptiveness on brand equity. J. Marketing Res. **56**(5), 862–878 (2019)
12. Luffarelli, J., Stamatogiannakis, A., Yang, H.: The visual asymmetry effect: An interplay of logo design and brand personality on brand equity. J. Marketing Res. **56**(1), 89–103 (2019)
13. Ma, N., Volkov, A., Livshits, A., Pietrusinski, P., Hu, H., Bolin, M.: An universal image attractiveness ranking framework. In: WACV (2019)
14. Min, E., Guo, X., Liu, Q., Zhang, G., Cui, J., Long, J.: A survey of clustering with deep learning: from the perspective of network architecture. IEEE Access **6**, 39501–39514 (2018)
15. Neumann, J., Samet, H., Soffer, A.: Integration of local and global shape analysis for logo classification. Patt. Recog. Lett. **23**(12), 1449–1457 (2002)
16. Sage, A., Agustsson, E., Timofte, R., Van Gool, L.: Logo synthesis and manipulation with clustered generative adversarial networks. In: CVPR (2018)
17. Stringhini, G., et al.: Follow the green: growth and dynamics in twitter follower markets. In: Internet Measurement Conference (2013)
18. Su, H., Gong, S., Zhu, X.: WebLogo-2M: scalable logo detection by deep learning from the web. In: ICCVW (2017)
19. Sundar, A., Noseworthy, T.J.: Place the logo high or low? Using conceptual metaphors of power in packaging design. J. Marketing **78**(5), 138–151 (2014)
20. Tang, H., Chen, K., Jia, K.: Unsupervised domain adaptation via structurally regularized deep clustering. In: CVPR (2020)
21. Wang, J., et al.: Logo-2k+: a large-scale logo dataset for scalable logo classification. In: AAAI (2020)
22. Zhan, X., Xie, J., Liu, Z., Ong, Y.S., Loy, C.C.: Online deep clustering for unsupervised representation learning. In: CVPR (2020)
23. Zhao, W.: A brief analysis on the new trend of logo design in the digital information era. In: ESSAEME (2017)

MediTables: A New Dataset and Deep Network for Multi-category Table Localization in Medical Documents

Akshay Praveen Deshpande⬤, Vaishnav Rao Potlapalli⬤, and Ravi Kiran Sarvadevabhatla(✉)⬤

Centre for Visual Information Technology, International Institute of Information Technology, Hyderabad (IIIT-H), Hyderabad 500032, India
ravi.kiran@iiit.ac.in
https://github.com/atmacvit/meditables

Abstract. Localizing structured layout components such as tables is an important task in document image analysis. Numerous layout datasets with document images from various domains exist. However, healthcare and medical documents represent a crucial domain that has not been included so far. To address this gap, we contribute MediTables, a new dataset of 200 diverse medical document images with multi-category table annotations. Meditables contains a wide range of medical document images with variety in capture quality, layouts, skew, occlusion and illumination. The dataset images include pathology, diagnostic and hospital-related reports. In addition to document diversity, the dataset includes implicitly structured tables that are typically not present in other datasets. We benchmark state of the art table localization approaches on the MediTables dataset and introduce a custom-designed U-Net which exhibits robust performance while being drastically smaller in size compared to strong baselines. Our annotated dataset and models represent a useful first step towards the development of focused systems for medical document image analytics, a domain that mandates robust systems for reliable information retrieval. The dataset and models can be accessed at https://github.com/atmacvit/meditables.

Keywords: Document analysis · Table localization · Healthcare · Medical · Semantic segmentation · Instance segmentation

1 Introduction

Document tables have been an efficient and effective technique for communicating structured information. With the advent of digital media, most document tables exist in PDF documents. Consequently, there have been efforts to develop algorithms for machine-based detection and understanding of tabular content from such media.

© Springer Nature Switzerland AG 2021
E. H. Barney Smith and U. Pal (Eds.): ICDAR 2021 Workshops, LNCS 12916, pp. 112–124, 2021.
https://doi.org/10.1007/978-3-030-86198-8_9

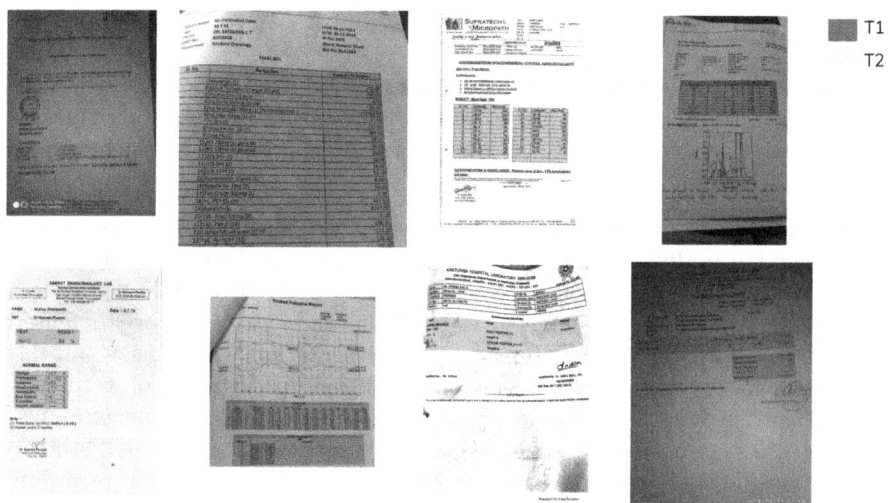

Fig. 1. Samples from our MediTables dataset. T1 and T2 represent two different kinds of table annotations. The diversity of the medical documents and table configurations is visible in the figure.

To understand and develop efficient, accurate algorithms for table localization and understanding, diverse datasets of document images were created and have been made available to the research community. In general, tables appear in varying formats and layouts which thwart heuristic approaches. Of late, deep learning [12] has proven to be a powerful mechanism to obtain state of the art results on many computer vision tasks, including table understanding.

While the existing datasets cover a number of domains, documents related to healthcare and medical domain are conspicuously absent. We seek to address this gap by contributing a new, annotated dataset. The documents in this dataset pose challenges not encountered in other datasets. Thus, this dataset adds to the diversity of the dataset pool.

Most of the available document datasets tend to contain similar levels of illumination and capture quality. Due to qualitative differences between existing datasets and medical documents, we also discover that pre-trained, deep-learning based table localization models trained on existing datasets do not generalize sufficiently on documents from the medical domain. Therefore, we also introduce a customized deep network for table localization in medical domain documents.

Specifically, we make the following contributions:

– A medical document image dataset called MediTables with annotations for two different types of tables.
– A deep learning model for table localization in medical document images.

The dataset and models can be accessed at https://github.com/atmacvit/meditables.

2   Related Work

Approaches for table localization characterize the problem either as a detection problem (i.e., identify axis-aligned bounding boxes for tables) or as a segmentation problem (i.e., obtain pixel-level labeling for tables). We review the literature related to these two approaches below.

Table Detection: There has been substantial work done in the wider context of table detection in document images and PDFs. Many of the earlier approaches were heuristic-based approaches. Ha et al. [7] proposed Recursive X-Y cut algorithm which recursively decomposes the document in blocks which are then used to build a X-Y tree which is later used for segmentation. Kieninger et al. [11] proposed a system called T-Recs which uses bottom up clustering of word entities and word geometries into blocks for segmentation of tables. Yildiz et al. [25] introduced a system known as pdf2table which uses a tool called pdf2html to convert the PDF file to its XML counterpart containing information regarding absolute location of text chunks. This XML along with certain heuristic rules is then used to perform table localization. Fang et al. [1] also showed a method which works on PDF documents. They use a four stepped approach for performing table detection. First, they parse PDF files and perform layout analysis. Separators and delimiting information is then mined to localize tables.

Hao et al. [8] propose a method where table regions are initially selected based on some heuristic rules and later, a convolutional neural network is trained. This is used to classify the selected regions into tables and non-tables. Huang et al. [9] use a modified YOLO framework for table localization with post processing step for additional performance improvement. Schreiber et al. [20] use a fine-tuned version of popular object detection framework, Faster RCNN [4], to detect tables in document images. Siddiqui et al. [21] propose a framework called DeCNT. In this approach, they combine a deformable CNN model with RCNN or an FCN instead of a conventional CNN. Gilani et al. [3] perform various distance transforms on the original image to create an alternative representation which is subsequently fed into an Faster R-CNN model for detecting tables.

Table Segmentation: Yang et al. [24] propose a multi-modal Fully Convolutional Network which segments documents images into various page elements (text, charts, tables) using both image information as well as underlying text in the image. Kavasidis et al. [10] model table localization as a semantic segmentation problem and use a Fully Convolutional Network pretrained on saliency detection datasets to develop visual cues similar to those of tables and graphs. The predictions of the network are further refined using Conditional Random Fields (CRFs). CascadeTabNet [14] is an Mask-RCNN based network trained to localize table regions in reasonably structured documents (Fig. 2).

3   MediTables Dataset

We introduce MediTables, a dataset of 200 multi-class annotated medical document images. The document images were scraped from various sources on the

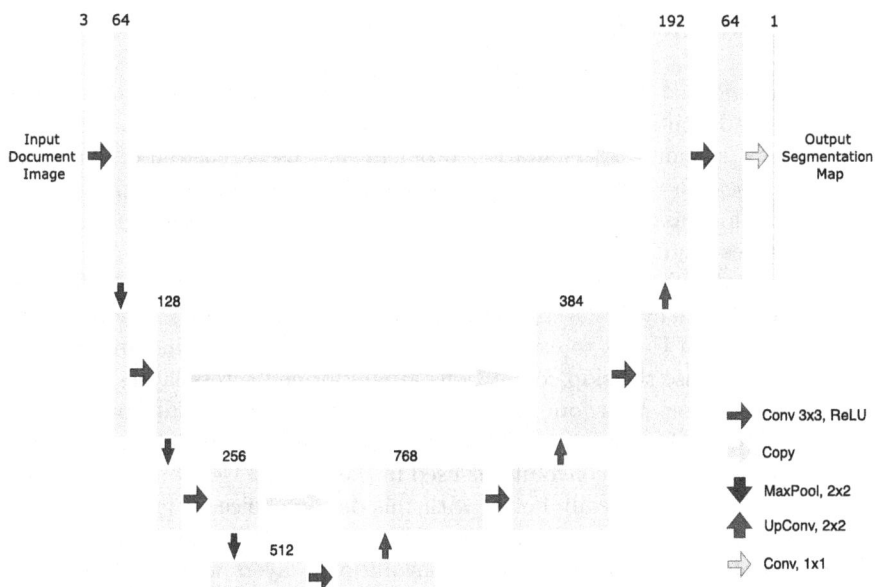

Fig. 2. The architecture for our modified U-Net.

Table 1. Table coverage by type in MediTables.

Type	# of documents				# of tables			
	Total	Train	Validation	Test	Total	Train	Validation	Test
T1 and T2	79	60	7	12	–	–	–	–
T1 only	73	43	15	15	190	126	29	35
T2 only	48	27	8	13	140	100	15	25
Total	200	130	30	40	330	226	44	60

internet. Our dataset contains a wide range of medical document images with variety in capture quality, layouts, skew, occlusion and illumination. They are a good representation of prevalent healthcare, medical images such as pathology, diagnostic and hospital-related reports. There are two kinds of annotated tables in our dataset (see Fig. 1):

- The first kind of table, hereafter referred to as T1, are tables which follow a conventional layout and tend to have some sort of demarcation between rows or between columns.
- The other kind of table, hereafter referred to as T2, consists of formatted data in key-value format which are usually found for fields such as names, identification numbers, addresses, age, etc.

The inclusion of such annotations can facilitate efficient retrieval of crucial meta-information from medical document images. Additional statistics related to the dataset can be seen in Table 1.

4 The Modified U-Net Deep Network

Due to the informal nature of document capture and consequent distortions induced, table layouts in our dataset are non-rectangular. Therefore, we model the task as a semantic segmentation problem and use a custom-designed U-Net [18] to localize tables. We chose a U-Net based model due to its success for segmentation tasks trained on small datasets. U-Net uses skip connections, which improves gradient flow and allows stable weight updates.

The original U-Net [18] has four skip connections across the network and requires cropping to meet the size demands for establishing skip connections. We modify the original U-Net to exclude the cropping in the cropping and copying step. This is because the skip connections are performed across layers that have the same spatial size. Thus, our modified U-Net has only three skip connections. Another difference is that we use a single convolution layer per down-sampling layer compared to double convolutions used in the original U-Net since we found performance to be empirically better with this design choice.

The model consists of a contraction and an expansion section. The contraction section of the model consists of 4 convolutional layers with 3×3 filters of $64, 128, 256$ and 512 channels successively. Each layer is followed by a Rectified Linear Unit (ReLU) activation layer. The convolutions are all single stride with a padding size of 1. The outputs of each convolution layer are max-pooled using a 2×2 kernel. The expansion section consists of 3 convolutional layers which perform upsampling using a 2×2 kernel with a padding of 1. After each upsampling step, features from the contraction section that have the same spatial dimensions as the current set of features are concatenated. The output, which has the same spatial dimensions as the input image, is obtained by applying 1×1 convolution filters on the feature output of the expansion section.

5 Experiment Setup

For our experiments, we consider multiple popular table localization datasets as described below.

5.1 Datasets

Marmot: The Marmot table recognition dataset consists of 2000 PDF document images with diverse page layouts and tabular formats. The Marmot layout analysis of fixed layout documents dataset consists of 244 clean document images and comprises of 17 labels for fragments in a document. From this, we selected 400 relevant images with tables.

UNLV: The UNLV dataset [22] contains 2889 scanned document images from sources such as newspapers and business letters. The resolution for such images are 200 to 300 DPI. We use the subset of 427 images containing tables for experiments.

Table 2. Performance comparison of original U-Net and modified U-Net evaluated on validation set of MediTables.

Model	Loss	IOU (%)	PPA (%)	F1 (%)
Original U-Net	$\mathcal{L}_{BCE}$	71.57	86.31	81.26
	$\mathcal{L}_{IoU}$	74.78	88.74	85.10
	$\mathcal{L}_{BCE} + \mathcal{L}_{IoU}$	**76.21**	89.39	86.20
Modified U-Net	$\mathcal{L}_{BCE}$	73.20	88.20	83.90
	$\mathcal{L}_{IoU}$	75.72	89.60	85.40
	$\mathcal{L}_{BCE} + \mathcal{L}_{IoU}$	75.81	**90.06**	**87.08**

UW3: The dataset consists of 1600 scanned skew-corrected document images. We have selected 120 document images, which contain atleast one table in them.

ICDAR Datasets: We use 124 documents from the ICDAR 2013 table detection competition [5]. Additionally, we used 549 document images with tables from ICDAR 2017 Page Object Detection Dataset [2].

TableBank: Recently, researchers from Beihang University and Microsoft Research Asia collected the largest document image based dataset with tables, TableBank [13]. It consists of 417,000 labeled tables and clean source documents.

To augment the datasets, we performed various standard augmentations on images such as Gaussian blurring, rotation, salt and pepper noise, Poisson noise and affine transformation. The augmentations were randomly applied to obtain a dataset consisting of 52,482 images.

5.2   Training and Implementation Details

To begin with, we trained our modified U-Net model using combined data from four existing datasets (Sect. 5.1). The resulting model was fine-tuned on the training set of 130 images sourced from our MediTables dataset.

All images and corresponding label map targets were resized to 512×512. For the optimization of our network, we used the popular Adam optimizer with a learning rate of 5×10^{-4}, with a corresponding mini-batch size of 16. The training was conducted in two phases. In the first phase of training, we used per-pixel binary cross-entropy loss ($\mathcal{L}_{BCE} = -y * log(\bar{y}) - (1 - y) * log(1 - \bar{y})$) where y is the ground-truth label, $\bar{y}$ is the prediction), for 15 epochs. In the second phase (i.e., 16th epoch onwards), we included the logarithmic version of IoU loss [16]

$$\mathcal{L}_{IoU} = -ln \left(\frac{X \cap \hat{X}}{X \cup \hat{X}} \right) \tag{1}$$

where $\hat{X}$ is the predicted image mask and X is the corresponding ground-truth label mask. In contrast with per-pixel cross-entropy loss, IoU loss optimizes for the table regions in a more direct manner and turns out to be crucial for overall

Table 3. Performance comparison between pre-trained and fine-tuned models.

(a) Performance of models trained only on existing document datasets and evaluated on test set of MediTables (PPA is not defined for detection models).

Model	IOU (%)	PPA (%)	F1 (%)
TableBank [13]	89.27 ± 18.21	NA	90.26 ± 17.24
YOLO-v3 [17]	19.85 ± 08.10	NA	17.44 ± 07.09
pix2pixHD [23]	21.06 ± 02.44	90.24 ± 06.13	42.37 ± 06.32
CascadeTabNet [14]	83.08 ± 19.00	95.70 ± 07.00	93.05 ± 01.00
Modified U-Net (Ours)	21.32 ± 16.16	71.14 ± 22.03	32.44 ± 21.10

(b) Performance of models pre-trained on existing document datasets and fine-tuned on MediTables (PPA is not defined for detection models).

Model	IOU (%)	PPA (%)	F1 (%)
TableBank [13]	95.15 ± 04.84	NA	97.10 ± 01.36
YOLO-v3 [17]	45.48 ± 25.47	NA	45.34 ± 11.41
pix2pixHD [23]	88.16 ± 06.91	97.61 ± 01.51	97.63 ± 00.74
CascadeTabNet [14]	89.61 ± 17.82	97.16 ± 05.26	95.07 ± 08.00
Modified U-Net (Ours)	$\mathbf{96.77 \pm 02.03}$	$\mathbf{99.51 \pm 00.21}$	$\mathbf{99.48 \pm 00.12}$

performance, as we shall see shortly. The final loss $\mathcal{L}$, used during the second phase of training, is a weighted combination of the aforementioned losses, i.e. $\mathcal{L} = \lambda_1 \mathcal{L}_{BCE} + \lambda_2 \mathcal{L}_{IoU}$ where $\lambda_1 = 1, \lambda_2 = 20$. The network is trained for a total of 58 epochs.

Finally, we combined training, validation sets and re-trained the model with the hyper-parameters and stopping criteria determined from the validation set experiments. The final model is evaluated on a disjoint test set (40 images). The entire training was performed using four Nvidia GeForce GTX 1080 Ti GPUs.

6   Experiments and Analysis

For all experiments, we compare performance using the standard measures – Intersection over Union (IoU) [19], Per pixel Average (PPA) [15], F-1 (Dice) score [6] – averaged over the evaluation set. The IoU is calculated corresponding to the tabular part of the document image, and the resulting nearest tabular mask from the model's output. The F-1 score is calculated using the conventional Dice coefficient formula for tabular regions of the document and the corresponding model outputs.

Modified v/s Original U-Net: To begin with, we compared the performance of original and modified U-Net. As Table 2 shows, our modified U-Net has a small but significant performance advantage over the original version, justifying its choice. The relatively smaller size of our modified U-Net (Table 4).

Table 4. Comparison of models by number of parameters

Model	# parameters (M = million)
TableBank [13]	17M
YOLO-v3 [17]	65M
pix2pixHD [23]	188M
CascadeTabNet [14]	83M
Original U-Net [14]	8M
Modified U-Net (Ours)	3.5M

As mentioned previously, the task of table localization can be viewed either as a table detection task or a table segmentation task. Consequently, we compared its performance in these two task settings by customized training of two popular object detection models and three semantic segmentation models.

Detection Models: For our experiments, we used TableBank [13], an open-source table detector trained on 163,417 MS Word documents and 253,817 LaTeX documents. We fine-tuned TableBank directly using the MediTables training set. We also trained a YOLO-v3 [17] object detection model. Unlike TableBank training, we followed the protocol used for our proposed model (i.e. pre-training on existing document datasets) (Sect. 5.2).

Segmentation Models: We trained pix2pixHD [23], a popular image pixel-level image translation model. Keeping the relatively small size of our dataset in mind, we trained a scaled down version. In addition, we trained CascadeTab-Net [14], a state-of-the-art segmentation model developed specifically for table segmentation. As a postprocessing step, we performed morphological closing on the results from segmentation approaches for noise reduction and filling holes in the output masks.

As a preliminary experiment, we examined performance when the models pre-trained on existing document datasets were directly evaluated on the MediTables test set (i.e. without any fine-tuning). As Table 3a shows, recent models developed specifically for table localization (CascadeTabNet [14], TableBank [13]) show good performance. Our modified U-Net's performance is relatively inferior, likely due to its inability to bridge the domain gap between existing datasets and medical domain documents directly.

Upon fine-tuning the document dataset pre-trained models data from the domain (i.e. MediTables dataset), a very different picture emerges (Table 3b). Our proposed approach (modified U-Net) outperforms strong baselines and existing approaches, including the models customized for table localization. We hypothesize that this is due to the ability of our modified U-Net to judiciously utilize the within-domain training data to close the gap between pre-trained setting and fine-tuned setting in an effective manner. Another important observation is that the deviation from average in our model is typically the smallest compared to other models. Finally, it is important to note that superior performance has been achieved by our model even though it is drastically smaller in size compared to customized and state of the art baselines (see Table 4).

Fig. 3. Table localization results for various models for images with highest IoU score using our model's predictions.

Performance Comparison by Table Type: The previous experiments focused on evaluation for all tables. To examine performance by table type (T1, T2), the previous 2-class (table, background) formulation was replaced by a 3-class prediction setup (T1, T2 and background). As Table 5 shows, our modified U-Net once again performs the best across table types and measures.

Figure 3 shows qualitative results on images with the highest IoU score as per our model's predictions. The superior quality of our results is evident. A similar set of results can be viewed by table type in Fig. 4. Images with the lowest IoU score as per our model prediction can be viewed in Fig. 5. These represent the most challenging images. As mentioned previously, a high degree of skew and small footprint of the table in the image generally affect the performance from our model. However, even for these images, the quality of results is quite acceptable.

Table 5. Per-table type performance of models pre-trained on existing document datasets and fine-tuned on MediTables (PPA is not defined for detection models)

Model	T1			T2		
	IOU (%)	PPA (%)	F1 (%)	IOU (%)	PPA (%)	F1 (%)
TableBank [13]	92.59 ± 11.27	NA	95.75 ± 00.07	84.90 ± 16.20	NA	90.95 ± 10.30
YOLO-v3 [17]	30.88 ± 02.83	NA	47.18 ± 14.47	56.38 ± 07.58	NA	72.10 ± 06.19
pix2pixHD [23]	85.42 ± 01.65	92.74 ± 06.82	92.13 ± 06.38	92.67 ± 01.34	99.10 ± 00.89	96.19 ± 00.71
CascadeTabNet [14]	92.66 ± 10.87	93.22 ± 10.37	95.84 ± 06.39	93.44 ± 10.21	93.76 ± 10.22	96.29 ± 06.14
Modified U-Net (Ours)	**95.48 ± 03.82**	**99.08 ± 00.91**	**97.69 ± 01.96**	**94.30 ± 00.30**	**99.62 ± 00.37**	**97.06 ± 01.61**

Fig. 4. Tables T1 (green) and T2 (red) segmentation results of three semantic segmentation models and two object detection models on the testing set of MediTables. Note that predictions (colors) may also be incorrect in terms of table type labels (T1,T2) in some instances. (Color figure online)

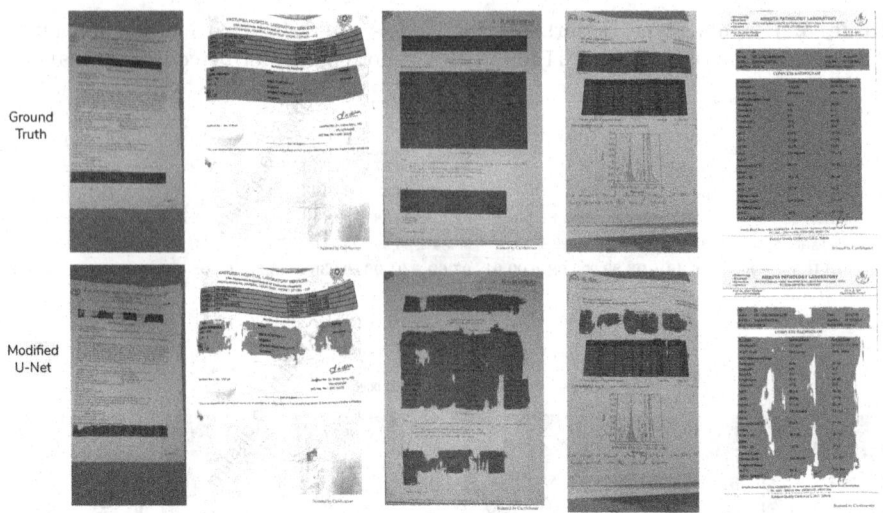

Fig. 5. Table localization results for various models for images with lowest IoU score using our model's predictions.

7   Conclusion

In this paper, we have presented a dataset for diverse healthcare and medical document images. We hope that our efforts encourage the community to expand our dataset and build upon our findings to enable richer understanding of medical document images. Given its distinct nature, we also expect our dataset to be considered along with existing datasets when benchmarking new table localization approaches in future.

We have also proposed a compact yet high performing approach for localizing and categorizing tables in medical documents. The performance of the proposed approach is greatly facilitated by our choice of using a segmentation network (as opposed to a detection network), the skip-connectivity for enhanced gradient flow and by our choice of losses and training procedure. Our model has the potential to operate as the first step in a processing pipeline for understanding tabular content in medical documents. Another significant advantage is the compact size of our model, making it potentially attractive for deployment on mobile and embedded devices.

References

1. Fang, J., Gao, L., Bai, K., Qiu, R., Tao, X., Tang, Z.: A table detection method for multipage pdf documents via visual seperators and tabular structures. In: 2011 International Conference on Document Analysis and Recognition, pp. 779–783. IEEE (2011)

2. Gao, L., Yi, X., Jiang, Z., Hao, L., Tang, Z.: ICDAR 2017 competition on page object detection. In: 2017 14th IAPR International Conference on Document Analysis and Recognition (ICDAR), vol. 1, pp. 1417–1422. IEEE (2017)

3. Gilani, A., Qasim, S.R., Malik, I., Shafait, F.: Table detection using deep learning. In: 2017 14th IAPR International Conference on Document Analysis and Recognition (ICDAR), vol. 1, pp. 771–776. IEEE (2017)

4. Girshick, R.: Fast R-CNN. In: Proceedings of the IEEE International Conference on Computer Vision, pp. 1440–1448 (2015)

5. Göbel, M., Hassan, T., Oro, E., Orsi, G.: ICDAR 2013 table competition. In: 2013 12th International Conference on Document Analysis and Recognition, pp. 1449–1453. IEEE (2013)

6. Goyal, M., Yap, M.H., Hassanpour, S.: Multi-class semantic segmentation of skin lesions via fully convolutional networks. arXiv preprint arXiv:1711.10449 (2017)

7. Ha, J., Haralick, R.M., Phillips, I.T.: Recursive XY cut using bounding boxes of connected components. In: Proceedings of 3rd International Conference on Document Analysis and Recognition, vol. 2, pp. 952–955. IEEE (1995)

8. Hao, L., Gao, L., Yi, X., Tang, Z.: A table detection method for pdf documents based on convolutional neural networks. In: 2016 12th IAPR Workshop on Document Analysis Systems (DAS), pp. 287–292. IEEE (2016)

9. Huang, Y., et al.: A yolo-based table detection method. In: 2019 International Conference on Document Analysis and Recognition (ICDAR), pp. 813–818. IEEE (2019)

10. Kavasidis, I., et al.: A saliency-based convolutional neural network for table and chart detection in digitized documents. In: Ricci, E., Rota Bulò, S., Snoek, C., Lanz, O., Messelodi, S., Sebe, N. (eds.) ICIAP 2019. LNCS, vol. 11752, pp. 292–302. Springer, Cham (2019). https://doi.org/10.1007/978-3-030-30645-8_27

11. Kieninger, T., Dengel, A.: The T-Recs table recognition and analysis system. In: Lee, S.-W., Nakano, Y. (eds.) DAS 1998. LNCS, vol. 1655, pp. 255–270. Springer, Heidelberg (1999). https://doi.org/10.1007/3-540-48172-9_21

12. LeCun, Y., Bengio, Y., Hinton, G.: Deep learning. Nature **521**(7553), 436 (2015)

13. Li, M., Cui, L., Huang, S., Wei, F., Zhou, M., Li, Z.: TableBank: table benchmark for image-based table detection and recognition. arXiv preprint arXiv:1903.01949 (2019)

14. Prasad, D., Gadpal, A., Kapadni, K., Visave, M., Sultanpure, K.: CascadeTabNet: an approach for end to end table detection and structure recognition from image-based documents (2020)

15. Prusty, A., Aitha, S., Trivedi, A., Sarvadevabhatla, R.K.: Indiscapes: instance segmentation networks for layout parsing of historical Indic manuscripts. In: 2019 International Conference on Document Analysis and Recognition (ICDAR), pp. 999–1006. IEEE (2019)

16. Rahman, M.A., Wang, Y.: Optimizing intersection-over-union in deep neural networks for image segmentation. In: Bebis, G., et al. (eds.) ISVC 2016. LNCS, vol. 10072, pp. 234–244. Springer, Cham (2016). https://doi.org/10.1007/978-3-319-50835-1_22

17. Redmon, J., Farhadi, A.: Yolov3: an incremental improvement. arXiv preprint arXiv:1804.02767 (2018)

18. Ronneberger, O., Fischer, P., Brox, T.: U-Net: convolutional networks for biomedical image segmentation. In: Navab, N., Hornegger, J., Wells, W.M., Frangi, A.F. (eds.) MICCAI 2015. LNCS, vol. 9351, pp. 234–241. Springer, Cham (2015). https://doi.org/10.1007/978-3-319-24574-4_28

19. Sarvadevabhatla, R.K., Dwivedi, I., Biswas, A., Manocha, S.: Sketchparse: towards rich descriptions for poorly drawn sketches using multi-task hierarchical deep networks. In: Proceedings of the 25th ACM international conference on Multimedia, pp. 10–18 (2017)

20. Schreiber, S., Agne, S., Wolf, I., Dengel, A., Ahmed, S.: DeepDeSRT: deep learning for detection and structure recognition of tables in document images. In: 2017 14th IAPR International Conference on Document Analysis and Recognition (ICDAR), vol. 1, pp. 1162–1167. IEEE (2017)

21. Siddiqui, S.A., Malik, M.I., Agne, S., Dengel, A., Ahmed, S.: DeCNT: deep deformable CNN for table detection. IEEE Access 6, 74151–74161 (2018)

22. Taghva, K., Nartker, T., Borsack, J., Condit, A.: UNLV-ISRI document collection for research in OCR and information retrieval 3967 (2000)

23. Wang, T.C., Liu, M.Y., Zhu, J.Y., Tao, A., Kautz, J., Catanzaro, B.: pix2pixhd: high-resolution image synthesis and semantic manipulation with conditional GANs

24. Yang, X., Yumer, E., Asente, P., Kraley, M., Kifer, D., Lee Giles, C.: Learning to extract semantic structure from documents using multimodal fully convolutional neural networks. In: Proceedings of the IEEE Conference on Computer Vision and Pattern Recognition, pp. 5315–5324 (2017)

25. Yildiz, B., Kaiser, K., Miksch, S.: pdf2table: a method to extract table information from pdf files. In: IICAI, pp. 1773–1785 (2005)

Online Analysis of Children Handwritten Words in Dictation Context

Omar Krichen$^{(\boxtimes)}$, Simon Corbillé, Eric Anquetil, Nathalie Girard, and Pauline Nerdeux

Univ Rennes, CNRS, IRISA, 35000 Rennes, France
{omar.krichen,simon.corbille,eric.anquetil,nathalie.girard,
pauline.nerdeux}@irisa.fr

Abstract. This paper presents a method for fine analysis of children handwriting on pen-based tablets. This work is in the context of the P2IA project, funded by the French government, which aims at designing a virtual notebook in order to foster handwriting learning for primary school pupils. In this work, we consider the task of analysing handwritten words in the context of a dictation exercise. This task is complex due to different factors: the children do not master yet the morphological aspects of handwriting, nor do they master orthography or translating phonetic sounds to actual graphemes (parts of word). In order to tackle this problem, we extend **to the context of dictation exercises** an analysis engine that was developed previously to deal with copying exercises. Two strategies were developed, the first one is a baseline approach and relies on double child input: the pupil types the word on a virtual keyboard after writing it with the stylus, thus the prior knowledge of the written word will drive the engine analysis. The second one relies on a single input: the child handwritten strokes. To drive the analysis, the strategy consists in generating hypotheses that are phonetically similar to the dictated instruction, which will act as probable approximations of the written word (sequence of letters), to cover potential orthographic mistakes by the pupil. To assist the learning process of the pupils, the engine returns different types of real-time feedbacks, that depend on the confidence of the analysis process (confident assessment on errors, warning, or reject).

Keywords: Handwriting recognition · Online handwriting · Digital learning

1 Introduction

This work is part of an Innovation and Artificial Intelligence Partnership (P2IA[1]), which supports the construction of solutions serving fundamental learning in French and mathematics in cycle 2 (CP, CE1, CE2). Here we are interested

---

[1] https://eduscol.education.fr/1911/partenariat-d-innovation-et-intelligence-artificielle-p2ia.

© Springer Nature Switzerland AG 2021
E. H. Barney Smith and U. Pal (Eds.): ICDAR 2021 Workshops, LNCS 12916, pp. 125–140, 2021.
https://doi.org/10.1007/978-3-030-86198-8_10

in defining a solution to help elementary school students in learning spelling. We propose a method to automatically analyse the production of handwritten words, written on pen tablets, in the context of dictation exercises.

The pedagogical foundation of this work lies in several studies that demonstrate the positive impact of using educational systems in the classroom, especially by using pen based tablets. In a critical review study [10], the authors reported that, among 12 highly trustworthy studies, 9 observed positive learning outcomes for the pupils, whereas 3 observed no difference in learning outcomes between the tablet setup and traditional pen and paper one. Moreover, the authors of [9] demonstrate that providing prompt feedback, which is facilitated with digital tools usage, is a key factor in improving learning performance.

In this context, defining a solution to help learning spelling involves different tasks. The first is to recognise children's handwritten words, and the second is to understand potential mistakes to give children appropriate feedback for each type of mistake.

The task of children handwritten words analysis is an open challenge. Indeed, even if deep learning based methods have made great progress in handwritten words recognition [13,14], most of them are targeted to adult data, and not suited to cope with children distorted handwriting. This work is an extension of previous project called IntuiScript, where the objective was to help preschool children learn to write. To achieve this goal, copy exercises have been designed, where the word instruction is displayed to the child who must reproduce it on the interface. The scientific challenge was to finely analyse the *handwriting quality*, in terms of letters shape, direction and order, in order to provide children with feedback on improving their writing skills. As the word instruction was displayed, it served as prior knowledge (groundtruth) to guide the process of recognising and analysing the child's written words and thus limit misinterpretation. The challenge was to deal with the degraded nature of handwriting (incorrect letter shapes). This previous work, with positive pedagogical results presented in [1], was transferred to "Learn & Go company" and integrated into the "Kaligo" solution, now used in French and English schools. Since this handwriting analysis method is based on knowledge of the word instruction and its display to the child, it is not robust enough to reliably extract letter-level segmentation when the child does not write the expected letters, which happens when the word is dictated (without display). As an extension of IntuiScript, the pedagogical objective of the P2IA project is to help children acquire orthographic knowledge, *i.e.* to learn graphemes and phonemes. The target population is primary school children who have acquired prior handwriting skills and, dictation/spelling exercises are proposed, such as the dictated instruction is heard but not seen by the child. The scientific challenge is to design an intelligent tutoring system [12] that is able to **provide orthographic feedback** to the child.

In a dictation context, the analysis task is more complex since the engine does not know what the child has written. We are faced with orthographic and phonetic errors, since the child only hears the instruction, as well as morphological errors. For clarity purposes, we define three important notions that will be present all along this paper:

- The instruction: the dictated word that the child has to correctly spell/reproduce;
- The handwriting/handwritten strokes: what the child actually wrote using the tablet stylus;
- The groundtruth: the letters sequence corresponding to the child production.

Figures 1, 2, 3 illustrates examples of the errors we encounter in this context. The instruction (or dictated word) is written in the box with a black border.

The Fig. 4 illustrates an example of the wide variety of pupils orthographic and morphological errors when the instruction "mes" (my in French) is dictated. As a consequence, the groundtruth is not available since its is likely to be unrelated to the dictated instruction.

Fig. 1. Morphological error on the shape of "o", orthographic/phonetic error (missing "s" at the end of the word)

Fig. 2. Morphological error on the shape of the letter "d", orthographic error: substitution of "eux" by "e"

Fig. 3. Orthographic errors: substitution of "c"-by "qu", likewise for-"mm" and "m"

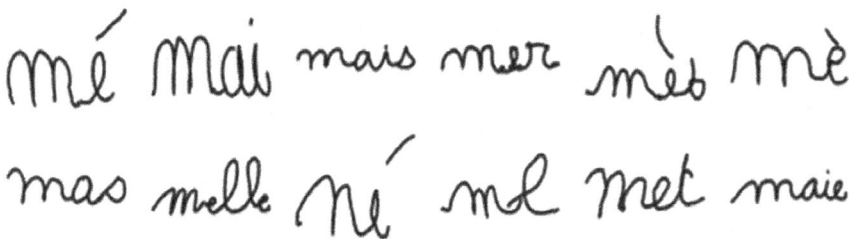

Fig. 4. Examples of pupils errors for the word "mes"

To address this challenge, **we propose a phonetic and morphological analysis strategy in a dictation context**. This strategy is based on the extension of the IntuiScript existing analysis engine suited to the copy context. This approach will be based on two independent modes:

– Double input strategy: in addition to writing the dictated word with the pen, the child also uses the tablet virtual keyboard to reproduce his/her production. This straightforward strategy provides the analysis engine with the necessary groundtruth prior knowledge to interpret the word correctly. This represents an intermediate solution that allows us to have an ideal baseline for the engine performance.
– Single input, phonetic hypotheses generation strategy: in order to be free from the user typed input and to cover the eventual errors made by the pupils, we integrate a phonetic engine, which role is to generate, given an instruction, phonetically similar pseudo-words (same sound as the instruction). Guided by these hypotheses, the analysis engine will try to predict the *actual* groundtruth.

As children are in a learning process, one of the challenges is to provide relevant feedback in real time to the child and to be as precise and exact as possible. So we need to minimise feedback errors, but we also need to moderate the details of the comments with the confidence of the analysis. Therefore, **we define a moderation strategy** (feedback generation mechanism), **based on the confidence of the analysis engine**. We evaluate these contributions on children data, 1087 words collected in the classroom.

The paper is organised as follows. The existing analysis engine is presented in Sect. 2. Sect. 3 describes the engine extension and adaption to the dictation context, whereas Sect. 4 illustrates the typology of the generated feedback. Experiments are presented in Sect. 5. Conclusion and future works are given in Sect. 6.

2   Existing Copying Analysis Engine

In this section, we present the main principles of the existing analysis engine, that was designed for a copy context. The Fig. 5 illustrates the analysis workflow.

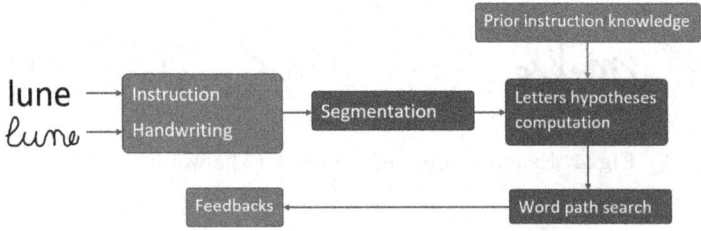

Fig. 5. Workflow of analysis engine in copy context

As discussed earlier in the paper, the inputs of the engine are the instruction and the pupil handwriting. The word analysis is divided into multiple steps as follows.

2.1 Segmentation

The segmentation process is based on two steps. First, the online signal is segmented into primary element by extracting all possible cutting points around the significant descending areas [2]. Second, as illustrated in Fig. 6 for the word "juste", a **segmentation lattice** is constructed. The first level of the lattice/graph is built from the primary segmentation by associating ascending areas with a descending area. The second level is made by merging two nodes from the first level and so on for the next levels. Then, the goal is to find the path in this lattice which corresponds to the character decomposition of the word written by the pupil. In Fig. 6, the best path is highlighted in green, whereas the explicit segmentation result is in the right top corner of the figure. This explicit segmentation is needed to analyse the letters in context, and provide precise feedback. This is not possible with current deep neural network approaches.

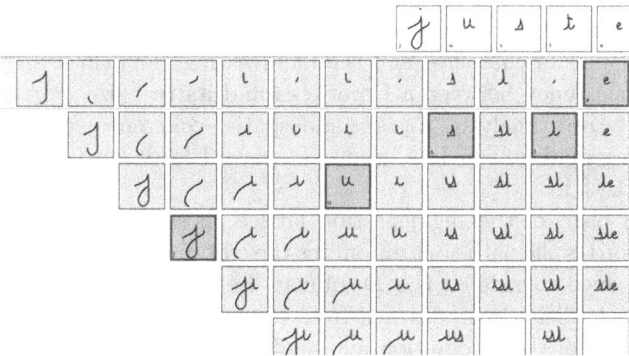

Fig. 6. Segmentation graph for the word "juste" (Color figure online)

More details regarding to the construction of the segmentation lattice can be found in [3].

2.2 Letter Hypothesis Computation

The next step is to compute letter hypotheses for each node of the segmentation lattice. First of all, a **recognition score** is computed with Evolve classifier [4], based on fuzzy inference [5]. These letter hypotheses are filtered in order to keep only the ones corresponding to expected letters in the instruction, as well as the ones with the best scores. This is the **recognition step**. Afterwards, an **analysis score** is computed for each hypothesis with a confidence based classifier [6], which use a intra and inter class scoring to deal with confusion between letters. Only the n *best* letters hypotheses belonging to the expected word and whose score is superior to a defined threshold are selected as valid hypotheses and the other ones are discarded. If there is no letter hypothesis verifying these two conditions, the letter with the best recognition score is kept as the sole valid hypothesis of the node. This is the **analysis step**.

This two-step process allows the information contained in the instruction to guide the selection of letter hypotheses. However, this approach is only suited if the pupil does not commit any orthographic error. That is why **this approach is suited to a copy context**, where the child sees the instruction to be reproduced. In a dictation context, this approach has to be adapted to the fact that the child does not know necessarily the instruction spelling.

2.3   Best Segmentation Path Search

The computation of the letters hypotheses is the entry point of the **word segmentation paths** search within the segmentation lattice. A word-level analysis score is calculated using the filtering of the node-level analysis scores for each possible path. This is combined with an n-gram and a spatial coherence scores. The **n-gram** score is related to the presence of bigram or letters of the word instruction. The **spatial coherence score** is calculated with the letters hypotheses of the paths, models of character and the handwriting. That makes it possible to check the consistency between a hypothesis and its real size. For more details, see [3]. This score (analysis, n-gram and spatial coherence) provides a metric on the writing quality, which is used as a reward feedback for the children in previous works [1].

The final path of the segmentation lattice chosen by the analyser as the recognised word is the path that minimises the edit distance with the instruction. The edit distance considered is a **Demerau Leveinshtein edit distance** [7], with optimised edition costs learned from the letters analysis (to deal with the confusion errors of the recognition and analysis process). This best path is not necessarily the one that maximises the analysis score. This way of retrieving the best path is well suited to the **copy context**, but becomes obsolete when the prior knowledge of groundtruth becomes unknown in the dictation context.

In this section, we have presented the principles of the analysis engine, specifically how the knowledge of the expected word, in a copy context, guides the analysis process. In the next section, we will present our new contribution to adapt this engine to a dictation context.

3   Adaption of the Engine in a Dictation Context

Since we are in a dictation context, the impact of the instruction in the guidance of the analysis process becomes obsolete when the child makes orthographic errors. To deal with this problem, we have designed two strategies that allow the engine to overcome this new challenge.

3.1   Double Input, Baseline Strategy

The first strategy to adapt the engine to this new context is a straightforward one: after the completion of the handwritten production, the system asks the child to enter with the keyboard, what he has written. As a consequence, the engine has an explicit knowledge of the groundtruth, which is the *childtyping* (what the child typed). To illustrate the key role of this prior knowledge, Figs. 7 and 8 present the analysis results of a pupil's handwritten word in two modes: using only the instruction as prior knowledge, and using the childtyping as prior knowledge. The difference in the two modes lies in the analysis results of the "a" node. Since the instruction is to write the word *mes*, the letters "m", "e", and "s" will guide the analysis process in the letter hypotheses computation step. Even if the letter "a" is the best ranked hypothesis for the highlighted node, it will be discarded due to the analysis filters and another hypothesis, "e", will be considered. It is clear that using the explicit groundtruth as prior knowledge solves this problem.

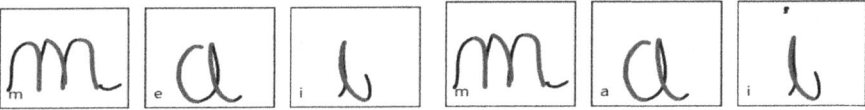

Fig. 7. Analysis driven by the Instruction (mes), recognised: mei

Fig. 8. Analysis driven by Childtyping (mai), recognised: mai

This intermediary strategy provides a baseline of how the engine would perform in ideal conditions. Surprisingly, the teachers associated to the project estimated that asking the child to type what he wrote could also have pedagogical benefits.

3.2   Phonetic Hypotheses Generation Strategy

The second strategy aims to be free from the user defined groundtruth and to predict it given only the instruction and the handwritten strokes. It is based on the integration of a phonetic hypotheses generation module to the analysis workflow. This module is based on the **Phonetisaurus engine** [8], a stochastic Grapheme to Phoneme (G2P) WFST (Weighted Finite State Transducer). This WFST is based on the principle of joint sequences [11] to align graphemes sequences with their corresponding phonemes in the learning phase. An N-Gram model is generated from the aligned joint sequences and transformed into a WFST model. This G2P model is then able to predict the pronunciation of a new word. To be adapted to our problematic, the G2P model is combined with a P2G (Phoneme to Grapheme) model so that the output of the combined model, given a new word, is a set of phonetically similar pseudo-words. We choose to generate, for each instruction, the 50 best hypotheses according to the Phonetisaurus engine ranking (for more details, see [8]). The Fig. 9 illustrates the new **phonetic analysis chain**.

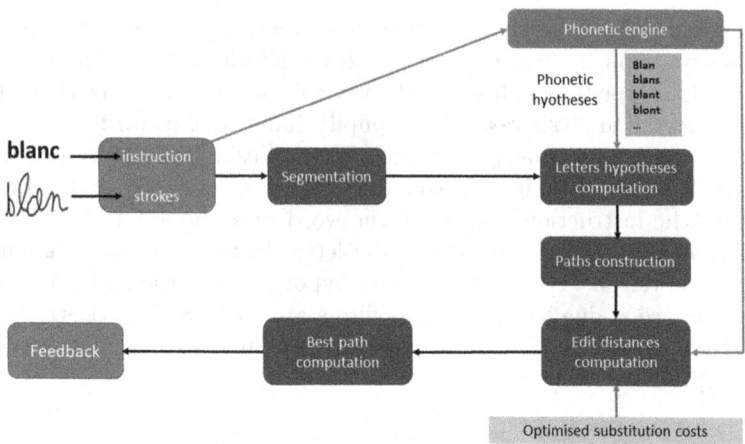

Fig. 9. Phonetic analysis chain workflow

A) Phonetic Hypotheses as Prior Knowledge in the Analysis Process.
The first adaptation of the analysis chain lies in modifying the filters that are
used for computing the letter hypotheses. We have explained in the last section
that for each segmentation node, only the letters that belong to the instruc-
tion/groundtruth are kept as valid hypotheses in the node.

Since the grountruth is unknown, the new analysis chain is guided by all the
generated phonetic hypotheses from the instruction (dictated word).

For the example illustrated in Figs. 7 and 8, instead of having "mes" (the
instruction) or "mai" (the child-typed groundtruth) as a the expected sequence,
the engine will have {mes, mais, mai, met, med...}. This means that neither "e"
nor "a" will be discarded *a priori* by the analysis filters. The filter criteria is
that the letter belongs to one of the expected sequences and that the analysis
score of the letter is amongst the n best scores. Another impact of having these
phonetic hypotheses guiding the analysis is that other hypotheses will be taken
into account, such as the letter "d", as shown in the Fig. 10. As a consequence,
"e" here is discarded since it is not among the n best hypotheses.

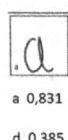

a 0,831

d 0,385

Fig. 10. Analysis scores of the "a" segmentation node

The word paths construction is the same as in the basic analysis chain. The
difference is in the computation of the n-gram scores of each path, since all bi-
grams of all phonetic hypothesis are included in the computation of this score.

B) Phonetic Hypotheses as Best Path Decision Criteria. We have presented in Sect. 2 the edition distance computation between the analyser word hypotheses, and the instruction/groundtruth, that enables the analysis engine to retrieve the handwritten word. This decision criterion is still suited to the *double input strategy*, since the child types the written word on the keyboard. However, this strategy **becomes obsolete without the prior knowledge of the groundtruth**. To tackle this problem, we compute the *phonetic correspondence* of each word segmentation path generated by the engine. By phonetic correspondence, we mean the phonetic hypothesis that has the minimum edition distance with the segmentation path. The edition score of each path will then depend on the generated phonetic hypotheses, as well as the optimised letter substitution costs learned by the analyser. As an example of this process, in Fig. 11, we can see that two segmentation paths (A and C in the figure) have a minimal edition distance score (0), since they are equal to two phonetic hypotheses, with "alors" being the dictated word, *i.e.* the instruction. We can also see that the edition score of "alxr" (B in the figure) is equal to 0.67, this number represents the optimised substitution cost of "o" with "x". This is an interesting example since there is a path that corresponds to the instruction (alors). However the groundtruth is equal to "alor", which means that the pupil made an orthographic mistake. It is clear that if the engine relied on the instruction to guide the interpretation process, it would have made an analysis error by choosing "alors" as the best path.

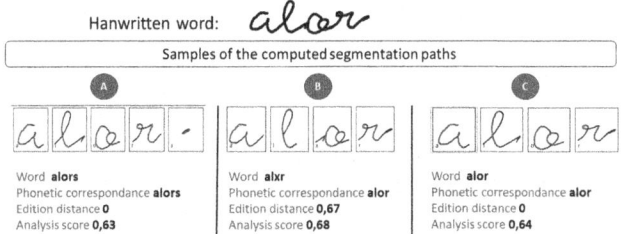

Fig. 11. Segmentation paths with their phonetic correspondences

In this particular case, the edition distance is not sufficient since we have two competing paths having the same edition score. To solve this issue, we include the analysis score in the decision criteria since this score reflects of the handwriting quality and enables the engine to discriminate between competing paths. **The phonetic analysis score is defined as follows.**

$$Score(path) = \frac{1}{1 + |EditScore(Path)|} * 0.3 + analysisScore(path) * 0.7$$

With this phonetic analysis score, the best path returned by the engine corresponds, in this example, to the groundtruth "alor" since the third path (C in the figure) has the highest analysis score of the two competing paths.

The phonetic correspondence is also important because it reflects the confidence of the phonetic analysis engine. Since there is no prior knowledge of the groundtrouth, this phonetic correspondence is a relatively efficient approximation.

C) Optimisation of the Phonetic Analysis Chain. Using all the letters from the phonetic hypotheses as prior knowledge in the letter hypotheses computation step enables to consider more possibilities in the path construction step, compared to the restrictive aspect of the filters in the copying analysis engine or with the double input strategy. However, one drawback of this phonetic strategy is that the correct letter hypothesis can sometimes be skipped if favour of others, as shown in the example in Fig. 12. For the word "rien", when all the phonetic hypotheses letters are added to the analysis filters, the analysis engine is not able to retrieve the groundtruth, whereas it is found when the analysis filters are restricted to "r", "i", "e", and "n".

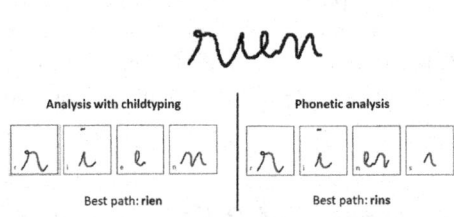

Fig. 12. Degradation of the analysis performance by adding phonetic hypotheses

To tackle this issue, we integrate the notion of **analysis competition**, illustrated in Fig. 13. This competition is between two analysis instances in order to find the best path, such as the first analysis (basic analysis in the figure) is guided by the instruction, and the second one (phonetic analysis) is guided by the phonetic hypotheses related to the expected word. If the basic analysis best path is equal to the expected word, its phonetic analysis score is computed and compared to the phonetic analysis score of the phonetic analysis best path. The path with the highest score is chosen as the final analysis result. If in the basic analysis best path is different from the expected path, the phonetic analysis best path is considered as the final result. This process enables the engine to retrieve a portion of the correctly written words that were misinterpreted before. We will study the impact of this optimisation in Sect. 5.

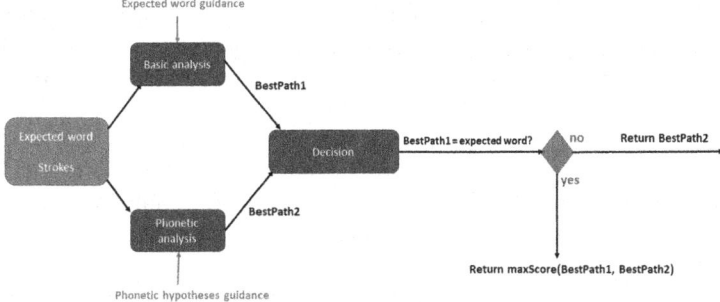

Fig. 13. Competition between analysis instances

After having presented our two strategies related to the orthographic analysis of pupils handwritten words, we present in the next section the feedback generation approach.

4   Feedback Typology

The objective of feedback generation by the system is to make the pupil aware of his/her orthographic errors. They are related to the difference between what was dictated and what was written. Therefore, these feedbacks highlight, in the pupil handwriting, the eventual insertions, deletions, substitutions of letters and accents. Table 1 illustrates some examples of such generated feedbacks. The red colour highlights a wrong insertion, whereas orange highlights a substitution.

Table 1. Feedback examples for dictated words

Instruction	Recognized	Error	Feedback
belle	belle	none	_belle_
alors	allore	insertion substitution	_allor_
céréale	cerèâle	Accents	_cereâle_

- **High confidence**: the recognised word is equal to the childtyping if the first strategy is activated, or is equal to its phonetic correspondence if the phonetic analysis strategy is activated $\implies$ the feedbacks are returned with a high degree of confidence. The feedbacks in Table 1 are *high confidence feedbacks*.
- **Medium confidence**: there is one letter that distinguishes between the recognised word and the childtyping/phonetic correspondence $\implies$ the system generates a warning on the uncertain zone/letter and other feedbacks are returned with a lesser confidence degree.

– **Reject**: there is more than one letter that distinguishes the recognised word and the childtyping/phonetic correspondence, no fine feedback is given to the child: the system informs the pupil that it was not able to analyse the production.

Table 3 presents two examples of medium confidence feedback and reject. For the first example, the engine recognises *alard* instead of the groundtruth *alord*, while the instruction is *alors*. The blue feedback on the "o" corresponds to a warning directed to the pupil, whereas the substitution of "s" by "d" is highlighted in grey, since there is a lesser degree of confidence in this feedback. For the second example, the phonetic analysis engine recognised *lemjeur*, which is completely unrelated phonetically to the instruction *bonjour*, or any of its phonetic hypotheses. Therefore, the system does not provide any fine feedback (Table 2).

Table 2. Examples of "medium confidence" feedback and reject

Dictated	Ground trouth	Recognized	Confidence	Feedback
alors	alord	alard	Medium	*alord*
bonjour	lezour	lmjeur	reject	*layour*

In this section, we have presented the feedbacks typology and our defined strategy to cope with analysis uncertainty. We present in the next section the performance of the analysis engine, as well as the pertinence of the generated feedback to the children.

5   Results

In this section, we will base our evaluation of the dictation adapted engine on a data-set of 1078 pseudo-words collected from children that use the system in the classroom. Due to GDRP restrictions on children's private data, we are not yet able to share this dataset publicly today. Table 3 presents some samples of this data-set enriched by the engine feedbacks.

5.1   Analysis Results

Table 4 presents the performance of each analysis mode/strategy on the test set. By analysis performance, we mean the correct segmentation and recognition of the handwritten strokes that lead automatically to a correct feedback. The fact that the ground truth is already available in the analysis with double input mode allows this strategy to have the best analysis rate by far (80.6%). It is interesting however to note that even with this ideal setting baseline, the engine has still an error rate of 19.4%, which demonstrates the complexity of the task in

Table 3. Feedback examples for dictated words: green (H) → correct high confidence feedback, green (M) → correct medium confident feedback, red (H) → error in high confidence feedback, blue (R) → feedback reject

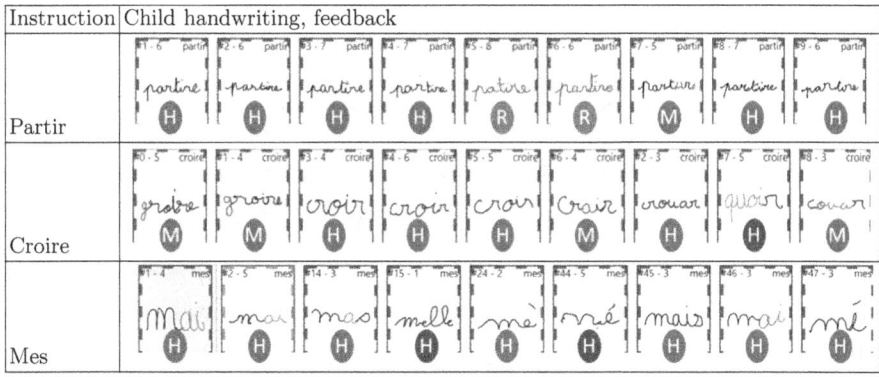

hand. The phonetic analysis strategy achieves a significantly lower recognition rate (69.4 %), however by far better than the existent copying analysis strategy (with only the instruction as prior knowledge). The gap between phonetic analysis and double input analysis can be explained by the fact that there are a lot of incorrectly written productions that are not phonetically similar to the instruction, which renders the phonetic guidance obsolete. The competition between analysis instances **in the optimised phonetic analysis strategy** results in a gain of 2% (71.4% recognition rate).

Table 4. Analysis performance of each strategy

Approach	Correctly analysed	Analysis rate
Copying analysis	633	58,72%
Double input analysis (childtyping)	869	80.6%
Phonetic analysis	748	69.4%
Optimised phonetic analysis	770	71.4%

In any case, there is room for improvement, as in optimising the recognition engines used to identify letter hypotheses, by using the amount of data that is being collected to improve the letter models. Moreover, the phonetic analysis chain can be improved by relearning the substitution costs and optimising the phonetic hypotheses generation process.

Given that there is some uncertainty in the interpretation robustness, its it important that the feedback generation strategy minimises the analysis engine errors.

5.2 Feedback Results

Table 5 presents the feedback generation results for the baseline double input analysis mode (three first columns) and for the optimized phonetic analysis mode (three last columns).

Table 5. Feedback generation pertinence

Confidence	Analysis with double input			Optimised phonetic analysis		
	Ratio	Errors	Errors rate	Ratio	Errors	Errors rate
High	857 (79.4%)	2	0.2%	819 (75.9%)	111	13.5%
Medium	137 (12.7%)	0	0%	85 (7.9%)	0	0%
Reject	88 (8.16%)	0	0%	174 (16.1%)	0	0%
Total feedback	994 (92.2%)	2	0.2%	904 (83.8%)	111	12.2%

We can see for the baseline double input, that even though the analysis rate is "only" 80 %, the feedback error ratio (total feedback errors/total feedback) is limited to 0.2%. This is due to the fact that most of the words that were not correctly analysed were either rejected or considered as medium confidence feedback. We can therefore conclude that the defined feedback statuses and the feedback generation decision criteria enable the system to have a good performance in the context of the analysis with double input.

Moreover, it is clear from the table that using the phonetic correspondence of the recognised word as a criterion for feedback generation is not as precise as using the baseline double input strategy, since it is an approximation of the ground truth. We can also see that there is more reject (174 versus 88), which is explained by the fact that we did not yet find a way to deal with children production that are phonetically incoherent with the production. In any case, we do observe the same improvement on the error ratio (feedback error ratio = 12.2% whereas analysis error ratio = 28.6%), which is encouraging. Finally, the possible improvements discussed relating to the analysis process would have a big impact on the feedback pertinence.

6 Conclusion

In this paper, we present an original approach for the orthographic analysis of children handwritten words in a dictation context. This approach is based on the extension of an existing analysis engine that was suited for copying exercises. Dictation exercises are more challenging since the child only hears the word he/she has to reproduce. As a consequence, we are faced with more morphological and orthographic errors. We defined two strategies to cope with this challenge. This first intermediary approach puts the user in the analysis loop, as the pupil has to type the word he/she has written on the keyboard after the production competition. This explicit groundtruth is then used as prior knowledge to drive the handwriting analysis process and to retrieve the written word. The second strategy aims to add fluidity to the interaction and to be free from the user defined groundtruth and is based on the generation of phonetically similar hypotheses for each instruction, that can cover a wide range of orthographic errors. We can consider that each phonetic hypothesis is a probable approximation of the groundtruth. We adapted the analysis process and the paths search decision criteria to cope with the fact that the ground truth is unknown. The experiments showed the improvement of the system performance with the integration of these new strategies, and the pertinence of the feedback generated to the pupils. Our future works consists in extending this analysis engine to the interpretation short sentences.

Acknowledgement. "P2IA" is funded by the French government. We would like to tank the project partners from Learn & Go, the University of Rennes 2, LP3C lab, INSA Rennes, University of Rennes 1 and IRISA lab. Additionally, parts of these works were supported by LabCom "Scripts and Labs" funded by the French National Agency for Research (ANR).

References

1. Bonneton-Botté, N.: Can tablet apps support the learning of handwriting? An investigation of learning outcomes in kindergarten classroom. Comput. Educ. 38 (2020)
2. Anquetil, E., Lorette, G.: On-line handwriting character recognition system based on hierarchical qualitative fuzzy modeling. In: Progress in Handwriting Recognition, pp. 109–116. World Scientific, New York (1997)
3. Simonnet, D., Girard, N., Anquetil, E., Renault, M., Thomas, S.: Evaluation of children cursive handwritten words for e-education. Pattern Recogn. Lett. **121**, 133–139 (2018)
4. Almaksour, A., Anquetil, E.: Improving premise structure in evolving Takagi-Sugeno neuro-fuzzy classifiers. Evolv. Syst. **2**, 25–33 (2011)
5. Takagi, T., Sugeno, M.: Fuzzy identification of systems and its applications to modeling and control. IEEE Trans. Syst. Man Cybern. **SMC-15.1**, pp. 116–132 (1985)
6. Simonnet, D., Anquetil, E., Bouillon, M.: Multi-criteria handwriting quality analysis with online fuzzy models. Pattern Recogn. **69**, 310–324 (2017)
7. Damerau, F.J.: A technique for computer detection and correction of spelling errors. Commun. ACM **7**, 171–176 (1964)
8. Novak, J., Dixon, P., Minematsu, N., Hirose, K., Hori, C., Kashioka, H.: Improving WFST-based G2P conversion with alignment constraints and RNNLM N-best rescoring (2012)
9. , Chickering, A.W., Zelda, F.G.: Seven principles for good practice in undergraduate education. AAHE Bull. **39**(7), 3–7 (1987)
10. Haßler, B., Major, L., Hennessy, S.: Tablet use in schools: a critical review of the evidence for learning outcomes. J. Comput. Assist. Learn. **32**, 139–156 (2016)

11. Bisani, M., Ney, H.: Joint-sequence models for grapheme-to-phoneme conversion. Speech Commun. **50**, 434–451 (2008)
12. Nkambou, R., Mizoguchi, R., Bourdeau, J.: Advances in Intelligent Tutoring Systems. Studies in Computational Intelligence, vol. 308. Springer, Heidelberg (2010). https://doi.org/10.1007/978-3-642-14363-2
13. Sheng, H., Schomaker, L.: Deep adaptive learning for writer identification based on single handwritten word images. Pattern Recogn. **88**, 64–74 (2008)
14. Kang, L., Rusinol, M., Fornes, A., Riba, P., Villegas, M.: Unsupervised writer adaptation for synthetic-to-real handwritten word recognition. In: The IEEE Winter Conference on Applications of Computer Vision, pp. 3502–3511 (2020)

A Transcription Is All You Need: Learning to Align Through Attention

Pau Torras[✉], Mohamed Ali Souibgui, Jialuo Chen, and Alicia Fornés

Computer Vision Center, Computer Science Department,
Universitat Autònoma de Barcelona, Barcelona, Spain
pau.torras@e-campus.uab.cat, {msouibgui,jchen,afornes}@cvc.uab.cat

Abstract. Historical ciphered manuscripts are a type of document where graphical symbols are used to encrypt their content instead of regular text. Nowadays, expert transcriptions can be found in libraries alongside the corresponding manuscript images. However, those transcriptions are not aligned, so these are barely usable for training deep learning-based recognition methods. To solve this issue, we propose a method to align each symbol in the transcript of an image with its visual representation by using an attention-based Sequence to Sequence (Seq2Seq) model. The core idea is that, by learning to recognise symbols sequence within a cipher line image, the model also identifies their position implicitly through an attention mechanism. Thus, the resulting symbol segmentation can be later used for training algorithms. The experimental evaluation shows that this method is promising, especially taking into account the small size of the cipher dataset.

Keywords: Handwritten symbol alignment · Hand-drawn symbol recognition · Sequence to Sequence · Attention models

1 Introduction

Historical ciphered manuscripts have recently attracted the attention of many researchers [6], not only for their own historical value, but also because of the challenges related to the transcription, decryption and interpretation of their contents. Indeed, many of these ciphers apply different techniques to hide their content from plain sight, for example, by using invented symbol alphabets. An example of a ciphered manuscript[1] is illustrated in Fig. 1.

Transcribing the sequence of symbols in the manuscript is the first step in the decryption pipeline [9]. However, machine learning-based recognition methods require annotated data, which is barely available. Indeed, an accurate labelling (e.g. annotation at symbol level) is desired, since it can then be used for training symbol classification, segmentation, spotting methods, etc. But, the few expert transcriptions are often available at paragraph or line level. For this reason, we propose to align each transcribed symbol with its representation in the

---

[1] https://cl.lingfil.uu.se/~bea/copiale/.

© Springer Nature Switzerland AG 2021
E. H. Barney Smith and U. Pal (Eds.): ICDAR 2021 Workshops, LNCS 12916, pp. 141–146, 2021.
https://doi.org/10.1007/978-3-030-86198-8_11

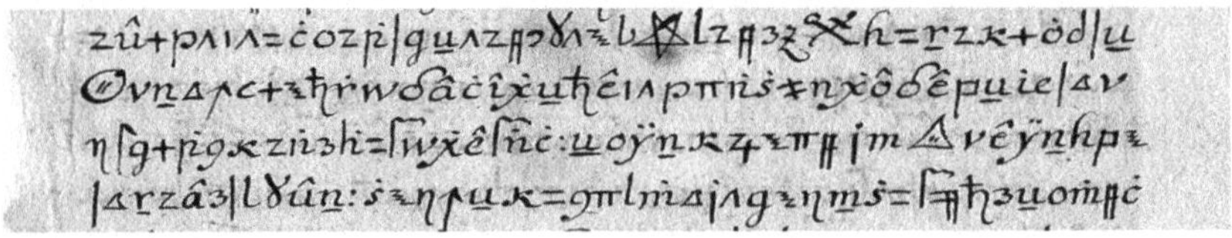

Fig. 1. An example of the Copiale ciphered manuscript, related to an 18th-century German secret society, namely the "oculist order".

manuscript image by using an attention-based Seq2Seq model [4], which implicitly infers the position of relevant visual features for every character output step.

The rest of the paper is organized as follows. First, in Sect. 2 we delve into relevant alignment methods present in the literature. We describe our approach in detail in Sect. 3, and the experiments in Sect. 4. Finally, in Sect. 5 we present some future work avenues and a few closing words.

2   Related Work

Many approaches exist for the task of alignment, which vary depending on the nature of the aligned manuscript. An example of domain-specific alignment can be found in Riba *et al.* [7], which consists of an image-to-image alignment using Dynamic Time Warping for detecting variations in music score compositions without the need of transcriptions. Similarly, Kassis *et al.* [5] use Siamese Neural Networks to align two handwritten text images with the same contents but different writing style.

The image-to-text alignment has been also researched. For example, Romero *et al.* [8] use Hidden Markov Models (HMM) and a dynamic programming algorithm to find candidate transcriptions of text lines and align them to the ground truth sequence. Fischer *et al.* [3] use HMMs and a first recognition pass.

A combination of both approaches is proposed by Ezra *et al.* [2], where they overfit a recogniser on the input data and generate a synthetic version of the image, which is aligned. They also use the OCR output and edit distances between said output and the ground truth for better performance.

3   Proposed Method

In this section we describe our architecture. We have used the Seq2Seq model with an attention mechanism proposed by Kang *et al.* [4] for HTR (Handwritten Text Recognition). We have adapted this technique for the task of alignment and performed several modifications related to the way attention masks are presented to improve their accuracy and flexibility for aligning cipher symbols.

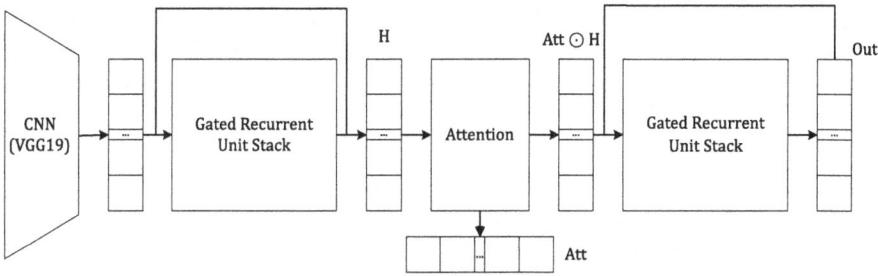

Fig. 2. Representation of the Seq2Seq model and the placement of attention within the pipeline.

3.1 Sequence to Sequence Model

The Seq2Seq model is an Encoder-Decoder architecture, which means that it processes an input set of vectors sequentially, generates an intermediate representation from them and then generates an output sequence based on said representation. The addition of an attention mechanism makes it possible for the intermediate representation to contain an unset number of vectors, since the model learns to assert the relevance of each of them and conditioning on those most useful in the current step. Our Seq2Seq model, depicted in Fig. 2, has a VGG19 convolutional network with its last max pooling layer removed as the first step in the pipeline, which accepts a $800 \times 64 \times 3$ px image of a text line as input. This generates a $50 \times 4 \times 512$-element (flattened to 50×2048) representation, which the Encoder, a stack of Gated Recurrent Units (GRU), further annotates into the hidden state H. Then, for each Decoder inference step, a vector Att of dimension 50 is computed through an attention mechanism. The input for the Decoder, another GRU stack, is the Hadamard product between the Att vector and the hidden state H, concatenated with the previously inferred symbol.

Our hypothesis is that there is a direct correlation between the position of a highly active attention mask and the position of the associated inferred symbol in the output text sequence, which enables us to perform alignment when learning to recognise lines.

3.2 Attention Mask Tuning

When applying the original Seq2Seq model in our data we found a major limitation, which was the fact that the attention mechanism can only provide a discrete set of fixed positions of a set width, since the attention mask is a 50-element vector that represents relevant areas in an 800px wide image. This made segmenting narrow characters or long sequences very difficult. Moreover, since the output of the attention mechanism is the result of a Softmax layer, no more than one attention band per character has a significant value.

Thus, we improved the model by treating the attention mask as a histogram and fitting a Normal Distribution onto it in order to find the position and width of the character in a more precise manner. Thus, every character mask is computed as:

$$m_{low} = \mu - kc_l\sigma \quad m_{high} = \mu + kc_h\sigma + 1 \tag{1}$$

where m_{low}, m_{high} are the lower and higher bounds of the character mask, μ is the mean of the histogram, σ is the standard deviation, c_l and c_h are a distribution skewness correction factor and k is a scaling factor to account only a set fraction of the standard deviation.

In this work, the k value was set to 0.5 after a tuning process in order to have only the most relevant samples of the distribution within the set boundaries. The skewness correction factors were computed as the ratio between the sum of all bins before or after the mean divided by the total sum of bins (excluding the highest valued one). Note also that both m_{low} and m_{high} need to be converted from the attention mask coordinate space into the image coordinate space.

Finally, since the model's only input is an image, the quality of the alignment relies on the model's capacity to produce a good output sequence of tokens. And given that our goal is not recognising the image, but instead aligning the associated transcription, the final mask prediction can be corrected by finding the shortest edit path between the output and ground truth sequences using Levenshtein's algorithm to remove unnecessary masks or adding padding when required. This underlying assumption considering the shortest edit path corresponds to the sequence of mistakes that a model has actually made, is quite strong, but we found that it prevents (or, at the very least, alleviates) misalignment in the majority of cases.

4   Experiments and Results

In this section we present the experiments performed to assess the viability of our model. We trained the Seq2Seq recogniser using line-level samples from the Copiale cipher with early stopping at 30 epochs with no Symbol Error Rate (SER) improvement on validation. Table 1 includes relevant information about the dataset and the model's hyperparameters.

Table 1. Relevant training information for experiment reproducibility.

Optimiser	Adam	Training samples	649
Learning rate (LR)	$3 \cdot 10^{-4}$	Validation samples	126
LR checkpoints	@ 20, 40, 60, 80, 100 epochs	Test samples	139
LR sigma	0.5	Dataset classes	126
Loss function	Cross-entropy	Avg. line length	42

For comparison, we used as baseline a learning-free method [1], which segments the line into isolated symbols based on connected components analysis.

Table 2. Experimental results, considering different Intersection over Union (IoU) thresholds. Metrics are Precision (Prec.), Recall (Rec.), F_1 score and Average Precision (AP). Symbol Error rate of the classifier is 0.365.

Exp.	IoU t = 0.25				IoU t = 0.50				IoU t = 0.75			
	Prec.	Rec.	F_1	AP	Prec.	Rec.	F_1	AP	Prec.	Rec.	F_1	AP
Baseline	0.31	0.30	0.31	0.12	0.25	0.24	0.24	0.07	0.14	0.16	0.14	0.02
Ours	0.94	0.89	0.91	0.84	0.59	0.55	0.57	0.34	0.10	0.09	0.10	0.001

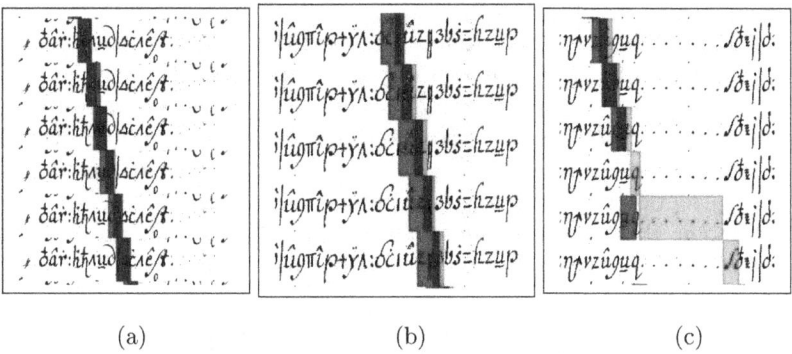

(a)   (b)   (c)

Fig. 3. Qualitative Results. Model's prediction in red, ground truth in green and the intersection of both in blue color. Each successive line within the image is a time step. These are fragments of a longer alignment sequence, cut for readability purposes. (a), (b) and (c) are examples of output quality patterns.

First, connected components are extracted, and then, grouping rules are join the components that likely belong to the same symbol. Then, the components are aligned to the sequence of transcribed symbols.

Quantitative results are shown on Table 2. Segmentation accuracy is evaluated as the Intersection over Union (IoU): the percentage of masks whose ground truth and prediction intersect in a ratio of t over the union of both areas. As it can be seen, our approach surpasses the baseline method in most scenarios.

The analysis of these results shows three general patterns:

– **Correct alignment** (Fig. 3a): There is a considerable proportion of cases with an overall correct alignment with limited error. Perfect masks are however rare, with some degree of error on the sides being relatively frequent.
– **Slight misalignment** (Fig. 3b): Mostly caused due to incorrect edit paths chosen after the recognition algorithm, narrow symbols or very long sequences, which cause the attention masks to be broader in comparison.
– **Misalignment** (Fig. 3c): Incorrect alignment when encountering very rare symbols or high output SER, mostly due to limited training data.

Finally, we note that we tested the potential for bootstrapping training under the same parameters with synthetic samples. We created a 40.000 sample dataset using segmented symbols from real Copiale pages. We trained the model with random-width lines with each symbol appearing under a uniform distribution and fine-tuned it real samples. However, results did not improve, which we attribute to characters in synth lines being broader than their real counterparts, which caused attention masks to skip symbols.

5 Conclusion

We have proposed an alignment method based on Seq2Seq models. Our method shows encouraging results given the small dataset. The main hindrances are the difficulty for training a very accurate model and the need of further mask processing in order to be able to find bounding boxes correctly. Thus we believe that, by refining the Levenshtein algorithm including confidence data to choose the right edits or modifying the attention mechanism to have more than one high activation mask, results might improve. It is also worth to explore adding supervised attention mask training to avoid having to tune masks after recognition, since some character-level annotated samples are available for the data we are working with and we might boost the performance further.

Acknowledgement. This work has been supported by the Swedish Research Council, grant 2018-06074, DECRYPT – Decryption of Historical Manuscripts, the Spanish project RTI2018-095645-B-C21 and the CERCA Program / Generalitat de Catalunya.

References

1. Baró, A., Chen, J., Fornés, A., Megyesi, B.: Towards a generic unsupervised method for transcription of encoded manuscripts. In: DATeCH, pp. 73–78 (2019)
2. Ezra, D.S.B., Brown-DeVost, B., Dershowitz, N., Pechorin, A., Kiessling, B.: Transcription alignment for highly fragmentary historical manuscripts: the dead sea scrolls. In: ICFHR, pp. 361–366 (2020)
3. Fischer, A., Frinken, V., Fornés, A., Bunke, H.: Transcription alignment of Latin manuscripts using hidden Markov models. In: HIP, pp. 29–36 (2011)
4. Kang, L., Toledo, J.I., Riba, P., Villegas, M., Fornés, A., Rusiñol, M.: Convolve, attend and spell: an attention-based sequence-to-sequence model for handwritten word recognition. In: Brox, T., Bruhn, A., Fritz, M. (eds.) GCPR 2018. LNCS, vol. 11269, pp. 459–472. Springer, Cham (2019). https://doi.org/10.1007/978-3-030-12939-2_32
5. Kassis, M., Nassour, J., El-Sana, J.: Alignment of historical handwritten manuscripts using Siamese neural network. In: ICDAR, vol. 1, pp. 293–298 (2017)
6. Megyesi, B., et al.: Decryption of historical manuscripts: the decrypt project. Cryptologia **44**(6), 545–559 (2020)
7. Riba, P., Fornés, A., Lladós, J.: Towards the alignment of handwritten music scores. In: GREC, pp. 103–116 (2015)
8. Romero-Gómez, V., Toselli, A.H., Bosch, V., Sánchez, J.A., Vidal, E.: Automatic alignment of handwritten images and transcripts for training handwritten text recognition systems. In: DAS, pp. 328–333 (2018)
9. Souibgui, M.A., Fornés, A., Kessentini, Y., Tudor, C.: A few-shot learning approach for historical ciphered manuscript recognition. In: ICPR, pp. 5413–5420 (2021)

Accurate Graphic Symbol Detection in Ancient Document Digital Reproductions

Zahra Ziran$^{(\boxtimes)}$ ⓘ, Eleonora Bernasconi ⓘ, Antonella Ghignoli ⓘ,
Francesco Leotta ⓘ, and Massimo Mecella ⓘ

Sapienza Università di Roma, Rome, Italy
{zahra.ziran,eleonora.bernasconi,antonella.ghignoli,
francesco.leotta,massimo.mecella}@uniroma1.it

Abstract. Digital reproductions of historical documents from Late Antiquity to early medieval Europe contain annotations in handwritten graphic symbols or signs. The study of such symbols may potentially reveal essential insights into the social and historical context. However, finding such symbols in handwritten documents is not an easy task, requiring the knowledge and skills of expert users, i.e., paleographers. An AI-based system can be designed, highlighting potential symbols to be validated and enriched by the experts, whose decisions are used to improve the detection performance. This paper shows how this task can benefit from feature auto-encoding, showing how detection performance improves with respect to trivial template matching.

Keywords: Paleography · Graphic symbol detection · Image processing · Machine learning

1 Introduction

A huge number of historical documents from Late Antiquity to early medieval Europe do exist in public databases. The NOTAE project (NOT A writtEn word but graphic symbols) is meant to study graphic symbols, which were added by authors of these documents with several different meanings. This task is very different though from processing words and letters in natural language as the symbols that we look for can be orthogonal to the content, making contextual analysis useless.

Labeling document pictures with positions of graphic symbols even in an unsupervised manner requires the knowledge of domain experts, paleographers in particular. Unfortunately, this task does not scale up well considering the high number of documents. This paper proposes a system that helps curators to

This research is part of the project *NOTAE: NOT A writtEn word but graphic symbols*, which has received funding from the European Research Council (ERC) under the European Union's Horizon 2020 research and innovation program (Advanced Grant 2017, GA n. 786572, PI Antonella Ghignoli). See also http://www.notae-project.eu. Copyright 2021 for this paper by its authors.

The original version of this chapter was revised: chapter 12 made as open access. The correction to this chapter is available at
https://doi.org/10.1007/978-3-030-86198-8_34

E. H. Barney Smith and U. Pal (Eds.): ICDAR 2021 Workshops, LNCS 12916, pp. 147–162, 2021.
https://doi.org/10.1007/978-3-030-86198-8_12

identify potential candidates for different categories of symbols. Researchers are then allowed to revise the annotations in order to improve the performance of the tool in the long run.

A method for symbol detection has already been proposed in the context of the NOTAE project [1]. Here, the authors have created a graphic symbols database and an identification pipeline to assist the curators. The symbol engine takes images as input, then uses the database objects as queries. It detects symbols and reduces noise in the output by clustering the identified symbols. Before this operation, the user is required to decide the binarization threshold from a prepared selection.

The approach proposed in this paper has several differences with the original one. First of all, in this new version of the tool, we rely on OPTICS [2], instead of DBSCAN [3], for clustering purposes. OPTICS uses the hyper-parameters MaxEps, and MinPts almost the same way as DBSCAN, but it distinguishes cluster densities on a more continuous basis. In contrast, DBSCAN considers only a floor for cluster density and filters noise by identifying those objects that are not contained in any cluster. In addition, our proposed pipeline implements an algorithm that sorts the objects of a cluster by confidence scores and selects the top match. So, the pipeline can control the number of predictions over different types of graphic symbols.

Moreover, the first tool has shown an high number of false positive, which are difficult to filter out. Here, we show how automatic detection of symbols can benefit from feature auto-encoding, showing how detection performance improves with respect to trivial template matching.

The paper is organized as it follows. Section 2 summarizes prior work on document analysis and digital paleography tools, metric learning, and graphic symbols spotting. Section 3 present our data cleaning process and image pre-processing tailored to the specific data domain, i.e., ancient documents. Section 4 covers the inner details of the proposed method. Section 5 show experimental results. Finally, Sect. 6 concludes the paper with a final discussion.

2 Related Work

In [7], the authors discuss the recent availability of large-scale digital libraries, where historians and other scientists can find the information they need to help with answering their research questions. However, as they state, researchers are still left with their traditional tools and limitations, and that is why they propose two new tools designed to address the need for document analysis at scale. Firstly, they consider a tool to match handwritings which is applied to documents that are fragmented and collected across tens of libraries. They also note the shortcomings of computer software in recommending matching scores without providing persuasive and satisfactory reasoning for researchers, as the ground truth is itself the subject of study and active research. Secondly, they mention a paleographic classification tool that recommends matching styles and dates with a given writing fragment. According to them, it seems like paleography researchers are interested in the why of recommender systems outputs as much as they value their accuracy.

Variational auto-encoders (VAE) [8] train a model to generate feature embeddings under the assumption that samples of a class should have the same embeddings. In [9], given the intra-class variance such as illumination and pose, the authors challenge that assumption. They believe that minimizing a loss function risks over-fitting on training data by ignoring each class's essential features. Moreover, by minimizing the loss function, the model could learn discriminative features based on intra-class variances. Also, they illustrate how the model struggles to generalize as samples from different classes but with the same set of intra-class variances cluster at the central part of the latent space. In addition to the KL-divergence [10] and reconstruction loss terms as in prior work, they add two metric-based loss terms. One of the new terms helps minimize the distance between samples of the same class in the presence of intra-class variations. The other new loss term prevents intra-class variances, which different classes might share, overpower essential features representing representational value.

Their framework, deep variational metric learning (DVML), disentangles class-specific discriminative features from intra-class variations. Furthermore, per their claim, it significantly improves the performance of deep metric learning methods by experimenting on the following datasets: CUB-200-2011, Cars196, and Stanford Online Products. In this work, we sample from the latent space by calculating the Kaiming-normal function, also known as He initialization [11], and we use that as epsilon to relate the mean and variance of the data distribution.

In [12], the authors focus on the problem of symbol spotting in graphical documents, architectural drawings in particular. They state the problem in the form of a paradox, as recognizing symbols requires segmenting the input image. The segmentation task should be done on a recognized region of interest. Furthermore, they want a model that works on digital schematics and scanned documents where distortions and blurriness are natural. Moreover, they also aim to build an indexing system for engineers and designers who might want to access an old document in an extensive database with a given symbol drawing that could only partially describe the design. Having those considerations in mind, the authors then propose a vectorization technique that builds symbols as a hierarchy of more basic geometric shapes. Then, they introduce a method for tiling the input document picture in a flexible and input-dependent manner. Their approach approximates arcs with low poly segments [13,14], and puts constraints on subsets of line segments such as distance ratios, angles, scales, etc. This way, they can model slight variations in the way that symbol queries build full representational graphs.

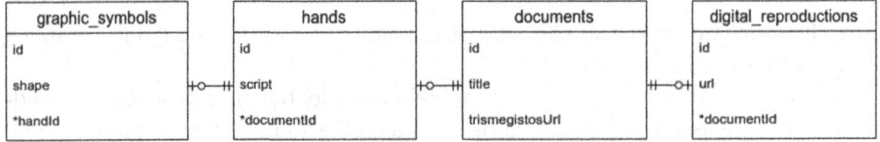

Fig. 1. Related tables in the NOTAE database

3   Data Preprocessing

3.1   Scraping Public Databases

With their expertise and knowledge of the domain, the NOTAE curators have gathered a database to find documents, hands, symbols, and digital reproductions, among much other useful information. The database tables are connected as a knowledge graph [15] (see Fig. 1). For example, a symbol present inside a script has an associated hand. The hand, in turn, comes from a document with an identification number. Then, we can get the digital reproduction of where the symbol comes from using the document ID.

3.2   Cleaning Duplicates

One of the implicit assumptions in dataset design is that sample images are unique. Scraped data is not clean, and it is likely to have duplicates. Web pictures come from different sources with different sizes and compression algorithms for encoding/decoding. So, comparing an image against the rest of the dataset to determine if it is a duplicate will not work most of the time. Using coarser features instead of raw pixels can destroy many trivial details that are not noticeable to the human eyes. Moreover, it is observed to be the case that some digital reproductions of the same picture have been ever so slightly cropped across public databases. So, the features need to be invariant under minor variations in data distribution but different enough between two unique pictures. The difference hash (dHash) algorithm [16] processes images and generates fixed-length hashes based on visual features. dHash has worked outstanding for our use case. In particular, generating 256-bit hashes and then using a relative Hamming distance threshold of 0.25 detects all duplicates. Among duplicate versions of scraped data, we chose the one with a higher resolution. By comparing image hashes against each other, we managed to clean the scraped data and thus create two datasets of unique samples such as graphic symbols and digital reproductions.

Figure 2 in particular shows the difference with respect to the quality of obtained results.

3.3   Binarization

Digital reproductions contain various supports such as papyrus, wooden tablets, slate, and parchment. In addition, due to preservation conditions and the passage of time, parts of the documents have been lost, and we deal with partial

Database A   **Database B**

Fig. 2. The effect of the proposed solution. On the left the result without cleaning duplicates, on the right after the cleaning operation.

observations of ancient texts and symbols. Accordingly, a pre-processing step seems necessary to foreground the handwritten parts and clear the background of harmful features and noise. Then, the issue of what threshold works best for such a diverse set of documents surfaces. In that regard, we follow the prior work [1] and hand-pick one value out of the five prepared threshold values that are input dependent. We find the first two of the threshold values by performing K-means clustering on the input image and then choosing the red channel, which is the most indicative value. Next, we calculate the other three thresholds as linear functions of the first two (taking the average, for example).

Template matching works on each color channel (RGB) separately, and so it returns three normalized correlation values. Consequently, the proper peak-finding function should take the average of them in order to find the location of the most probable box (see further in Sect. 4.2 for more on peak-finding in template matching). However, since document pictures have a wide range of supports with various colors and materials, using color images is optimal, whereas binary images work the best. First, we remove the background using the selected threshold value. Next, we apply the erosion operator to remove noise and the marginal parts further. Finally, we fill the foreground with the black color to get the binary image. In our experiments, the binarization step has proven to be at least an order of magnitude more effective in reducing false positives, compared to when we tried color images.

3.4   Dataset Design

The simple baseline begins with the binarization of document pictures and template matching using the NOTAE graphic symbols database. These two steps make for an end-to-end pipeline already and identify graphic symbols with a given picture (see Fig. 3.) Next, we split our preferred set of unique binarized digital reproductions into three different subsets: train, validation, and test. The partitioning ratio is 80% training data and 10% for each test and validation subsets.

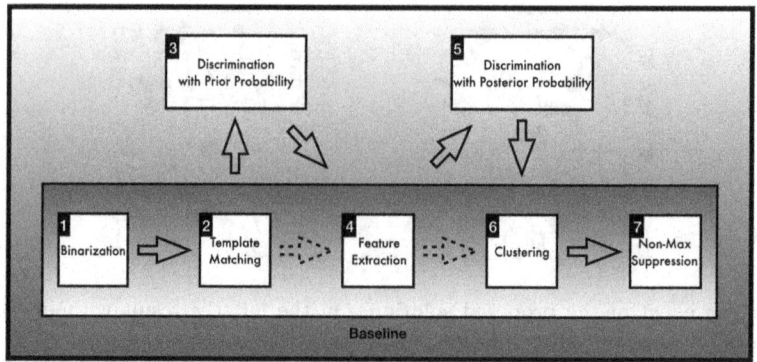

Fig. 3. Identification pipeline

3.5   Initial Symbol Clustering

As discussed in the introduction, in this new version of the annotation tool we moved from DBSCAN to OPTICS for symbol clustering. A description of how OPTICS forms denser clusters follows. First, it defines core-distance as the minimum distance within the Eps-neighborhood of an object such that it satisfies the MaxEps condition. In general, core-distance is less than MaxEps, and that is why there is a MaxEps rather than a fixed Eps in OPTICS. Then, it uses core-distance to define the reachability score as a function of one object concerning another. The reachability of object o with respect to a different object p is defined to be the maximum between either of two values: the core-distance of o or the distance between o and p. Reachability is instead not defined if objects o and p are not connected. Using a cluster expansion loop with the given core distances, OPTICS can reorder database objects between subclusters and superclusters where cluster cores come earlier and noise later. Object ordering plus reachability values prove to be much more flexible than a naive cluster-density condition in the way DBSCAN works.

4   Modeling Approach

Our target is to determine very particular graphic symbols in a digital reproduction and find their positions as smaller rectangles inside the picture frame. The NOTAE database supplies the templates we look for, so the simplest possible model can be a template matching algorithm. It takes a picture and a set of templates as inputs and returns a set of bounding boxes and the confidence scores assigned to each one of them as outputs. Then, one could select the final predictions from the top of the boxes sorted based on their scores. However, in practice, we observed that the simple model also returns too many false positives, bounding boxes with relatively high confidence scores but contain non-symbols. Moreover, the rate of false positives increases as a linear function of the database size. This inefficiency in naive template matching poses a problem since the NOTAE system design relies on the growth of the database for making its suggestions brighter. So, template matching is a simple and fast model for identifying graphic symbols in a document picture, but it has relatively limited precision.

In the identification pipeline, the template matching step is done for every graphic symbols database object. For one object (template), the algorithm returns a field of correlation densities over the input document picture, as many as the number of pixels in the given picture. So we select the one with the maximum score as the final match. Also, template matching uses five different sizes of each object. The scales range from 5% of the picture width up to 20% because that is about the size of the symbols in documents. Hence, the first step of the pipeline produces five bounding boxes per database object.

After template matching is over, we can recover some precision by way of updating confidence scores. Fast template matching is possible by transforming visual data from the spatial dimension to the frequency dimension. One can ignore some high-frequency features to speed up the process and then transform the results back to the spatial dimension. In [17] Fourier transform does so to reduce computation complexity. However, once the algorithm has queried the database and is done with its prediction process, then we can afford to update the confidence scores using a more computationally complex approach that would be quite infeasible right from the beginning. We engineer visual features for both database objects and regions of interest (ROI) for that purpose, as template matching predicts. Suppose u and v are two such features extracted using a method of our choice (u represents a template while v represents an identification ROI, for example.) If we find the lengths of these feature vectors then it becomes easy to see how similar they are:

$$correlation = \frac{<u,v>}{|u| \cdot |v|},$$

$<\cdot,\cdot>$ denotes the inner product on the vector space of features, $|\cdot|$ denotes the length of a vector and the correlation is in the closed interval $[-1, 1]$.

Due to reasons that will be discussed later in this section, we can build features in a particular latent space to preserve the save the same metric from the previous step. In fact, we propose to build a discriminator that uses the correlation between features to update the prediction probabilities and prune away false positives.

We already identified potential candidates for graphic symbols in a document picture, then discriminated against some of them based on engineered features, and finally, filtered outliers based on size. However, all those steps pertain to more individual and local symmetries rather than considering what an ensemble of identifications has in common. That is where clustering and unsupervised classification comes into play and further reduce false positives. Using the same engineered features, be it histogram of oriented gradients (HOG) or learned embeddings, a clustering algorithm can group the identified symbols into one cluster and label the rest of the identifications as noise. In this last major step to improve the results, global symmetries are the main deciding factor as to whether a bounding box should be in the graphic symbols group or not. In the clustering step, individual boxes relate to each other via a distance function. Setting a minimum neighborhood threshold, clusters of specific densities can form, as discussed in the previous section. At the end of every promising identification pipeline, they apply a non-maximum suppression algorithm. In overlapping bounding boxes, those with lower confidence scores are removed in favor of the top match. See Fig. 3 for a representation of our identification pipeline.

4.1   Updating Identification Probabilities

Suppose T, F, M and D be events: T as the event that a box is true positive, F as the event that a box is false positive, M as the event that the template Matching model labels a box as true positive, and D is the event that the Discriminator model labels a box as true positive. Also, suppose MD be the event that both the template Matching and Discriminator models label a box as true positive.

Please note that the sum of prior probabilities should be equal to one.

$$P(T) + P(F) = 1.$$

Next, let's appeal to the Bayes theorem. In the symbol identification task, write down the posterior probabilities of such events occurring:

$$P(T|MD) = \frac{P(MD|T) \cdot P(T)}{P(MD|T) \cdot P(T) + P(MD|F) \cdot P(F)},$$

or, in an equivalent way:

$$P(T|MD) = \frac{P(MD|T) \cdot P(T)}{P(MD)}.$$

First, the template matching model acts on the graphics symbols database. The input document picture is implicit here as it stays constant throughout the pipeline. Then, the template matching model returns one match per pixel in the document picture. A suitable cut-off threshold as a hyper-parameter will reduce the number of symbols based on the confidence scores. So, we only select the top match for each database object (template). Next, the discriminator model acts on the top matches. Furthermore, thus the template-matching model and the discriminator model participate in a function composition at two different levels of abstraction. In this composition, template matching works with raw pixels, whereas discrimination works with high-level embedding vectors.

$$updated\ scores = Discriminate \circ Match(database),$$

where $\circ$ denotes the function composition by first applying $Match$ and then $Discriminate$ on the database, object by object.

If we assume that events M and D are independent (or slightly correlated), then we can say that they are conditionally independent given T.

$$P(MD|T) = P(M|T) \cdot P(D|T)$$

Therefore the updated probability will be:

$$P(T|MD) = \frac{P(M|T) \cdot P(D|T) \cdot P(T)}{P(MD)}$$

Performing some computation to simplify the posterior probability:

$$P(T|MD) = \frac{P(M|T) \cdot P(DT)}{P(MD)}$$

$$P(T|MD) = \frac{P(M|T) \cdot P(T|D) \cdot P(D)}{P(MD)}$$

$$P(T|MD) = \frac{P(M|T) \cdot P(T|D)}{Q(1,2)},$$

where $Q(1,2) = \frac{P(MD)}{P(D)}$.
Since $1 = P(T|MD) + P(F|MD)$, therefore:

$$1 = \frac{P(M|T) \cdot P(T|D) + P(M|F) \cdot P(F|D)}{Q(1,2)}.$$

Now, it is obvious that

$$Q(1,2) = P(M|T) \cdot P(T|D) + P(M|F) \cdot P(F|D)$$

And that conclusion implies that the updated probability is as follows:

$$P(T|MD) = \frac{P(M|T) \cdot P(T|D)}{P(M|T) \cdot P(T|D) + P(M|F) \cdot P(F|D)} \tag{1}$$

Q: Where do we get the value $P(M|T)$ from? **A:** The Average Recall (AR) of the template matching function gives the value for $P(M|T)$. It is the probability that the fast template matching algorithm identifies a symbol given that it is a true symbol. **Q:** Where do we get the value $P(T|D)$ from? **A:** The Average Precision (AP) of the discriminator function gives the value for $P(T|D)$. It is the probability that a symbol is true given that the discriminator model has labeled it positive. **Q:** What does $P(M|F)$ mean? **A:** It is the probability that the template-matching model identifies a symbol given that it is negative. **Q:** What does $P(F|D)$ mean? **A:** It is the probability that a symbol is false given that the discriminator model has labeled it positive.

The template matching model produces potential bounding boxes in a digital reproduction with the graphic symbols database. Next in the pipeline, we use an attention mechanism to discriminate for the boxes that are more likely to be true with the given digital reproduction. The discriminator is indifferent to the location of the query symbol and only cares about whether the matching box is

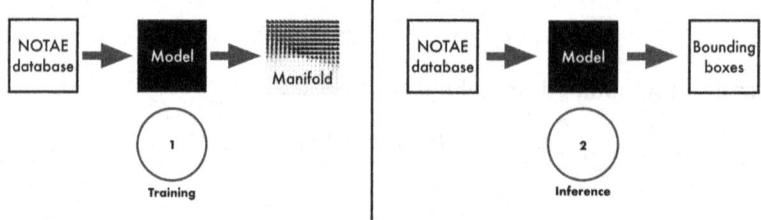

Fig. 4. Filtering noise with low overhead as the inference has lower latency.

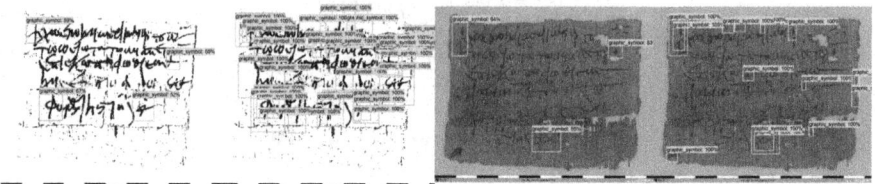

Fig. 5. Identifications on the left side and ground truth on the right.

similar to it or not. Therefore, the discrimination step is an image classification task in essence. Figure 4 shows how the two steps, symbol matching, and classification, share the same database objects. The discrimination model, step 5 in Fig. 3, introduces a posterior probability function $P(T|D)$ and assigns to every box a value from -1 to 1. The sequential update of information now changes first to consider event M and then update with event D.

Finally, we can normalize the discrimination confidence score by adding one unit and dividing it by 2 to get a correct probability value in $[0, 1]$, formally known as an affine transformation. Next, we use it to replace the score from the template matching step. The posterior probability $P(T|MD)$ is correlated to the scores coming from both steps: template matching and discrimination. The rest of the pipeline will work the same (see Fig. 5). In the following subsection, we are going to use this result to focus on the feature engineering that maximizes $P(T|D)$, that is, the true positive rate has given the second event, discrimination.

4.2   Latent Clustering

By now, we have established the probability that a graphic symbol is true given that the discriminator model has labeled it positive works based on a correlation between the source symbol and the target identification. As indicated, we need to look more closely at the choice of metric and distance functions. Because the more accurate we are in determining the actual distance between two objects, the better we can reason about if the two objects in question are related and why.

Suppose the distribution of the graphic symbols database is described by manifold M. Here, we do not assume any structure beyond that there is a probability $p(x)$ that we discover a given object x in it. Except for maybe a smooth frame at x for applying convolutional filters. Since it is a complex manifold, as is the case with most objects in the real world, it could be intractable to explain with a reasonable amount of information. Therefore, we defer to a latent manifold $\tilde{M}$ which is finite-dimensional and could potentially explain the most important aspects that we care about in objects from M. What we need here is a map, such as ϕ, from manifold M into manifold $\tilde{M}$ such that our choice of metric in the latent manifold $\tilde{M}$ results in a predictable corresponding metric in the original manifold M.

Accordingly, we could reason unseen objects knowing that for every input in the domain of graphic symbols manifold, there will be a predictable output in the co-domain of the latent manifold. Predictable in the sense that our metric in the latent space would work as expected. In this context, the encoder model plays the part of the inverse of a smooth map. It maps objects from the pixel space onto the latent space.

$$Encode : pixel\ space \mapsto latent\ space$$

Suppose that p and v are vector representations of an ROI (inside a document picture) and a graphic symbol, respectively. Next, we define a few smooth maps for computing the probabilities of our modeling approach.

$$P(M|T) := \arg\max_{i,j} Match(p_{i,j}, v),$$

given by

$$Match(p_{i,j}, v) = p_{i,j} * v = <\hat{p_{i,j}}, \hat{v}>,$$

The inner product between normalized elements from the template matching sliding window at (i, j) of the input picture and normalized database elements makes sense if both vector spaces are of the same actual dimension. Here i and j are the maximum arguments of the term on the right, which reflect our process of selecting the top match based on confidence scores. We take the maximum value among the inner products so that it corresponds to the most probable location in the document picture.

$$Discriminate(p_{i,j}, v) := P(T|D),$$

given by

$$Discriminate(p_{i,j}, v) = <Encode(p_{i,j}), Encode(v)>.$$

For taking symbols from the pixel space to the latent space (embeddings), we can use the encoder part of a variational auto-encoder (VAE) model. We trained a VAE model on the graphic symbols database in a self-supervised manner to get the embeddings of unseen symbols. The model uses a deep residual architecture (ResNet18 in Fig. 6) [18] and the bottleneck in this neural network would be the latent layer where the features are sampled from.

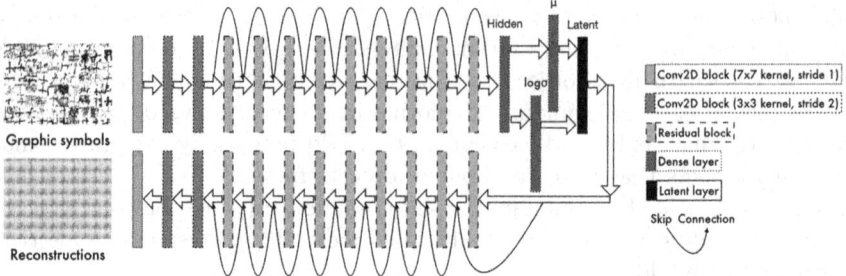

Fig. 6. The VAE encodes graphic symbols, upper row, and decodes them, lower row.

4.3 Optimization Objective

The loss function should look like the following equation since according to equation (1) from earlier in this section, we want the training objective to minimize $P(M|F)$ and $P(F|D)$ while maximizing $P(M|T)$ and $P(T|D)$ (up to a proxy function.)

$$\mathcal{L} = \alpha \cdot reconstruction + \beta \cdot [KL\ divergence],$$

where, α and β are hyper-parameters in $\mathbb{R}$. The reconstruction loss term above is the mean square error of the input image and its decoded counterpart. A point in the latent space should be similar to a sample from the normal distribution if we want the model to learn a smooth manifold. When the latent distribution and the normal distribution are the most alike, the KL-divergence loss term should be approximately equal to zero. Adding the relative entropy loss term to the loss function justifies our assumption on the learned manifold being a smooth one.

5 Quantifying Model Performance

In order to perform evaluation, it is helpful to imagine the annotation tool as a generic function that maps elements from an input domain to the output. In our case, in particular, we want to map tuples of the form (document_picture, symbol) to a bounding box array. As evaluation method, we employed mean Average Precision (mAP) [5], which outputs the ratio of true symbols over all of the identified symbols.

Additionally, we annotated the dataset using the Pascal VOC [6] format in order to evaluate the system using well-established tools.

We used an object detection model by the moniker CenterNet ResNet50 V2 512×512 [19], which was pre-trained on the MS COCO 17 dataset [20]. It is a single-stage detector that has achieved 29.5% mAP with COCO evaluation tools. In order to repurpose it for our work, we generated annotations for 183 unique digital reproductions using our pipeline and then fine-tuned the object detection model on the annotated data. It is not so easy to measure how helpful our approach is using offline training as the model outputs have to be first justified

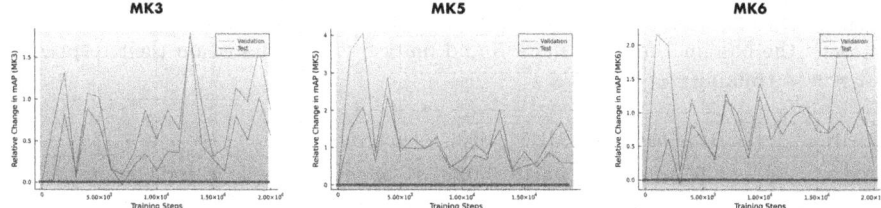

Fig. 7. Improvements in mAP validate the pipeline. The horizontal line is the baseline.

by the model and then interpreted and validated by domain experts. Therefore, the evaluation protocol in this section merely focuses on the coherency and accuracy of the results. The different variants of the identified symbols datasets are partitioned with different ratios and random seeds, so they also serve as a multi-fold testing apparatus. This section considers improvements in precision since it is normal for symbol spotting methods to perform well in terms of recall.

In the spirit of an iterative pipeline design, we generated seven different identified symbols datasets. Using roughly the identical digital reproductions and graphic symbols validated our data and modeling approach. For the baseline, we bypassed steps 3 and 5 in the pipeline (Fig. 3) and also used HOG features to have a model as close as possible to prior work [1]. Next, we used the encoder with a binary classifier and generated mark 3. This modification puts steps 3 and 5 of the pipeline into effect. We have compared the evaluation results of MK3 with that of the baseline model, which is about double the precision, suggesting the effectiveness of the discrimination step in improving the true positive rate. Mark 5 follows the same architecture as mark 3. However, it adds discrimination based on bounding box area and foreground density after discrimination with posterior probability, which further improved the results (compare the third and the fourth columns in Table 1).

Then, we modified the pipeline by training the encoder and hard-wiring a discriminator function to calculate posterior probabilities using cosine similarity. The object detection model trained on the identified symbols mark 6 dataset yielded new evaluation results. MK6 annotations look much better than their predecessors in a qualitative way. Interestingly, MK6 annotations seem to generalize well over different scales (see the bottom image in Fig. 5), as it is the first dataset among the series to identify small symbols as well.

The evaluation of MK3 was when we picked up on the trend that we could gain model performance by focusing more on the data rather than the model. By manually labeling the binarized version of the graphic symbols database, we excluded almost half of the objects as non-symbols to get to a dataset of 722 graphic symbols. So, we should attribute some of the improvements over the baseline model to the data cleaning process. That process called for training the auto-encoder model again with the clean data. Table 2 brings the final improvement rates over the baseline with MK3, MK5, and MK6. We added the validation set to Table 2 and Fig. 7 in order to show that our approach is not sensitive to the choice of hyper-parameters. Because test results are strongly correlated with

Table 1. Symbol identification performance results related to the identified symbols datasets: the baseline, mark 3, mark 5 and mark 6 (all evaluated on their respective test sets at training step 2000.)

Metric	Baseline	MK3	MK5	MK6	Comment
AP	**0.011**	**0.022**	**0.048**	**0.028**	AP at IoU = .50:.05:.95 (primary metric)
AP@IoU = .50	*0.047*	*0.101*	*0.148*	*0.102*	AP at IoU = .50 (PASCAL VOC metric)
AP@IoU = .75	0.000	0.003	0.015	0.012	AP at IoU = .75 (strict metric)
AP@small	0.009	0.000	0.000	0.022	AP for small objects: area $< 32^2$
AP@medium	0.016	0.021	0.028	0.033	AP for medium objects: $32^2 <$ area $< 96^2$
AP@large	0.000	0.051	0.083	0.025	AP for large objects: area $> 96^2$
AR@max=1	0.003	0.008	0.019	0.006	AR given 1 detection per image
AR@max=10	0.024	0.053	0.071	0.033	AR given 10 detections per image
AR@max=100	0.083	0.126	0.158	0.085	AR given 100 detections per image
AR@small	0.075	0.000	0.000	0.021	AR for small objects: area $< 32^2$
AR@medium	0.102	0.130	0.094	0.088	AR for medium objects: $32^2 <$ area $< 96^2$
AR@large	0.000	0.076	0.246	0.086	AR for large objects: area $> 96^2$

Table 2. Guiding the identification pipeline design by measuring the relative change in mAP, dataset to dataset.

Identified symbols dataset	mAP relative change (valid.)	mAP relative change (test)
Mark 3	69%	51%
Mark 5	**102%**	**119%**
Mark 6	80%	86%

validation. MK5 performs at least twice better than the baseline, and so it is a good candidate to replace it as a new baseline. So, we expect it to perform as well on unseen data. The following relation allows us to calculate the relative change in mAP:

$$relative\ change\ in\ mAP = \frac{proposed\ mAP - baseline\ mAP}{baseline\ mAP} \cdot 100\%.$$

Table 2 presents the relative change in mAP while Table 1 puts the main challenge metric into its proper context. As an illustration, mark 5 outperforms the baseline by 102% and 119% in the validation and test subsets, respectively.

6   Conclusions

In this paper, we have shown how the detection scores provided by fast template matching can be the key to annotate extensive databases in an efficient way. In previous work, the idea is that the bigger the database grows, the more brilliant the symbol engine gets. However, more significant databases also cause more false positives due to inefficiencies in template matching. In this work, we first removed duplicates and then hand-picked binarized versions of the scraped images. Then, a series of identified graphic symbols datasets to validate our hypotheses on data and modeling was designed. The confidence scores of symbol matching using a binary classifier where the discriminative features sampled from the latent space as an approximation of the original space updated. Next, we justified our assumptions about the effectiveness of our distance function in providing a metric for filtering false positives. Not only we managed to recover results from the baseline model, but also there was a significant improvement in model performance across validation and test subsets. Even though many false positives make it through the final stage of the pipeline, we illustrated how a trained detection model generalizes well on the annotated data and why it solves the paradox of segmenting for spotting or spotting for segmentation. Our approach applies to intelligent assistants for database curators and researchers. In a domain where labeled data is scarce, we have adopted evaluation metrics that enable researchers to quantify model performance with weakly labeled data.

The fact that modifications to the pipeline have a clear impact on model performance regarding the relative change in mAP helps define a reward function. Based on the behavior of model performance, we believe that the relative change in mAP could introduce a new term to the loss function. In future work, we would like to see agents that can use this metric to fill in the gaps between sparse learning signals from domain experts during interactive training sessions.

References

1. Boccuzzi, M., et al.: Identifying, classifying and searching graphic symbols in the NOTAE system. In: Ceci, M., Ferilli, S., Poggi, A. (eds.) IRCDL 2020. CCIS, vol. 1177, pp. 111–122. Springer, Cham (2020). https://doi.org/10.1007/978-3-030-39905-4_12

2. Ankerst, M., Breunig, M.M., et al.: OPTICS: ordering points to identify the clustering structure, ACM SIGMOD Rec. (1999). https://doi.org/10.1145/304181.304187

3. Ester, M., Kriegel, H.P., et al.: A density-based algorithm for discovering clusters in large spatial databases with noise. In: Proceedings of the 2nd International Conference on Knowledge Discovery and Data Mining, Portland, OR, pp. 226–231. AAAI Press (1996)

4. Huang, J., Rathod, V., et al.: Speed/accuracy trade-offs for modern convolutional object detectors. In: CVPR (2017)

5. COCO detection evaluation. https://cocodataset.org/#detection-eval. Accessed 17 Mar 2021

6. Everingham, M., Winn, J.: The PASCAL Visual Object Classes Challenge 2012 (VOC2012) Development Kit (2012)

7. Wolf, L., Potikha, L., Dershowitz, N., Shweka, R., Choueka, Y.: Computerized paleography: tools for historical manuscripts. In: 18th IEEE International Conference on Image Processing, pp. 3545–3548 (2011). https://doi.org/10.1109/ICIP.2011.6116481

8. Kingma, D.P., Welling, M.: Auto-Encoding Variational Bayes (2014). arXiv:1312.6114

9. Lin, X., Duan, Y., Dong, Q., Lu, J., Zhou, J.: Deep variational metric learning. In: Ferrari, V., Hebert, M., Sminchisescu, C., Weiss, Y. (eds.) ECCV 2018. LNCS, vol. 11219, pp. 714–729. Springer, Cham (2018). https://doi.org/10.1007/978-3-030-01267-0_42

10. Kullback, S., Leibler, R.A.: On information and sufficiency. Ann. Math. Statist. **22**(1) 79–86 (1951). https://doi.org/10.1214/aoms/1177729694

11. He, K., et al.: Delving deep into rectifiers: surpassing human-level performance on ImageNet classification. In: Proceedings of the IEEE International Conference on Computer Vision (2015)

12. Rusiñol, M., Lladós, J.: Symbol spotting in technical drawings using vectorial signatures. In: Liu, W., Lladós, J. (eds.) GREC 2005. LNCS, vol. 3926, pp. 35–46. Springer, Heidelberg (2006). https://doi.org/10.1007/11767978_4

13. Ramer, U.: An iterative procedure for the polygonal approximation of plane curves. Comput. Graph. Image Process. **1**(3), 244–256 (1972). https://doi.org/10.1016/S0146-664X(72)80017-0. ISSN 0146-664X

14. Douglas, D.H., Peucker, T.K.: Algorithms for the Reduction of the Number of Points Required to Represent a Digitized Line or its Caricature (2011). https://doi.org/10.1002/9780470669488.ch2

15. Bernasconi, E., et al.: Exploring the historical context of graphic symbols: the NOTAE knowledge graph and its visual interface. In: IRCDL 2021, pp. 147–154 (2021)

16. Krawetz, N.: Kind of Like That, In: The Hacker Factor Blog (2013). http://www.hackerfactor.com/blog/?/archives/529-Kind-of-Like-That.html. Accessed 29 June 2021

17. Lewis, J.P.: Fast Template Matching, In: Vision Interface 95, Canadian Image Processing and Pattern Recognition Society, Quebec City, Canada, pp. 120–123 (1995)

18. He, K., Zhang, X., et al.: Deep residual learning for image recognition (2015). arXiv:1512.03385

19. Duan, K., Bai, S., et al.: CenterNet: keypoint triplets for object detection (2019). arXiv:1904.08189

20. Lin, T.-Y., et al.: Microsoft COCO: common objects in context. In: Fleet, D., Pajdla, T., Schiele, B., Tuytelaars, T. (eds.) ECCV 2014. LNCS, vol. 8693, pp. 740–755. Springer, Cham (2014). https://doi.org/10.1007/978-3-319-10602-1_48

ICDAR 2021 Workshop on Camera-Based Document Analysis and Recognition (CBDAR)

CBDAR 2021 Preface

We are glad to welcome you to the proceedings of the 9th edition of the International Workshop on Camera-based Document Analysis and Recognition (CBDAR 2021). CBDAR 2021 builds on the success of the previous eight editions held in 2019 (Sydney, Australia), 2017 (Kyoto, Japan), 2015 (Nancy, France), 2013 (Washington DC, USA), 2011 (Beijing, China), 2009 (Barcelona, Spain), 2007 (Curitiba, Brazil), and 2005 (Seoul, South Korea).

CBDAR aims to move away from the comfort zone of scanned paper documents and to investigate the innovative ways of capturing and processing both paper documents and other types of human created information in the world around us, using cameras.

Since the first edition of CBDAR in 2005, the research focus has shifted many times, but CBDAR's mission to provide a natural link between document image analysis and the wider computer vision community, by attracting cutting edge research on camera-based document image analysis, has remained very relevant.

In this 9th edition of CBDAR we received eight submissions, coming from authors in four different countries (Brazil, India, Japan, and the USA). Each submission was carefully reviewed by three expert reviewers. The Program Committee of the workshop was comprised of 12 members. We would like to take this opportunity to thank all the reviewers for their meticulous reviewing efforts.

Taking into account the recommendations of the reviewers, we selected six papers for presentation in the workshop. This results in an acceptance rate of 75% for CBDAR 2021.

We would like to take this opportunity to thank the MDPI Journal of Imaging for sponsoring the best paper award and the CircularSeas project, co-financed by the Interreg Atlantic Area Program through the European Regional Development Fund, for supporting the successful organization of CBDAR 2021.

The participation of attendants from both academia and industry has remained an essential aspect of CBDAR. The program of this edition was carefully crafted to appeal to both. Apart from the presentation of new scientific work, we were pleased to include an invited talk.

We hope that the program of CBDAR 2021 attracted much interest in the community and that the participants enjoyed the workshop which, in order to facilitate participation, was held in a hybrid mode.

We are looking forward meeting again in person at a future CBDAR event.

September 2021

Sheraz Ahmed
Muhammad Muzzamil Luqman

Organization

Workshop Chairs

Sheraz Ahmed   DFKI, Kaiserslautern, Germany
Muhammad Muzzamil Luqman   L3i, La Rochelle University, France

Program Committee

Anna Zhu   Wuhan University of Technology, China
Cheng-Lin Liu   Chinese Academy of Sciences, China
C. V. Jawahar   IIIT, India
Dimosthenis Karatzas   Universitat Autonoma de Barcelona, Spain
Faisal Shafait   National University of Science and
                                  Technology, Pakistan
Joseph Chazalon   EPITA Research and Development Lab.,
                                  France
Kenny Davila   Universidad Tecnológica Centroamericana,
                                  Honduras
Masakazu Iwamura   Osaka Prefecture University, Japan
Muhammad Imran Malik   National University of Science and
                                  Technology, Pakistan
Nibal Nayef   MyScript, France
Petra Gomez-Krämer   L3i, La Rochelle University, France
Umapada Pal   Indian Statistical Institute, Kolkata, India

Inscription Segmentation Using Synthetic Inscription Images for Text Detection at Stone Monuments

Naoto Morita, Ryunosuke Inoue, Masashi Yamada$^{(\boxtimes)}$ (ID), Takatoshi Naka (ID), Atsuko Kanematsu (ID), Shinya Miyazaki (ID), and Junichi Hasegawa (ID)

School of Engineering, Chukyo University, Toyota, Japan
{t317076,t317009}@m.chukyo-u.ac.jp,
{myamada,t-naka,kanematsu,miyazaki,hasegawa}@sist.chukyo-u.ac.jp

Abstract. Stone monuments have historical value, and the inscriptions engraved on them can tell us about the events and people at the time of their installation. Photography is an easy way to record inscriptions; however, the light falling on the monument, the resulting shadows, and the innate texture of the stone can make the text in the photographs unclear and difficult to recognize. This paper presents a method for inferring pixel-wise text areas in a stone monument image by developing a deep learning network that can deduce the shape of kanji characters. Our method uses pseudo-inscription images for training a deep neural network, which is generated by synthesizing a shaded image representing the engraved text and stone texture image. Through experiments using a High Resolution Net (HRNet), we confirm that the HRNet achieves high accuracy in the task of inscription segmentation and that training with pseudo-inscription images is effective in detecting inscriptions on real stone monuments. Thus, synthetic inscription images can facilitate efficient and accurate detection of text on stone monuments, thereby contributing to further history research.

Keywords: Synthetic data · Inscription · Text detection · Deep learning

1 Introduction

Stone monuments have an important place in history research. The inscriptions engraved on them can vividly inform us about the events that occurred at the time of their installation and the life of people living at that time. For example, the "Three Stelae of Kozuke Province" in Japan is a group of historical monuments and an important source of information about cultural interactions in East Asia from the 7th to the 8th century. Photography has been used as an easy way to record these inscriptions. However, the light present at the time of photography, the resulting shadows created, and the basic texture of the stone can make the text in the photographs indistinct and difficult to recognize. This

© Springer Nature Switzerland AG 2021
E. H. Barney Smith and U. Pal (Eds.): ICDAR 2021 Workshops, LNCS 12916, pp. 167–181, 2021.
https://doi.org/10.1007/978-3-030-86198-8_13

paper proposes a method for inferring pixel-wise text areas from the inscription images of stone monuments using deep learning that would help to decipher the script engraved on the stone.

In an inscription image, the text area corresponds to the area on the image that has been engraved according to the shape of the text characters. The image on the left side of Fig. 1 is an actual inscription. Depending on how the light hits the engraved area, a shadow is cast over part of the engraved area; the brightness of the reflected light varies depending on the inclination of the engraved surface. In other words, the shading in the text area is not uniform. Hence, it is difficult to detect the text area using simple binarization, such as Otsu's method. Therefore, regarding the process as a semantic segmentation task would better infer the original shape of the text from the shading. Thus, the current paper refers to this task as inscription segmentation. The inscription segmentation method can infer the shape of the engraved text from the shape of the shadows and shading features, even if the shading of a text area is not uniform. The binary image on the right side of Fig. 1 represents an ideal inscription segmentation that was created manually by checking the inscription image visually. The white pixels indicate the text areas. The goal of this research is to develop a deep learning network that can make such inferences. If the image output from such a network is used in conjunction with the original photo, it can support the task of reading the text. It can also be used as a preprocessing method for automatic text recognition.

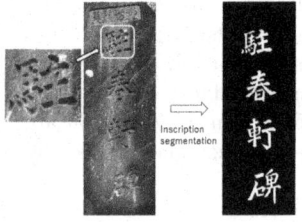

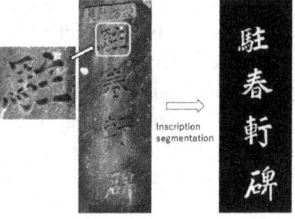

Fig. 1. (left) The engraved parts are not uniform in shade. (right) The white pixels indicate the text areas; This image was created manually by checking the inscription image visually.

Fig. 2. Factors making text obscure: dirt, stone texture, and weathering.

Our target is stone monuments in Japan that contain Chinese characters, 'kanji'. However, deep learning requires sufficient training data and datasets of stone monuments containing kanji are not yet available. In addition, it takes a lot of effort to create a dataset for training from images of actual stone monuments. Therefore, we use computer generated technology to generate pseudo-inscription images, which are then used for deep learning. Generally, it is necessary to learn the character features of all the classes in order to recognize individual characters in an inscription. There are more than 2,000 classes of kanji characters currently used in Japan, and more than 5,000 if we include characters used in older eras. However, for our purpose of detecting a text area, it is not necessary to learn the character features of all the classes because kanji characters are composed of basic strokes. If we can detect the individual strokes, the text area has been detected. Based on this assumption, we created 50 images containing 900 distinct kanji characters in total and generated pseudo-inscription images from them.

The procedure we perform to generate a pseudo-inscription image is as follows: We create a three-dimensional shape representing the engraved characters; we set a light source, render it to generate a shaded image, and synthesize the shaded image with a stone texture image prepared in advance. Using pseudo-inscription images as training data, we aim to obtain a model that is capable of detecting text areas. Note that in actual stone monuments, the factors that obscure the shading of the engraved text are not only the stone texture but also other factors that include dirt and weathering, as seen in Fig. 2. The pseudo-inscription images we generated did not take into account heavy dirt and weathering. In our experiments, we will determine how severe stains and weathering affect the results of inscription segmentation.

To the best of our knowledge, this is the first study that involves training a network using pseudo-inscription images and detecting text engraved on stone monuments. The main contribution of this study can be summarized as follows: (1) We propose a low-cost image synthesis technique suitable for dynamic pseudo-inscription image generation in the learning process. (2) We verify that the network trained with pseudo-inscription images outperforms conventional binarization methods in the task of inscription segmentation.

2   Related Work

The tasks of text detection can be categorized into the task of estimating the region surrounding the text and the segmentation task of detecting pixel-wise areas of individual characters. Baek et al. proposed CRAFT for the former task [2]. Document binarization and inscription segmentation belong to the latter task.

There are several studies that have attempted to perform text detection and text recognition on inscription images. Baht et al. proposed a method for restoring degraded inscription on ancient stone monuments in India using a binarization method based on phase-based feature maps and geodesic morphology [3]. Liu et al. proposed a method for obtaining the bounding box of characters engraved on oracle bones using a conventional model of object detection [12]. Using an entropy-based feature extraction algorithm, Qin et al. proposed another method for obtaining the bounding box of characters from a scene with stone monuments [17].

Binarization is a conventional approach to separate text and background from document images. In the case of historical documents, binarization is not an easy task due to the presence of faded text, ink bled through the page, and stains. Kitadai et al. proposed a deciphering support system using four basic binarization methods [10]. Peng et al. proposed a binarization method incorporating multi-resolution attention that achieved better accuracy than the best model in the ICDAR2017 competition [15,16]. In addition, methods using a two-dimensional morphological network [13] and cascading modular U-nets [9] have been proposed. The difference between inscription segmentation of stone monuments and binarization of document images tasks is that the text on a document image is black or gray with almost constant shading, whereas the text on stone monument images is incomplete and needs to be reconstructed based on shadows and shading. It is also necessary to deal with a variety of stone textures.

The objective of this study is inscription segmentation of Japanese stone monuments. Datasets that can be used for deep learning have not been created or are not publicly available. Therefore, in this study, we generate pseudo-inscription images and use them to train the network. There has been research to artificially generate images of text and use them to increase the accuracy of text recognition. Jaderberg et al. [7] generated text in various fonts, styles, and arrangements, and created a dataset of automatically generated text images that are difficult to recognize. They showed that a network trained on it is effective for text recognition in real scenes. Tensmeyer et al. proposed a DGT-CycleGAN that generated highly realistic synthetic text data [22]. However, these networks are not intended for inscriptions and do not consider engraved text.

The purpose of this study is to verify the effectiveness of our generated pseudo-data for inscription segmentation, select a representative deep network, and optimize it for this purpose. Some of the representative networks that have been proposed for semantic segmentation are a fully convolutional network (FCN), SegNet, Unet, FCDenseNet, DeepLabV3, and High Resolution Net (HRNet) [1,4,8,18–21]. HRNet, in particular, has achieved higher accuracy than the other networks. In this study, we use an HRNet in an optimized form.

3   Proposed Method

In this section, the method of generating pseudo-inscription images and the deep neural network used in this research are described.

3.1   Creating Images of Pseudo-inscription

The pseudo-inscription image is generated using the following procedure.

1. Generate the text image.
2. Generate the three-dimensional mesh data with the text engraved into it.
3. Generate a shaded text image by setting a light source and rendering it.
4. Blend with a stone texture image.

The text image is composed of the handwritten kanji characters found in a calligraphy instruction book [14] containing 1,000 different classes of characters written in three different calligraphy styles: standard ('kaisyo-cho'), cursive ('so-sho'), and running ('gyo-sho'). Of these, 50×18 kai-sho characters are extracted,

18 at a time, and transferred to a single image, making a total of 50 images. The text image is a binary image with a spatial resolution of 320×448. Each of the 50 images is numbered and denoted by $m_1, m_2, \cdots, m_{50}$. The text images created are shown in Fig. 3.

Fig. 3. Fifty binary text images including 18 distinct kanji characters per image, 18×50 distinct kanji characters in total.

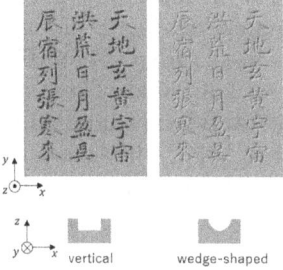

Fig. 4. Three-dimensional model of engraved text. Two types of engraving: vertical (left) and wedge-shaped (right).

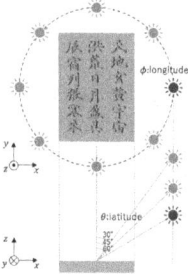

Fig. 5. Polar coordinates of light location $(r, \theta, \phi) : r = \infty, \theta = 30°, 45°, 60°,$ $\phi = 0°, 45°, \cdots, 315°$.

Next, we prepare a planar grid mesh with the same resolution as the text image, and move the mesh vertices corresponding to the interior of the characters in the $-z$-axis direction. This generates a mesh that represents the three-dimensional shape of the engraved character. Two types of engraving methods are adopted, as seen in Fig. 4. The first is vertical engraving, in which all the z-coordinates of the vertices inside the character are set to $z = -c$, where c is an arbitrary constant satisfying $c > 0$. The second is wedge-shaped engraving. For this, the binary text image is distance-transformed to find the distance d from the black pixel; the z-coordinates of the vertices inside the character are set to $z = -(ad)^{\frac{1}{3}}$, where a is an arbitrary constant satisfying $a > 0$. We determine the values of c and a as follows: The stroke width in the binary text images is about

10 pixels. Therefore, we set the value of c to 5, so that the engraving depth of the vertical type is to be half the stroke width. For wedge-shaped type, we set the value of a to 10, so that the maximum depth is to be 3.6 when a stroke width is 10.

Next, the three-dimensional mesh is loaded onto the renderer (Blender 2.83 LTS), parallel light sources are set, and a shaded image is generated. (r, θ, ϕ) is the polar coordinate representation of the position of the light source. All combinations of $r = \infty$, $\theta = 0°, 45°, \cdots, 315°$, and $\phi = 30°, 45°, 60°$ are rendered. This results in 24 shaded images with different shading from a single three-dimensional mesh. Figure 4 shows two examples of shaded images generated with the two different engraving methods. Figure 5 shows the 24 different light source directions (θ, ϕ). The number of the shaded images generated from a single text image m_k is 48 (2 types of engraving methods and 24 types of light source positions). The set of 48 images is denoted by $S_k = \{s_1^k, s_2^k, \cdots s_{48}^k\}$.

The next step synthesizes the shaded image with stone texture. Solid textures are often used to virtually reproduce the stone texture [11]. However, mapping a solid texture to an engraved three-dimensional mesh is computationally more expensive than synthesizing two-dimensional images. In this study, in order to perform data augmentation at a low cost, a pre-prepared image of the stone texture is combined with the shaded image generated earlier. Figure 6 shows the 18 different stone texture images that were actually used. These are free images that are publicly available as stone slab images. When blending the stone textures in the data augmentation method described below, they are randomly scaled to produce a synthetic image with diversity. In this section, we explain the synthesis method. The images used for synthesis are the shaded, stone texture, and binary text images.

s, t, and m represent the shaded, stone texture, and binary text images, respectively. $\overline{m}$ denotes the inverted image of m. The blended image is called a pseudo-inscription image, denoted by p and calculated as:

$$p = \overline{m} \odot s \odot t + \alpha m + (1 - \alpha) m \odot s \odot t, \tag{1}$$

where $s \odot t$ represents the Hadamard product (element-wise product) of the shaded and stone texture images, and $\overline{m} \odot s \odot t$ represents the non-text area, which is $s \odot t$. The text area is composed of a binary text image m and $s \odot t$ with a blend ratio of α. In this study, $0 \le \alpha \le 0.3$ is used. Figure 7 shows an example of a pseudo-inscription image p when $\alpha = 0.0, 0.1, 0.2, 0.3$. There are two reasons to blend the binary text image m into the text area. The first is because the engraved text area is often light gray in actual stone monuments. The second reason is to facilitate the learning process by providing samples in which it is difficult (e.g., $\alpha = 0.0$ in Fig. 7) and easy (e.g., $\alpha = 0.3$ in Fig. 7) to detect text areas during the training phase.

Fig. 6. Eighteen texture images of a stone slab.

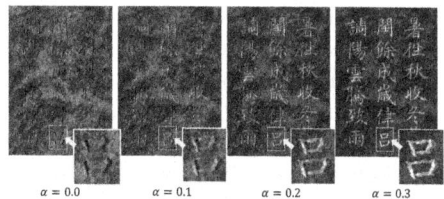

$\alpha = 0.0$ $\alpha = 0.1$ $\alpha = 0.2$ $\alpha = 0.3$

Fig. 7. Pseudo-inscription images created by blending shaded, texture, and binary text images: $\alpha = 0.0, 0.1, 0.2, 0.3$.

3.2   Network Structure

We use an optimized form of an HRNet. The input is a 3-channel image, and the output is a 1-channel image. The structure of the HRNet used in this study is shown in Fig. 8. The numbers below the feature map denote the numbers of channels.

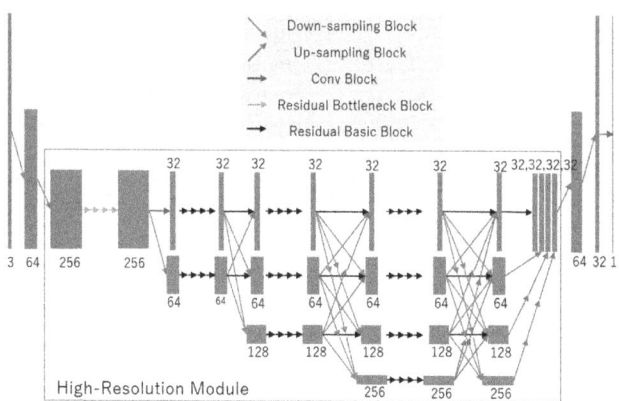

Fig. 8. Structure of HRNet.

This HRNet has two down-sampling blocks and two up-sampling blocks at the beginning and end, respectively, and a high-resolution module in between. Considering the memory limitation of the GPU used in this research, down-sampling is performed first before inputting to the high-resolution module. Down-sampling and up-sampling are designed to increase spatial resolution by a factor of $1/2$ and 2, respectively. The high-resolution module is composed of down-sampling blocks, up-sampling blocks, conv blocks, residual bottleneck blocks and residual basic blocks. The implementation details of these blocks are shown in Table 1. In the case of this implementation, the `HRNet have equal input and output resolutions when the input resolution is a multiple of 32.

The residual block used is not ResNet [6], but the one proposed in Deep Pyramidal Residual Networks [5]. The image input to the high-resolution module first passes through four bottleneck blocks. The image processed in the bottleneck block is input to both the conv and down-sampling blocks and is converted into a 32-channel image with the resolution maintained and a 64-channel image with the resolution halved. Thereafter, in the high-resolution module, the number of image channels is doubled for each down-sampling and halved for each up-sampling. At the transition to each resolution, down-sampling or up-sampling is repeated until the resolution reaches that of the destination image. At that point, the rectified linear unit (ReLU) function is applied after adding all the transitioned images before the ReLU function at the end of the down-sampling and up-sampling blocks. In addition, all residual blocks are basic blocks, except for the four bottleneck blocks immediately after the input to the module. The high-resolution module finally up-samples the low-resolution images to match the resolution of all the images, applies the ReLU function, and outputs their concatenation.

Table 1. Configuration of each block in HRNet.

Down-sampling Blk	Conv Blk	Residual Bottleneck Blk	Residual Basic Blk
conv k:4, s:2, p:1	conv k:3, s:1, p:1	BatchNorm	BatchNorm
BatchNorm	BatchNorm	conv k:1, s:1, p:0	conv k:3, s:1, p:1
ReLU	ReLU	BatchNorm	BatchNorm
	(last block of HRNet)	ReLU	ReLU
Up-sampling Blk	conv k:3, s:1, p:1	conv k:3, s:1, p:1	conv k:3, s:1, p:1
up conv k:4, s:2, p:1		BatchNorm	BatchNorm
BatchNorm		ReLU	
ReLU		conv k:1, s:1, p:0	
		BatchNorm	

4 Experiments

This section describes the two experiments that were conducted as part of this study. In the first experiment, we vary the number of data used for training and observe the relation between the number of training data and accuracy. We assume that it is not necessary to learn all classes of kanji characters for inscription segmentation because if the stroke of a character can be detected then characters that are combinations of strokes can also be detected. The validity of this assumption is verified through this experiment. In the second experiment, the accuracy of the proposed network (which is trained on pseudo-inscription images) in segmenting actual inscription images is measured. Thus, we verify whether the network trained with only pseudo-inscription images can detect the text of real inscriptions.

4.1   Training Data, Validation Data, and Data Augmentation

The input to the network is a pseudo-inscription image that is generated dynamically during the data augmentation process described as follows. The target is a binary text image.

In the training and validation phases, we first specify the range of shaded image sets $S_1, S_2, \cdots S_{50}$ to be used for inscription image generation. In this experiment, we specify 10 sets $S_1, S_2, \cdots, S_{10}$ for the validation data and n sets $S_{11}, S_{12}, \cdots S_{10+n}$ for the training data. For instance, for $n = 5$, the set of 5 from S_{11} to S_{15} are used for training. Each set $S_k = \{s_1^k, \cdots, s_{48}^k\}$ has 48 images. Therefore, a total of $N_{val} = 480$ shaded images are used for the validation phase and $N_{train} = 48n$ shaded images for the training phase.

Pseudo-inscription images to be input to the network are generated by the following data augmentation method.

1. Contrast stretch the shaded image s_i^k so that the minimum and maximum intensities are 0 and 255, respectively. Let s' be the image obtained by this processing.
2. Determine the crop size $(w, h) = (r_0, r_0)$ and crop position c_0 randomly. Obtain the cropped image from s', resize it to 192×192 so that the resolution is to be multiple of 32, and let s'' be the obtained image. Produce the cropped image from the binary text image m_k using the same crop size and crop position, resize it in the same way, and let m' be the obtained image.
3. Randomly select a stone texture image t_j. Randomly determine the crop size (r_1, r_1) and crop position c_1. Obtain the cropped image from t_j. Resize it to 192×192, and let t' be the obtained image.
4. Randomly determine the blending ratio α and synthesize s'', m', and t' as in Eq. 1 to generate a pseudo-inscription image p of size 192×192.
5. Randomly change the hue, saturation, and brightness of p.

Figure 9 shows 16 examples of the results of the aforementioned data augmentation process from step 1 to 4. The binary text image m', generated by step 2, is the target of the network. In the experiment, $96 \leq r_0, r_1 \leq 288$. This is to set the magnification of the scaling transformations from 2/3 to 2.

In the training and validation phases, we also add an image of the stone texture without text as a negative sample to the input data of the network. In this case, the target image is an image where all pixels are black. When the number of data for training and validation is N, the number of negative samples to be added is $0.4N$, which accounts for approximately 29% of the total.

4.2   Implementation Details

The loss function is a mean squared error. In the training phase, the learning rate is 10^{-4}. The optimization algorithm used is adaptive moment estimation. The batch size is 48 and the samples are sorted randomly. All experiments are conducted on Pytorch and run on a workstation with 3.5 GHz, 10-core CPU, 64 G RAM, GeForce RTX 2080 Ti, and Windows 10 OS. For $n = 40$, it takes about 5 h for the training and validation phases of 300 epochs.

Fig. 9. Pseudo-inscription images generated by data augmentation processing.

4.3 Results for Validation Data

The shaded image sets used in the training phase are $S_{11}, S_{12}, \cdots S_{10+n}$. In this experiment, we compare the values of the loss function during training for the five cases: $n = 1, 5, 10, 20,$ and 40. The number of classes of kanji characters appearing in the training data are 18, 90, 180, 360, and 720, respectively. To make the total number of iterations equal, the maximum number of epochs is set to 12000, 2400, 1200, 600, and 300, respectively. The data for verification is $S_1, S_2, \cdots S_{10}$, and the number of classes appearing in it is 180. These classes are different from those of characters for training. The loss during training is shown in Fig. 10. From left to right, these are loss graphs when $n = 1, 5, 10, 20,$ and 40. In each graph, the horizontal axis is the number of epochs and the vertical axis is the loss. The larger the difference in loss between the training and validation data, the smaller n becomes. These overfitting phenomena suggest that the number of classes of kanji characters in the training data is not sufficient. The loss to the validation data becomes smaller as n becomes larger, but there is little difference in the loss when n is 20 and 40. This indicates that the effect of increasing the number of classes in the training data is attenuated.

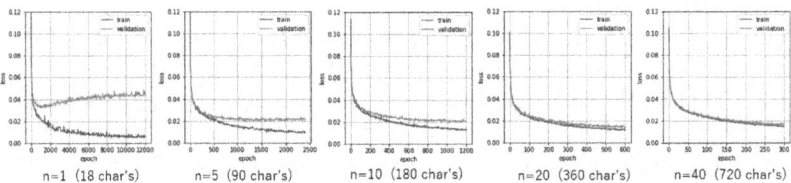

Fig. 10. Loss for training and validation data: From left to right, these are loss graphs when n is 1, 5, 10, 20, and 40.

Table 2 shows the precision, recall, and f-measure for the validation data. To calculate these, a single channel image output from the network is binarized with a threshold of 128. The precision, recall, and f-measure are defined as follows: $precision = \frac{TP}{TP+FP}$, $recall = \frac{TP}{TP+FN}$, and $f\text{-}measure = \frac{2 \cdot precision \cdot recall}{precision+recall}$, where TP, FP, and FN are the number of pixels with true positives, false positives, and false negatives, respectively. Furthermore, when $n = 40$, we train up to a maximum epoch of 500 so that the loss would converge. The bottom row of Table 2 shows the results for $n = 40$ with a maximum of 500 epochs.

Table 2. Accuracy for validation data.

Setting		HRNet		
n	Epochs	Precision	Recall	f-measure
1	12000	0.69	0.54	0.60
5	2400	0.85	0.74	0.79
10	1200	0.86	0.73	0.79
20	600	0.89	0.84	0.86
40	300	0.89	0.78	0.83
40	500	0.92	0.86	0.88

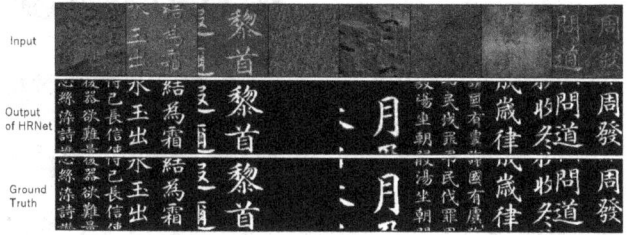

Fig. 11. Inference results for validation data: input (top row), output of HRNet (second row), and ground truth (third row).

The output of the network for the eight inputs of the validation phase is shown in Fig. 11. The first line is the input images before color jittering. The fourth input image from the left is a negative sample with only a stone texture without text. The second line is the outputs of HRNet and the third line is the targets. Through visual evaluation, our HRNet is able to perform sufficient segmentation.

4.4   Results for Real Inscription Images

Next, the results for actual stone monuments are described. The images of actual stone monuments in this experiment were taken at Hakodate Park and Koushoji Temple in Japan. The image of the actual stone monument is resized at 0.05 intervals from 0.1 to 1.8, for a total of 32 different magnifications. The result with the highest value of f-measure is used for evaluation.

Figure 12 shows the actual image of the stone monument, HRNet output, and ground truth. The ground truth was created through visual examination of the monument image; the parts that were difficult to see owing to various factors such as dirt were inferred by using knowledge of kanji. For the stone monument in (a), the text areas of "駐", "春", and "軒" are detected correctly. For the stone monument in (b), the over-detection of the stone texture is prevented, and the darker shadows in the stroke are detected. However, the entire stroke is not detected, and the output stroke is thin in some areas. For the stone monument in (c), text areas that are difficult to see with the naked eye are detected, but there is over-detection at the right edge and top of the output image. For the stone monument in (d), the text in the dirty area in the center is not sufficiently detected. In order to cope with this problem, it is necessary to create and train a pseudo-inscription image that reproduces various kinds of stains.

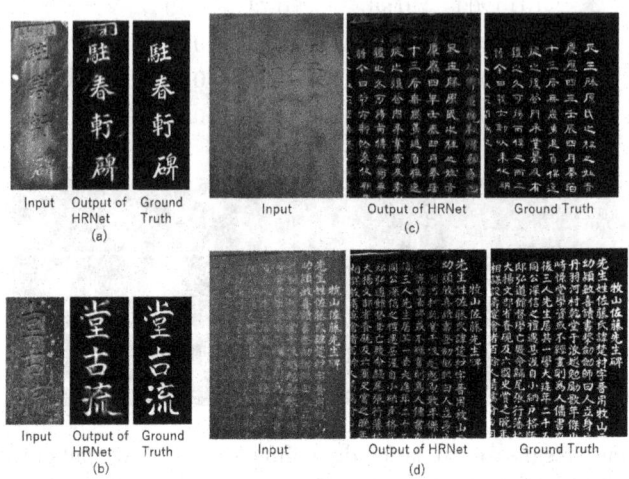

Fig. 12. Inference results for four real inscription images: input (left), output of HRNet (center), and ground truth (right).

The second column of Table 3 shows the precision, recall, and f-measure for the same examples as Fig. 12. The values are inferior to the accuracy for the pseudo-inscription images described earlier. This suggests the network might overfit pseudo-inscription images. However, we believe that the output results are valid enough for the purpose of assisting inscription reading.

Next, we compare the quantitative accuracy of our method with that of binarization methods without learning. Kitadai et al. proposed a deciphering support system [10] using the following four basic binarization methods: (1) converting to grayscale and thresholding, (2) thresholding each BGR component channel and obtaining their logical product, (3) thresholding the difference image from the representative background color, and (4) converting to hsv and thresholding the saturation component channel. In our experiment, two types of thresholding were used: Otsu's method and adaptive thresholding. In addition, the mean color of the input image was used in (3) instead of the representative background color. These four binarization methods and two types of thresholding were applied to the input image to obtain eight binarization results. The results with the highest f-measure among the eight binarization results are shown in the third column of Table 3. For all the four samples, the f-measure of HRNet is better than that of the conventional binarization methods.

Table 3. Accuracy achieved for real inscription images shown in Fig. 12 (a), (b), (c) and (d).

Input		HRNet			Binarization proposed in [10]		
Image	Size	*Precision*	*Recall*	*f-measure*	*Precision*	*Recall*	*f-measure*
a	160 × 480	0.72	0.85	0.78	0.14	0.62	0.23
b	96 × 256	0.74	0.82	0.78	0.16	0.55	0.25
c	576 × 800	0.61	0.73	0.66	0.10	0.44	0.17
d	572 × 768	0.91	0.50	0.65	0.70	0.55	0.62

Figure 13 shows the other three stone monuments and the output results. In (a), part of the character "女" in the second line from the left is illegible due to weathering. The pseudo-inscription image does not currently represent this kind of weathering; hence, the strokes in the weathered area are not detected in the output result. The stone monument in (b) is difficult to read because of the high contrast texture. Many text areas are not detected because of this. The inscription in (c) is written in a peculiar calligraphy style. Our pseudo-inscription image contains only characters in the standard calligraphy style, 'kai-sho', and may need to be adapted to a variety of calligraphy styles.

From these experimental results, we can confirm that inscription segmentation is possible by learning pseudo-inscription images and can be used to support reading. The accuracy for real stone monuments can be further improved by increasing the variation of pseudo-inscription images by increasing the variation of stone texture patterns, calligraphy styles, character placement, and character size. The proposed method assumes that the user can eventually select an appropriate image resolution. In order to reduce the human efforts, the network model needs to be improved to be robust to the variety of input image resolutions. In addition, a more rigorous quantitative evaluation using a dataset with more real images is needed to confirm the practicality of the method.

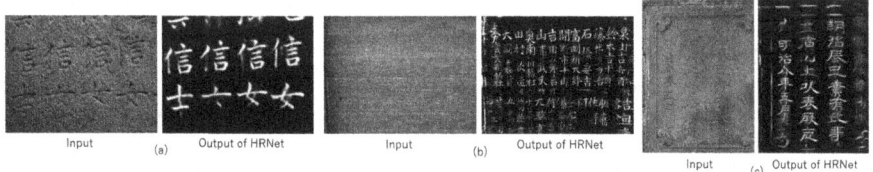

Fig. 13. Inference results for three real stone monuments including weathering (left), intense texture of shades (center), and peculiar calligraphy style (right).

5   Conclusion

In this paper, we proposed a method for generating pseudo-inscription images. The proposed method is a low-cost technique to synthesize shaded images and stone textures, and is suitable for dynamic pseudo-inscription image generation in the learning process. In the experiments, an HRNet was trained using this pseudo-inscription image. We confirmed that inscription segmentation does not necessarily require learning every class of kanji characters and that learning images containing 720 characters can be used for inscription segmentation of 180 other different characters. The network trained on the pseudo-inscription images was applied to the inscription segmentation of real stone monuments. The results showed that the network was capable of inscription segmentation and could be used to support reading. Thus, the proposed approach can contribute to history research involving text detection. However, in order to improve the accuracy for real images, it is necessary to increase the variation of pseudo-inscription images, and deal with dirt and weathering. In addition, in order to prove that every kanji character can be detected, we need to prepare a dataset including more than 5,000 kinds of kanji characters and verify the detection accuracy. A more rigorous quantitative evaluation using a dataset with more real images is also needed to confirm the practicality of the method. In future works, we plan to expand the dataset, and conduct these experiments.

Acknowledgement. This work was supported by JSPS KAKENHI Grant Number JP20H01304.

References

1. Badrinarayanan, V., Kendall, A., Cipolla, R.: SegNet: a deep convolutional encoder-decoder architecture for image segmentation. IEEE Trans. Pattern Anal. Mach. Intell. **39**(12), 2481–2495 (2017)
2. Baek, Y., Lee, B., Han, D., Yun, S., Lee, H.: Character region awareness for text detection. In: Proceedings of the IEEE/CVF Conference on Computer Vision and Pattern Recognition (CVPR), June 2019
3. Bhat, S., Seshikala, G.: Restoration of characters in degraded inscriptions using phase based binarization and geodesic morphology. Int. J. Recent Technol. Eng. **7**(6), 1070–1075 (2019)
4. Chen, L., Papandreou, G., Schroff, F., Adam, H.: Rethinking atrous convolution for semantic image segmentation. CoRR abs/1706.05587 (2017). http://arxiv.org/abs/1706.05587
5. Han, D., Kim, J., Kim, J.: Deep pyramidal residual networks. In: 2017 IEEE Conference on Computer Vision and Pattern Recognition (CVPR), pp. 6307–6315 (2017)
6. He, K., Zhang, X., Ren, S., Sun, J.: Deep residual learning for image recognition. In: 2016 IEEE Conference on Computer Vision and Pattern Recognition (CVPR), pp. 770–778 (2016)
7. Jaderberg, M., Simonyan, K., Vedaldi, A., Zisserman, A.: Synthetic data and artificial neural networks for natural scene text recognition. In: Workshop on Deep Learning, NIPS (2014)

8. Jégou, S., Drozdzal, M., Vazquez, D., Romero, A., Bengio, Y.: The one hundred layers Tiramisu: Fully convolutional DenseNets for semantic segmentation. In: 2017 IEEE Conference on Computer Vision and Pattern Recognition Workshops (CVPRW), pp. 1175–1183 (2017)

9. Kang, S., Iwana, B.K., Uchida, S.: Cascading modular u-nets for document image binarization. In: 2019 International Conference on Document Analysis and Recognition (ICDAR), pp. 675–680 (2019)

10. Kitadai, A., Saito, K., Hachiya, D., Nakagawa, M., Baba, H., Watanabe, A.: Support system for archaeologists to read scripts on Mokkans. In: 2005 International Conference on Document Analysis and Recognition (ICDAR), vol. 2, pp. 1030–1034 (2005)

11. Kopf, J., Fu, C.W., Cohen-Or, D., Deussen, O., Lischinski, D., Wong, T.T.: Solid texture synthesis from 2D exemplars. ACM Trans. Graph. (Proc. SIGGRAPH 2007) **26**(3), 2:1–2:9 (2007)

12. Liu, G., Xing, J., Xiong, J.: Spatial pyramid block for oracle bone inscription detection. In: Proceedings of the 2020 9th International Conference on Software and Computer Applications, pp. 133–140 (2020)

13. Mondal, R., Chakraborty, D., Chanda, B.: Learning 2D morphological network for old document image binarization. In: 2019 International Conference on Document Analysis and Recognition (ICDAR), pp. 65–70 (2019)

14. Ono, G.: Thousand Character Classic in Three Styles, Kai, Gyo and So. Maar-sha (1982). (in Japanese)

15. Peng, X., Wang, C., Cao, H.: Document binarization via multi-resolutional attention model with DRD loss. In: 2019 International Conference on Document Analysis and Recognition (ICDAR), pp. 45–50 (2019)

16. Pratikakis, I., Zagoris, K., Barlas, G., Gatos, B.: ICDAR 2017 competition on document image binarization (DIBCO 2017). In: 2017 14th IAPR International Conference on Document Analysis and Recognition (ICDAR), vol. 01, pp. 1395–1403 (2017)

17. Qin, X., Chu, X., Yuan, C., Wang, R.: Entropy-based feature extraction algorithm for stone carving character detection. J. Eng. **2018**(16), 1719–1723 (2018)

18. Ronneberger, O., Fischer, P., Brox, T.: U-Net: convolutional networks for biomedical image segmentation. In: Navab, N., Hornegger, J., Wells, W.M., Frangi, A.F. (eds.) MICCAI 2015. LNCS, vol. 9351, pp. 234–241. Springer, Cham (2015). https://doi.org/10.1007/978-3-319-24574-4_28

19. Shelhamer, E., Long, J., Darrell, T.: Fully convolutional networks for semantic segmentation. IEEE Trans. Pattern Anal. Mach. Intell. **39**(4), 640–651 (2017)

20. Sun, K., Xiao, B., Liu, D., Wang, J.: Deep high-resolution representation learning for human pose estimation. In: 2019 IEEE/CVF Conference on Computer Vision and Pattern Recognition (CVPR), pp. 5686–5696 (2019)

21. Sun, K., et al.: High-resolution representations for labeling pixels and regions. CoRR abs/1904.04514 (2019)

22. Tensmeyer, C., Brodie, M., Saunders, D., Martinez, T.: Generating realistic binarization data with generative adversarial networks. In: 2019 International Conference on Document Analysis and Recognition (ICDAR), pp. 172–177 (2019)

Transfer Learning for Scene Text Recognition in Indian Languages

Sanjana Gunna$^{(\boxtimes)}$ ⓘ, Rohit Saluja ⓘ, and C. V. Jawahar ⓘ

Centre for Vision Information Technology, International Institute of Information Technology, Hyderabad 500032, India
{sanjana.gunna,rohit.saluja}@research.iiit.ac.in, jawahar@iiit.ac.in,
https://github.com/firesans/STRforIndicLanguages

Abstract. Scene text recognition in low-resource Indian languages is challenging because of complexities like multiple scripts, fonts, text size, and orientations. In this work, we investigate the power of transfer learning for all the layers of deep scene text recognition networks from English to two common Indian languages. We perform experiments on the conventional CRNN model and STAR-Net to ensure generalisability. To study the effect of change in different scripts, we initially run our experiments on synthetic word images rendered using Unicode fonts. We show that the transfer of English models to simple synthetic datasets of Indian languages is not practical. Instead, we propose to apply transfer learning techniques among Indian languages due to similarity in their n-gram distributions and visual features like the vowels and conjunct characters. We then study the transfer learning among six Indian languages with varying complexities in fonts and word length statistics. We also demonstrate that the learned features of the models transferred from other Indian languages are visually closer (and sometimes even better) to the individual model features than those transferred from English. We finally set new benchmarks for scene-text recognition on Hindi, Telugu, and Malayalam datasets from IIIT-ILST and Bangla dataset from MLT-17 by achieving 6%, 5%, 2%, and 23% gains in Word Recognition Rates (WRRs) compared to previous works. We further improve the MLT-17 Bangla results by plugging in a novel correction BiLSTM into our model. We additionally release a dataset of around 440 scene images containing 500 Gujarati and 2535 Tamil words. WRRs improve over the baselines by 8%, 4%, 5%, and 3% on the MLT-19 Hindi and Bangla datasets and the Gujarati and Tamil datasets.

Keywords: Scene text recognition · Transfer learning · Photo OCR · Multi-lingual OCR · Indian languages · Indic OCR · Synthetic data

1 Introduction

Scene-text recognition or Photo-Optical Character Recognition (Photo-OCR) aims to read scene-text in natural images. It is an essential step for a wide

ⓒ Springer Nature Switzerland AG 2021
E. H. Barney Smith and U. Pal (Eds.): ICDAR 2021 Workshops, LNCS 12916, pp. 182–197, 2021.
https://doi.org/10.1007/978-3-030-86198-8_14

Fig. 1. Clockwise from top-left; "**Top:** Annotated Scene-text images, **Bottom:** Baselines' predictions (row-1) and Transfer Learning models' predictions (row-2)", from Gujarati, Hindi, Bangla, Tamil, Telugu and Malayalam. Green, red, and "_" represent correct predictions, errors, and missing characters, respectively. (Color figure online)

variety of computer vision tasks and has enjoyed significant success in several commercial applications [9]. Photo-OCR has diverse applications like helping the visually impaired, data mining of street-view-like images for information used in map services, and geographic information systems [2]. Scene-text recognition conventionally involves two steps; i) Text detection and ii) Text recognition. Text detection typically consists of detecting bounding boxes of word images [4]. The text recognition stage involves reading cropped text images obtained from the text detection stage or from the bounding box annotations [13]. In this work, we focus on the task of text recognition.

The multi-lingual text in scenes is a crucial part of human communication and globalization. Despite the popularity of recognition algorithms, non-Latin language advancements have been slow. Reading scene-text in such low resource languages is a challenging research problem as it is generally unstructured and appears in diverse conditions such as scripts, fonts, sizes, and orientations. Hence a large amount of dataset is usually required to train the scene-text recognition models. Conventionally, the synthetic dataset is used to deal with the problem since a large number of fonts are available in such low resource languages [13]. The synthetic data may also serve as an exciting asset to perform controlled experiments, e.g., to study the effect of transfer learning with the change in script or language text. We investigate such effects for transfer from English to two Indian languages in this work, i.e., Hindi and Gujarati. We also explore the transferability of features among six different Indian languages. We share 2500 scene text word images obtained from over 440 scenes in Gujarati and Tamil to demonstrate such effects. In Fig. 1, we illustrate the sample annotated images from our datasets, and IIIT-ILST and MLT datasets, and the predictions of our models. The overall methodology we follow is that we first generate the synthetic datasets in the six Indian languages. We describe the dataset generation process and motivate the work in Sect. 2. We then train the two deep neural networks we introduce in Sect. 3 on the individual language datasets. Subsequently, we apply transfer-learning on all the layers of different networks from one language to another. Finally, as discussed in Sect. 4, we fine-tune the networks on standard datasets and examine their performance on real scene-text images in Sect. 5. We finally conclude the work in Sect. 6. The summary of our contributions are as follows:

1. We investigate the transfer learning of complete scene-text recognition models i) from English to two Indian languages and ii) among the six Indian languages, i.e., Gujarati, Hindi, Bangla, Telugu, Tamil, and Malayalam.
2. We also contribute two datasets of around 500 word images in Gujarati and 2535 word images in Tamil from a total of 440 Indian scenes.
3. We achieve gains of 6%, 5%, and 2% in Word Recognition Rates (WRRs) on IIIT-ILST Hindi, Telugu, and Malayalam datasets in comparison to previous works [13,20]. On the MLT-19 Hindi and Bangla datasets and our Gujarati and Tamil datasets, we observe the WRR gains of 8%, 4%, 5%, and 3%, respectively, over our baseline models.
4. For the MLT-17 Bangla dataset, we show a striking improvement of 15% in Character Recognition Rate (CRR) and 24% in WRR compared to Bušta et al. [2], by applying transfer-learning from another Indian language and plugging in a novel correction RNN layer into our model.

1.1 Related Work

We now discuss datasets and associated works in the field of photo-OCR.

Works of Photo-OCR on Latin Datasets: As stated earlier, the process of Photo-OCR conventionally includes two steps: i) Text detection and ii) Text recognition. With the success of Convolutional Neural Networks (CNN) for object detection, the works have been extended to text detection, treating words or lines as the objects [12,27,37]. Liao et al. [10] extend such works to real-time detection in scene images. Karatzas et al. [8] and Bušta et al. [1] present more efficient and accurate methods for text detection. Towards reading scene-text, Wang et al. [30] propose an object recognition pipeline based on a ground truth lexicon. It achieves competitive performance without the need for an explicit text detection step. Shi et al. [21] propose a Convolutional Recurrent Neural Network (CRNN) architecture, which integrates feature extraction, sequence modeling, and transcription into a unified framework. The model achieves remarkable performances in both lexicon-free and lexicon-based scene-text recognition tasks. Liu et al. [11] introduce Spatial Attention Residue Network (STAR-Net) with spatial transformer-based attention mechanism to remove image distortions, residue convolutional blocks for feature extraction, and an RNN block for decoding the text. Shi et al. [22] propose a segmentation-free Attention-based method for Text Recognition (ASTER) by adopting Thin-Plate-Spline

(TPS) as a rectification unit. It tackles complex distortions and reduces the difficulty of irregular text recognition. The model incorporates ResNet to improve the network's feature representation module and employs an attention-based mechanism combined with a Recurrent Neural Network (RNN) to form the prediction module. Uber-Text is a large-scale Latin dataset that contains around $117K$ images captured from 6 US cities [36]. The images are available with line-level annotations. The French Street Name Signs (FSNS) data contains around $1000K$ annotated images, each with four street sign views. Such datasets, however, contain text-centric images. Reddy et al. [16] recently release RoadText-1K to introduce challenges with generic driving scenarios where the images are not text-centric. RoadText-1K includes 1000 video clips (each 10 s long at 30 fps) from the BDD dataset, annotated with English transcriptions [32].

Works of Photo-OCR on Non-Latin Datasets: Recently, there has been an increasing interest in scene-text recognition for non-Latin languages such as Chinese, Korean, Devanagari, Japanese, etc. Several datasets like RCTW ($12k$ scene images), ReCTS-25k ($25k$ signboard images), CTW ($32k$ scene images), and RRC-LSVT ($450k$ scene images) from ICDAR'19 Robust Reading Competition (RRC) exist for Chinese [23,25,33,35]. Arabic datasets like ARASTEC (260 images of signboards, hoardings, and advertisements) and ALIF ($7k$ text images from TV Broadcast) also exist in the scene-text recognition community [28,31]. Korean and Japanese scene-text recognition datasets include KAIST ($2,385$ images from signboards, book covers, and English and Korean characters) and DOST ($32k$ sequential images) [5,7]. The MLT dataset available from the ICDAR'17 RRC contains $18k$ scene images (around $1 - 2k$ images per language) in Arabic, Bangla, Chinese, English, French, German, Italian, Japanese, and Korean [15]. The ICDAR'19 RRC builds MLT-19 over top of MLT-17 to contain $20k$ scene images containing text from Arabic, Bangla, Chinese, English, French, German, Italian, Japanese, Korean, and Devanagari [14]. The RRC also provides $277k$ synthetic images in these languages to assist the training. Mathew et al. [13] train the conventional encoder-decoder, where Convolutional Neural Network (CNN) encodes the word image features. An RNN decodes them to produce text on synthetic data for Indian languages. Here an additional connectionist temporal classification (CTC) layer aligns the RNN's output to labels. The work also releases an IIIT-ILST dataset for testing that reports Word Recognition Rates (WRRs) of 42.9%, 57.2%, and 73.4% on $1K$ real images in Hindi, Telugu, and Malayalam, respectively. Bušta et al. [2] proposes a CNN (and CTC) based method for text localization, script identification, and text recognition. The model is trained and tested on 11 languages of MLT-17 dataset. The WRRs are above 65% for Latin and Hangul and are below 47% for the remaining languages. The WRR reported for Bengali is 34.20%. Recently, an OCR-on-the-go model and obtain the WRR of 51.01% on the IIIT-ILST Hindi dataset and the Character Recognition Rate (CRR) of 35% on a multi-lingual dataset containing 1000 videos in English, Hindi, and Marathi [20]. Around 2322 videos in these languages recorded with controlled camera movements like tilt, pan, etc., are additionally shared at https://catalist-2021.github.io/.

Table 1. Statistics of synthetic data. μ, σ represent mean, standard deviation.

Language	# Images	Train	Test	μ, σ word length	# Fonts
English	17.5M	17M	0.5M	5.12, 2.99	>1200
Gujarati	2.5M	2M	0.5M	5.95, 1.85	12
Hindi	2.5M	2M	0.5M	8.73, 3.10	97
Bangla	2.5M	2M	0.5M	8.48, 2.98	68
Tamil	2.5M	2M	0.5M	10.92, 3.75	158
Telugu	5M	5M	0.5M	9.75, 3.43	62
Malayalam	7.5M	7M	0.5M	12.29, 4.98	20

Transfer Learning in Photo-OCR: With the advent of deep learning in the last decade, transfer learning became an essential part of vision models for tasks such as detection and segmentation. [17,18]. The CNN layers pre-trained from the Imagenet classification dataset are conventionally used in such models for better initialization and performance [19]. The scene-text recognition works also use the CNN layers from the models pre-trained on Imagenet dataset [11,21,22]. However, to our best knowledge, there are no significant efforts on transfer learning from one language to another in the field of scene-text recognition, although transfer learning seems to be naturally suitable for reading low resource languages. We investigate the possibilities of transfer learning in all the layers of deep photo-OCR models.

2   Datasets and Motivation

We now discuss the datasets we use and the motivation for our work.

Synthetic Datasets: As shown in Table 1, we generate $2.5M$, or more, word images each in Hindi, Bangla, Tamil, Telugu, and Malayalam[1] with the methodology proposed by Mathew et al. [13]. For each Indian language, we use $2M$ images for training our models and the remaining set for testing. Sample images of our synthetic data are shown in Fig. 2. For English, we use the models pre-trained on the $9M$ MJSynth and $8M$ SynthText images [3,6]. We generate $0.5M$ synthetic images in English with over 1200 fonts for testing. As shown in Table 1, English has a lower average word length than Indian languages. We list the Indian languages in the increasing order of language complexity, with visually similar scripts placed consecutively, in Table 1. Gujarati is chosen as the entry point from English to Indian languages as it has the lowest word length among all Indian languages. Subsequently, like English, Gujarati does not have a top-connector line that connects different characters to form a word in Hindi and Bangla (refer to Fig. 1 and 2). Also, the number of Unicode fonts available in Gujarati is

---

[1] For Telugu and Malayalam, our models trained on $2.5M$ word images achieved results lower than previous works, so we generate more examples equal to Mathew et al. [13].

Fig. 2. Clockwise from top-left: synthetic word images in Gujarati, Hindi, Bangla, Tamil, Telugu, & Malayalam. Notice that a top-connector line connects the characters to form a word in Hindi or Bangla. Some vowels and characters appear above and below the generic characters in Indian languages, unlike English.

fewer than those available in other Indian languages. Next, we choose Hindi, as Hindi characters are similar to Gujarati characters and the average word length of Hindi is higher than Gujarati. Bangla has comparable word length statistics with Hindi and shares the property of the top-connector line with Hindi. Still, we keep it after Hindi in the list as its characters are visually dissimilar and more complicated than Gujarati and Hindi. We use less than 100 for fonts in Hindi, Bangla, and Telugu. We list Tamil after Bangla because these languages share similar vowels' appearance (see the glyphs above general characters in Fig. 2). Tamil and Malayalam have the highest variability in word length and visual complexity compared to other languages. Please note that we have over 150 fonts available in Tamil.

Real Datasets: We also perform experiments on the real datasets from IIIT-ILST, MLT-17, and MLT-19 datasets (refer to Sect. 1.1 for these datasets). To enlarge scene-text recognition research in complex and straight forward low-resource Indian Languages, we release 500 and 2535 annotated word images in Gujarati and Tamil. We crop the word images from 440 annotated scene images, which we obtain by capturing and compiling Google images. We illustrate sample annotated images of different datasets in Fig. 1. Similar to MLT datasets, we annotate the Gujarati and Tamil datasets using four corner points around each word (see Tamil image at bottom-right of Fig. 1). IIIT-ILST dataset has two-point annotations leading to an issue of text from other words in the background of a cropped word image as shown in the Hindi scene at the top-middle of Fig. 1.

Motivation: As discussed earlier in Sect. 1.1, most of the scene-text recognition works use the pre-trained Convolutional Neural Networks (CNN) layers for improving results. We now motivate the need for transfer learning of the com-

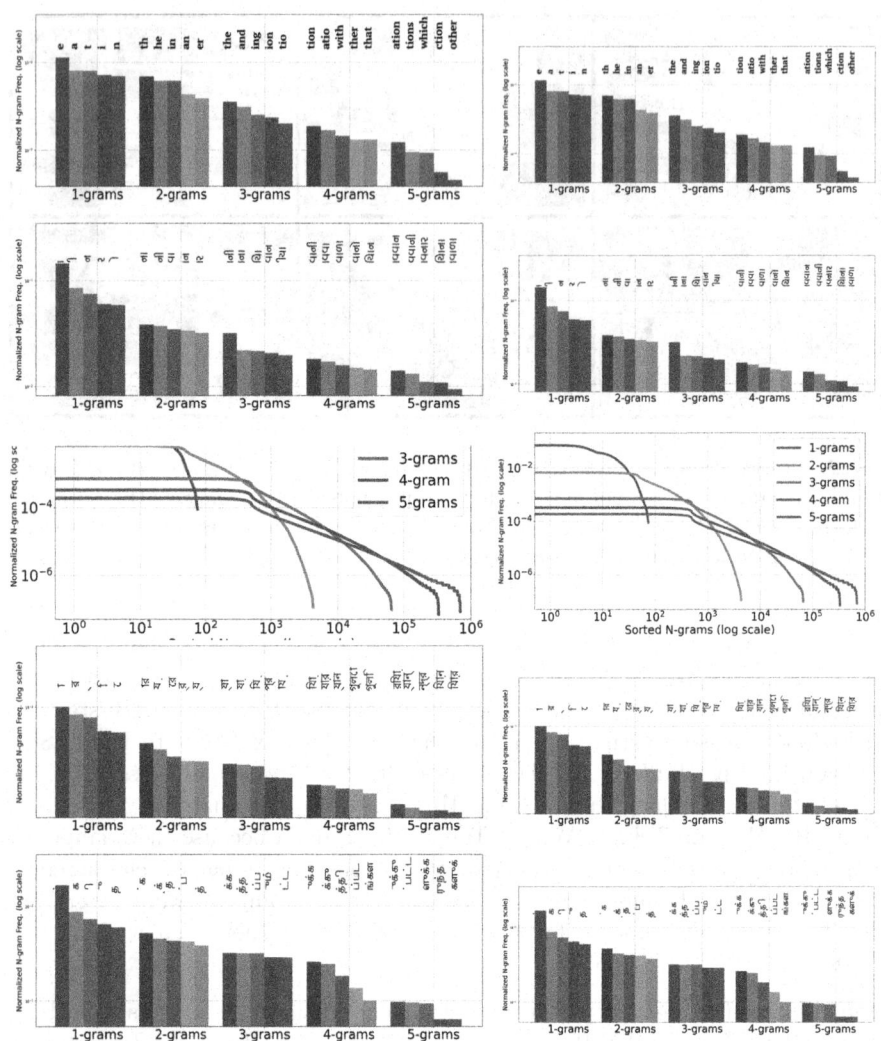

Fig. 3. Distribution of Char. n-grams ($n \in [1, 5]$) from $2.5M$ words in English, Gujarati, Hindi, Bangla, and Tamil (top to bottom): Top-5 (left) and All (right).

plete recognition models discussed in Sect. 1 and the models we use in Sect. 3 among different languages. As discussed in these sections, the Recurrent Neural Networks (RNNs) form another integral component of such reading models. Therefore, we illustrate the distribution of character-level n-grams they learn in Fig. 3[2] for the first five languages we discussed in the previous section (we notice that the last two languages also follow the similar trend). On the left, we show

---

[2] For plots on the right, we use moving average of 10, 100, 1000, 1000, 1000 for 1-grams, 2-grams, 3-grams, 4-grams, and 5-grams, respectively.

the frequency distribution of top-5 n-grams, ($n \in [1, 5]$). On the right, we show the frequency distribution of all n-grams with $n \in [1, 5]$. We use $2.5M$ words from each language for these plots. We consider both capital and small letters separately for English, as it is crucial for the text recognition task. Despite this, we note that top-5 n-grams are composed of small letters. The Indian languages, however, do not have small and capital letters like English. However, the total number of English letters (given that small letters are different from capitals) is of the same order as Indian languages. The x-values (≤ 100) for the drops in 1-gram plots (blue curves) of Fig. 3 also illustrates this. So it becomes possible to compare the distributions. Next, we note that most of the top-5 n-grams comprise vowels for all the languages. Moreover, the overall distributions are similar for all the languages. Hence, we propose that the RNN layers' transfer among the models of different languages is worth an investigation.

It is important to note the differences between the n-grams of English and Indian languages. Many of the top-5 n-grams in English are the complete word forms, which is not the case with Indian languages owing to their richness in inflections (or fusions) [29]. Also, note that the second and the third 1-gram for Hindi and Bangla in Fig. 3 (left), known as Halanta, is a common feature of top-5 Indic n-grams. The Halanta forms an essential part of joint glyphs or *aksharas* (as advocated by Vinitha et al. [29]). In Figs. 1 and 2, the vowels, or portions of the joint glyphs for word images in Indian languages, often appear above the top-connector line or below the generic consonants. All this, in addition to complex glyphs in Indian languages, makes transfer learning from English to Indian languages ineffective, which is detailed in Sect. 5. Thus, we also investigate the transferability of features among the Indic scene-text recognition models in the subsequent sections.

3   Models

This section explains the two models we use for transfer learning in Indian languages and a plug-in module we propose for learning the correction mechanism in the recognition systems.

CRNN Model: The first model we train is Convolutional-Recurrent Neural Network (CRNN), which is the combination of CNN and RNN as shown in Fig. 4 (left). The CRNN network architecture consists of three fundamental components, i) an encoder composed of the standard VGG model [24], ii) a decoder consisting of RNN, and iii) a Connectionist Temporal Classification (CTC) layer to align the decoded sequence with ground truth. The CNN-based encoder consists of seven layers to extract feature representations from the input image. The model abandons fully connected layers for compactness and efficiency. It replaces standard squared pooling with 1×2 sized rectangular pooling windows for 3rd and 4th max-pooling layer to yield feature maps with a larger width. A two-layer Bi-directional Long Short-Term Memory (BiLSTM) model, each with a hidden size of 256 units, then decodes the features. During the training phase, the CTC layer provides non-parameterized supervision to align the decoded predictions

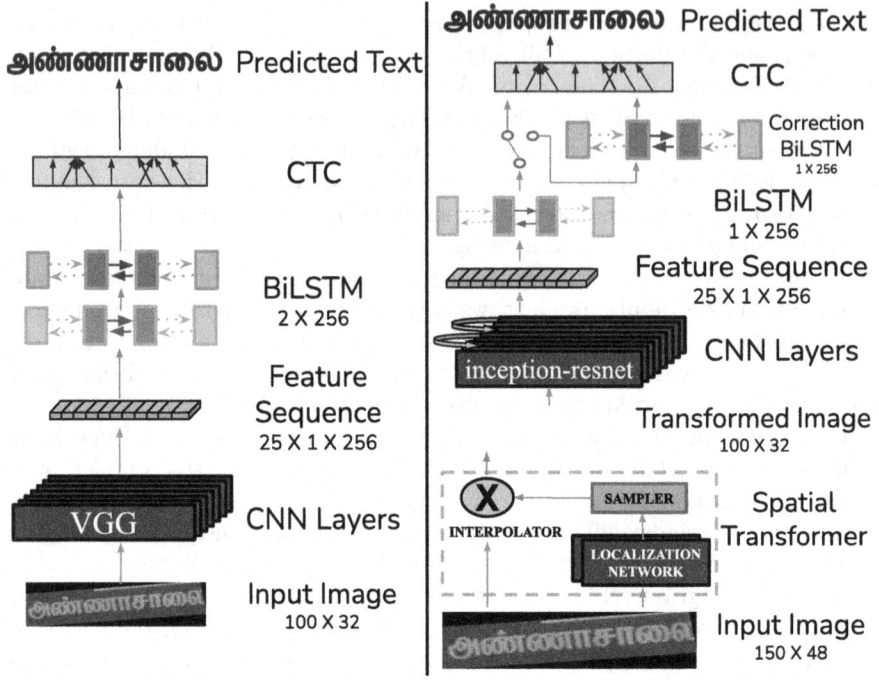

Fig. 4. CRNN model (left) and STAR-Net with a correction BiLSTM (right).

with the ground truth. The greedy decoding is used during the testing stage. We use the PyTorch implementation of the model by Shi et al. [21].

STAR-Net: As shown in Fig. 4 (right), the STAR-Net model consists of three components, i) a Spatial Transformer to handle image distortions, ii) a Residue Feature Extractor consisting of a residue CNN and an RNN, and iii) a CTC layer to align the predicted and ground truth sequences. The transformer consists of a spatial attention mechanism achieved via a CNN-based localization network, a sample, and an interpolator. The localizer predicts the parameters of an affine transformation. The sampler and the nearest-neighbor interpolator use the transformation to obtain a better version of the input image. The transformed image acts as the input to the Residue Feature Extractor, which includes the CNN and a single-layer BiLSTM of 256 units. The CNN used here is based on the inception-resnet architecture, which can extract robust image features required for the task of scene-text recognition [26]. The CTC layer finally provides the non-parameterized supervision for text alignment. The overall model consists of 26 convolutional layers and is end-to-end trainable [11].

Correction BiLSTM: After training the STAR-Net model on a real dataset, we add a correction BiLSTM layer (of size 1×256), an end-to-end trainable module, to the end of the model (see Fig. 4 top-right). We train the complete model again on the same dataset to implicitly learn the error correction mechanism.

Table 2. Results of individual CRNN & STAR-Net models on synthetic datasets.

Language	CRNN-CRR	CRNN-WRR	STAR-Net-CRR	STAR-Net-WRR
English	77.13	38.21	**86.04**	**57.28**
Gujarati	94.43	81.85	**97.80**	**91.40**
Hindi	89.83	73.15	**95.78**	**83.93**
Bangla	91.54	70.76	**95.52**	**82.79**
Tamil	82.86	48.19	**95.40**	**79.90**
Telugu	87.31	58.01	**92.54**	**71.97**
Malayalam	92.12	70.56	**95.84**	**82.10**

4   Experiments

The images, resized to 150×18, form the input of STAR-Net. The spatial transformer module, as shown in Fig. 4 (right), then outputs the image of size 100×32. The inputs to the CNN Layers of CRNN and STAR-Net are of the same size, i.e., 100×32, and the output size is $25 \times 1 \times 256$. The STAR-Net localization network has four plain convolutional layers with 16, 32, 64, and 128 channels. Each layer has the filter size, stride, and padding size of 3, 1, and 1, followed by a 2×2 max-pooling layer with a stride of 2. Finally, a fully connected layer of size 256 outputs the parameters which transform the input image. We train all our models on $2M$ or more synthetic word images as discussed in Sect. 2. We use the batch size of 16 and the ADADELTA optimizer for stochastic gradient descent (SGD) for all the experiments [34]. The number of epochs varies between 10 to 15 for different experiments. We test our models on 0.5 M synthetic images for each language. We use the word images from IIIT-ILST, MLT-17, and MLT-19 datasets for testing on real datasets. We fine-tune the Bangla models on 1200 training images and test them on 673 validation images from the MLT-17 dataset to fairly compare with Bušta et al. [1]. Similarly, we fine-tune only our best Hindi model on the MLT-19 dataset and test it on the IIIT-ILST dataset to compare with OCR-on-the-go (since it is also trained on real data) [20]. To demonstrate generalizability, we also test our models on 3766 Hindi images and 3691 Bangla images available from MLT-19 datasets [14]. For Gujarati and Tamil, we use 75% of word images to fine-tune our models and the remaining 25% for testing.

5   Results

In this section, we discuss the results of our experiments with i) individual models for each language, ii) the transfer learning from English to two Indian languages, and iii) the transfer learning from one Indian language to another.

Performance on Synthetic Datasets: It is essential to compare the results on synthetic datasets of different languages sharing common backgrounds, as it provides a good intuition about the difficulty in reading different scripts. In Tables 2

Table 3. Results of transfer learning (TL) on synthetic datasets. Parenthesis contain results from Table 2. TL among Indic scripts improves STAR-Net results.

Language	CRNN-CRR	CRNN-WRR	STAR-Net-CRR	STAR-Net-WRR
English → Gujarati	92.71 (**94.43**)	77.06 (**81.85**)	97.50 (**97.80**)	90.90 (**91.40**)
English → Hindi	88.11 (**89.83**)	70.12 (**73.15**)	94.50 (**95.78**)	80.90 (**83.93**)
Gujarati → Hindi	**91.98** (89.83)	73.12 (**73.15**)	**96.12** (95.78)	**84.32** (83.93)
Hindi → Bangla	91.13 (**91.54**)	70.22 (**70.76**)	**95.66** (95.52)	**82.81** (82.79)
Bangla → Tamil	81.18 (**82.86**)	44.74 (**48.19**)	**95.95** (95.40)	**81.73** (79.90)
Tamil → Telugu	87.20 (**87.31**)	56.24 (**58.01**)	**93.25** (92.54)	**74.04** (71.97)
Telugu → Malayalam	90.62 (**92.12**)	65.78 (**70.56**)	94.67 (**95.84**)	77.97 (**82.10**)

and 3, we present the results of our experiments with synthetic datasets. As noted in Table 2, the CRNN model achieves the Character Recognition Rates (CRRs) and Word Recognition Rates (WRRs) of i) 77.13% and 38.21% in English and ii) above 82% and 48% on the synthetic dataset of all the Indian languages (refer to columns 1 and 2 of Table 2). The low accuracy on the English synthetic test set is due to the presence of more than 1200 different fonts (refer Sect. 2). Nevertheless, using a large number of fonts in training helps in generalizing the model for real settings [3,6]. The STAR-Net achieves remarkably better performance than CRNN on all the datasets, with the CRRs and WRRs above 90.48 and 65.02 for Indian languages. The reason for this is spatial attention mechanism and powerful residual layers, as discussed in Sect. 3. As shown in columns 3 and 5 of Table 2, the WRR of the models trained in Gujarati, Hindi, and Bangla are higher than the other three Indian languages despite common backgrounds. The experiments show that the scripts in latter languages pose a tougher reading challenge than the scripts in former languages.

We present the results of our transfer learning experiments on the synthetic datasets in Table 3. The best individual model results from Table 2 are included in parenthesis for comparison. We begin with the English models as the base because the models have trained on over 1200 fonts and $17M$ word images as discussed in Sect. 2, and are generic. However, in the first two rows of the table, we note that transferring the layers from the model trained on the English dataset to Gujarati and Hindi is inefficient in improving the results compared to the individual models. The possible reason for the inefficiency is that Indic scripts have many different visual and slightly different n-gram characteristics from English, as discussed in Sect. 2. We then note that as we try to apply transfer learning among Indian languages with CRNN (rows 3–7, columns 1–2 in Table 3), only some combinations work well. However, with STAR-Net (rows 3–7, columns 3–4 in Table 3), transfer learning helps improve results on the synthetic dataset from a simple language to a complex language[3]. For Malayalam, we observe that the individual STAR-Net model is better than the one transferred from Telugu, perhaps due to high average word length (refer Sect. 2).

---

[3] We also discovered experiments on transfer learning from a tricky language to a simple one to be effective but slightly lesser than the reported results.

Table 4. Results on real datasets. FT indicates fine-tuned.

Language	Dataset	# Images	Model	CRR	WRR
Gujarati	Ours	125	CRNN	84.93	64.80
			STAR-Net	85.63	64.00
			STAR-Net Eng→Guj	78.48	60.18
			STAR-Net Hin→Guj	**88.47**	**69.60**
Hindi	IIIT-ILST	1150	Mathew et al. [13]	75.60	42.90
			CRNN	78.84	46.56
			STAR-Net	78.72	46.60
			STAR-Net Eng→Hin	77.43	44.81
			STAR-Net Guj→Hin	79.12	47.79
			OCR-on-the-go [20]	–	51.09
			STAR-Net Guj→Hin FT[a]	**83.64**	**56.77**
Hindi	MLT-19	3766	CRNN	86.56	64.97
			STAR-Net	86.53	65.79
			STAR-Net Guj→Hin	**89.42**	**72.96**
Bangla	MLT-17	673	Bušta et al. [2]	68.60	34.20
			CRNN	71.16	52.74
			STAR-Net	71.56	55.48
			STAR-Net Hin→Ban	72.16	57.01
			W/t Correction BiLSTM	**83.30**	**58.07**
Bangla	MLT-19	3691	CRNN	81.93	74.26
			STAR-Net	82.80	77.48
			STAR-Net Hin→Ban	**82.91**	**78.02**
Tamil	Ours	634	CRNN	**90.17**	70.44
			STAR-Net	89.69	71.54
			STAR-Net Ban→Tam	89.97	**72.95**
Telugu	IIIT-ILST	1211	Mathew et al. [13]	**86.20**	57.20
			CRNN	81.91	58.13
			STAR-Net	82.21	59.12
			STAR-Net Tam→Tel	82.39	**62.13**
Malayalam	IIIT-ILST	807	Mathew et al. [13]	**92.80**	73.40
			CRNN	84.12	70.36
			STAR-Net	91.50	72.73
			STAR-Net Tel→Mal	92.70	**75.21**

[a] Fine-tuned on MLT-19 dataset as discussed earlier. We fine-tune all the layers.

Performance on Real Datasets: Table 4 depicts the performance of our models on the real datasets. At first, we observe that for each Indian language, the overall performance of the individual STAR-Net model is better than the individual CRNN model (except for Gujarati and Hindi, where the results are very close). Based on this and similar observations in the previous section, we present the results of transfer learning experiments on real datasets only with the STAR-Net model[4]. Next, similar to the previous section, we observe that the transfer learning from English to Gujarati and Hindi IIIT-ILST datasets (rows 3

---

[4] We also tried transfer learning with CRNN; STAR-Net was more effective.

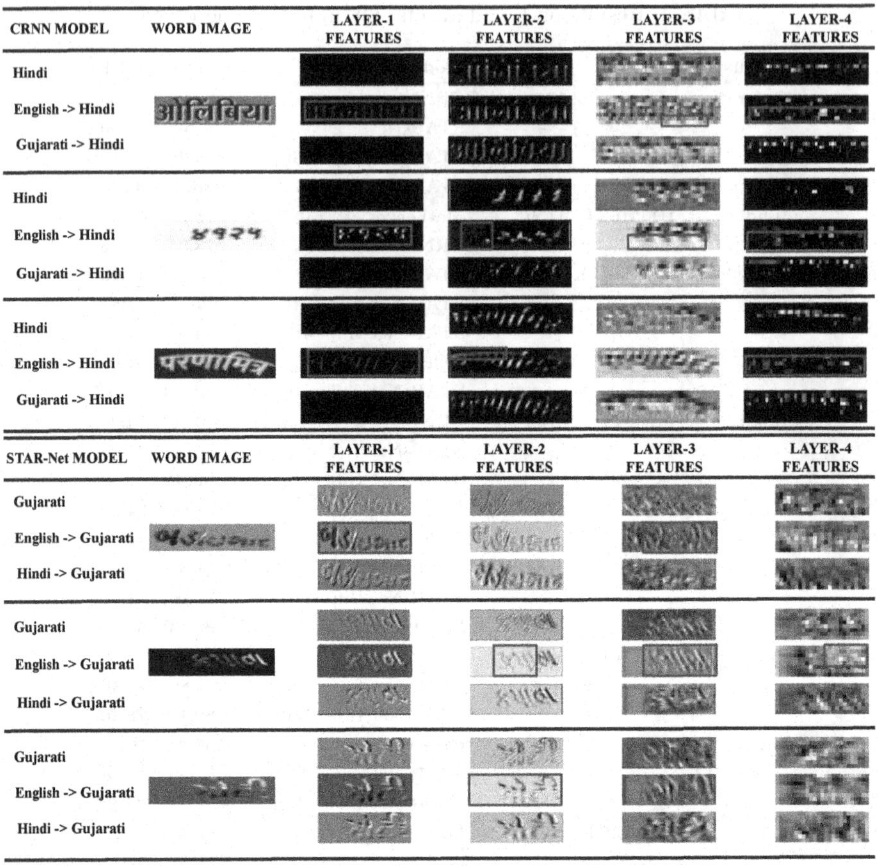

Fig. 5. CNN Layers visualization in the **Top:** CRNN models trained on Hindi, English→Hindi, and Gujarati→Hindi; and **Bottom:** STAR-Net models trained on Gujarati, English→Gujarati, and Hindi→Gujarati. Red boxes indicate the regions where the features for the model transferred from English are activated (as white), whereas the features from the other two models are not. (Color figure online)

and 8 in Table 4) is not as effective as individual models in these Indian languages (rows 2 and 7 in Table 4). Finally, we observe that the performance improves with the transfer learning from a simple language to a complex language, except for Hindi→Gujarati, for which Hindi is the only most straightforward choice. We achieve performance better than the previous works, i.e., Bušta et al. [1], Mathew et al. [13], and OCR-on-the-go [20]. Overall, we observe the increase in WRRs by 6%, 5%, 2% and 23% on IIIT-ILST Hindi, Telugu, and Malayalam, and MLT-17 Bangla datasets compared to the previous works. On the MLT-19 Hindi and Bangla datasets, we achieve gains of 8% and 4% in WRR over the baseline individual CRNN models. On the datasets we release for Gujarati and Tamil, we improve the baselines by 5% and 3% increase in WRRs. We present the qualitative results of our baseline CRNN models as well as best transfer learning models in Fig. 1. The green and red colors represent the correct predictions and errors, respectively. "_" represents the missing character. As can be seen, most of the mistakes are single-character errors.

Since we observe the highest gain of 23% in WRR (and 4% in CRR) for the MLT-17 Bangla dataset (Table 4), we further try to improve these results. We plug in the correction BiLSTM (refer Sect. 3) to the best model (row 18 of Table 4). The results are shown in row 19 of Table 4. As shown, the correction BiLSTM improves the CRR further by a notable margin of 11% since the BiLSTM works on character level. We also observe the 1% WRR gain, thereby achieving the overall 24% WRR gain (and 15% CRR gain) over Bušta et al. [1].

Features Visualization: In Fig. 5 for the CRNN model (top three triplets), we visualize the learned CNN layers of the individual Hindi model, the "English →Hindi" model, and the "Gujarati→Hindi" model. The red boxes are the regions where the first four CNN layers of the model transferred from English to Hindi are different from the other two models. The feature visualization again strengthens our claim that transfer from the English reading model to any Indian language dataset is inefficient. We notice a similar trend for the Gujarati STAR-Net models, though the initial CNN layers look very similar to word images (bottom three triplets in Fig. 5). The similarity also demonstrates the better learnability of STAR-Net compared to CRNN, as observed in previous sections.

6   Conclusion

We generated 2.5M or more synthetic images in six different Indian languages with varying complexities to investigate the language transfers for two scene-text recognition models. The underlying view is that the transfer of image features is standard in deep models, and the transfer of language text features is a plausible and natural choice for the reading models. However, we observe that transferring the generic English photo-OCR models (trained on over 1200 fonts) to Indian languages is inefficient. Our models transferred from one Indian language to another perform better than the previous works or the new baselines we created for individual languages. We, therefore, set the new benchmarks for scene-text recognition in low-resource Indian languages. The proposed Correction BiLSTM, when plugged into the STAR-Net model and trained end-to-end, further improves the results.

References

1. Bušta, M., Neumann, L., Matas, J.: Deep textspotter: an end-to-end trainable scene text localization and recognition framework. In: ICCV (2017)
2. Bušta, M., Patel, Y., Matas, J.: E2E-MLT - an unconstrained end-to-end method for multi-language scene text. In: Carneiro, G., You, S. (eds.) ACCV 2018. LNCS, vol. 11367, pp. 127–143. Springer, Cham (2019). https://doi.org/10.1007/978-3-030-21074-8_11

3. Gupta, A., Vedaldi, A., Zisserman, A.: Synthetic data for text localisation in natural images. In: Proceedings of the IEEE Conference on Computer Vision and Pattern Recognition, pp. 2315–2324 (2016)
4. Huang, Z., Zhong, Z., Sun, L., Huo, Q.: Mask R-CNN with pyramid attention network for scene text detection. In: WACV, pp. 764–772. IEEE (2019)
5. Iwamura, M., Matsuda, T., Morimoto, N., Sato, H., Ikeda, Y., Kise, K.: Downtown Osaka scene text dataset. In: Hua, G., Jégou, H. (eds.) ECCV 2016, Part I. LNCS, vol. 9913, pp. 440–455. Springer, Cham (2016). https://doi.org/10.1007/978-3-319-46604-0_32
6. Jaderberg, M., Simonyan, K., Vedaldi, A., Zisserman, A.: Synthetic data and artificial neural networks for natural scene text recognition. In: Workshop on Deep Learning, NIPS (2014)
7. Jung, J., Lee, S., Cho, M.S., Kim, J.H.: Touch TT: scene text extractor using touchscreen interface. ETRI J. **33**(1), 78–88 (2011)
8. Karatzas, D., et al.: ICDAR 2015 competition on robust reading. In: 2015 ICDAR, pp. 1156–1160. IEEE (2015)
9. Lee, C.Y., Osindero, S.: Recursive recurrent nets with attention modeling for OCR in the wild. In: CVPR, pp. 2231–2239 (2016)
10. Liao, M., Shi, B., Bai, X., Wang, X., Liu, W.: Textboxes: a fast text detector with a single deep neural network. In: AAAI, pp. 4161–4167 (2017)
11. Liu, W., Chen, C., Wong, K.Y.K., Su, Z., Han, J.: STAR-Net: a spatial attention residue network for scene text recognition. In: BMVC, vol. 2 (2016)
12. Liu, X., Liang, D., Yan, S., Chen, D., Qiao, Y., Yan, J.: FOTS: fast oriented text spotting with a unified network. In: Proceedings of the IEEE Conference on Computer Vision and Pattern Recognition, pp. 5676–5685 (2018)
13. Mathew, M., Jain, M., Jawahar, C.: Benchmarking scene text recognition in Devanagari, Telugu and Malayalam. In: ICDAR, vol. 7, pp. 42–46. IEEE (2017)
14. Nayef, N., et al.: ICDAR2019 robust reading challenge on multi-lingual scene text detection and recognition-RRC-MLT-2019. In: 2019 International Conference on Document Analysis and Recognition (ICDAR), pp. 1582–1587. IEEE (2019)
15. Nayef, N., et al.: Robust reading challenge on multi-lingual scene text detection and script identification - RRC-MLT. In: 14th ICDAR, vol. 1, pp. 1454–1459. IEEE (2017)
16. Reddy, S., Mathew, M., Gomez, L., Rusinol, M., Karatzas, D., Jawahar, C.: RoadText-1K: text detection & recognition dataset for driving videos. In: 2020 IEEE International Conference on Robotics and Automation (ICRA), pp. 11074–11080. IEEE (2020)
17. Ren, S., He, K., Girshick, R., Sun, J.: Faster R-CNN: towards real-time object detection with region proposal networks. arXiv preprint arXiv:1506.01497 (2015)
18. Romera, E., Alvarez, J.M., Bergasa, L.M., Arroyo, R.: ERFnet: efficient residual factorized ConvNet for real-time semantic segmentation. IEEE Trans. Intell. Transp. Syst. **19**(1), 263–272 (2017)
19. Russakovsky, O., et al.: Imagenet large scale visual recognition challenge. Int. J. Comput. Vis. **115**(3), 211–252 (2015)
20. Saluja, R., Maheshwari, A., Ramakrishnan, G., Chaudhuri, P., Carman, M.: OCR On-the-Go: robust end-to-end systems for reading license plates and street signs. In: 15th IAPR International Conference on Document Analysis and Recognition (ICDAR), pp. 154–159. IEEE (2019)
21. Shi, B., Bai, X., Yao, C.: An end-to-end trainable neural network for image-based sequence recognition and its application to scene text recognition. IEEE Trans. Pattern Anal. Mach. Intell. **39**(11), 2298–2304 (2016)

22. Shi, B., Yang, M., Wang, X., Lyu, P., Yao, C., Bai, X.: ASTER: an attentional scene text recognizer with flexible rectification. IEEE Trans. Pattern Anal. Mach. Intell. **41**, 2035–2048 (2018)
23. Shi, B., et al.: ICDAR2017 competition on reading Chinese text in the wild (RCTW-17). In: 14th ICDAR, vol. 1, pp. 1429–1434. IEEE (2017)
24. Simonyan, K., Zisserman, A.: Very deep convolutional networks for large-scale image recognition. arXiv preprint arXiv:1409.1556 (2014)
25. Sun, Y., et al.: ICDAR 2019 competition on large-scale street view text with partial labeling - RRC-LSVT. In: 2019 International Conference on Document Analysis and Recognition (ICDAR), pp. 1557–1562. IEEE (2019)
26. Szegedy, C., Ioffe, S., Vanhoucke, V., Alemi, A.: Inception-v4, Inception-Resnet and the impact of residual connections on learning. In: Proceedings of the AAAI Conference on Artificial Intelligence, vol. 31 (2017)
27. Tian, Z., Huang, W., He, T., He, P., Qiao, Yu.: Detecting text in natural image with connectionist text proposal network. In: Leibe, B., Matas, J., Sebe, N., Welling, M. (eds.) ECCV 2016, Part VIII. LNCS, vol. 9912, pp. 56–72. Springer, Cham (2016). https://doi.org/10.1007/978-3-319-46484-8_4
28. Tounsi, M., Moalla, I., Alimi, A.M., Lebouregois, F.: Arabic characters recognition in natural scenes using sparse coding for feature representations. In: 13th ICDAR, pp. 1036–1040. IEEE (2015)
29. Vinitha, V., Jawahar, C.: Error detection in indic OCRs. In: 2016 12th IAPR Workshop on Document Analysis Systems (DAS), pp. 180–185. IEEE (2016)
30. Wang, K., Babenko, B., Belongie, S.: End-to-end scene text recognition. In: ICCV, pp. 1457–1464. IEEE (2011)
31. Yousfi, S., Berrani, S.A., Garcia, C.: ALIF: a dataset for Arabic embedded text recognition in TV broadcast. In: 2015 13th International Conference on Document Analysis and Recognition (ICDAR), pp. 1221–1225. IEEE (2015)
32. Yu, F., Xian, W., Chen, Y., Liu, F., Liao, M., Madhavan, V., Darrell, T.: BDD100K: a diverse driving video database with scalable annotation tooling. arXiv preprint arXiv:1805.04687, vol. 2, no. 5, p. 6 (2018)
33. Yuan, T., Zhu, Z., Xu, K., Li, C., Mu, T., Hu, S.: A large Chinese text dataset in the wild. J. Comput. Sci. Technol. **34**(3), 509–521 (2019)
34. Zeiler, M.: ADADELTA: an adaptive learning rate method, p. 1212 (December 2012)
35. Zhang, R., et al.: ICDAR 2019 robust reading challenge on reading Chinese text on signboard. In: 2019 International Conference on Document Analysis and Recognition (ICDAR), pp. 1577–1581. IEEE (2019)
36. Zhang, Y., Gueguen, L., Zharkov, I., Zhang, P., Seifert, K., Kadlec, B.: Uber-text: a large-scale dataset for optical character recognition from street-level imagery. In: SUNw: Scene Understanding Workshop-CVPR, vol. 2017 (2017)
37. Zhou, X., et al.: East: an efficient and accurate scene text detector. In: CVPR, pp. 2642–2651 (2017)

How Far Deep Learning Systems for Text Detection and Recognition in Natural Scenes are Affected by Occlusion?

Aline Geovanna Soares⓪, Byron Leite Dantas Bezerra$^{(\boxtimes)}$⓪,
and Estanislau Baptista Lima⓪

University of Pernambuco, Recife, Brazil
{ags4,ebl2}@ecomp.poli.br, byron.leite@upe.br
http://ppgec.upe.br

Abstract. With the rise of deep learning, significant advances in scene text detection and recognition in natural images have been made. However, the severe impact threat to the algorithm's performance caused by occlusion still consists of an open issue due to the lack of consistent real-world datasets, richer annotations, and evaluations in the specific occlusion problem. Therefore, unlike previous works in this field, our paper addresses occlusions in scene text recognition. The goal is to evaluate the effectiveness and efficiency of existing deep architectures for scene text detection and recognition in various occlusion levels. First, we investigated state-of-the-art scene text and recognition methods and evaluated these current deep architectures performances on ICDAR 2015 dataset without any generated occlusion. Second, we created a methodology to generate large datasets of scene text in natural images with ranges of occlusion between 0 and 100%. From this methodology, we produced the ISTD-OC, a dataset derivated from the ICDAR 2015 database to evaluate deep architectures under different levels of occlusion. The results demonstrated that these existing deep architectures that have achieved state-of-the-art are still far from understanding text instances in a real-world scenario. Unlike the human vision systems, which can comprehend occluded instances by contextual reasoning and association, our extensive experimental evaluations show that current scene text recognition models are inefficient when high occlusions exist in a scene. Nevertheless, for scene text detection, segmentation-based methods, such as PSENet and PAN, are more robust in predicting higher levels of occluded texts. In contrast, methods that detect at the character level, such as CRAFT, are unsatisfactory to heavy occlusions. When it comes to recognition, attention-based methods that benefit contextual information have performed better than CTC-based methods.

Keywords: Occlusion · Text scenes · Deep learning

Supported by Foundation for the Support of Science and Technology of the State of Pernambuco (FACEPE), Coordenação de Aperfeiçoamento de Pessoal de Nível Superior - Brasil (CAPES) - Finance Code 001, and CNPq - Brazilian agencies.

E. H. Barney Smith and U. Pal (Eds.): ICDAR 2021 Workshops, LNCS 12916, pp. 198–212, 2021.
https://doi.org/10.1007/978-3-030-86198-8_15

1   Introduction

Text is one of humanity's most influential inventions and has played an essential role in human life, used for communication and information transfer, so far from ancient times. The text has rich and precise semantic information embodied that is very useful in a wide range of computer vision-based applications such as robotics, image search, automotive assistance, text localization, text identification, and so on [3,15]. Therefore, text detection and recognition in natural scenes have become vital and active research topics in computer vision and document analysis. One of the primary reasons for this is the rapid developments in camera-based applications on portable devices such as smartphones and tablets, which have facilitated the acquisition and processing of large numbers of images with text every day [11]. Due to the increase in high-performance and low-cost image capturing devices, natural scene text recognition (STR) applications rapidly expand and become more popular. Therefore, in the past few years, several techniques have been explored to solve scene text recognition. Although several remarkable breakthroughs have been made in the pipeline, text recognition in natural images is still challenging, caused by the significant variations of scene text in color, font, orientations, languages, spatial layout, uncontrollable background, camera resolution, partial occlusion, motion blurriness, among others problems [19,28].

In the last few years, deep learning methods, notably Convolutional Neural Network (CNN), have become popular in the computer vision community by substantially advancing the state-of-the-art of various tasks such as image classification, object detection, but also in many other practical applications, including scene text recognition [25]. Recently, with excellent performance on generic visual recognition, CNN has become one of the most attractive methods for scene text recognition, and many methods have been proposed, for instance, PSENet [25], EAST [30], PAN [26], and CRAFT [3]. Such methods have obtained state-of-the-art results in terms of accuracy in benchmarks such as ICDAR 2015 [10], MSRA-TD500 [27] and COCO-Text [22]. However, some of these existing methods failed in some complex cases, such as partial occlusion, arbitrarily shaped and curved texts, which are difficult to represent with a single rectangle. Furthermore, in several studies such as those carried out by Raisi et al. [19], partial occlusion appears as an open issue. It is a thick, contiguous, and spatially additive noise that partially hides one object from the other and represents a severe threat to a pattern recognition system's performance, hindering the observation and reconstruction of information in a reliable manner [29]. Also, few researchers focus on overcoming this challenge instead of improving scene text recognition performance by extracting more robust and effective visual features [1,17].

In this paper, we address the issue of occlusions in scene text identification. We first investigated and evaluated the effectiveness of four popular deep architectures for scene text detection, i.e., PAN [26], CRAFT [3], PSENet [25], EAST [30], and others four for scene text recognition task, i.e., ROSETTA [4], RARE [21], CRNN [20], and STAR-Net [12] on ICDAR 2015 dataset without any generated occlusion. Second, we present as a contribution of this work a methodology to implement occlusion and generate large datasets of occluded scene text in natural images from benchmarks such as ICDAR 2015. This brand new dataset, named Incidental Scene Text Dataset - Occlusion (ISTD-OC), contains 15000 images for detection and 65450 cropped word images for recognition, with levels of occlusions in an interval that ranges from zero to a hundred percent. We also streamline the research over the ISTD-OC dataset by conducting a comparative evaluation on the eight state-of-the-art text identification algorithms reviewed, a baseline reference for future research in the specific occlusion problem.

The goal is to provide a dataset with different occlusion levels to explore how far these existing deep architectures perceive texts in different occlusion levels. To present the results achieved through our analysis, the remaining parts of this paper are arranged as follows: In Sect. 2, we briefly summarize the state-of-the-art deep models for scene text detection and recognition, and take a look at some scene text datasets like ICDAR 2015 dataset, warning the need of specific data for a treat the occlusion problem. In Sect. 3, we show how the occlusion methodology was created to generate the ISTD-OC dataset. Section 4 indicate some evaluation protocols and a discussion about the algorithms' performances. Finally, in Sect. 5, we synthesize our contribution and present future work ideas.

2   Literature Review

Scene text detection and recognition in natural images have received much attention during the past decades in the computer vision community. This section introduces existing scene text recognition methods, categorized into two main categories: text detection and text recognition in natural scenes. We briefly review both classical machine learning approaches and deep learning approaches from recent years. Besides, benchmark datasets will also be presented.

2.1   Text Detection in Natural Scenes

Text detection is still a popular and active research area in the computer vision field. Text detection aims to determine text from the input image, often represented by a bounding box [11].

Most previous methods are usually combining handcraft feature extraction techniques and machine learning models. These features are used to discriminate region or non-region of text scene image, which can be based in sliding window and connected component methods [11,13]. Mishra et al. [14] developed a standard sliding window model to detect characters' potential locations in natural scenes using standards sliding window methods and character aspect ratio prior. Connected component-based methods first extract candidate components from the image using, for instance, similar properties such as color clustering, boundaries, or textures and then filter out non-text components using manually designed rules or automatically trained classifiers on handcrafted features [11,19]. There are two usual methods, i.e., stroke width transform (SWT) and maximally stable extremal regions (MSER), which are more efficient and robust than sliding windows schemes [11]. However, the mentioned classical methods require multiple complicated sequential steps, demanding and repetitive pre-processing and post-processing steps to not quickly propagate errors to other levels. Furthermore, these methods might fail in front of challenging situations such as detecting text under low illumination, text with multiple connected characters, and occlusion [11,13,19].

On the other hand, most current methods have explored deep learning models to detect texts in natural scenes [15]. Since deep learning techniques have advantages over the traditional models, simplifying pipelines and expanding the capacity of the system of generalization, becoming gradually the mainstream [19,30]. The last deep learning-based text detection methods were inspired by object detection structure and can be classified as bounding-box regression-based and segmentation-based approaches [19].

Bounding-box regression-based methods for text detection aim to predict the candidate bounding boxes regarding text, such as usually is done with an object [19]. In EAST [30], a highly simplified pipeline and efficient to perform inference at real-time speed are proposed. FCN is applied to detect text regions directly without using the steps of candidate aggregation and word partition, and then NMS is used to detect word or line text. The model prediction occurs through rotated boxes or quadrangles of words or text lines at each point in the text region.

Segmentation-based methods in text detection aim to classify text regions according to pixel level in images [19]. Wang et al. [25] proposed the progressive scale expansion network (PSENet) to detect adjacent text instances better—the algorithm aims to find kernels with multiple scales and separate text instances close to each other accurately. In CRAFT [3], character affinity maps were used to connect detected characters into a single word and a weakly supervised framework to train a character-level detector. However, this model requires many training images, which increases the time processing and can be challenging for platforms with limited resources. Wang et al. [26] proposed an efficient and accurate arbitrary-shaped text detector, named Pixel Aggregation Network (PAN), which is featured by a low computational-cost segmentation and learnable post-processing. The architecture can introduce multi-level information to guide better segmentation.

2.2 Text Recognition in Natural Scenes

Text recognition in natural scenes is an essential task in computer vision, with many applications. However, text recognition in natural images is still challenging [5]. Therefore, to address these problems, researchers have proposed numerous methods for this task in a few years in the face of extensive studies.

In the past two decades, most classical methods follow bottom-up or top-down approaches to classify characters are linked up into words or to recognize words directly [19]. For example, in [24] HOG features are first extracted from each sliding window, then a pre-trained nearest neighbor or SVM classifier is applied to classify the input characters word image. However, in bottom-up schemes, the low representation capability of handcrafted features cannot achieve either sufficient recognition accuracy and building models that can handle text recognition in the wild. Using top-down methods requires that all input images are inside the word-dictionary dataset to better recognition rate [19].

Recent advances in deep neural network architectures boosted the performance of scene text recognition models [28] bringing improvements in hardware systems, automatic feature learning, generalization, and the capacity of running complex algorithms in real-time [5].

To translate a text instance in computer-readable strings sequences, one of the main approaches is segmentation-free methods. The idea is mapping the entire text instance image into a target string sequence directly, through an encoder-decoder framework, thus, avoiding character segmentation [5]. These models generally contain image processing stages, feature representation, sequence modeling, and prediction [5]. Two major approaches are CTC-based and attention-based. For instance, [4,12,23] have used connectionist temporal classification (CTC) [7] for prediction of character sequences.

The development of many attention-based methods improved STR models with applications to 2D prediction problems. The implicit language model's construction applies implicit attention automatically to enhance in-depth features in the decoding process [5]. Besides, they also work to improve contextual understanding, as in STAR-Net.

Both CTC and attention-based approaches have limitations and advantages. For instance, the attention-based approaches can achieve higher recognition accuracy on isolated word recognition tasks but perform worse on sentence recognition than CTC-based approaches. Therefore, the right prediction methods differ according to different application scenarios and constraints [5,11,13].

2.3 Datasets of Text in Natural Images

Most research suggests that deep learning approaches need many data to demonstrate good performances [28]. However, labeling and collecting training data is usually costly [6]. Two main categories of datasets include those with real-world images and those containing synthetic images [28]. Internet images, street view, google images, and google glasses are some kinds of images in these datasets [28]. For instance, SVT is collected from Google Street View and contains 647 images for evaluation and 245 for training. Many of the images are noisy, blurry, or low-resolution and contain perspective projections augmented by annotations for numbers and text, respectively [3]. SynthText (ST) [8] is an example of the synthetically generated dataset and was initially designed for scene text detection.

For all mentioned above detection and recognition approaches, it is heeded that the primary purpose is to achieve better performances in challenges such as arbitrary shape, adjacent instances, and languages but none of the methods proposed in state of the art focuses on images where a part of the text is missing due to occlusion. Baek et al. [2] highlight how each of the works differs in constructing and using their datasets and investigate the bias caused by the inconsistency when comparing performance between different works. The study shows that prior works have used different sets of training datasets and combined real-world and synthetic data and call into question whether these inconsistencies are due to the proposed module's contribution or better or more extensive training data. Thus, in order of a fair comparison with the work done in the literature, we propose in this work a methodology to apply partial occlusions in different levels to the benchmark dataset ICDAR 2015 [10], intending to evaluate baseline algorithms under text occlusion images without inconsistencies and disparity.

ICDAR 2015 Dataset. The ICDAR15 dataset, named "Incidental Scene Text," was created for the Robust Reading Competition of ICDAR 2015. It is commonly used as a reference for the assessment of text detection or recognition schemes. The detection part has 1,500 images in total that consists of 1,000 training and 500 testing images for detection, and the recognition part consists of 4468 images for training and 2077 images for testing. This irregular dataset contains real-world images of text on signboards, books, posters, and other objects with world-level axis-aligned bounding box annotations captured by Google Glasses. Moreover, text instances are blurred or skewed since they are acquired without users' prior preference or intention. Thus, many are noisy, rotated, and low resolution and complex background conditions.

3   ISTD-OC Dataset

There are some ways to generate occlusions in scenes from the real world, such as [18]. Our work collected 1500 images for detection and 4468 images for recognition from the ICDAR 2015 dataset, where the core idea is to deliver images with a range of partial occlusion in a certain degree of randomness.[1]

To do this, we developed a systematic methodology in which, for each image, word instances were occluded differently, as shown in Fig. 1. The prediction was made considering the proportion of the text size to the original image. Although the methodology suggests implementing regular occlusions in the form of a rectangle, the proposal is sufficient for the initial analysis of current models' generalization ability in the face of occlusion.

Therefore, occlusion corresponds to a random part of the original image overlaid on the text, and its generated by a Gaussian distribution. In this way, minimal texts are prevented from receiving very large occlusions and vice versa. The occlusion level fluctuates between 0 and 100%, intending to analyze the techniques graphically and at what level of occlusion performance deteriorates.

---

[1] https://github.com/alinesoares1/ISTD-OC-Dataset.

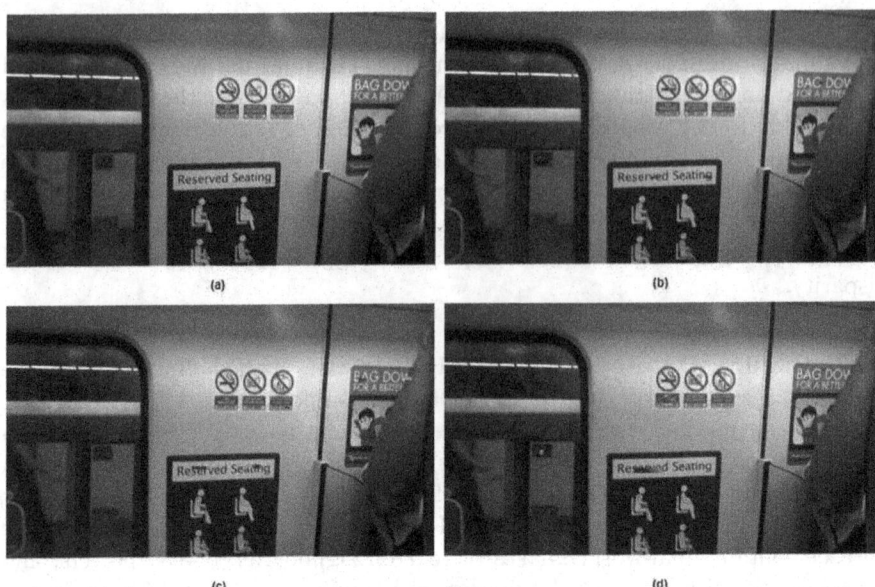

Fig. 1. An illustrative example of the occlusion generation based on the ICDAR 2015 dataset; (a) is an original image from ICDAR 2015 dataset with no occlusions, (b), (c) and (d) are samples of occlusion generation in 30%, 50% and 90% levels, respectively.

The equation that determines the size of the occlusion can be seen below. W and H represent the width and height of the image, respectively. In addition, the width and height of the bounding box are represented by w and h. Alpha (α) is a parameter that indicates how much weight is given to the size of the box for the page; in our experiments, we used 0.55. Rand is a random value generated by the Gaussian distribution of an average of 10% and a standard deviation of 5%, which defines the degree of occlusion.

$pbwW$ is the relationship between the width of the text box and the image, and $pbhH$ is the relationship between the height of the text box and the height of the image. This means the fraction between w and W, as well as h and H, respectively. The occlusion of the text region must be bw wide and bh height, given by:

$$bw = \alpha \times w \times rand + (1 - \alpha) \times w \times pbwW \qquad (1)$$

$$bh = \alpha \times h \times rand + (1 - \alpha) \times h \times pbhH \qquad (2)$$

A sample of the occlusion generation can be seen in Fig. 2, with the size of the occlusion defined, after texts instances detected from the original image (see Fig. 2a and 2b), any other part of the image is pasted with the dimensions of the occlusion over the region of the text, as shown in Fig. 2c. However, to make the task of identifying occluded texts a little more complex, in Fig. 2d–e, we added noise to the RGB channels of the occlusion, so it is possible to prevent models from learning patterns based on equal regions to locate occlusions.

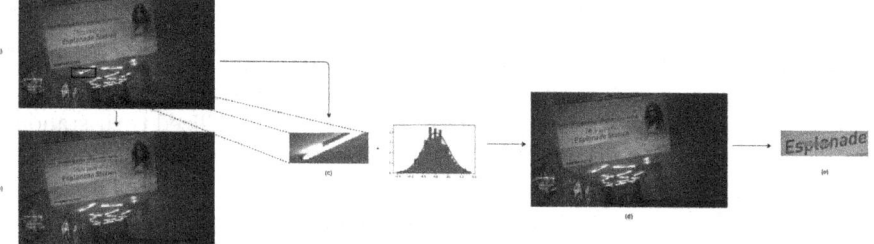

Fig. 2. Sample of occlusion generation in ICDAR 2015 dataset.

It is interesting to note that even using the ICDAR 2015 dataset as a reference to assess the occlusion problem and generate occluded text images; once the text region is located through the coordinates, the occlusion generation methodology allows different levels of obstruction to be applied to the text of any other desired dataset. The choice of this specific benchmark as a source is due only to its recurrent application in the analysis of baseline algorithms, avoiding possible inconsistencies during the evaluation process.

4   Experimental Evaluation

Our work aims to evaluate the effectiveness of the existing deep architectures to perform scene text detection and recognition in the proposed dataset with different occlusion levels.

In this work, CRAFT [3], EAST [30], PAN [26], and PSENet [25] were used for text detection, due to the recent results achieved by these models on this task [19]. It is worth mentioning that for each method, except EAST and PSENet, we used the corresponding pre-trained model directly from the authors' GitHub page trained on the ICDAR15 dataset. For both EAST and PSENet, we trained the algorithm on ICDAR15 according to the authors' code. For testing, the ISTD-OC has been used. This evaluation strategy allows to analyze the impact of occlusions on these techniques and avoids unbiased evaluation.

For scene text recognition schemes, the deep learning-based techniques chosen are based on the Baek et al. [2] study, which has introduced a unified four-stage STR framework that most existing STR models match. The inconsistencies of training and evaluation datasets and the performance gap results were discussed in their work. Therefore, models such as CRNN [20], RARE [21], STAR-Net [12] and ROSETTA [4] were re-implemented with the proposed framework and under consistent settings. The best model found in each technique shows competitive performance to state-of-the-art methods in ICDAR 2015 dataset. All recognition models have been trained on combination of SynthText [8] and MJSynth [9] datasets. For evaluating, 2077 occluded cropped word instances images from ISTD-OC were used. The experiments on the ISTD-OC dataset were performed approximately one hundred times for every model evaluated, that is, ten times in each interval with a step of 10%, between 0 and 100% of occlusion.

4.1   Evaluation Metrics

Text Detection. Commonly models who evaluate the ICDAR dataset use the protocols for performing quantitative comparison among the text detection techniques. To quantify the chosen methods' performance, we utilized their standard evaluation metrics: Precision (P) and Recall (R) metrics used in the information retrieval field. Precision and Recall are based on using the ICDAR15 intersection over union (IoU) metric [10], which is obtained for the jth ground-truth and ith detection bounding box as follow:

$$IoU = \frac{Area(G_j \cap D_i)}{Area(G_j \cup D_i)} \tag{3}$$

and the threshold for counting the correct detection is $IoU \geq 0.5$. Furthermore, H-mean or F1-score were also used as follow:

$$F1 - Score = 2 \times \frac{P \times R}{P + R} \tag{4}$$

Text Recognition. Based on the most common evaluation metrics in STR systems, the word error rate (WER) and the character error rate (CER) were adopted in this work. These metrics indicate the amount of text that the applied model did not recognize correctly. CER is defined as the minimum number of editing operations at the character level, considering the respective ground truth. WER is specified in the same way, but when it comes to words [16].

4.2   Evaluation of Text Detection Approaches

Figures 3, 4 and 5 illustrates the detection performance of the selected state-of-the-art text detection methods, namely, CRAFT [3], PSENet [25], PAN [26] and EAST [30] related to precision, f1-score and recall metric, respectively. In the Figs. 3a, 4a and 5a, we show box plots to depict the lack of robustness of the models facing occlusion and how it affects the performance of these text detection algorithms on the scene.

Initially, we focus on these models' capability in terms of generalization, i.e., how a trained model on one dataset can detect and recognize challenging text instances on "other" datasets. From these graphs, it is also possible to infer the variability, median, outliers, and other information from the data obtained. Comparing the CRAFT approach's performance in images without occlusions and high-level occluded samples, are obtained a performance decline of 37% in precision, as shown in Fig. 3b. In contrast, PSENet, the second-best model evaluated on ICDAR15 with high occlusions in precision, had the worst f1-score value decreasing performance with a decline of 40% compared from images non-occluded, as we can see in Fig. 4b.

Besides that, even PAN performing 70% of precision for ranges of occlusion between 10% and 50%, when it comes to recall, EAST and PSENet performed better, as shown in Fig. 5b. Moreover, it has been observed that the decline of performance increases significantly when text instances are affected not only by occlusion but by a combination of incidental and diversified text detection, i.e., low-resolution images, complex backgrounds.

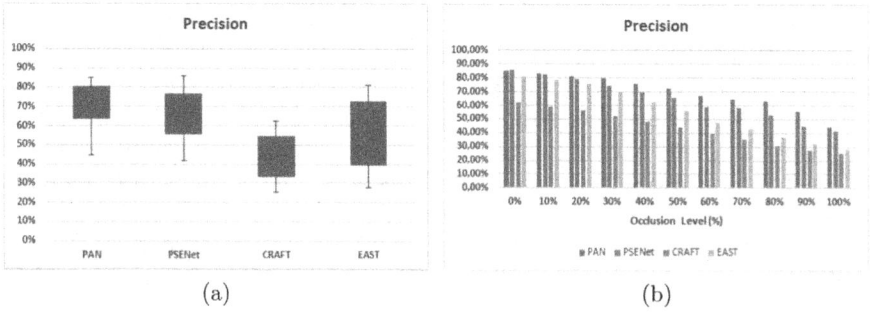

Fig. 3. Detection performance of deep state-of-the-art models based on precision metric.

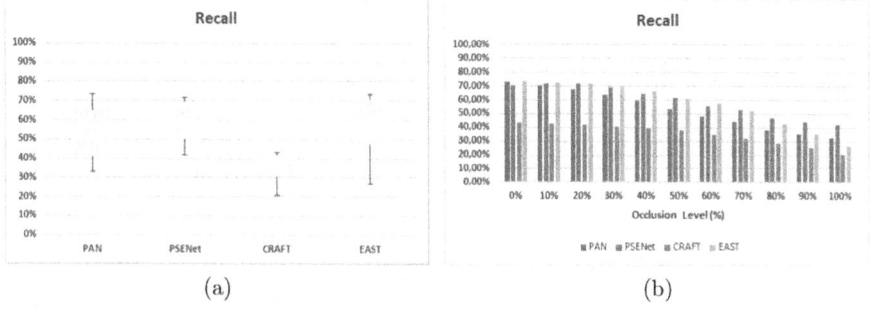

Fig. 4. Detection performance of deep state-of-the-art models based on recall metric.

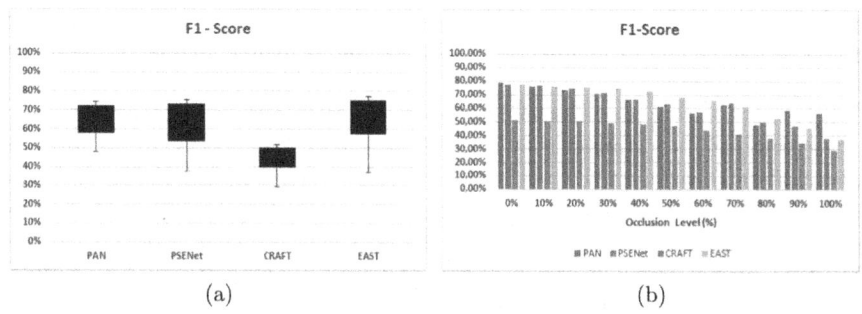

Fig. 5. Detection performance of deep state-of-the-art models based on F1-score metric.

Figure 6 presents qualitative results of our evaluation for each model. Even though both PAN and PSENet approaches offer better robustness in detecting occlusions due to their capacity of learning information gradually, these method's performances are still far from perfect, significantly when occlusion levels exceed 70%. Moreover, although significant progress has been made with deep learning and these models have proven generalizability to many scene text datasets even with synthetic data, they fail to adapt to varying inputs, such as text instances with occlusion. The CRAFT model uses character-level ground truth in existing real-world level datasets synthetically to compensate for the lack of character annotation. However, it can not infer enough information about word-level occlusion, and it is computationally costly as it requires many data to generate the affinity maps. The EAST model works based on an object detection approach and needs handcrafted information about orientation, anchor design, etc., limiting the efficiency of the model.

4.3   Evaluation of Text Recognition Approaches

We evaluated our generated ISTD-OC dataset with occlusion, as shown in Fig. 8, in some scene text recognition benchmarks, i.e., ROSETTA [4], RARE [21], CRNN [20], and STAR-Net [12]. Figures 7a and 7b summarizes the comparative results in terms of word and character error rate for these models. It can be seen from these Figures that all methods performed several accuracy declines in the presence of high occlusions, although models that contain a rectification module have been able to perform a little better on occlusions compared to other models. The STAR-Net, for example, presents a rectification module in their feature extraction stage for spatially transforming text images, and RARE combines a Spatial Transformer Network (STN) and a Sequence Recognition Network (SRN).

In the STN, an input image is spatially transformed into a rectified image, i.e., the STN produces an image that contains regular text since a text recognizer works best when its input images contain tightly-bounded regular text. Besides, the module combination proposed by Baek et al. cites Analysis for RARE, which attributes an attention module for prediction and a BiLSTM for the sequential modeling stage. It allows the model to outperform the CTC-based modules combinations, such as CRNN and ROSETTA, because attention methods better handle the alignment problem in the irregular text compared to CTC-based methods.

It is worth noting that even though all investigated algorithms are pre-trained only with synthetic datasets, they can still handle manageable occlusion levels in natural scenes, based on the word error rate showed in Fig. 7b. Furthermore, in general, attention-based methods that benefit from a deep backbone for feature extraction and transformation network for rectification have performed better than CTC-based methods, as in CRNN, STAR-Net, and ROSETTA. Besides that, we believe that contextual information is significant for feature representation in scene text recognition methods since focusing on informative regions in images helps generate discriminative features. For example, when humans have

Fig. 6. Qualitative evaluation of text detection state-of-the-arte models on ISTD-OC. Each row presents a sample of results for PSENet (a), EAST (b), CRAFT (c) and PAN (d) in levels of 20%, 40% and 80% of occlusion.

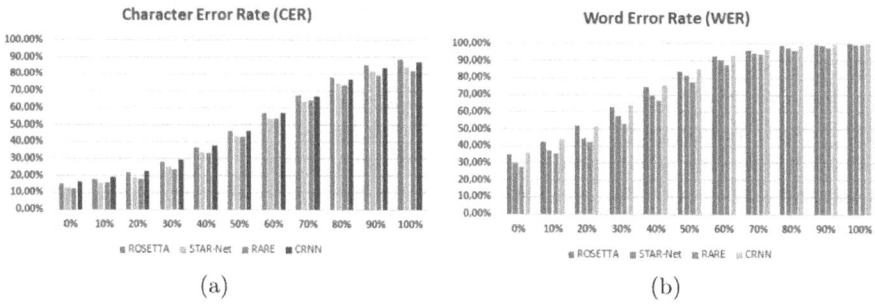

Fig. 7. Performance of the state-of-the-art text recognition models based on the CER and WER metrics.

to identify a text region in a scene image, no matter how complicated the background clutters are, the localization of the text instance is subserved by local processes sensitive to local features and a global process of retrieving structural context.

Fig. 8. Samples of cropped word instances from ISTD-OC dataset under 10% (a), 40% (b), 60% (c) and (d) 70% occlusion levels.

5   Conclusion and Future Work

In this work, we investigated state-of-the-art deep architectures for scene text detection and recognition in the case of occlusion. Consequently, and as a contribution of this work, we proposed a systematic methodology to generate occlusion, which resulted in the ISTD-OC dataset. The released dataset is designed to examine and comparing benchmark models' ability to handle different occlusion levels.

The experimental results suggest that these existing deep architectures for scene text detection and recognition are far from the human visual system's ability to read occluded texts in natural daily-life situations.

However, the research has significant importance in the scene text recognition field, referencing more studies in complex and diverse scenes. Despite contributions, this work is only a small step towards recognizing occluded texts in natural scenes robustly. We should improve the way occlusion is generated in future works, representing more real-world scenarios. Moreover, we will investigate the causes of failure of these existing models for scene text detection and recognition under different occlusion levels to build a single and lightweight network for end-to-end scene text detection and recognition able to handle different levels of occlusion.

References

1. Adak, C., Chaudhuri, B.B., Blumenstein, M.: Impact of struck-out text on writer identification. In: 2017 International Joint Conference on Neural Networks (IJCNN), pp. 1465–1471. IEEE (2017)
2. Baek, J., Han, Y., Kim, J.O., Lee, J., Park, S.: What is wrong with scene text recognition model comparisons? Dataset and model analysis (2019)
3. Baek, Y., Lee, B., Han, D., Yun, S., Lee, H.: Character region awareness for text detection (2019)

4. Borisyuk, F., Gordo, A., Sivakumar, V.: Rosetta: large scale system for text detection and recognition in images. In: Proceedings of the 24th ACM SIGKDD International Conference on Knowledge Discovery & Data Mining, pp. 71–79 (2018)
5. Chen, X., Jin, L., Zhu, Y., Luo, C., Wang, T.: Text recognition in the wild: a survey. ACM Comput. Surv. (CSUR) **54**, 1–35 (2020)
6. Efimova, V., Shalamov, V., Filchenkov, A.: Synthetic dataset generation for text recognition with generative adversarial networks, vol. 1143315, p. 62 (2020). https://doi.org/10.1117/12.2558271
7. Graves, A., Fernández, S., Gomez, F., Schmidhuber, J.: Connectionist temporal classification: labelling unsegmented sequence data with recurrent neural networks. In: Proceedings of the 23rd International Conference on Machine Learning, pp. 369–376 (2006)
8. Gupta, A., Vedaldi, A., Zisserman, A.: Synthetic data for text localisation in natural images. In: Proceedings of the IEEE Computer Society Conference on Computer Vision and Pattern Recognition, December 2016, pp. 2315–2324 (2016). https://doi.org/10.1109/CVPR.2016.254
9. Jaderberg, M., Simonyan, K., Vedaldi, A., Zisserman, A.: Synthetic data and artificial neural networks for natural scene text recognition. arXiv preprint arXiv:1406.2227 (2014)
10. Karatzas, D., et al.: ICDAR 2015 competition on robust reading. In: 2015 13th International Conference on Document Analysis and Recognition (ICDAR), pp. 1156–1160. IEEE (2015)
11. Lin, H., Yang, P., Zhang, F.: Review of scene text detection and recognition. Arch. Comput. Methods Eng. **27**(2), 433–454 (2020). https://doi.org/10.1007/s11831-019-09315-1
12. Liu, W., Chen, C., Wong, K.Y.K., Su, Z., Han, J.: Star-net: a spatial attention residue network for scene text recognition. In: BMVC, vol. 2, p. 7 (2016)
13. Long, S., He, X., Yao, C.: Scene text detection and recognition: the deep learning era. Int. J. Comput. Vis. **129**, 1–24 (2020)
14. Mishra, A., Alahari, K., Jawahar, C.: Scene text recognition using higher order language priors (2012)
15. Mittal, A., Shivakumara, P., Pal, U., Lu, T., Blumenstein, M., Lopresti, D.: A new context-based method for restoring occluded text in natural scene images. In: Bai, X., Karatzas, D., Lopresti, D. (eds.) DAS 2020. LNCS, vol. 12116, pp. 466–480. Springer, Cham (2020). https://doi.org/10.1007/978-3-030-57058-3_33
16. Neto, A.F.D.S., Bezerra, B.L.D., Toselli, A.H.: Towards the natural language processing as spelling correction for offline handwritten text recognition systems. Appl. Sci. **10**(21), 7711 (2020). https://doi.org/10.3390/app10217711
17. Nisa, H., Thom, J.A., Ciesielski, V., Tennakoon, R.: A deep learning approach to handwritten text recognition in the presence of struck-out text. In: International Conference Image and Vision Computing New Zealand (December 2019). https://doi.org/10.1109/IVCNZ48456.2019.8961024
18. Qi, J., et al.: Occluded video instance segmentation. arXiv preprint arXiv:2102.01558 (2021)
19. Raisi, Z., Naiel, M.A., Fieguth, P., Jun, C.V.: Text detection and recognition in the wild: a review, pp. 13–15 (2020)
20. Shi, B., Bai, X., Yao, C.: An end-to-end trainable neural network for image-based sequence recognition and its application to scene text recognition. IEEE Trans. Pattern Anal. Mach. Intell. **39**(11), 2298–2304 (2016)

21. Shi, B., Wang, X., Lyu, P., Yao, C., Bai, X.: Robust scene text recognition with automatic rectification. In: Proceedings of the IEEE Conference on Computer Vision and Pattern Recognition, pp. 4168–4176 (2016)

22. Veit, A., Matera, T., Neumann, L., Matas, J., Belongie, S.: Coco-text: dataset and benchmark for text detection and recognition in natural images. arXiv preprint arXiv:1601.07140 (2016)

23. Wang, J., Hu, X.: Gated recurrent convolution neural network for OCR. In: Advances in Neural Information Processing Systems, vol. 30, pp. 335–344 (2017)

24. Wang, K., Babenko, B., Belongie, S.: End-to-end scene text recognition. In: 2011 International Conference on Computer Vision, pp. 1457–1464. IEEE (2011)

25. Wang, W., et al.: Shape robust text detection with progressive scale expansion network. In: Proceedings of the IEEE Computer Society Conference on Computer Vision and Pattern Recognition, June 2019, pp. 9328–9337 (2019). https://doi.org/10.1109/CVPR.2019.00956

26. Wang, W., et al.: Efficient and accurate arbitrary-shaped text detection with pixel aggregation network. In: Proceedings of the IEEE International Conference on Computer Vision, October 2019, pp. 8439–8448 (2019). https://doi.org/10.1109/ICCV.2019.00853

27. Yao, C., Bai, X., Liu, W., Ma, Y., Tu, Z.: Detecting texts of arbitrary orientations in natural images. In: 2012 IEEE Conference on Computer Vision and Pattern Recognition, pp. 1083–1090. IEEE (2012)

28. Yuan, T.L., Zhu, Z., Xu, K., Li, C.J., Hu, S.M.: Chinese text in the wild. arXiv preprint arXiv:1803.00085 (2018)

29. Zhao, F., Feng, J., Zhao, J., Yang, W., Yan, S.: Robust LSTM-autoencoders for face de-occlusion in the wild. IEEE Trans. Image Process. **27**, 778–790 (2016)

30. Zhou, X., et al.: EAST: an efficient and accurate scene text detector (2015)

CATALIST: CAmera TrAnsformations for Multi-LIngual Scene Text Recognition

Shivam Sood[1]([✉]) [iD], Rohit Saluja[2] [iD], Ganesh Ramakrishnan[1] [iD], and Parag Chaudhuri[1] [iD]

[1] IIT Bombay, Mumbai, India
{ssood,ganesh,paragc}@cse.iitb.ac.in
[2] IIIT Hyderabad, Hyderabad, India
rohit.saluja@research.iiit.ac.in
https://catalist-2021.github.io

Abstract. We present a CATALIST model that 'tames' the attention (heads) of an attention-based scene text recognition model. We provide supervision to the attention masks at multiple levels, *i.e.*, line, word, and character levels while training the multi-head attention model. We demonstrate that such supervision improves training performance and testing accuracy. To train CATALIST and its attention masks, we also present a synthetic data generator ALCHEMIST that enables the synthetic creation of large scene-text video datasets, along with mask information at character, word, and line levels. We release a real scene-text dataset of $2k$ videos, CATALIST$_d$ with videos of real scenes that potentially contain scene-text in a combination of three different languages, namely, English, Hindi, and Marathi. We record these videos using 5 types of camera transformations - (i) *translation*, (ii) *roll*, (iii) *tilt*, (iv) *pan*, and (v) *zoom* to create transformed videos. The dataset and other useful resources are available as a documented public repository for use by the community.

Keywords: Scene text recognition · Video dataset · OCR in the wild · Multilingual OCR · Indic OCR · Video OCR

1 Introduction

Reading the text in modern street signs generally involves detecting the boxes around each word in the street signs and then recognizing the text in each box. Reading street signs is challenging because they often appear in various languages, scripts, font styles, and orientations. Reading the end-to-end text in scenes has the advantage of utilizing the global context in street signs, enhancing the learning of patterns. One crucial factor that separates a character-level OCR system from an end-to-end OCR system is reading order. Attention is thus needed to locate the initial characters, read them, and track the correct reading order in the form of change in characters, words, lines, paragraphs, or columns (in multi-column texts). This observation forms the motivation for our work.

© Springer Nature Switzerland AG 2021
E. H. Barney Smith and U. Pal (Eds.): ICDAR 2021 Workshops, LNCS 12916, pp. 213–228, 2021.
https://doi.org/10.1007/978-3-030-86198-8_16

Fig. 1. Sample video frames from CATALIST$_d$

Obtaining large-scale multi-frame video annotations is a challenging problem due to unreliable OCR systems and expensive human efforts. The predictions obtained on videos by most OCR systems are fluctuating, as we motivate in Sect. 3. The fluctuations in the accuracy of the extracted text may also be due to various external factors such as partial occlusions, motion blur, complex font types, distant text in the videos. Thus, such OCR outputs are not reliable for downstream applications such as surveillance, traffic law enforcement, and cross-border security system.

In this paper, we demonstrate that the photo OCR systems can improve by guiding the attention masks based on the orientations and positions of the camera. We improve an end-to-end attention-based photo-OCR model on continuous video frames by taming the attention masks in synthetic videos and on novel controlled datasets that we record for capturing possible camera movements.

We begin by motivating our work in Sect. 3. We base a video scene-text recognition model (referred to as CATALIST) on partly supervised attention. Like a teacher holding a lens through which a student can learn to read on a board, CATALIST exploits supervision for attention masks at multiple levels (as shown in Fig. 3). Some of the attention masks might be interpreted as covering different orientations in frames during individual camera movements (through separate masks). In contrast, others might focus on the line, word, or character level reading order. We train CATALIST using synthetic data generated using a non-trivial extension of SynthText [6]. The extension allows for the generation of text videos using different camera movements while also preserving character-level information. We describe the CATALIST model which 'tames' the attention (heads) in Sect. 4.1. We demonstrate that providing direct supervision to attention masks at multiple levels (*i.e.*, line, word, and character levels) yields improvement in the recognition accuracy.

To train CATALIST and its attention masks, we present a synthetic data generator ALCHEMIST[1] that enables the synthetic creation of large scene-text video datasets, along with mask information at character, word and line levels. We describe the procedure to generate synthetic videos in Sect. 4.2.

We also present a new video-based real scene-text dataset, CATALIST$_d$ in Sect. 4.3. Figure 1 shows the sample video frames of the dataset. We create these videos using 5 types of camera transformations - (i) *translation*, (ii) *roll*, (iii) *tilt*, (iv) *pan*, and (v) *zoom*. We provide the dataset and experimental details in Sect. 5. We summarize the results in Sect. 6 and conclude the work in Sect. 8.

2   Related Work

We now introduce the approaches to tackle various issues in the field of photo OCR. Works specific to text localization are proposed by Gupta et al. [6]. Liao et al. [11,13] augments such work to real-time detections in the end-to-end scenes. Karatzas et al. [9] and Bušta et al. [3] present better solutions in terms of accuracy and speed. The problem of scene-text spotting, however, remains complicated owing to variations in illumination, capturing methods and weather conditions. Moreover, the movement of the camera (or objects containing text) and motion blur in videos can make it harder to recognize the scene-text correctly. There has been a rising interest in end-to-end scene-text recognition in images over the last decade [2,3,9,10,16]. Recent text-spotters by Bušta et al. [3,4] include deep models that are trained end-to-end but with supervision at the level of text as well as at the level of words and text-boxes. The two recent breakthroughs in this direction, which work directly on complete scene images without supervision at the level of text boxes, are:

1. STN-OCR by Bartz et al. [2]: A single neural network for text detection and text recognition. The model contains a spatial transformation network that encodes the input scene image. It then applies a recurrent model over the encoded image features to output a sequence of grids. Combining the grids and the input image returns the series of word images present in the scene. Another spatial transformer network process the word images for recognition. This work does not need supervision at the level of detection.
2. Attention-OCR by Wojna et al. [19]: This work employs an inception network (proposed by Szegedy et al. [18]) as an encoder and an LSTM with attention as a decoder. The work is interesting because it does not involve any cropping of word images but works on the principle of soft segmentation through attention. The attention-OCR model performs character-level recognition directly on the complete scene image thus utilizing the global context while reading the scene. This model has an open-source TensorFlow (a popular library for deep learning by Abadi et al. [1]) implementation.

---

[1] ALCHEMIST stands for synthetic video generation in order to tame **A**ttention for **L**anguage (line, word, character, *etc.*) and other camera-**CH**ang**E**s and co**M**binat**I**ons for **S**cene **T**ext.

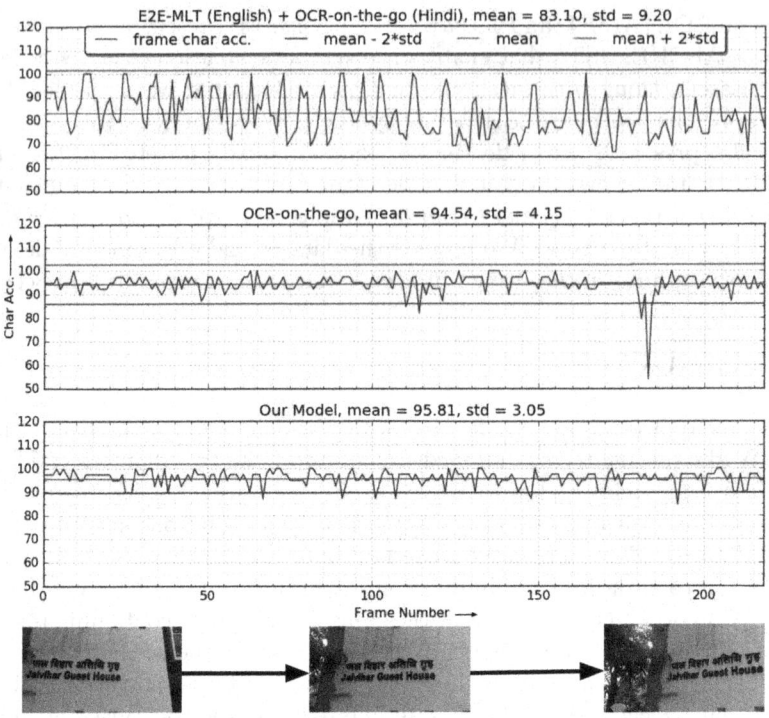

Fig. 2. Frame wise accuracy of 3 text-spotters on a simple video exhibiting *pan*

Both these works experiment on French Street Name Signs (FSNS) dataset, on which Attention-OCR performs the best. The Attention-OCR model also outperforms another line-level segmentation-based method (refer to work by Smith et al. [17]) on the FSNS dataset. Recently, the OCR-on-the-go model outperforms these models on the FSNS dataset using a multi-head attention mechanism [15]. In this work, we set new benchmarks for reading Indian street signs in a large number of video frames. The FSNS dataset contains around $10M$ images annotated with end-to-end transcriptions similar to ours. Different large-scale datasets are available in English. Uber-Text by Zhang et al. [21] include over $0.1M$ images annotated at line-level, captured from 6 US cities. Reddy et al. [14] annotate 1000 video clips from the BDD dataset [20] at line-level. We provide end-to-end transcriptions for our dataset similar to FSNS. Additionally, we also share noisy annotations at word-level and paragraph-level for each frame.

3   Motivation

We motivate our work of training the scene-text spotting models on the real (as well as synthetic) videos captured via continuous camera movements. Various end-to-end scene-text spotters, such as the ones proposed by Bušta et al. [3,4], train on synthetic as well as augmented real data to cover different capturing perspectives/orientations. The problem, however, is that during the training phase, such models do not exploit all the continuous perspectives/orientations captured by the camera movement (or scene movement). Thus the OCR output fluctuates when tested on all/random video frames. Also, to deploy such models on real-time videos, two scenarios may occur. Firstly, the multi-frame consensus is desirable to improve OCR accuracy or interactive systems. Secondly, since it is computationally expensive to process each frame for readability, it is not possible to verify the quality of the frame to be OCR-ed. In any of these scenarios, the recognition system needs to work reasonably well on continuous video frames.

We present the frame level accuracy of E2E-MLT proposed by Bušta et al. [4] on an 8 s video clip with a frame size of 480×260 in the first plot of Fig. 2 (with sample frames shown at the bottom). Since the model does not work for Hindi, we recognize the Hindi text using *OCR-on-the-go* model [15]. As shown, the E2E-MLT model produces the most unstable text on a simple video (from the test dataset) with the average character accuracy of 83.1% and the standard deviation of 9.20. The reason for this is that E2E-MLT, which does not train on continuous video frames, produces extra text-boxes on many of them during the detection phase. Thus extra noise characters or strings are observed during recognition. For instance, the correct text "Jalvihar Guest House" appears in 18 frames, the text "Jalvihar arG Guest House" appears in 10 frames, and the text "Jalvihar G Guest House" appears in 9 frames. The text "Jalvihar G arGu Guesth R H House" appears in one of the frame.

The instability in the video text, however, reduces when we use the *OCR-on-the-go* model by Saluja et al. [15] to read these video frames. As shown in the second plot of Fig. 2, we achieve the (higher) average character accuracy of 94.54% and (lower) standard deviation of 4.15. This model works on the principle of end-to-end recognition and soft detection via unsupervised attention. The instability further reduces, as shown in the third plot of Fig. 2, when we train our CATALIST model on the continuous video datasets proposed in this work.

4   Methodology

We use end-to-end attention-based encoder-decoder model proposed by Wojna et al. [19]. For better inference of attention masks, and improved recognition, we use the multi-head version of this model, proposed by Saluja et al. [15]. In Fig. 3, we present the CATALIST model, that uses multi-task learning to update attention masks. Each mask is updated based on two loss functions. For end-to-end supervision, we use cross-entropy loss. To train attention heads, we use dice loss [12] between the predicted masks and the segmented masks obtained using text-boxes from SynthText proposed by Gupta et al. [6]. We also transform the synthetic images, along with text-boxes, to form videos which we describe in the end of this section.

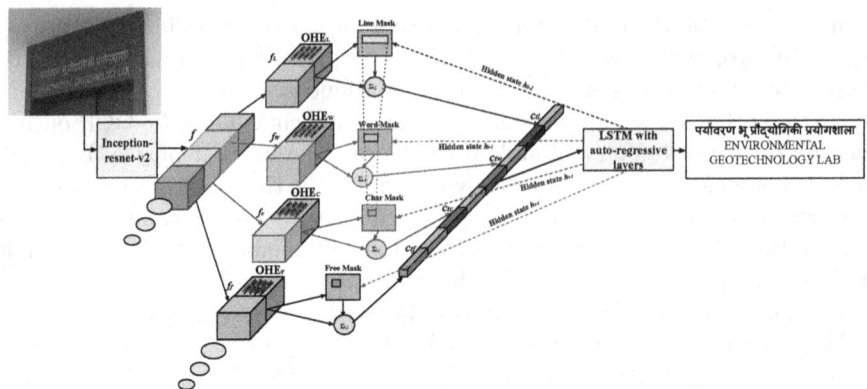

Fig. 3. CATALIST tames attention mask at multiple levels of granularity. The first three masks, namely line, word, and char mask, are supervised. The remaining attention masks are set free. Figure shows the first four attention masks.

4.1   The CATALIST Model

As shown in Fig. 3, the powerful inception-based encoder proposed by Szegedy et al. [18], which performs multiple convolutions in parallel, enhances the ability to read the text at multiple resolutions. We extract the features f from the input image using the inception-based encoder. Moreover, the multi-head attention mechanism in our model exploits: i) the splits of feature f into f_L, f_w, f_c, $f_f{}^2$, etc. (refer Fig. 3), ii) one-hot-encoded (OHE) vectors (OHE_L, OHE_w, OHE_c, OHE_f, etc.[3]) for both x and y coordinates of each feature split, iii) hidden layer at the previous decoding step (h_{t-1}) of an LSTM (decoder). To learn the attention at multiple levels of granularity, we provide supervision to the first three masks in the form of the line, word, and character level segmented binary images. The remaining masks are set free to assist/exploit end-to-end recognition/supervision. Thus we refer to the first three of them as line mask, word mask, and char mask in Fig. 3. We also hard-code the word mask to remain inside the line mask, and the character mask to remain inside the word mask. The context vectors (c_L, c_w, c_c, c_f, etc.), which are obtained after applying the attention mechanism, are fed into the LSTM to decode the characters in the input image.

It is important to note that for each input frame, the features f and splits remain fixed, whereas the attention masks move in line with the decoded characters. Thus, we avoid using simultaneous supervision for all the character masks (or word masks or line masks) in a frame. Instead, we use a sequence of masks (in the form of segmented binary images) at each level for all the video frames.

---

[2] f_L represents the features used for producing line masks, f_w represents features used for word masks, f_c represents features used for character masks, and f_f represents features used for free attention masks.

[3] for the corresponding features f_L, f_w, f_c, f_f, etc.

We accomplish this by keeping the word-level as well as the line-level segmented images constant and moving the character level segmented images while decoding the characters in each word. Once the decoding of all the characters in a word is complete, the word level segmented image moves to the next word in the line, and the character level image keeps moving as usual. Once the model has decoded all the characters in a line of text, the line (and word) level segmented image moves to the next line, and the character level segmented image continues to move within the word image.

4.2   The ALCHEMIST Videos

We generate synthetic data for training the attention masks (as well as the complete model) using our data generator, which we refer to as ALCHEMIST. ALCHEMIST enables the synthetic creation of large scene-text video datasets. ALCHEMIST overlays synthetic text on videos under 12 different transformations described in the next section. By design, we preserve the information of the transformation performed, along with information of the character, word, and line positions (as shown in Fig. 7). This information in the synthetic data provides for fairly detailed supervision on the attention masks in the CATALIST model. We build ALCHEMIST as an extension of an existing fast and scalable engine called SynthText proposed by Gupta et al. [6].

Methodology: According to pinhole camera model, a (2-d) point x (in homogeneous coordinate system) of image captured by a camera is given by Eq. 1.

$$x = K[R|t]X \tag{1}$$

Here K is the intrinsic camera matrix, R and t are rotation and translation matrices respectively, and X is a (3-d) point in *real world coordinates* in an homogeneous coordinate system.

Fig. 4. For videos with camera *pan*, we find Homography between the corners of a rectangle and 4 points equidistant from them (which form one of the blue trapeziums). (Color figure online)

For generating synthetic videos, we first select a fixed crop within the synthetic image (as denoted by the green rectangle in Fig. 5). We then warp the corners of the crop by finding a planar homography matrix H (using algorithm given by Hartley et al. [7]) between the corner coordinates and four points equidistant from corners (direction depends on the kind of transformation as

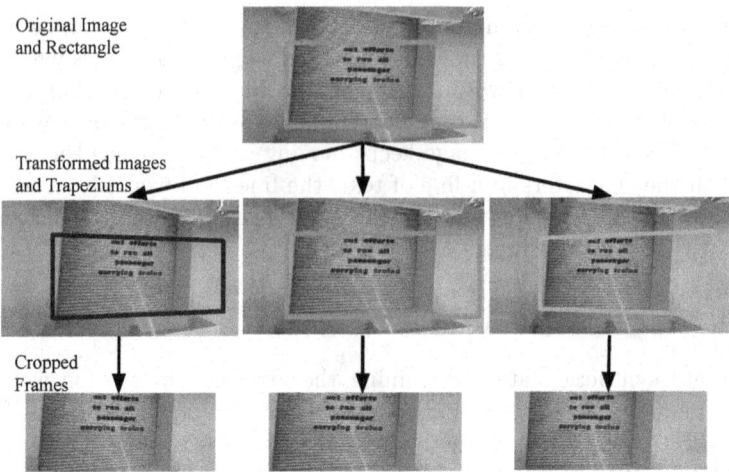

Fig. 5. Generating video with camera *pan* (3 frames at the bottom for dark-blue, green and light-blue perspectives respectively) from an image (at the top). (Color figure online)

explained later). For Fig. 4 (and Fig. 5), we find the planar homography matrix H between corners of one of the blue trapezium and the green rectangle. Thus, instead of a 2D point x in the homogeneous coordinate system as explained earlier, we get a translated point x_{new} defined in Eq. 2:

$$x_{new} = HK[R|t]X \tag{2}$$

Here, H is the known homography. The above equation is simplified from the equation below:

$$x_{new} = KT[R|t]X = KTK^{-1}K[R|t]X \tag{3}$$

Here T is the unknown transformation matrix. We then warp the complete image using H and crop the rectangular region (refer green rectangle in Fig. 5), to obtain the video frames. To find all the homography matrices for a video with camera *pan*, we consider the corners of the trapezium moving towards the rectangle corners. Once the homography matrix becomes the *identity matrix*, we move the corners of the trapezium away from the rectangle in the opposite direction to the initial flow (to form the mirrors of the initial trapeziums, e.g. light-blue trapezium in Fig. 5).

The process for generating videos with camera *tilt* is similar to that of *pan*. The only difference is that the trapeziums in videos with camera *tilt* have vertical sides as parallel (as shown in Fig. 6a) whereas the trapeziums in videos with camera *pan* have horizontal sides as parallel. For the videos with camera *roll*, we utilize the homography matrices between the corners of the rectangles rotating around the text center and the base (horizontal) box, as shown in Fig. 6b.

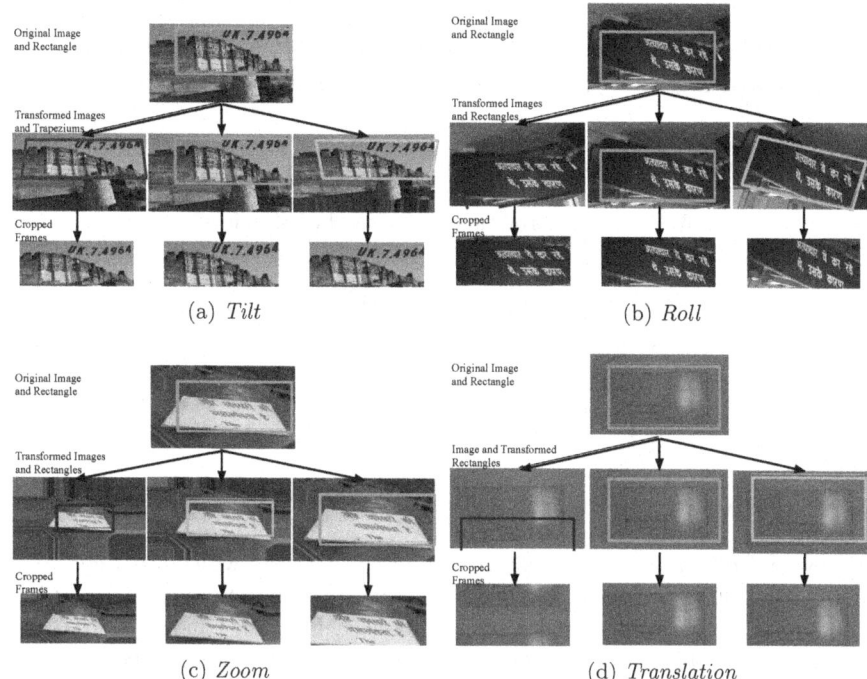

Fig. 6. Generating video with camera (a) *tilt*, (b) *roll*, (c) *zoom* and (d) *translation* *(frames at the bottom)*

For videos with camera *translation*, we use the regions a moving rectangle beginning from one text boundary to the other and generate the frames, as shown in Fig. 6d. We make sure that the complete text, with rare partial occlusion of boundary characters, lies within each frame of the videos.

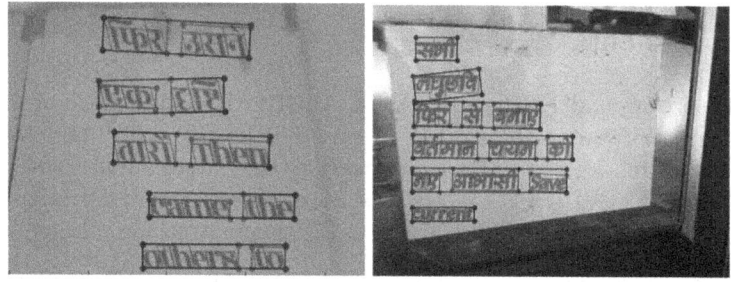

Fig. 7. Sample frames from the synthetic videos with multi-level text-boxes (Color figure online)

We also use the homography H to transform the multi-level text-boxes in the cropped image. Figure 7 depicts sample video frames with text-boxes at the line, word, and character[4] levels – shown in blue, green, and red, respectively.

Table 1. Distribution of videos in the CATALIST$_d$ dataset

S.No.	Transformation type	Number of videos
1.	Translation	736
2.	Roll	357
3.	Tilt	387
4.	Pan	427
5.	Zoom	402

4.3   The CATALIST$_d$ Videos

We now present a new video-based scene-text dataset, which we refer to as CATALIST$_d$. Every video in CATALIST$_d$ contains scene-text, potentially in a combination of three different languages, namely, English, Hindi, and Marathi. For each such scene-text, we create 12 videos using 12 different types of camera *transformations*, broadly categorized into 5 groups:- (i) four types of *translation*, that could be left, right, up and down, (ii) two types of *roll*, including clockwise and anti-clockwise, (iii) two types of *tilt* which could be up-down or down-up motion, (iv) two types of *pan*, that is left-right and right-left, and (v) two types of *zoom* which could be in or out. We use a camera with a tripod stand to record all these videos to have a uniform control.

We summarize the distribution of different types of videos in Table 1. It is important to note that there are four types of translations, whereas there are only two types for all other transformations. We capture these videos at 25 fps with a resolution of 1920 × 1080.

5   Experiments

We synthesize around 12000 videos using ALCHEMIST data generator, which we use only for training the models. We use 50 Unicode fonts[5] and 18 license plate fonts[6] to render text in these videos. Here the duration and frame-rate for each video are 5 s and 25 fps, respectively. Moreover, we record a total of around $2k$ real videos (uniformly divided across 12 camera transformations) using a

---

[4] For Devanagari (the script used for Hindi and Marathi), we carefully consider the boxes at the level of joint-glyphs instead of characters since rendering characters individually (to obtain character level text-boxes) hamper glyph substitution rules that form the joint glyphs in Devanagari.

[5] http://indiatyping.com/index.php/download/top-50-hindi-unicode-fonts-free.

[6] https://fontspace.com/category/license%20plate.

camera mounted over tripod stand for CATALIST$_d$ dataset. The setup allows smooth camera movements for *roll, tilt, pan* and *zoom*. We record the horizontal translation videos with the camera and tripod moving on a skateboard. Other translation videos, which exhibit top to bottom and reverse movements, have jitter because our tripod does not allow for smooth translation while recording such videos. We use a train: test split of 75:25, and carefully avoid letting any testing labels (as well as redundancy of the scenes) be present in the training data. We additionally record around $1k$ videos using handheld mobile phones and use them for training the models. Finally, we also make use of the 640 videos shared by Saluja et al. [15]. We refer to the complete training dataset described above as CATALIST$_{ALL}$ in the next sections.

Fig. 8. A sample video frame from ICDAR'15 competition with text-boxes sorted using our algorithm

We further add the ICDAR'15 English video dataset of 25 training videos (13,450 frames) and 24 testing videos (14,374 frames) by Karatzas et al. [9] to the datasets. For each frame in the ICDAR'15 dataset, we first cluster the text-boxes into paragraphs and then sort the paragraph text-boxes from top-left to bottom-right. A sample video frame with the reading order mentioned above and the text-boxes sorted using our algorithm are shown in Fig. 8. We visually verify that the reading order remains consistent throughout their appearance and disappearance in the videos. The reading order, changes when a new piece of text appears in the video or an old piece of text disappears from the video.

Although we record the controlled videos with a high resolution of 1920×1080, we work with the frame size of 480×260 for all videos owing to the more limited size of the videos captured on mobile devices, as well as for to reduce training time on a large number of video frames. To take care of resolution as well as to remove the frames without text, we extract the 480×260 sized clips containing the mutually exclusive text regions in the videos from the ICDAR'15 dataset. Features of size $14 \times 28 \times 1088$ are extracted from the mixed-6a layer of inception-resnet-v2 [18]. The maximum sequence length of the labels is 180, so we unroll the LSTM decoder for 180 steps. We train all the models for 15 epochs.

Table 2. Test accuracy on different datasets.

S. No.	Training model	Training data	Test data	Char. Acc.	Seq. Acc.
1.	OCR-on-the-go (8 free masks)	OCR-on-the-go[a]		35.00 [15]	1.30
2.	CATALIST model (8 free masks)	CATALIST$_{ALL}$[b]	OCR-on-the-go 200 test videos	65.50	7.76
3.	CATALIST model (3 superv., 5 free masks)	CATALIST$_{ALL}$		**68.67**	**7.91**
4.	CATALIST model (8 free masks)	CATALIST$_{ALL}$	491 CATALIST$_d$ videos	**73.97**	6.50
5.	CATALIST model (3 superv., 5 free masks)			73.60	**7.96**
6.	CATALIST model (8 free masks)	CATALIST$_{ALL}$	24 ICDAR'15 Competition videos	34.37	**1.70**
7.	CATALIST model (3 superv., 5 free masks)			**35.48**	0.72

[a]640 real videos + 700k synthetic images.
[b]3.7k real videos + 12k synthetic videos.

6   Results

We now present the results of the CATALIST model on the different datasets described in the previous section. It is important to note that we use a single CATALIST model to jointly train on all the datasets (CATALIST$_{ALL}$) at once.

Results on the OCR-on-the-go Dataset. In the first three rows of Table 2, we show the results on the test data used for *OCR-on-the-go* model by Saluja et al. [15]. The first row shows the results of this work. As shown in row 2, there is a dramatic improvement in character accuracy by 30.50% (from 35.0%

to 65.5%) as well as sequence accuracy by 6.46% (1.30% to 7.76%), due to proposed CATALIST model as well as the ALCHEMIST and CATALIST$_d$ datasets we have created. Adding the multi-level mask supervision to the CATALIST model further improves the accuracies by 3.17% (from 65.50% to 68.67%) and 0.15% (from 7.76% to 7.91%).

Results on the CATALIST$_d$ Dataset. As shown in the fourth and fifth row of Table 2, the gain of 1.46% (6.50 to 7.96) is observed in the sequence accuracy of the CATALIST model, when we use the mask supervision. We, however, observe a slight gain of 0.37% in character level accuracy when all the masks are set free (*i.e.*, trained without any direct supervision).

Results on the ICDAR'15 Competition Dataset. We observe a gain of 1.11% (from 34.37% to 35.48%) in character-level accuracy on the ICDAR'15 competition dataset due to mask supervision. The end-to-end sequence accuracy for this dataset is as low as 1.70% for the model with all free masks and further lowers (by 0.98%) for the model with the first 3 masks trained using direct semantic supervision. We observe that the lower sequence accuracy for this dataset is due to the complex reading order in the frames.

7   Frame-Wise Accuracies for all Transformations

In Fig. 2, we presented the frame-level accuracy of E2E-MLT (with the Hindi text recognized using *OCR-on-the-go* model), *OCR-on-the-go* model, and the present work on an 8 s video exhibiting *pan*. In this section, we present the frame-level accuracy of the above mentioned text-spotters for the other transformations: *roll*, *zoom*, *tilt*, and *translation*. The accuracy plots for a video with 88 frames (at 25 fps) exhibiting *roll* (clockwise) is shown in Fig. 9a. We use the formulae in Eq. 4 for calculating the character accuracy taking noise characters into consideration.

$$Accuracy = 100 * \frac{length(GT) - edit\_distance(P, GT)}{length(GT)} \tag{4}$$

Here, GT denotes the ground truth sequence and P is the predicted sequence. For some of the frames (with large amounts of transformations), predicted sequence contains a lot of noise characters. As a result, the *edit_distance* between predicted sequence and ground truth sequence may go higher than the length of ground truth sequence. Thus we get negative accuracy for some of the frames in Fig. 9a. As shown, our model has the highest mean and lowest standard deviation for this video as well. It demonstrates the importance of the CATALIST model trained with mask supervision on continuous video frames. Furthermore, it is essential to note that all the models perform poorly at the start of this video due to larger amounts of rotation as compared to the later parts of the video.

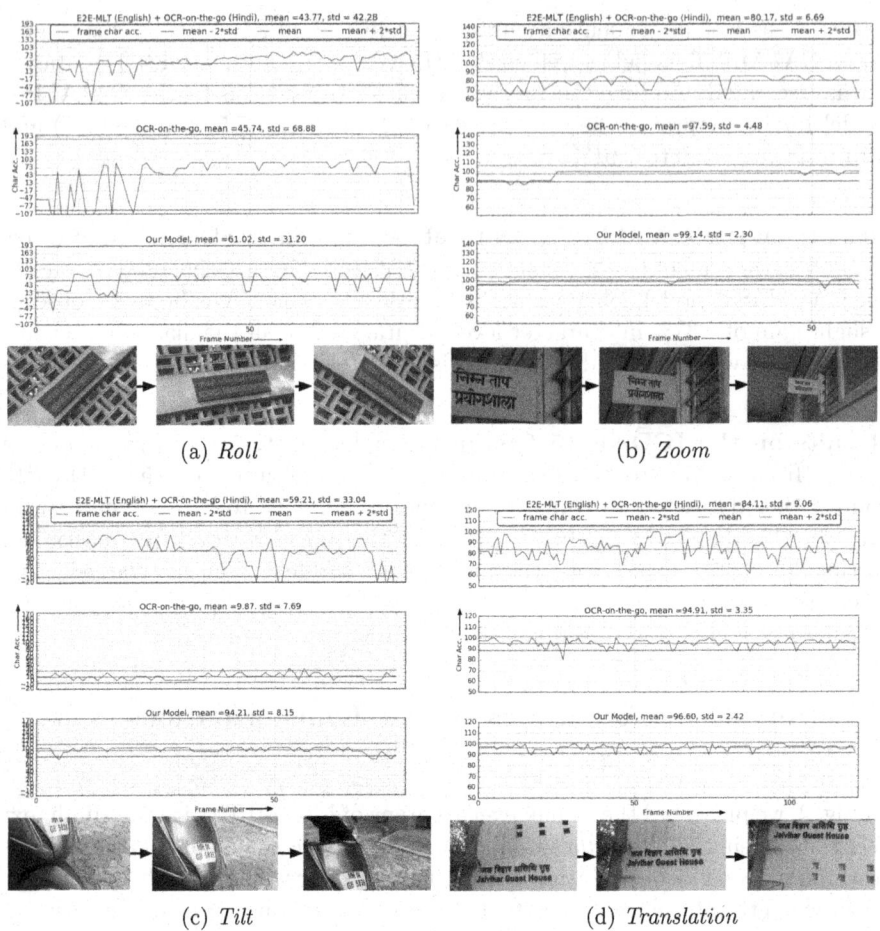

Fig. 9. Frame-wise accuracy of 3 text-spotters on videos exhibiting (a) *roll*, (b) *zoom*, (c) *tilt* and (d) *translation*

In Fig. 9b, we present similar plots for a video with 58 frames exhibiting *zoom* (out). The signboard in the video only contains Hindi text. The E2E-MLT model, however, outputs some English characters due to script mis-identification. Owing to this, the overall accuracy of the topmost plot (E2E-MLT + *OCR-on-the-go*) in Fig. 9b is most unstable. Our model again achieves the highest mean and lowest standard deviation across all the video frames.

The plots for a video with 75 frames exhibiting *tilt* (up-down) is shown in Fig. 9c. As shown, contrary to other figures, the *OCR-on-the-go* model performs poorly on this video. The reason for this is that the model perhaps overfits to its license plates dataset. E2E-MLT generalizes well with respect to *OCR-on-the-go* model, however, our model has the highest average accuracy. In Fig. 9d, we present similar plots for a video with 121 frames exhibiting *translation* (upward).

As discussed earlier in Sect. 5, the video clips recorded with the vertical camera movements in the setup possess jitter because the tripod does not allow for smooth translation while recording such videos. Our model, however, outputs the text with the highest accuracy and lowest standard deviation for the video we present in Fig. 9d.

8 Conclusion

In this paper, we presented CATALIST, a multi-task model for reading scene-text in videos and ALCHEMIST, a data generator that produces the videos from text images. These synthetic videos mimic the behaviour of videos captured with five different camera movements. We also presented the CATALIST$_d$ dataset of around two thousand real videos recorded with the camera movements mentioned above. By training the CATALIST model on both real and synthetic videos, we set new benchmarks for the task of reading multi-lingual scene-text in Hindi, Marathi, and English. The multi-level mask supervision improved either character or sequence (or both) accuracy on three different datasets with varying complexities.

9 Future Work

The camera movement information in CATALIST$_d$ dataset is ideal for Capsule Network [5,8]. Unlike conventional CNNs, capsule networks are viewpoint invariant. The transformation information can help capsules with video scene-text detection by helping the network learn about camera movements.

Acknowledgment. We thank Shubham Shukla for dataset collection and annotation efforts.

References

1. Abadi, M., et al.: Tensorflow: a system for large-scale machine learning. In: 12th {$USENIX$} Symposium on Operating Systems Design and Implementation (OSDI 2016), pp. 265–283 (2016)
2. Bartz, C., Yang, H., Meinel, C.: STN-OCR: a single neural network for text detection and text recognition. arXiv preprint arXiv:1707.08831 (2017)
3. Bušta, M., Neumann, L., Matas, J.: Deep textspotter: an end-to-end trainable scene text localization and recognition framework. In: International Conference on Computer Vision (2017)
4. Bušta, M., Patel, Y., Matas, J.: E2E-MLT - an unconstrained end-to-end method for multi-language scene text. In: Carneiro, G., You, S. (eds.) ACCV 2018. LNCS, vol. 11367, pp. 127–143. Springer, Cham (2019). https://doi.org/10.1007/978-3-030-21074-8_11
5. Duarte, K., Rawat, Y.S., Shah, M.: Videocapsulenet: a simplified network for action detection. arXiv preprint arXiv:1805.08162 (2018)

6. Gupta, A., Vedaldi, A., Zisserman, A.: Synthetic data for text localisation in natural images. In: Proceedings of the IEEE Conference on Computer Vision and Pattern Recognition, pp. 2315–2324 (2016)

7. Hartley, R., Zisserman, A.: Multiple View Geometry in Computer Vision. Cambridge University Press, Cambridge (2003)

8. Hinton, G.E., Sabour, S., Frosst, N.: Matrix capsules with EM routing. In: International Conference on Learning Representations (2018)

9. Karatzas, D., et al.: ICDAR 2015 competition on robust reading. In: 2015 13th International Conference on Document Analysis and Recognition (ICDAR), pp. 1156–1160. IEEE (2015)

10. Karatzas, D., et al.: ICDAR 2013 robust reading competition. In: 2013 12th International Conference on Document Analysis and Recognition, pp. 1484–1493. IEEE (2013)

11. Liao, M., Shi, B., Bai, X., Wang, X., Liu, W.: Textboxes: a fast text detector with a single deep neural network. In: AAAI, pp. 4161–4167 (2017)

12. Milletari, F., Navab, N., Ahmadi, S.A.: V-net: fully convolutional neural networks for volumetric medical image segmentation. In: 2016 Fourth International Conference on 3D Vision (3DV), pp. 565–571. IEEE (2016)

13. Liao, M., Shi, B., Bai, X.: TextBoxes++: a single-shot oriented scene text detector. CoRR abs/1801.02765 (2018)

14. Reddy, S., Mathew, M., Gomez, L., Rusinol, M., Karatzas, D., Jawahar, C.: RoadText-1K: text detection & recognition dataset for driving videos. In: 2020 IEEE International Conference on Robotics and Automation (ICRA), pp. 11074–11080. IEEE (2020)

15. Saluja, R., Maheshwari, A., Ramakrishnan, G., Chaudhuri, P., Carman, M.: Robust end-to-end systems for reading license plates and street signs. In: 2019 15th IAPR International Conference on Document Analysis and Recognition (ICDAR), pp. 154–159. IEEE (2019)

16. Shi, B., Bai, X., Yao, C.: An end-to-end trainable neural network for image-based sequence recognition and its application to scene text recognition. IEEE Trans. Pattern Anal. Mach. Intell. $39(11)$, 2298–2304 (2017)

17. Smith, R., et al.: End-to-end interpretation of the French street name signs dataset. In: Hua, G., Jégou, H. (eds.) ECCV 2016, Part I. LNCS, vol. 9913, pp. 411–426. Springer, Cham (2016). https://doi.org/10.1007/978-3-319-46604-0_30

18. Szegedy, C., Vanhoucke, V., Ioffe, S., Shlens, J., Wojna, Z.: Rethinking the inception architecture for computer vision. In: Proceedings of the IEEE Conference on Computer Vision and Pattern Recognition, pp. 2818–2826 (2016)

19. Wojna, Z., et al.: Attention-based extraction of structured information from street view imagery. In: 2017 14th IAPR International Conference on Document Analysis and Recognition (ICDAR), vol. 1, pp. 844–850. IEEE (2017)

20. Yu, F., et al.: BDD100K: a diverse driving video database with scalable annotation tooling. arXiv preprint arXiv:1805.04687, vol. 2, no. 5, p. 6 (2018)

21. Zhang, Y., Gueguen, L., Zharkov, I., Zhang, P., Seifert, K., Kadlec, B.: Uber-text: a large-scale dataset for optical character recognition from street-level imagery. In: SUNw: Scene Understanding Workshop-CVPR, vol. 2017 (2017)

DDocE: Deep Document Enhancement with Multi-scale Feature Aggregation and Pixel-Wise Adjustments

Karina O. M. Bogdan[1]([⊠])⬡, Guilherme A. S. Megeto[1]⬡, Rovilson Leal[1]⬡,
Gustavo Souza[1]⬡, Augusto C. Valente[1]⬡, and Lucas N. Kirsten[2]⬡

[1] Instituto de Pesquisas Eldorado, Campinas, SP, Brazil
{karina.bogdan,guilherme.megeto,rovilson.junior,
gustavo.souza,augusto.valente}@eldorado.org.br
[2] HP Inc., Porto Alegre, RS, Brazil
lucas.nedel.kirsten@hp.com

Abstract. Digitizing a document with a smartphone might be a difficult task when there are suboptimal environment lighting conditions. The presence of shadows and insufficient illumination may reduce the quality of the content, such as its readability, colors or other aesthetic aspect. In this work, we propose a lightweight neural network to enhance photographed document images using a feature extraction that aggregates multi-scale features and a pixel-wise adjustment refinement step. We also provide a comparison with different methods, including methods not originally proposed for document enhancement. We focused on aesthetics aspects of the images, for which we used traditional image quality assessment (IQA) metrics and others based on deep learning models of human quality perception of natural images. Our deep document enhancement (DDocE) method was able to lessen the negative effects of different artifacts, such as shadows and insufficient illumination, while also maintaining a good color consistency, resulting in a better final enhanced image than the ones obtained with other methods.

Keywords: Deep learning · Document enhancement · Image enhancement

1 Introduction

The popularity, multitasking purpose, and processing power of smartphones have been leading to a substitution of dedicated hardware for different tasks and a concentration of applications within these devices. For the task of digitizing an image, a dedicated hardware such as a flatbed scanner provides a highly controlled lighting environment. On the other hand, using a mobile phone to provide a high-quality-scanned aspect of a photographed document might be a difficult task when there are suboptimal environment lighting conditions, resulting in

© Springer Nature Switzerland AG 2021
E. H. Barney Smith and U. Pal (Eds.): ICDAR 2021 Workshops, LNCS 12916, pp. 229–244, 2021.
https://doi.org/10.1007/978-3-030-86198-8_17

images with artifacts caused by inadequate illumination, presence of shadows, or blurriness due to camera focus.

Many existing approaches for image enhancement focus on natural images, specially to solve the problem of low-light environments and degraded images [8, 20, 26, 30]. Natural image enhancement shares challenging aspects with document enhancement mostly due to the environment lighting influence. However, there are aspects that are specially crucial for the document enhancement task (e.g., the presence of shadows in text regions, and crumpled paper) since they can interfere in the readability of the text. Moreover, the enhancement must preserve the color regions and provide a uniform enhancement over each region of the document image to avoid undesired effects, for instance, on regions that are less affected by the artifacts.

Techniques for document enhancement include traditional image processing algorithms [5, 25] and machine learning approaches [10, 13, 17, 24]. The enhancement can focus on the document geometry, which can be distorted, i.e., not flat considering the camera perspective, requiring techniques to unwarp the deformed document through rectification [18, 23]; or focus on removing scanning artifacts included in the image capturing process, such as effects of an uneven illumination over the document [5, 10, 13, 17, 24]. The latter can also consider a document image with a crumpled remaining aspect, but in this case, the capture process is taken with the unwarped version of the document.

As aforementioned, one of the most challenging aspects of document enhancement can rise from non-uniform illumination over the captured document. But, although shadow removal methods [17, 24] have a similar objective regarding the elimination of shadows and uneven illumination, they mostly do not enhance other aesthetic aspects of the document, such as colors and remaining artifacts of crumpled paper. For instance, Lin et al. [17] train their method on shadow/shadow-free document pairs that were captured in the same environment, as their objective is purely shadow removal.

In this work, we are not only interested in removing shadows and uneven illumination artifacts, but also enhancing the photographed image to be as close as possible to its original digital version, or a high quality artifact-free scanned image, before printing the document. An example of the task and results of state-of-the-art models are presented in Fig. 1. We address the task of document enhancement through a method that aggregates features from multiple scales producing enhancement adjustments at the pixel level to be applied cumulatively in a refinement process, based on Guo et al. [8]. The proposed method is designed in an end-to-end fashion, being, to the best of our knowledge, one of the first methods to this end based on deep learning that is trained from scratch, i.e., without requiring pre-trained models. As shown in the experiments section, the method is not only able to deal with a wide range of documents including colored documents, but also produces a lightweight solution for the task.

Our main contributions include the following. We propose a lightweight enhancement model, named DDocE, for photographed documents that uses a feature extraction module based on a multi-scale feature aggregation approach

| Input | Original image | Our model | HP Inc. [13] | Fan [5] | Zero-DCE [8] full-reference |

Fig. 1. Example of document enhancement results. The method of Fan [5] and our model present the best results, whereas the HP Inc. [13] and the Zero-DCE [8] model over enhance the image, specifically the top region of the apple, for which Zero-DCE loses the color saturation of the region.

that expands the pixel neighborhood view, which is demonstrated to be an important factor to deal with colored documents. We present an investigation of other state-of-the-art models, which were not originally proposed for the problem, to build a strong comparison and possible baseline for document enhancement. We provide an extensive set of experiments using a robust evaluation protocol that considers complementary IQA metrics, including metrics based on the human perceptual assessment of natural images, to assess the quality of the produced enhancements. Moreover, we also propose a simple data augmentation technique to improve the enhancements.

2   Related Work

Enhancement of Natural Images. The enhancement of natural images has an aesthetic objective of improving a digital image to be more pleasant to human perception considering different characteristics such as color balance, contrast, brightness, sharpness, etc. [7]. Works such as [2,8,14,19,20,30] aim to automate the human effort to provide partial or full image enhancement. But although all these models tackle the image enhancement problem, they focus on specific branches of the task. For instance, the enhancement can focus on improving the aesthetic aspect that is highly affected [2,8,14] or not [19,20,30] by environment factors included in the capture process.

Regarding the models that focus on improving environment artifacts included in the digital images, Islam et al. [14] proposed a generative model for underwater image enhancement, while Guo et al. [8] and Atoum et al. [2] proposed models for low-light exposure image enhancement. Moreover, Guo et al. [8] focused on global enhancement by pixel-wise adjustments applied over a refinement process to remove the underexposed aspect of the image, and Atoum et al. [2] focused on color enhancement by proposing a color-wise attention model that separates the enhancement of the lightness and color components.

Enhancement of Document Images. The goal of document enhancement is to improve the digital image quality of a document by removing artifacts or restoring any degradation effect generated by the capturing process (e.g., using a mobile camera or a flatbed scanner) to facilitate human readability, which could aid in the extraction and recognition of textual content by an Optical Character Recognition (OCR) system, but mostly aiming to restore its original digital appearance [6,10,11].

Regarding situations of controlled environments (e.g., controlled illumination) or simpler images (e.g., black and white), traditional image processing techniques may be sufficient to remove noise and enhance a region of interest, using techniques such as histogram equalization and binarization [10]. However, for colored documents and environments with inadequate lighting, binarization, for example, will affect the document appearance which will no longer resemble the original digital version. For those cases, a more robust document enhancement approach is required. To that end, Fan [5] proposed a watershed-based segmentation model with traditional image processing techniques, such as color, illumination, and sharpening corrections. On the other hand, He et al. [10] proposed a deep learning model that learns binarization maps to be used in a refinement process of a model that learns degradation patterns in the document images. Finally, HP Inc. [13] proposed a method that first tries to identify the document content through segmentation and then uses this segmentation mask to enhance the document image using an ensemble of modern deep learning architectures.

Image Quality Assessment (IQA). In order to evaluate the quality of a digital image, enhancement approaches mostly rely on image quality assessment (IQA) metrics to evaluate the aesthetic aspect of their improved images [3,4, 21,28,31]. Regarding the specific task of document image quality assessment (DIQA), some metrics were also proposed, such as the work of Hussain et al. [12] that proposed a no-reference metric, which is different from the scenario considered here, since we expect to compare our enhanced image to a reference image; and the work of Alaei et al. [1] that proposed a model based on local and global texture similarity to predict image quality on a data set considering only JPEG image compression distortions applied on a small set of reference images[1].

3   Multi-scale Feature Aggregation for Document Enhancement

We propose a deep learning approach that aggregates multi-scale and contextual features from the input document image and generates enhancement curves to be applied at pixel level in a refinement process. First, the model extracts features from the input image with a squeeze and expand architecture that uses an intermediate multi-scale context aggregation. The enhancement features obtained in this first process are then used to predict pixel-wise curves to improve the photographed image and obtain an *artifact-free* (i.e., enhanced) version of it. This

---

[1] No open source implementation was found to reproduce the proposed metric.

process uses a refinement procedure where, at each step, one pixel-wise enhancing curve is applied on the previous enhanced image, being the last step the one generating the final enhanced image. An overview of our model is presented in Fig. 2. The following outlines each component of our model.

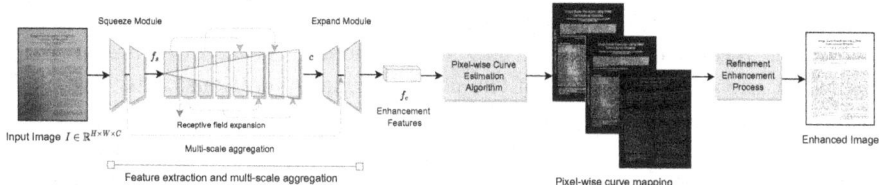

Fig. 2. Overview of the model. From an input image I we first squeeze and generate features f_s. Then, we pass f_s through the multi-scale feature aggregation generating features c. From c, the expand module generates the enhancement features f_e that are used to estimate the n pixel-wise curves with same shape as I. These curves are applied cumulatively over the input image in order to enhance it. To visualize the pixel-wise adjustments in the figure, we considered all RGB adjustments for $n = 3$ curves altogether.

Squeeze-Expand Modules. The squeeze and expand modules work as an encoding and decoding process. The squeeze module downsamples the input image $I \in \mathbb{R}^{H \times W \times C}$, such that C is the number of color channels, into a feature vector $f_s \in \mathbb{R}^{H' \times W' \times C_s}$, such that $H' \leq H$, $W' \leq W$. Hence, this first module decreases the resolution of the input image for the following *multi-scale aggregation* module, which processes the features maintaining its resolution. The features f_s are then used on the multi-scale feature aggregation module to generate the context and aggregated features $c \in \mathbb{R}^{H' \times W' \times 2C_s}$. The expand module then upscales the feature vector c up to $f_e \in \mathbb{R}^{H \times W \times C_e}$. Therefore, we upscale c up to the original height and width of the input image I to generate the pixel-wise adjustments in the curve estimation step. The squeeze module is composed of two convolutional layers using kernel size 3×3 with stride 2. The expand module is composed of two transposed convolutional layers using kernel size 3×3 with stride 2 where $C_e = 32$.

Multi-scale Feature Aggregation. In order to aid in the identification of camera-captured artifacts, such as noise and shadows, one can consider different regions of the pixel neighborhood in order to capture more context and alleviate the effort of the algorithm to differentiate artifacts from real parts of the original document content. This expanded view of the pixel neighborhood can even enable a proper enhancement of particular regions that are more affected than others (e.g., non-uniform illumination over the document).

Towards a global enhancement that is still capable of dealing with different local artifacts, we propose to use a feature aggregator that encloses multi-scale features from the input image I. These multi-scale features are extracted by expanding the view of the input features pixel neighborhood by augmenting the receptive field of the convolutional operations through dilated convolutions. We extract these features with a module based on the work of Yu et al. [29] for semantic segmentation, which originally expands the receptive field for feature extraction from 3×3 up to 67×67 of the features resolution, such that each expansion step represents a different scale of the receptive field.

In our feature extraction process of the multi-scale aggregation module, we expand the receptive field up to 69×69 and follow the architecture of Yu et al. [29] for the dilation factors of the convolutional operations. This module is added between our *squeeze-expand modules* and uses as input the features $f_s \in \mathbb{R}^{H' \times W' \times C_s}$, from the squeeze module, to aggregate multi-scale information and generate a contextual feature vector $c \in \mathbb{R}^{H' \times W' \times 2C_s}$. Therefore, we maintain the features resolution in c, but change the receptive field over the network. The features c are then used by the expand module, which generates f_e. The final architecture of the multi-scale feature aggregation is composed of eight blocks of one convolution and one Leaky Relu activation function ($\alpha = 0.2$) operation following the structure of Table 1 with skip connections between initial and mid-level layers as illustrated in Fig. 2.

Table 1. Architecture of the multi-scale feature aggregation module that encloses and processes different scales of the input features f_s.

Layer	1	2	3	4	5	6	7	8	Output
Convolution	3×3	3×3	3×3	3×3	3×3	3×3	3×3	3×3	–
Dilation factor	1	1	2	3	8	16	1	1	–
Receptive field	3×3	5×5	9×9	17×17	33×33	65×65	67×67	69×69	–
Output channels	C_s	C_s	C_s	C_s	C_s	C_s	C_s	C_s	$2C_s$
Input channels	C_s	C_s	C_s	C_s	C_s	$2C_s$	$2C_s$	$2C_s$	$2C_s$
Skip connection (concatenated layer)	–	–	–	–	1	2	3	4	–

Pixel-Wise Enhancement Process. From the aggregated features and its posterior expansion that generates features f_e, we estimate n pixel-wise enhancement curves to be applied over the mobile photographed document in a refinement process. The pixel-wise curve estimation process produces curve maps $\mathcal{A}_i \in \mathbb{R}^{H \times W \times C} \; \forall \; i \in \mathbb{N}, 1 \leq i \leq n$ from f_e, where each curve is applied at its corresponding iteration i of the refinement process. Each curve estimation model, which generates $\mathcal{A}_i$ from f_e, is implemented using a convolutional layer of kernel size 3×3 and stride 1 followed by a hyperbolic tangent activation function.

We adapted the refinement process of Guo et al. [8], which was initially proposed for low-light exposure of natural images with no target reference, to a full-reference scenario, where we considered the original digital image of the document (i.e., the PDF version) as target (ground truth). The refinement process generates an image $E_i \in \mathbb{R}^{H \times W \times C}$, such that $1 \leq i \leq n$, after a sequence of operations, defined in Eq. 1, applied over the previous enhancement. The result at each step is defined by:

$$E_i = E_{i-1} + \mathcal{A}_i E_{i-1}(1 - E_{i-1}), \tag{1}$$

where E_0 is the normalized input image $I \in [0,1]$, and $\mathcal{A}_i \in \mathbb{R}^{H \times W \times C}$ is a trainable matrix that contains coefficients $\alpha_{kl} \in [-1,1]$, where $0 \leq k \leq H$ and $0 \leq l \leq W$, that controls the magnitude and exposure level of each pixel in the second term of Eq. 1. The second term of Eq. 1 works as a highlight-and-diminish operation for the enhanced image E_{i-1} as it is encouraged to learn how to uncover low-light exposure or shadow regions and how to remove the noise in the image with the full-reference loss function. Therefore, after the last cumulative enhancement step, we generate the final enhanced image $\hat{I}$, such that $\hat{I} = E_n$.

The loss function used to train the model consists of the L1 full-reference loss between the image enhanced by our model $\hat{I} \in \mathbb{R}^{H \times W \times C}$ and the expected enhancement or ground truth $I_{GT} \in \mathbb{R}^{H \times W \times C}$:

$$\mathcal{L} = \lambda \|I_{GT} - \hat{I}\|_1 \tag{2}$$

where λ is a weight[2] for the enhancement comparison.

4   Experiments

Implementation Details. We empirically set $\lambda = 50$ and $n = 3$, and considered RGB images ($C = 3$). The networks were trained from scratch with the Adam [16] optimizer with a learning rate of 10^{-4} ($\beta_1 = 0.9$). The weights of our models were initialized from a normal distribution $\mathcal{N}(0, 0.02)$ and the models were trained up to 50 epochs. We reduced the learning rate by a factor of 0.1 after 5 epochs of no significant improvements in the validation loss, and, similarly, stopped the training after 30 epochs of no significant improvements in the validation loss. All convolutional layers use a padding to the input that ensures that the output has the same shape as the input. A dropout of 0.05 is applied in the first two layers of the multi-scale aggregation module. All models were implemented with TensorFlow 2.3.0.

Datasets. For our experiments, we used a private dataset composed of several photographed types of documents such as: plain text, magazines, articles, advertising documents, flyers, and receipts; in black and white or including colors; and also with variations of shadows. Since such type of data is scarce and difficult to obtain, for the train set we considered as ground truth the enhancements obtained by the method of Fan [5], which were manually chosen to prevent the addition of poor enhancements as ground-truth. In the test set, we used pairwise images composed of mobile-captured document images and its original digital form (i.e., the PDF version) as the ground-truth images. Hence, we considered

---

[2] Although we considered only one loss, we used an adaptive optimizer with this setup. Therefore, we maintained this definition here for reproducibility purposes.

two sets of data: the train set composed of 674 pairs of images, the photographed document and the Fan [5] enhanced image; and the test set composed of 198 pairs of images, the photographed document and the original digital image. Overall, all image pairs were manually chosen in order to build a dataset composed of ground-truth enhancements of high quality and that presented a good level of improvement compared to the raw (photographed) image.

In order to augment the data, we performed random horizontal and vertical flips, and extracted 256×256 overlapping patches from the training images with a sliding window technique with a stride of 128. Additionally, we evaluated a sampling method with variable window size to increase the pixel neighborhood of the image patches, which we referred to as *pyramid sampling*. The pyramid setup considered several windows sizes with patches of size: 256×256 (stride 128), 512×512 (stride 256), 1024×1024 (stride 512), and the full image. Since the document images contain large blank regions, we also apply a Laplacian operator over the patches to discard samples below an absolute sum of gradient threshold of $9 \cdot 10^5$.

Evaluation Metrics. For the evaluation protocol, we considered the following full-reference IQA metrics: peak signal-to-noise ratio (PSNR), multi-scale structural similarity (MS-SSIM) [28], PieAPP [21], WaDIQaM [3], LPIPS [31], and DISTS [4] – where the last four metrics are perceptual metrics based on deep learning. The quantitative results were calculated considering a standard resolution of 512×512 for all test images. Therefore, for testing, we considered the full images with this fixed resolution, which is different from the data augmentation applied on the train set that considered patches.

Baseline Models. We compare our model with six baselines methods: U-Net [22], Pix2Pix [15], Pix2PixHD [27], Fan [5], HP Inc. [13][3], and Zero-DCE [8]. The first three were proposed to perform semantic segmentation and image-to-image translation, and the last three for image enhancement tasks. Although the first three methods were not originally proposed for enhancement tasks, they are competitive or state-of-the-art models and some of them were adapted to similar tasks [10]. Therefore, to strengthen our evaluation protocol, we adapted the original task of these models to perform full-reference document enhancement, which considers both the photographed document and the corresponding original image of the document to be available during training. For fair comparison, we trained the following models from scratch with the same dataset used to train our model: U-Net [22], Pix2Pix [15], Pix2PixHD [27], and Zero-DCE [8]. Moreover, we used the pre-trained model of HP Inc. [13] that was trained with a dataset composed partially by the dataset considered in this work.

4.1 Quantitative Evaluation

The quantitative comparison between our model and the baselines for document enhancement is presented in Table 2. Although the first group of baselines were

---

[3] Code and pre-trained models kindly provided by the authors.

Table 2. Quantitative results. First group of results correspond to models not originally proposed for document enhancement but that were adapted to it. The second group corresponds to models originally proposed for image enhancement: for document (Fan [5] and HP Inc. [13]) and natural images (Zero-DCE [8]), respectively. Best results for each metric are highlighted in boldface. # Indicates the round up number of parameters.

Model	# P.	PSNR↑	MS-SSIM↑	PieAPP↓	WaDIQaM↑	LPIPS↓	DISTS↓
U-Net [22]	2M	15.9 ± 2.7	0.81 ± .07	**0.89 ± 0.74**	0.56 ± .10	0.166 ± .05	0.196 ± .05
Pix2Pix [15]	57M	16.2 ± 2.7	0.81 ± .08	1.18 ± 0.81	0.55 ± .13	0.168 ± .05	0.191 ± .05
Pix2PixHD [27]	113M	15.7 ± 2.7	0.79 ± .09	0.95 ± 1.12	0.51 ± .12	0.206 ± .06	0.236 ± .06
Fan [5]	–	15.1 ± 3.4	0.80 ± .08	1.00 ± 0.94	0.58 ± .08	0.163 ± .05	0.162 ± .05
HP Inc. [13]	5M	**16.9 ± 3.0**	**0.84 ± .07**	1.49 ± 0.88	0.57 ± .10	0.157 ± .05	0.169 ± .04
Zero-DCE [8] full-reference	79k	15.9 ± 3.0	0.82 ± .08	1.10 ± 0.84	0.58 ± .09	0.165 ± .05	0.186 ± .04
Our model	595k	16.1 ± 2.7	0.82 ± .07	1.09 ± 0.86	0.58 ± .09	0.157 ± .05	0.167 ± .05
Our model w/ pyramid	595k	16.1 ± 3.0	0.83 ± .07	1.08 ± 0.79	**0.63 ± .06**	**0.143 ± .05**	**0.147 ± .04**

not originally proposed for the full-reference document enhancement problem, they achieved competitive results. Between them, the Pix2PixHD [27] generated the worst result. However, this is a promising result considering that the model does not use a pixel-wise reconstruction loss (e.g., L1 or L2) as the other ones. Overall, the U-Net [22] was considered the best model of the first group since it achieved the best result in four of the six IQA metrics. Nevertheless, despite the results of six IQA metrics, an assertive conclusion on this first group is difficult to be made as some results are very close to each other.

Regarding the second group of models, which were originally proposed for image enhancement, we first noticed that the algorithm proposed by Fan [5] produced competitive results despite being a traditional approach for the problem and based only on image processing techniques. Overall, the work of HP Inc. [13] generated the best results considering the PSNR and MS-SSIM metrics and close values to our best results. We also included the results of the Zero-DCE [8] model adapted to our full-reference task. Although our refinement process algorithm is based on Guo et al. [8], we achieved better results, which could indicate the relevance of our proposed feature extraction based on multi-scale aggregation. Moreover, our model generates a lightweight solution considering the approximated number of parameters of each model evaluated.

4.2 Qualitative Evaluation

We present examples of qualitative results in Figs. 3, 4, and 5. From these figures we can observe that there are particular types of documents that are challenging for all models, such as the first document in Fig. 3 for which all models missed the gray dog positioned at the center of the document, probably mistaking this region for a shadow or scanning artifact. The result with the U-Net [22] model preserves residual pixels of that part, but still lacks the proper enhancement of the region. Regarding documents with colored regions, some models presented

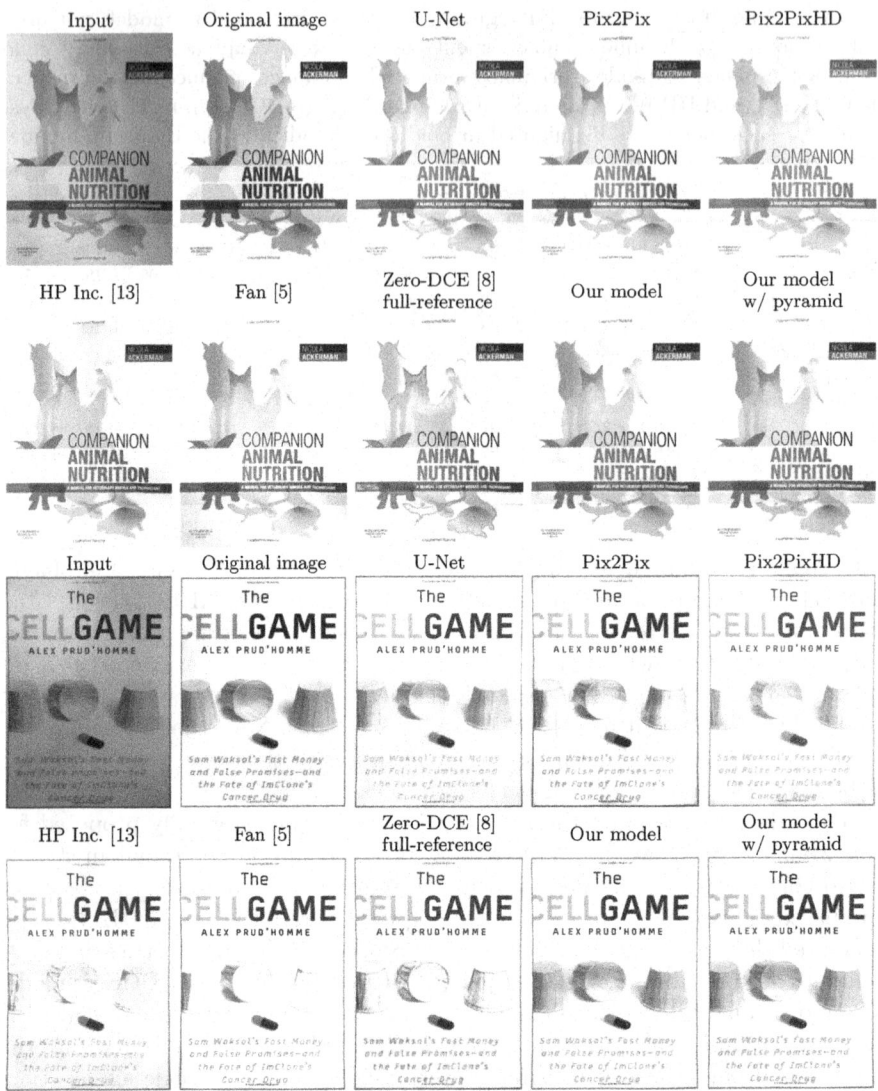

Fig. 3. Qualitative results of each model evaluated for documents with large colored regions.

a better filling of existing colored regions in the enhanced image by preserving the color saturation of the areas. Examples include the head of the orange dog in the first document of Fig. 3 and the cups in the second document in the same figure, for which our model achieved the best qualitative results.

Another challenging scenario is the one presented in Fig. 4, which includes heavily shaded regions. For the first document in Fig. 4, with the shade concentrated on the bottom-left corner area, our model generated the best results but

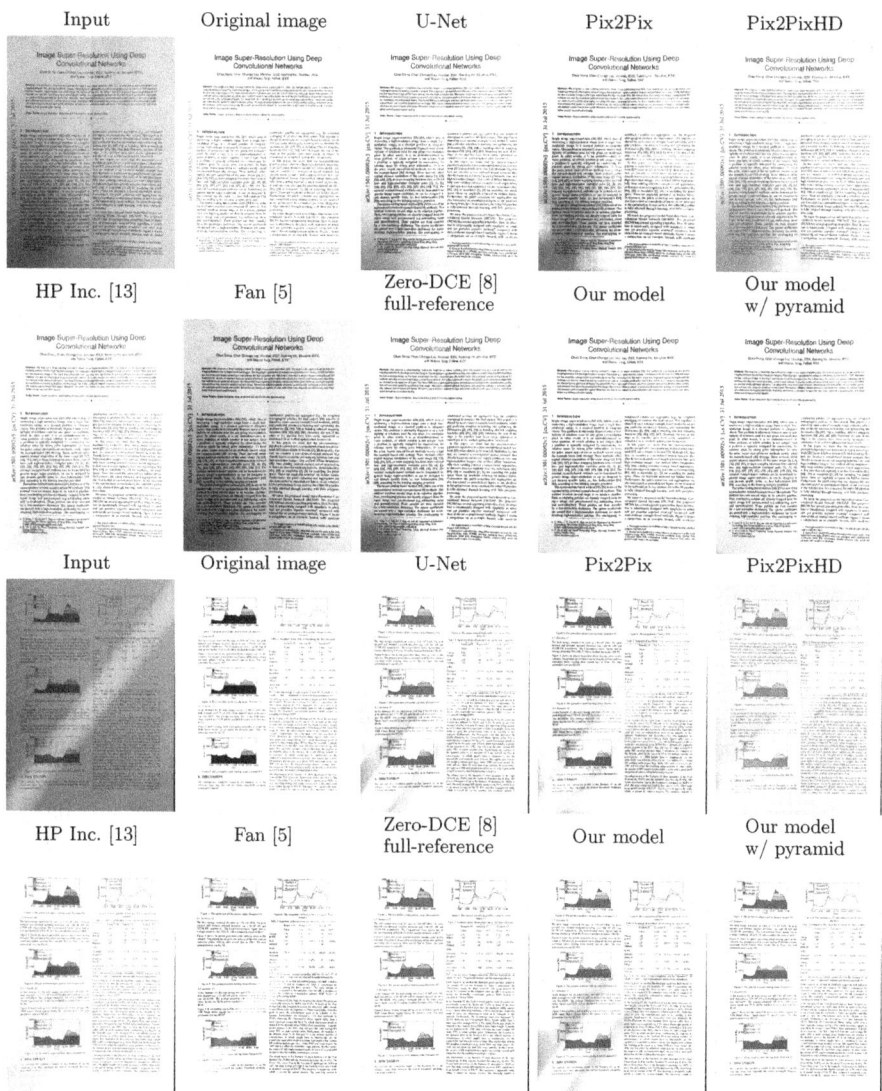

Fig. 4. Qualitative results of each model evaluated for documents with large areas of text and shade artifacts.

still kept a residual of artifacts in the corner of the shaded region. The model of HP Inc. [13] also generated a good result, but left traces of residual artifacts which are more alleviated in ours. On the other hand, for the second document, the best results were achieved with Fan [5] and our model with the pyramid sampling strategy. Although both results leave residual artifacts in the enhanced image, mostly by whitening the region without the shade, our model obtained the best results by diminishing the shade effect and maintaining a uniform aspect

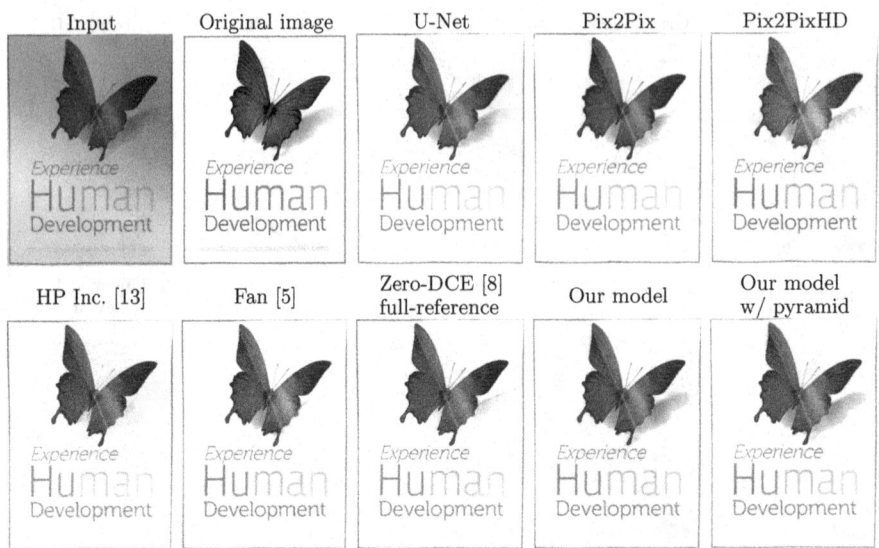

Fig. 5. Qualitative results of each model evaluated for a document with colored regions and with a shade that can be mislead as scanning artifacts.

in the enhancement. Nevertheless, there are parts of the document that can be incorrectly interpreted as a scanning artifact, such as the shade in the document of Fig. 5, for which only our model was capable of correctly enhancing it by maintaining the butterfly shadow. Lastly, considering the results of our model with different sampling strategies, we noticed that the dense sampling provided by the pyramid strategy, which samples large views of the image adding more context information, generated better results mostly aiding in the removal of residual scanning artifacts.

4.3 Ablation Studies

Feature Extraction. Since our main contribution lies on the feature extraction module, we provide a comparison of different models for this part. We start by considering the feature extractor of Guo et al. [8] as we use their pixel-wise refinement process. Then, we also considered the architecture of Hamaguchi et al. [9] for the multi-scale aggregation module, which uses a different setup for the dilation factors of the convolutional operations. Moreover, we also considered variations of our model by removing the squeeze and expand modules, and by considering a traditional convolutional operation (dilation factor of 1) instead of the setup considered for the multi-scale aggregation module presented in Sect. 3.

A comparison between the results of the ablation study is presented in Fig. 6 and the quantitative results in Table 3 (first and second groups of results). One aspect that we noticed in almost all results with colored documents is that other approaches for the feature extraction generally obtain results where the colored

Table 3. Ablation experiments including: first group - variations in the feature extractor of our model, second group - variations of our model; and third group - in the refinement process. Best results for each metric are highlighted in boldface.

Model	PSNR↑	MS-SSIM↑	PieAPP↓	WaDIQaM↑	LPIPS↓	DISTS↓
Our model w/ Zero-DCE [8] extractor	15.7 ± 3.0	0.81 ± .08	0.94 ± .84	0.58 ± .07	0.163 ± .05	0.163 ± .04
Multi-scale aggregation w/ Hamaguchi et al. [9]	16.0 ± 2.7	0.81 ± .08	1.02 ± .88	0.59 ± .08	0.161 ± .05	0.167 ± .04
Multi-scale aggregation w/ Hamaguchi et al. [9] w/o squeeze-expand	15.9 ± 2.9	0.82 ± .08	1.01 ± .85	0.59 ± .07	0.160 ± .05	0.162 ± .04
Our model w/o squeeze-expand	16.1 ± 2.9	0.82 ± .07	**0.92 ± .80**	0.60 ± .07	0.154 ± .05	0.162 ± .04
Multi-scale aggregation (dilation factor of 1)	16.1 ± 2.8	0.81 ± .08	1.03 ± .86	0.59 ± .08	0.158 ± .05	0.159 ± .04
Our model w/ $n = 1$	14.2 ± 2.0	0.81 ± .07	1.21 ± .98	0.46 ± .12	0.178 ± .05	0.231 ± .05
Our model w/ $n = 5$	16.1 ± 2.8	0.82 ± .08	1.02 ± .81	0.59 ± .09	0.155 ± .05	0.166 ± .04
Our model w/ $n = 8$	**16.1 ± 2.7**	0.82 ± .07	1.07 ± .88	0.58 ± .09	0.157 ± .05	0.169 ± .04
Our model ($n = 3$)	**16.1 ± 2.7**	0.82 ± .07	1.09 ± .86	0.58 ± .09	0.157 ± .05	0.167 ± .05
Our model ($n = 3$) w/ pyramid	16.1 ± 3.0	**0.83 ± .07**	1.08 ± .79	**0.63 ± .06**	**0.143 ± .05**	**0.147 ± .04**

regions are poorly filled with their true colors, although other aspects such as shadows are correctly enhanced. Even the variations of our model result in poorly filled regions, which could indicate the importance of the squeeze, multi-scale aggregation, and expand modules altogether. Nonetheless, regions of text with colored background are still a challenge for our model as the background region sometimes is overly enhanced generating an overly brightened aspect where parts of the colored region lose its saturation.

Refinement Process. To understand how learning less or more adjustments really impact the results, we also evaluated the number of refinements. Guo et al. [8] had the best results for low-light enhancement of natural images with $n = 8$ refinements. Hence, we evaluate this number along with $n = 1$ and $n = 5$, for which we present the results in the third group of Table 3. From these results, we observe that one iteration alone is not enough to enhance the document images. On the other hand, five and eight refinements seem to obtain similar results with our default setup of $n = 3$. However, some qualitative results presented in Fig. 6 show us that the difference between the refinements from $n = 3$ up to $n = 8$ is hardly noticed as the enhanced results look very similar.

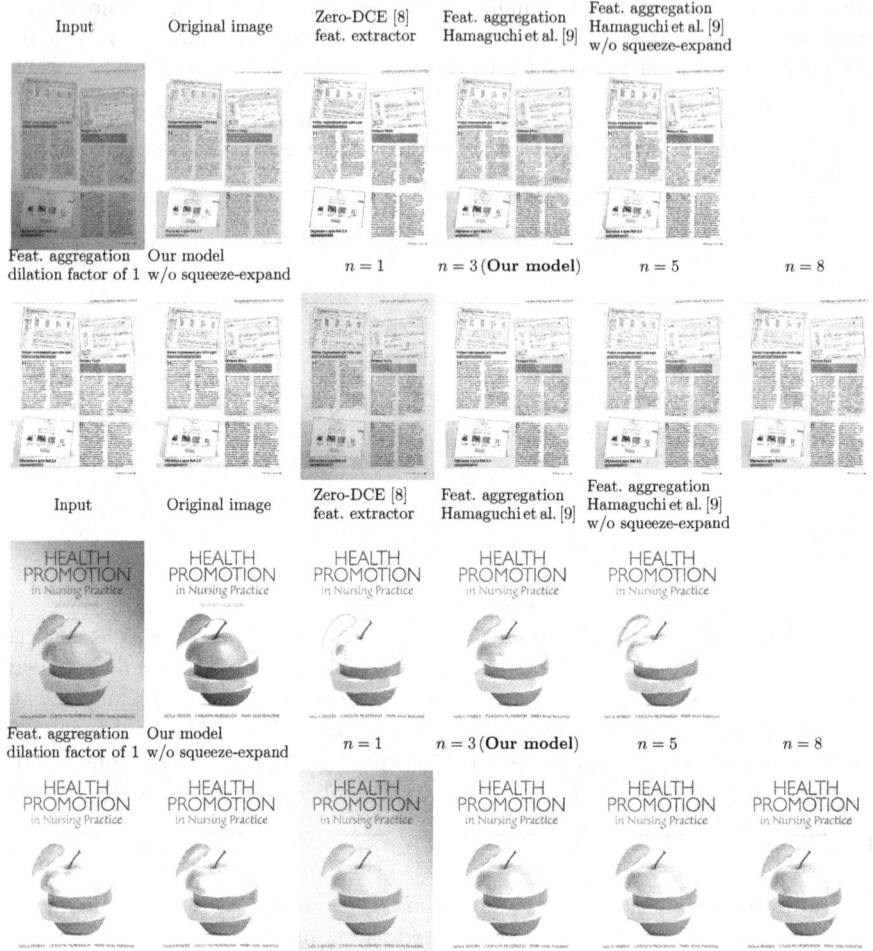

Fig. 6. Comparison between the results from the ablation studies: variations in the feature extractor of our model, variations of our model; and in the refinement process.

5   Conclusions

We have presented a model to enhance photographed document images based on a multiple scale aggregation procedure. The DDocE model first extracts enhancement features from the input image, which are then used in a pixel-wise enhancement and refinement process. We include in our evaluation protocol models not originally proposed for the document enhancement task, for which we made adaptations to consider a full-reference enhancement problem. We show through an extensive set of experiments the role of each component of our model indicating the importance of them altogether. Although there are still some open challenging scenarios, such as the proper enhancement of background colored regions with text and complex shadows, we show that our model generates promising results maintaining the content and improving the visual quality of the input photographed image, besides having the advantage of being a lightweight model.

Acknowledgments. This work was done in cooperation with HP Inc. R&D Brazil, using incentives of the Brazilian Informatics Law (Law n°. 8.2.48 of 1991). The authors would like to thank Sebastien Tandel, Erasmo Isotton, Rafael Borges, and Ricardo Ribani.

References

1. Alaei, A., Conte, D., Blumenstein, M., Raveaux, R.: Document image quality assessment based on texture similarity index. In: IAPR Workshop on Document Analysis Systems (DAS), pp. 132–137 (2016)
2. Atoum, Y., Ye, M., Ren, L., Tai, Y., Liu, X.: Color-wise attention network for low-light image enhancement. In: CVPRW, pp. 2130–2139 (2020)
3. Bosse, S., Maniry, D., Müller, K., Wiegand, T., Samek, W.: Deep neural networks for no-reference and full-reference image quality assessment. IEEE TIP **27**(1), 206–219 (2018)
4. Ding, K., Ma, K., Wang, S., Simoncelli, E.P.: Image quality assessment: unifying structure and texture similarity. CoRR abs/2004.07728 (2020)
5. Fan, J.: Enhancement of camera-captured document images with watershed segmentation. In: CBDAR, pp. 87–93 (2007)
6. Farrahi Moghaddam, R., Cheriet, M.: A variational approach to degraded document enhancement. IEEE TPAMI **32**(8), 1347–1361 (2010)
7. Lv, F., Li, Y., Lu, F.: Attention guided low-light image enhancement with a large scale low-light simulation dataset. CoRR abs/1908.00682 (2020)
8. Guo, C., et al.: Zero-reference deep curve estimation for low-light image enhancement. In: CVPR, pp. 1780–1789 (2020)
9. Hamaguchi, R., Fujita, A., Nemoto, K., Imaizumi, T., Hikosaka, S.: Effective use of dilated convolutions for segmenting small object instances in remote sensing imagery. In: WACV, pp. 1442–1450 (2018)
10. He, S., Schomaker, L.: DeepOtsu: document enhancement and binarization using iterative deep learning. Pattern Recognit. **91**, 379–390 (2019)
11. Hedjam, R., Cheriet, M.: Historical document image restoration using multispectral imaging system. Pattern Recognit. **46**(8), 2297–2312 (2013)
12. Hussain, M., Wahab, A.W.A., Idris, Y.I.B., Ho, A.T., Jung, K.H.: Image steganography in spatial domain: a survey. Signal Process.: Image Commun. **65**, 46–66 (2018)
13. INC, H.: A workflow for document enhancement through content segmentation and multiple enhancements. Technical Disclosure Commons (2020). https://www.tdcommons.org/dpubs_series/3119
14. Islam, M.J., Xia, Y., Sattar, J.: Fast underwater image enhancement for improved visual perception. IEEE Robot. Autom. Lett. **5**(2), 3227–3234 (2020)
15. Isola, P., Zhu, J.Y., Zhou, T., Efros, A.A.: Image-to-image translation with conditional adversarial networks. In: CVPR, pp. 5967–5976 (2017)
16. Kingma, D.P., Ba, J.: Adam: a method for stochastic optimization. In: ICLR (2015)

17. Lin, Y.H., Chen, W.C., Chuang, Y.Y.: BEDSR-Net: a deep shadow removal network from a single document image. In: CVPR, pp. 12905–12914 (2020)
18. Ma, K., Shu, Z., Bai, X., Wang, J., Samaras, D.: DocUNet: document image unwarping via a stacked U-Net. In: CVPR, pp. 4700–4709 (2018)
19. Moran, S., Marza, P., McDonagh, S., Parisot, S., Slabaugh, G.: DeepLPF: deep local parametric filters for image enhancement. In: CVPR, pp. 12826–12835 (2020)
20. Moran, S., McDonagh, S., Slabaugh, G.: CURL: neural curve layers for global image enhancement. CoRR abs/1911.13175 (2020)
21. Prashnani, E., Cai, H., Mostofi, Y., Sen, P.: PieAPP: perceptual image-error assessment through pairwise preference. In: CVPR, pp. 1808–1817 (2018)
22. Ronneberger, O., Fischer, P., Brox, T.: U-Net: convolutional networks for biomedical image segmentation. In: Navab, N., Hornegger, J., Wells, W.M., Frangi, A.F. (eds.) MICCAI 2015, Part III. LNCS, vol. 9351, pp. 234–241. Springer, Cham (2015). https://doi.org/10.1007/978-3-319-24574-4_28
23. Tian, Y., Narasimhan, S.G.: Rectification and 3D reconstruction of curved document images. In: CVPR, pp. 377–384 (2011)
24. Wang, B., Chen, C.L.P.: An effective background estimation method for shadows removal of document images. In: ICIP, pp. 3611–3615 (2019)
25. Wang, B., Chen, C.L.P.: Local water-filling algorithm for shadow detection and removal of document images. Sensors **20**(23), 6929 (2020)
26. Wang, R., Zhang, Q., Fu, C.W., Shen, X., Zheng, W.S., Jia, J.: Underexposed photo enhancement using deep illumination estimation. In: CVPR, pp. 6842–6850 (2019)
27. Wang, T.C., Liu, M.Y., Zhu, J.Y., Tao, A., Kautz, J., Catanzaro, B.: High-resolution image synthesis and semantic manipulation with conditional GANs. In: CVPR, pp. 8798–8807 (2018)
28. Wang, Z., Simoncelli, E.P., Bovik, A.C.: Multiscale structural similarity for image quality assessment. In: The Thirty-Seventh Asilomar Conference on Signals, Systems Computers, vol. 2, pp. 1398–1402 (2003)
29. Yu, F., Koltun, V.: Multi-scale context aggregation by dilated convolutions. In: ICLR (2016)
30. Zamir, S.W., et al.: Learning enriched features for real image restoration and enhancement. CoRR abs/2003.06792 (2020)
31. Zhang, R., Isola, P., Efros, A.A., Shechtman, E., Wang, O.: The unreasonable effectiveness of deep features as a perceptual metric. In: CVPR, pp. 586–595 (2018)

Handwritten Chess Scoresheet Recognition Using a Convolutional BiLSTM Network

Owen Eicher$^{(\boxtimes)}$, Denzel Farmer, Yiyan Li, and Nishatul Majid

Fort Lewis College, Durango, CO, USA
{oeicher,dfarmer,yli,nmajid}@fortlewis.edu

Abstract. Chess players while playing in Over-the-Board (OTB) events use scoresheets to record their moves by hand, and later the event organizers digitize these sheets for an official record. This paper presents a framework for decoding these handwritten scoresheets automatically using a convolutional BiLSTM neural network designed and trained specifically to handle chess moves. Our proposed network is pretrained with the IAM handwriting dataset [1] and later fine-tuned with our own Handwritten Chess Scoresheet (HCS) dataset [2,3]. We also developed two basic post-processing schemes to improve accuracy by cross-checking moves between the two scoresheets collected from the players with white and black pieces. The autonomous post-processing involves no human input and achieves a Move Recognition Accuracy (MRA) of 90.1% on our test set. A second semi-autonomous algorithm involves requesting user input on certain unsettling cases. On our testing, this approach requests user input on 7% of the cases and increases MRA to 97.2%. Along with this recognition framework, we are also releasing the HCS dataset which contains scoresheets collected from actual chess events, digitized using standard cellphone cameras and tagged with associated ground truths. This is the first reported work for handwritten chess move recognition and we believe this has the potential to revolutionize the scoresheet digitization process for the thousands of chess events that happen each day.

Keywords: Chess scoresheet recognition · Offline handwriting recognition · Convolutional BiLSTM network · Latin handwriting recognition · Handwritten chess dataset

1 Introduction

Chess is one of the most popular board games in the world, with approximately 605 million regular players (and more playing every day) [4]. During an Over-the-Board (OTB) chess event, both players in each match are generally required to record their own and their opponent's moves on a chess recording sheet, or scoresheet, by hand. These moves are usually recorded in a Standard Algebraic Notation (SAN). Figure 1, captured from an OTB chess tournament shows how players write moves during a live match along with the scoresheets collected from

© Springer Nature Switzerland AG 2021
E. H. Barney Smith and U. Pal (Eds.): ICDAR 2021 Workshops, LNCS 12916, pp. 245–259, 2021.
https://doi.org/10.1007/978-3-030-86198-8_18

the players with white and black pieces. These scoresheets contain a complete move history along with player information and match results. They are useful to keep as official game records and to settle any disputes that may arise, hence their popularity in competitive settings. Moves are not recorded digitally in an attempt to prevent computer assisted cheating, which is a big issue in the world of chess. The scoresheet system, while secure, leaves event organizers with sometimes hundreds of scoresheets which must be manually digitized into Portable Game Notation (PGN). This process requires extensive time and labor, as organizers must not only type each move, but in the case of a conflict or illegible writing, play out the game on a chess board to infer moves. As an alternative to this legacy approach, we propose a deep neural network architecture to perform automatic conversion from pictures of these scoresheets to a PGN text file. In most practical cases, a cellphone camera is an extremely convenient way of collecting data, much more so than a flat-bed scanner. Keeping this in mind, we developed our detection system, from data acquisition to training and testing, to perform digitization on cellphone camera photos.

Fig. 1. Captured from an Over-the-Board (OTB) chess tournament (left), highlighting scoresheets from the players with white (middle) and black pieces (right).

In theory, chess moves could be recognized with a standard offline Latin handwriting recognition system, since chess universally uses Latin symbols. But this generic approach has its shortcomings. Chess scoresheets offer many distinguishing features that we can leverage to significantly increase move recognition accuracy, including:

- Two scoresheets for each match are available since both players write on their own copy of all the moves played. These copies can be cross referenced for validation.
- Chess moves use a much smaller character set than the entire Latin alphabet, allowing only 31 characters as opposed to the 100+ ASCII characters.
- Traditional post-processing techniques (spell checking, Natural Language Processing or NLP, etc.) do not apply, instead chess moves can be ruled valid or invalid based on SAN syntax errors or move illegality dictated by the game rules.

– Handwritten moves are contained inside well defined bounding boxes (Fig. 2) with some natural degrees of shift, therefore the process of individual move segmentation is much simpler than in unconstrained handwriting.

Furthermore, the vast majority of chess moves are 2–5 characters long with only a few rare exceptions, and none are longer than 7 characters. These differences between the generic Latin script and handwritten chess moves make it worthwhile to approach this problem separately, while invoking the traditional wisdom of offline handwriting recognition. To accomplish this, we pretrain a deep neural network on an existing Latin handwriting dataset, IAM [1] and later fine-tune our model with redefined classes and network size adjustments. We also use post-processing algorithms to restrict output and improve accuracy, since valid chess moves allow only a limited set of letters, numbers, and symbols (Table 1 and 2) in specific positions (e.g., moves cannot start with a number, cannot end with a character, etc.). At this time of writing, no other work has been reported for a handwritten chess scoresheet recognition framework. Services for digitizing chess scoresheets such as Reine Chess [5] currently exist, but they require games be recorded on their proprietary scoresheets with very specific formats. They cannot be applied to existing scoresheet formats, and would require tournaments to alter their structure, causing a variety of problems. Scoresheet-specific solutions also offer no solution to retroactively digitize scoresheets and cannot be easily applied to other documents. The scope of this project is to create a general offline handwriting recognition framework, which in the future may be applicable to a variety of documents, with chess scoresheets featured as a single application. Furthermore, there was previously no publicly available chess scoresheet dataset that we could use for our network training. Therefore, we accumulated our own database, the Handwritten Chess Scoresheet (HCS) dataset [2,3]. This dataset contains scoresheets from actual chess events digitized with a cellphone camera and tagged with associated ground truths. This is a publicly accessible dataset to encourage further research.

Handwriting recognition generally takes one of two approaches: either character spotting followed by isolated character recognition, or full-word recognition. The first approach works better for scripts with large alphabets and other complicated attributes like fusion of characters [6]. Our framework uses the later approach, as it does not require character level ground truth location tagging which makes data preparation much simpler. The script in our problem is also not complex enough to warrant character spotting. We use a Recurrent Neural Network (RNN) which can take sequence input and predict sequence outputs to solve this problem. Specifically, we use the Bidirectional Long-Short Term Memory (BiLSTM) variant of RNN, as basic RNNs are susceptible to vanishing gradients. In contrast, Long Short-Term Memory (LSTM) cells retain information seen earlier in the sequence more effectively, and bidirectional LSTMs (BiLSTM) are able to make predictions using information in both the positive and negative time directions. This approach has proven to be extremely powerful in word recognition: using a variety of datasets for training, Bruel et al. achieved a 0.6% Character Error Rate (CER) for English text trained on several datasets [7].

Shkarupa et al. achieved 78% word-level accuracy in classifying unconstrained Latin handwriting on the KNMP Chronicon Boemorum and Stanford CCCC datasets [8]. Dutta et al. was able to achieve a 12.61% Word Error Rate (WER) when recognizing unconstrained handwriting on the IAM dataset [9]. Alternatively, gated recurrent unit (GRU) networks can function as lighter versions of LSTMs, since they behave similarly but use fewer trainable parameters [10]. Although they are faster to train, in most cases LSTMs outperforms GRUs in terms of recognition accuracy, which is a key concern for a framework to be reliable for practical use. Different segmented text recognition approaches for problems like bank check recognition, signature verification, tabular or form based text identification, etc. can be relevant to our chess move recognition problem. One notable demonstration for recognizing segmented handwritten text was presented by Su et al. using two RNN classifiers with Histogram of Oriented Gradient (HOG) and traditional geometric features to obtain 93.32% accuracy [11].

Table 1. Standard Algebraic Notation (SAN) for chess moves. Short and long castles (not included in the table) are denoted by 'O-O' and 'O-O-O' respectively.

Piece	Disambiguating File/Rank	If Capture	Destination File	Destination Rank	If Promotion	If Check or Mate
Pawn ~ (none) King ~ K Queen ~ Q Rook ~ R Bishop ~ B Knight ~ N	Either from a - h or from 1 - 8 for disambiguating File or Rank respectively	X	One choice from a - h	One choice from 1 - 8	= followed by the Piece Q, R, B or N	Check ~ + Mate ~ #

This paper presents an end-to end system for offline chess scoresheet recognition with a convolutional BiLSTM neural network as an alternative to the existing inefficient method of manual digitization. Each handwritten move is extracted from its scoresheet during pre-processing and passed through the network to generate a prediction. An autonomous post-processing algorithm outputs a final prediction for the text box after cross-checking with the game's second scoresheet. Alternatively, we present a semi-autonomous post-processing algorithm that significantly increases accuracy to near human-level by requesting user input for especially difficult handwriting samples and invalid entries. Even the semi-autonomous process can save hundreds of hours for event organizers and drastically reduce human labor required for digitizing chess games.

2   Offline Chess Scoresheet Recognition

2.1   Preprocessing

A chess scoresheet has predefined boxes where players enter moves during a game. While there are different styles of scoresheets, each that we encountered

Table 2. Example chess moves in SAN format. Dis. or Disambiguation file or rank is used only when multiple piece can move to the same Destination Square.

Sample move	Move description	Piece	Dis. File	Dis. Rank	Capture	Dest. Square	Promote	Check mate
Nf3	Knight to f3	N	~	~	~	f3	~	~
R1f4	Rank 1 Rook fo f4	R	~	1	~	f4	~	~
Bxe5	Bishop takes e5	B	~	~	x	e5	~	~
Rdf8	d-file Rook to f8	R	d	~	~	f8	~	~
Qh4e1	h4 Queen to e1	Q	h	4	~	e1	~	~
e8=Q	Pawn to e8 promotes to Queen	~	~	~	~	e8	=Q	~
e4	Pawn to e4	~	~	~	~	e4	~	~
Qxf7#	Queen takes f7 Checkmate	Q	~	~	x	f7	~	#
Bxc3+	Bishop takes c3 Check	B	~	~	x	c3	~	+
dxe5	d-file Pawn takes e5	~	d	~	x	e5	~	~
O-O	Short Castle	~	~	~	~	~	~	~
O-O-O	Long Castle	~	~	~	~	~	~	~

ordered moves in essentially the same way: four columns, each representing 30 moves, with columns and rows separated by solid grid lines. To train our network to recognize these moves, we developed an algorithm that isolates each move based on those grid lines. First we convert our RGB images to gray-scale, and then to binary images with an adaptive threshold using Otsu's method. Then we use two long, thin kernels (one horizontal and one vertical) with sizes relative to input image dimensions, and morphological operations (erosion followed by dilation) with those kernels to generate an image containing only grid lines. With this simplified image, we use a border following algorithm [12] to generate a hierarchical tree of contours. Each contour is compressed into 4 points, storing only corners of each quadrilateral. Any contour which is significantly larger or smaller than the size of a single move-box (again, calculated relative to the total image size) can be ignored. The final contours are sorted based on their positions relative to one another, and each is labelled by game, move and player. Finally, we crop each move-box with a padding on the top and bottom of 15% and 25% respectively, since written moves overflow the bottom of their box more often and more severely than the top. This process is displayed in Fig. 2. We did not pad box sides because chess moves are short and the players rarely need to cross the side boundaries. This method of pre-processing is nearly agnostic to scoresheet style, and will work with any scoresheet style which includes 4-columns and solid grid lines.

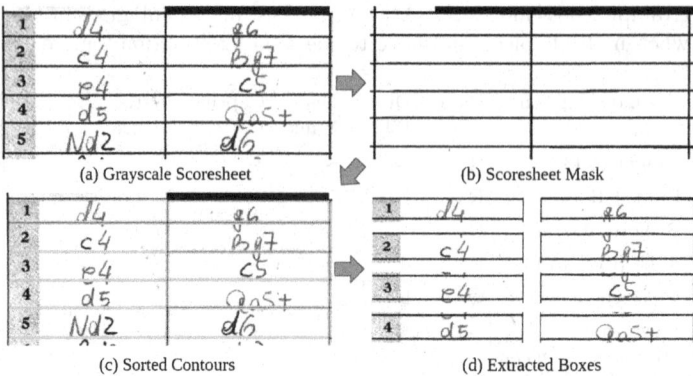

Fig. 2. Stages of move-box extraction: (a) grayscale image, (b) morphological operations generate a mask, (c) contour detection and sorting from the mask, and (d) sorted contours cropping with top and bottom padding of 15% and 25%.

2.2   BiLSTM Neural Network Architecture

We present a convolutional BiLSTM network for recognizing handwritten chess moves, as outlined in [13]. The network includes 10 layers, separated into 3 functional groups. The first 7 layers are convolutional and take in the gray-scale image of a single move-box, scaled to a resolution of 64 × 256 pixels. These layers act as feature extractors and convert image data into a feature map. This feature map removes unnecessary information and creates a sequence input for the recurrent layers. The recurrent model itself consists of 2 BiLSTM layers, which take the feature map and generate a sequence output to a final dense network. The BiLSTM neural network is a modification on the traditional RNN which takes advantage of hidden memory units to better process sequence data. The final dense network converts LSTM sequence output into a loss matrix. Afterwards, a Connectionist Temporal Classification (CTC) loss function [14] is applied which converts this matrix into a string. The CTC loss function and decoder better estimate the network loss than more general loss functions and allows for simple repeat errors common to the LSTM layers to be corrected. This loss matrix encodes a string as a vector of one-hot encoded character vectors, so the final matrix has dimensions of the number of allowed characters by maximum output length (plus a small padding on output length for blank characters).

First we pretrained our network with the IAM dataset [1], one of the most used collections of unconstrained handwritten English sentences. Approximately 86,000 words of this dataset were used to pretrain our network so that it could learn the overall Latin script and later be modified for chess move recognition. Figure 3 shows several sample images from the IAM dataset. This process of transfer learning is a widely used technique which can achieve faster and better network training, especially useful when working with a small dataset. While pretraining, the network uses 2 BiLSTM layers with 256 hidden units and a dense network which outputs a loss matrix of 81 × 31. These dimensions allow pretraining on the IAM dataset which has a total of 80 allowed character classes (entire Latin alphabet, numerals and frequently used symbols) and a maximum word length of 27.

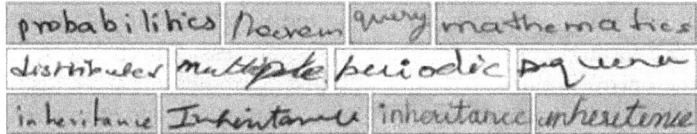

Fig. 3. Sample images from the IAM dataset used for pretraining.

Afterwards, we restructured this pretrained network for chess move recognition. The SAN format uses 30 allowed characters with a maximum possible move length being 7. This particular move recording standard is the most widely used chess notation in the world adopted by almost all the chess international organizations, including the United States Chess Federation (USCF) and the Fédération Internationale des Échecs (FIDE), which overseas all world-class competitions. To match with the SAN system, we reduced the BiLSTM layers from 256 hidden units to 64, as well as re-tuned the final dense layer to output a loss matrix of 30×12 instead of 81×31. Figure 4 shows the network with both the pretraining and fine-tuning structures. The maximum length from the output is slightly overestimated, allowing the network to have additional characters to compensate for CTC loss 'blank' characters and double spotting of a single character (which occurs frequently with LSTM layers). Since the SAN system does not include any valid chess moves with adjacent repetition, these extra characters are completely removed during the decoding phase.

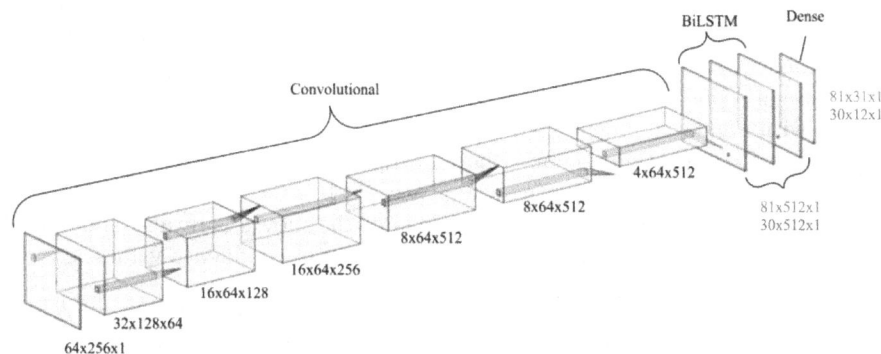

Fig. 4. Layer graph of the BiLSTM network. Orange and green values show the dimensions for the pretraining network and our chess move recognition network respectively. Pooling, batch normalization, and reshaping layers not pictured. (Color figure online)

The convolutional BiLSTM has a variety of advantages over other handwriting recognition structures. The LSTM works with sequence input and output data, allowing for variable length moves, which is crucial for a chess game (for example, both 'e4' and 'Qh4xe1' are valid algebraic notation moves). Unlike other character-specific handwriting detection networks, a LSTM-based network does not require labeled bounding boxes around individual characters, which involves a much more intensive ground-truth tagging process. The network's consideration of the entire input image and not just a single character also allows it to take advantage of context clues, which are often extremely important when decoding chess moves. If the network can gain a general understanding of the syntax of algebraic notation (e.g., that the first characters of a move are often in the form 'Piece' 'a-h' '1-8', e.g. Nf3), then it can make much more accurate inferences about the ground truth of an ambiguous character. The bidirectional structure of the BiLSTM layers make inference easier, allowing the network to leverage context from the entire input as it reads.

2.3   Post-processing

Given the availability of two unique scoresheets for each game (one written by each player) and the fixed structure of the algebraic notation for every chess move, we developed two simple post-processing strategies:

Autonomous System: The autonomous system relies on cross-checking between white and black scoresheets and spotting invalid notations to improve accuracy. Here, invalid notation refers to a syntactic error rather than an illegal move with respect to game rules. The system compares two predictions for each move along with their confidence values. The confidence value is calculated as the exponential of negative CTC loss between the raw network output and its decoded string. The system then makes the following logical decisions:

- If the predictions agree, the system accepts that prediction regardless of their confidence values.
- In case of a prediction conflict where both of the moves are valid, the prediction with higher confidence score is accepted.
- In case of a prediction conflict where both of the moves are invalid, the prediction with higher confidence score is accepted.
- In case of a prediction conflict where one of the moves is invalid and one is valid, the valid prediction is accepted regardless of confidence values.

For example, if the network prediction is 'NG4' (invalid) for one sheet and 'Ng4' (valid) for the other, the final prediction becomes 'Ng4', regardless of their confidence scores. Alternatively, if both moves were valid (e.g. 'Ng4' and 'Ne4') or both invalid (e.g. 'NG4' and 'NG8'), the system then accepts the one with the higher confidence regardless of their validity. After all predictions of a game, both scoresheets are presented to the user with confident predictions labelled in blue and lower-confidence predictions highlighted in pink as shown in Fig. 10.

We chose a confidence threshold of 95% in this autonomous approach. This allows the user to take a brief look after the network generated results and spot obvious issues.

Semi-autonomous System: To increase accuracy even further, we can leverage the user on certain moves which are more likely to be wrong. This allows significant improvements to accuracy with minimal interruption to the user. As with autonomous post-processing, we take in two sets of predictions with their respective confidence values, one for the scoresheet written by each player. We classify each prediction in each set as **confident** if it meets a confidence threshold of 90%, and **valid** if it follows correct SAN. If a move is both confident and valid, we additionally label it as **certain**, as we are more certain that it accurately represents what the player wrote. Otherwise it is either invalid or low-confidence, so we label it **uncertain**. In the following cases, the algorithm makes a decision without interruption:

- If both predictions are the same and at least one is **certain**, then that prediction is accepted.
- If predictions are different, but one is **certain** and the other is **uncertain** then the **certain** prediction is accepted.

While most prediction pairs fall into one of these two categories, if neither applies then interrupts are handled as follows:

- **Strong Conflict:** If predictions are different and both **certain**, then the user is prompted for a manual entry.
- **Weak Conflict:** If predictions are different and both **uncertain**, then the user is prompted for a manual entry.
- **Weak Match:** If predictions are the same but neither is **certain**, then the user is also prompted.

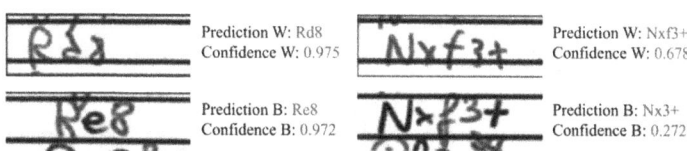

Fig. 5. Example cases where the user is interrupted for a manual entry in semi-autonomous post processing system: **Strong Conflict** (left) where both predictions are **certain** and **Weak Conflict** (right) where predictions are **uncertain**.

Figure 5 gives examples of **strong** and **weak** conflict cases. The **strong** conflict example involves two **certain** predictions, 'Rd8' and 'Re8', each valid and confident (with confidences of 97%). **Strong** conflicts occur most often when one of the players makes a mistake in writing. These cases are impossible to correct without going through a move-by-move game analysis. The **weak** conflict example shows predictions of '$Nxf3+$' and 'Nx3+', each with low confidence values (less than 70%). Both of these moves are considered **uncertain**:'Nxf3+' is associated with a low confidence value and 'Nx3+' is both low in confidence as well as an invalid SAN. Thus through our semi-autonomous system, this causes a user interrupt to ensure a successful PGN conversion. When the user is interrupted, we display both full scoresheets along with the network's predictions and ask the user to input the correct move with a user interface as shown in Fig. 6.

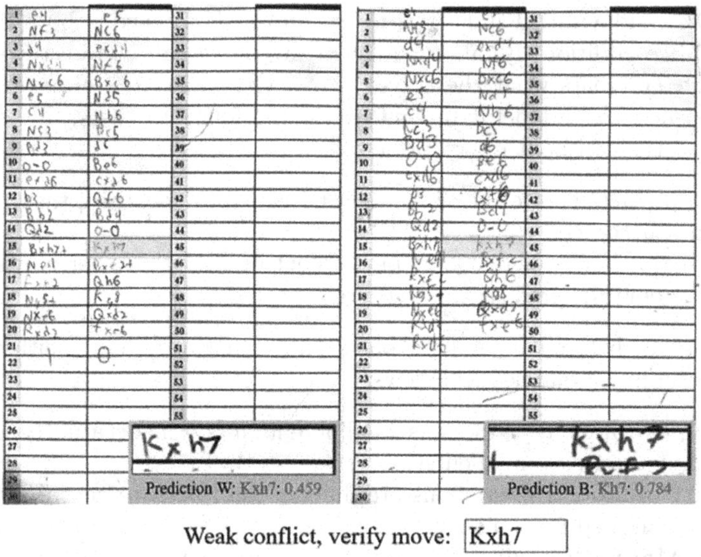

Weak conflict, verify move: Kxh7

Fig. 6. The semi-autonomous user interface, displaying both scoresheets from a game with the pertinent moves highlighted, along with a zoomed view of the move-boxes and the network's guess for each.

2.4   Data Augmentation

Fig. 7. Original image (top left) and ten augmented images. Ground truth: Nb4

Our HCS dataset is small for training a deep neural network at this first release and therefore it was necessary to increase the training set size by data augmentation. By applying meaningful image transformations to each move-box image, it is possible to simulate a larger dataset which almost always ensures a more generally trained network. We generated 10 augmented images for each sample using one of the following transformations and range randomly:

- Rotation (between - 10° to 10°),
- Horizontal scaling (between - 20% to 20%) and
- Horizontal shear (between - 15° to 15°).

Figure 7 shows a few sample of these augmented images. More complex data augmentation techniques are available and could have been used potentially to push the network performance even further, but we kept this process simple with just the basic techniques.

3   The Handwritten Chess Scoresheet (HCS) Dataset

There is no publicly available chess scoresheet dataset at the time of this research so it was necessary to develop our own, the HCS dataset [2,3], to train and test our network. This dataset consists of 158 total games consisting of 215 pages of chess scoresheet images digitized using a standard cellphone camera in natural lighting conditions. These images are tightly cropped and a standard corner detection based transformation is applied to eliminate perspective distortion. The headers and footers were also cropped out from each image in order to maintain player anonymity. The scoresheets were collected from actual chess events; they were not artificially prepared by volunteers. Despite posing challenges for training, identifying handwriting "in the wild" has more generality than identifying text from artificial, pristine images. The data is therefore very diverse, consisting of many different handwriting and scoresheet styles, varieties in ink colors, natural occurrences of crossed-off, rewritten and out-of-the box samples, a few of which are shown in Fig. 9.

Each scoresheet in the dataset contains 120 text boxes (60 boxes for each player). However, many of these boxes are empty since most chess matches last for far fewer than 60 moves. Omitting the empty boxes, there are approximately 13,810 handwritten chess moves at the first release of our dataset. Some of these scoresheets came with a photocopied or carbon copied version as well. These were included in the dataset since such variations of images provide a natural form of data augmentation which can be useful for the training process. We manually created the ground truth version of each game and stored in a text file, sample shown in Fig. 8. Both the images and the ground truth text files are stored with a naming convention given by:

$$[game\#]\_[page\#]\_[move\#]\_[white/black].png/txt$$

The HCS dataset [2,3] is public and free to use for researchers who want to work with similar problems. This can be accessed at *https://sites.google.com/view/chess-scoresheet-dataset* or *http://tc11.cvc.uab.es/datasets/HCS_1*.

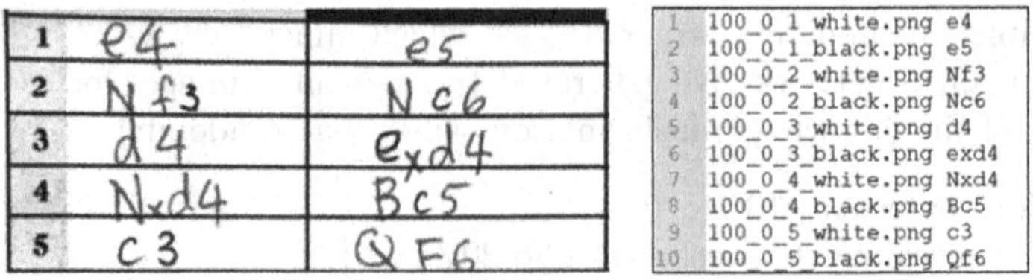

Fig. 8. Raw images (left) and ground truth labels (right) from the HCS dataset.

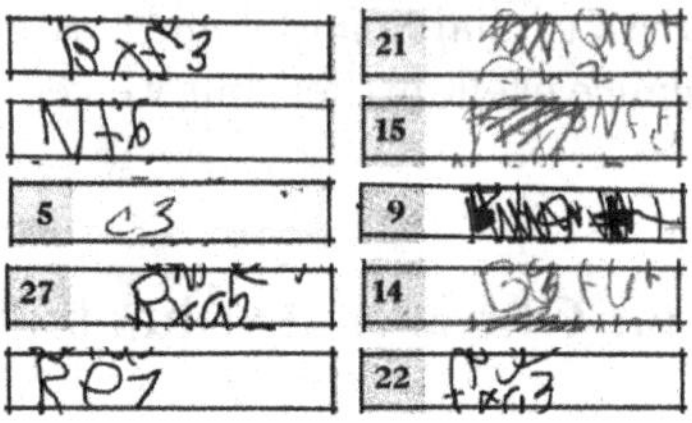

Fig. 9. Clear samples (left) and messy samples (right) from the HCS dataset.

4   Training and Results

Our BiLSTM network is trained with 2,345 unique box-label pairs from 35 unique pages along with their photocopied or carbon copied versions if available. This generated a train set size of 4,706 image pairs. We perform 10:1 data augmentation to expand this size to 47,060 image pairs. These data are separated into batches of 32 and trained with a learning rate of 0.0005 for 10 epochs. For testing, we used 7 games or 14 scoresheets, which translates into 828 testing image pairs. This test set was composed of data from writers/players unseen by the network during training. Not all of the games in the HCS dataset have unique (not photocopied) white and black player copies, which is fine for the network training, but not testing. Our post-processing pipeline uses a comparison framework from two unique copies, so the test set is carefully chosen with the games where both white and black player scoresheets are available.

Table 3. Character Recognition Accuracy (CRA), Move Recognition Accuracy (MRA), and Interruption Rate are for various levels of post-processing.

Post-processing	CRA	MRA	Interruption Rate
None	92.1%	82.5%	None
Autonomous	95.5%	90.1%	None
Semi-autonomous	98.1%	97.2%	7.00%

Our results, outlined in Table 3 demonstrate not just the effectiveness of our network, with a raw move recognition accuracy of 82.5%, but also the effectiveness of our post-processing schemes. The autonomous post-processing system increases this accuracy by 9.2%, translating to 6 prediction errors for the average 30 move game. Figure 10 shows an example of output prediction for a partial game, with each move labelled and low confidence moves highlighted. This presentation allows the user to quickly spot and check for any prediction errors.

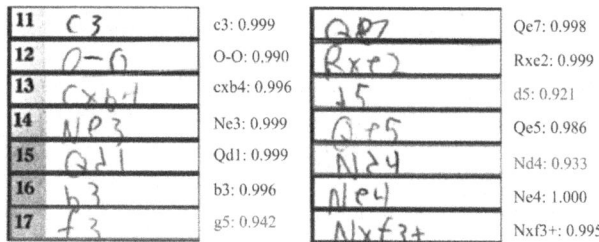

Fig. 10. The network output of a scoresheet portion as presented to the user. Low confidence predictions ($< 95\%$) are presented in pink for further inspection.

Leveraging the user with the semi-autonomous system improves the accuracy by a further 7.9% (17.8% from the raw prediction), while only requesting user input on 7% of moves. This means that for a 30 move game a user will be prompted roughly 4 times on average and these minor interrupts reduce errors to only a couple of moves. The accuracy for the semi-autonomous algorithm is much higher, with a move error rate of 2.8%. This is at least comparable or sometimes even better than actual human error rate that can occur naturally from pure manual entries. Many of these errors are caused when players accidentally write different moves, and one of those moves is above the confidence threshold while the other is not. This means that the user is not prompted for the move, and whichever has a higher confidence is selected, which may or may not be correct, but is currently always counted as an error. This error is common, so our true accuracy may be slightly higher than the measured accuracy of 97.2%. Since our current ground truth values are based on the handwriting itself, and not what was played in the game, we may be outputting a prediction which is correct to the game, but not to the handwriting of one scoresheet.

Errors from both autonomous and semi-autonomous systems are primarily caused by incorrect prediction, when the network does not accurately recognize an individual move. These errors generally fall into one of three categories: rare move structures, illegible writing, or information lost due to cropping. Rare or less frequent moves such as pawn promotions, moves involving disambiguation files or ranks or even long castles introduce training data imbalance. While easily recognizable to a human, our small dataset includes very few (or sometimes no) examples of these move structures, so the network makes a simpler and shorter prediction. A larger dataset or selectively introducing such less frequent moves could counteract this data imbalance and therefore reduce this error case. Messy, illegible writing which includes crossed-out or cramped letters (examples shown in Fig. 9) often causes single move errors, however most of these are solved with our post-processing algorithm. Both the autonomous and semi-autonomous algorithms cross-check both player's scoresheet, one of which is likely to be written in a better way. Finally, many players do not write a move entirely within its move-box, and this missing information also causes prediction errors.

5   Conclusion

We live in an era where deep learning can be applied to a vast number of complex problems, and yet surprisingly little work has been done on handwritten chess move recognition. A handwriting recognition framework is particularly useful with chess since players are bound to record moves by hand in order to prevent computer aided cheating. Chess is one of the most popular board game today and its popularity is growing faster than ever. Here, we present an attempt to save thousands of man-hours currently being spent by event organizers to transform chess scoresheets into standard PGN files everyday. While saving a substantial amount of manpower, our framework makes the record keeping process extremely convenient. Chess scoresheet digitization has a variety of uses from post-game engine analysis to efficient game-publishing, and in most professional and serious events, it is mandatory to keep these records. Therefore, we firmly believe our presented approach can have a big influence and potentially revolutionize the chess event management system practiced today.

Starting from an extremely limited number of data samples we were able to achieve an accuracy of above 90% with no human interaction needed at all. This autonomous approach presents the resultant scoresheet to the user with highlights for the low confidence predictions. Our semi-autonomous approach increased accuracy to near 98% with an average of only 4 interrupts for manual entry required per game. Although this performance is comparable to actual human level error, even this 2% error rate can cause issues in practical cases. This small gap to achieve a perfect result could be further reduced by increasing the training data size and including rare occurrence moves like pawn promotions or moves with disambiguation files/ranks. Along with that, a sophisticated post-processing system could be implemented which not only identifies SAN syntax errors, but also can detect illegal moves by assessing the current board position from the move history. With tools from graph theory it could be possible to determine the fewest number of moves requiring correction in order to have a top-to-bottom valid game - this could take the system accuracy very close to 100%. Another possibility is to utilize the most played moves in a certain position in the post-processing, which could be especially useful in common chess openings. Thanks to the rapid growth in online chess, lots of open source databases are available to get this information. In order to encourage further research and development in this area we are releasing the HCS dataset to the public.

Acknowledgements. We want to express our sincerest gratitude to Mr. George Lundy, FIDE National Arbitor (US) at Chandra Alexis Chess Club, for his enthusiasm and support in this research.

References

1. Marti, U.-V., Bunke, H.: The IAM-database: an English sentence database for offline handwriting recognition. Int. J. Doc. Anal. Recogn. **5**(1), 39–46 (2002). https://doi.org/10.1007/s100320200071
2. Eicher, O., Farmer, D., Li, Y., Majid, N.: Handwritten Chess Scoresheet dataset (2021). https://sites.google.com/view/chess-scoresheet-dataset
3. Eicher, O., Farmer, D., Li, Y., Majid, N.: Handwritten Chess Scoresheet Dataset (HCS) (2021). http://tc11.cvc.uab.es/datasets/HCS_1
4. Chess.com. How popular is chess? (2020). https://www.chess.com/news/view/how-popular-is-chess-8306
5. Sudharsan, R., Fung, A.: Reine chess (2020). https://www.reinechess.com/
6. Majid, N., Smith, E.H.B.: Segmentation-free Bangla offline handwriting recognition using sequential detection of characters and diacritics with a faster R-CNN. In: 2019 International Conference on Document Analysis and Recognition (ICDAR), pp. 228–233. IEEE (2019)
7. Breuel, T.M., Ul-Hasan, A., Al-Azawi, M.A., Shafait, F.: High-performance OCR for printed english and fraktur using LSTM networks. In: 2013 12th International Conference on Document Analysis and Recognition, pp. 683–687 (2013)
8. Shkarupa, Y., Mencis, R., Sabatelli, M.: Offline handwriting recognition using LSTM recurrent neural networks. In: The 28th Benelux Conference on Artificial Intelligence (2016)
9. Dutta, K., Krishnan, P., Mathew, M., Jawahar, C.: Improving CNN-RNN hybrid networks for handwriting recognition. In: 2018 16th International Conference on Frontiers in Handwriting Recognition (ICFHR), pp. 80–85 (2018)
10. Cho, K., et al.: Learning phrase representations using RNN encoder-decoder for statistical machine translation, arXiv preprint arXiv:1406.1078 (2014)
11. Su, B., Zhang, X., Lu, S., Tan, C.L.: Segmented handwritten text recognition with recurrent neural network classifiers. In: 2015 13th International Conference on Document Analysis and Recognition (ICDAR). IEEE, pp. 386–390 (2015)
12. Suzuki, S.: Topological structural analysis of digitized binary images by border following. Comput. Vis. Graph. Image Process. **30**(1), 32–46 (1985)
13. Shi, B., Bai, X., Yao, C.: An end-to-end trainable neural network for image-based sequence recognition and its application to scene text recognition. IEEE Trans. Pattern Anal. Mach. Intell. **39**(11), 2298–2304 (2016)
14. Graves, A., Schmidhuber, J.: Offline handwriting recognition with multidimensional recurrent neural networks. In: Proceedings of the Advances in Neural Information Processing Systems, Vancouver, Canada, vol. 545, p. 552 (2009)

ICDAR 2021 Workshop on Arabic and Derived Script Analysis and Recognition (ASAR)

ASAR 2021 Preface

We are glad to welcome you to the proceedings of the 4th International Workshop on Arabic and Derived Script Analysis and Recognition (ASAR 2021), which was held on September 16, 2021, in Lausanne, Switzerland, in conjunction with ICDAR (15th International Conference on Document Analysis and Recognition). The workshop was organized by the REGIM Lab (University of Sfax, Tunisia).

The ASAR workshop provides an excellent opportunity for researchers and practitioners at all levels of experience to meet colleagues and to share new ideas and knowledge about Arabic and derived script document analysis and recognition methods. The workshop enjoys strong participation from researchers in both industry and academia.

In this 4th edition of ASAR we received 14 submissions, coming from authors in 10 different countries. Each submission was reviewed by three expert reviewers. The Program Committee of the workshop comprised 21 members, who generated a total of 42 reviews. We would like to take this opportunity to thank the Program Committee members and sub-reviewers for their meticulous reviewing efforts!

Taking into account the recommendations of the Program Committee members, we selected nine papers for presentation in the workshop, resulting in an acceptance rate of 64%.

This edition of ASAR included a keynote speech by Sourour Njah and Houcine Boubaker from the University of Sfax, Tunisia, on the "Beta Elliptic Model for On-line Handwriting Generation and Analysis".

We would also like to thank the organizations that have supported us, especially the ICDAR organizers and the REGIM Lab, University of Sfax, Tunisia.

We hope that you enjoyed the workshop!

Adel M. Alimi
Bidyut Baran Chaudhur
Fadoua Drira
Tarek M. Hamdani
Amir Hussain
Imran Razzak

Organization

General Chairs

Adel M. Alimi — University of Sfax, Tunisia
Bidyut Baran Chaudhur — Indian Statistical Institute, Kolkata, India
Fadoua Drira — University of Sfax, Tunisia
Tarek M. Hamdani — University of Monastir, Tunisia
Amir Hussain — Edinburgh Napier University, UK
Imran Razzak — Deakin University, Australia

Program Committee

Alireza Alaei — Southern Cross University, Australia
Mohamed Ben Halima — University of Sfax, Tunisia
Syed Saqib Bukhari — German Research Center for Artificial
Intelligence (DFKI), Germany
Haikal El Abed — German International Cooperation
(GIZ) GmbH, Germany
Jihad El-Sana — Ben Guion University of the Negev, Israel
Najoua Essoukri Ben Amara — University of Sousse, Tunisia
Jaafaral Ghazo — Prince Mohammad Bin Fahd University,
Saudi Arabia
Afef Kacem Echi — Laboratoire LATICE, Tunisia
Slim Kanoun — University of Sfax, Tunisia
Driss Mammass — Ibn Zohr University, Morocco
Ikram Moalla — University of Sfax, Tunisia
Volker Märgner — Technische Universität Braunschweig,
Germany
Mark Pickering — University of New South Wales, Australia
Samia Maddouri Snoussi — University of Jeddah, Saudi Arabia
Daniel Wilson-Nunn — The Alan Turing Institute, UK

RASAM – A Dataset for the Recognition and Analysis of Scripts in Arabic Maghrebi

Chahan Vidal-Gorène[1,4]([⊠])[iD], Noëmie Lucas[2][iD], Clément Salah[3,5],
Aliénor Decours-Perez[4], and Boris Dupin[4]

[1] École Nationale des Chartes – Université Paris Sciences & Lettres,
65 rue Richelieu, 75003 Paris, France
chahan.vidal-gorene@chartes.psl.eu
[2] GIS Moyen-Orient et mondes musulmans – UMS 2000 (CNRS/EHESS),
96 boulevard Raspail, 75006 Paris, France
noemie.lucas@ehess.fr
[3] Sorbonne-Université, Faculté des Lettres, 21 rue de l'école de médecine,
75006 Paris, France
[4] Calfa, MIE Bastille, 50 rue des Tournelles, 75003 Paris, France
{alienor.decours,boris.dupin}@calfa.fr
[5] Institut d'Histoire et Anthropologie des Religions, Faculté de Théologie et Sciences
des Religions, Université de Lausanne, 1015 Lausanne, Switzerland
clement.salah@unil.ch

Abstract. The Arabic scripts raise numerous issues in text recognition and layout analysis. To overcome these, several datasets and methods have been proposed in recent years. Although the latter are focused on common scripts and layout, many Arabic writings and written traditions remain under-resourced. We therefore propose a new dataset comprising 300 images representative of the handwritten production of the Arabic Maghrebi scripts. This dataset is the achievement of a collaborative work undertaken in the first quarter of 2021, and it offers several levels of annotation and transcription. The article intends to shed light on the specificities of these writing and manuscripts, as well as highlight the challenges of the recognition. The collaborative tools used for the creation of the dataset are assessed and the dataset itself is evaluated with state of the art methods in layout analysis. The word-based text recognition method used and experimented on for these writings achieves CER of 4.8% on average. The pipeline described constitutes an experience feedback for the quick creation of data and the training of effective HTR systems for Arabic scripts and non-Latin scripts in general.

Keywords: Arabic Maghrebi scripts · Dataset · Manuscripts · Layout analysis · HTR · Crowdsourcing

This work was carried out with the financial support of the French Ministry of Higher Education, Research and Innovation. It is in line with the scientific focus on digital humanities defined by the Research Consortium Middle-East and Muslim Worlds (GIS MOMM). We would also like to thank all the transcribers and people who took part in the hackathon and ensured its successful completion.

E. H. Barney Smith and U. Pal (Eds.): ICDAR 2021 Workshops, LNCS 12916, pp. 265–281, 2021.
https://doi.org/10.1007/978-3-030-86198-8_19

1   Introduction

The automatic analysis of handwritten documents has become a classic preliminary step for numerous digital humanities projects that benefit from the mass digitization policy of heritage institutions. Following the competitions organized in recent years, at ICFHR and ICDAR notably, several robust architectures for layout analysis of historical documents have been developed [8], whose application to non-Latin script documents provide equivalent results [10,14]. The HTR architectures specialized on a type of document or on a hand also achieve a very high recognition score, even though the literature is mostly Latin script based, as well as the proven pipelines composed of character-level HTR and post-processing [7]. The non-Latin, cursive and RTL writings, like the Arabic scripts, remain an open problem in digital humanities with a wide variety of approaches [11]. Although specialized databases have emerged in recent years (see *infra* Sect. 2.1), they are often focused on the layout [6,10] and on common documents and writings, leaving out numerous under-resourced written traditions.

We are presenting a new dataset for the analysis and the recognition of handwritten Arabic documents, the first dataset focused on the writings called "Maghrebi scripts", also known as "Western scripts", or "round scripts". This term encompasses a variety of styles that have common characteristics and are poorly represented in digital humanities. These scripts dating back to the 10th century have been widely used however in the Islamic West – al-Andalus and North-Africa –, as well as sub-Saharan Africa until the 20th century[1]. They have numerous specificities that differentiate them from the classical problems met for Arabic handwritten character recognition. The rounded shape of these scripts may be explained by the writing tool used, that is qalams made from large reed straws cut in half lengthwise, with a pointed nib and not a biseled one as in the Islamic East [3]. Therefore, the Maghrebi scripts constitute a family of rounded scripts that share a number of characteristics, first of all very rounded loops, that can be seen in the manuscripts in the present dataset (see *infra* Sect. 2.3). The main characteristics[2] of the scripts are displayed in Table 1.

The dataset, resulting from a collaborative hackathon held from January to April 2021, intends to cover a large spectrum of the handwritten production in Maghrebi scripts. The choice for a multilevel annotation (semantic and baseline annotation, word-level and character-level transcription) aims to offer to the scientific community a comprehensive dataset dedicated to the creation and evaluation of complete HTR pipelines, from layout analysis to text recognition for this written tradition. After a short presentation of the related work for the datasets, we propose a complete description of the manuscripts, the annotations and the editorial choices made for the transcription. The creation of the dataset

---

[1] The history and the origins of these scripts have been an important scientific open debate [3,4]. The most recent works, in particular those of U. Bongianino, have foregrounded the different itineraries (from books to qurans, from al-Andalus to the Maghreb) followed by these writings between the 10th and the 13th century [4].

[2] Characteristics are taken from U. Bongianino [4]; theoretical realizations are taken from the article of N. Van de Boogert upon which U. Bongianino draws [13].

has also benefited from a collaborative and semi-automatic work, with architectures dedicated to non-Latin scripts, that we assess with the view to replicate for other under-resourced languages.

Table 1. Some characteristics and realizations of Maghrebi scripts

Letters	Characteristics	Theoretical realization	Examples from mss ARA.1977, ARA.609 and ARA.417
bā' tā' ṯā' fā'	(i) **Isolated position:** concave form – (ii) **Final position:** closing denticle in the shape of an inverted comma	ب / ـب	وركب لصاحب مغلوـــ / وركب لصاحب مغلوب
dāl ḏāl	**Isolated, median and final positions:** concave downstroke and final downward spur (*dāl kāfiyya*)	د / ـد	جزيد دنانير الديـان / يزيد دنانير الديان
dāl ḏāl	**Final position:** marked semicircular descender, resembling the letters *rā'* and *zā'*	ـر	بعرد بغزما محمد / بعد فقد محمد
sīn šīn ṣād ḍād qāf nūn	**Final position:** exaggerated semi-circular descenders, often described as 'swooping' or 'plunging', stretching below the following word	ن / ـں	من ماله كان بن عبد
ṣād ḍād ṭā' ẓā	Oval or semi-circular body and lack of denticle	ط	سبط الاصطلاح قسنطينة / سبط الاصطلاح قسنطينة
'ayn ġayn	**Initial position:** oversized curl	عـ	عايشه عم عادة / عايشه عم عادة
kāf	**Initial and median positions:** semicircle topped by a diagonal stroke	كـ / ـكـ	وكتب كذلك كثير / وكتب كذلك كثير
mīm	**Final and isolated positions:** long curved tail in two variants (concave or convex)	م / م	ايام تقدم اسلام / ايام تقدم اسلام
hā' tā' marbūṭa	**Isolated position:** drawn in the shape of a '6', sometimes inverted	٥ / ه	يذكره هاذه ايـاه / يذكره هاذه اياه

2   Dataset and Arabic Maghrebi Manuscripts of the BULAC

2.1   Existing Resources for Arabic Scripts

Following the competitions organized in recent years, at ICFHR and ICDAR, several datasets for Arabic written documents analysis have emerged. Such is the case of RASM2018 [6], focused on the Arabic scientific manuscripts on Qatar Digital Library, which is used to evaluate the layout analysis of documents in Arabic scripts. The 100 images of the dataset are annotated at different levels: region, polygons, lines and text. The Arabic scripts, however, present unique challenges for text-region detection and baseline detection. Therefore, in order to include these specificities, RASM2018 has been considerably expanded by BADAM [10] with a dataset focused on the Arabic scripts, comprising 400 images annotated at the text-region and baseline level.

These datasets cover a wide-ranging production of texts in non-Maghrebi Arabic scripts to enable the training of dedicated models [10]. More generally, there are other smaller or more specialized datasets, like HADARA80P [12] and VMH-HD [9], annotated at the region and word level, like KERTAS [2], dedicated to manuscripts datation, or like WAHD [1], dedicated to writer identification. Aside from the targeted tasks and the languages concerned, these datasets shed light on the variety of existing perspectives for handwriting, either inspired by the Latin languages, or by word-based approach. A FCN followed by a post-processing for baseline extraction gives robust results even on the most complex layouts [10,14]. Manuscripts in Arabic Maghrebi scripts are largely excluded from the datasets.

2.2   Dataset Composition

RASAM is available under an open license[3]. It comprises 300 annotated images extracted from three manuscripts selected among the collections of the Bibliothèque Universitaires des LAngues et Civilisations (BULAC)[4]. The images of the dataset are in JPEG format and have varying resolutions from 96 DPI to 400 DPI. Experiments are carried out on the hackathon results (v1.0, 297 images). Dataset has been expanded in June 2021 (v1.1, 300 images, includes minor corrections).

Thus, two manuscripts of the dataset have been chosen among the 150 Arabic manuscripts available online (MS.ARA.1977 and MS.ARA.609); the third and last manuscript (MS.ARA.417) of the corpus has been digitalized at our request. The variety of topics, the representative type of the Maghrebi script, as well as the diversity of layouts has informed our choice of manuscripts to annotate and

---

[3] https://github.com/calfa-co/rasam-dataset.

[4] The BULAC holds the second biggest fund of Arabic manuscripts in France (2.458 identified documentary units). BULAC collections contains a substantial proportion of the manuscripts copied in Maghrebi script. 150 Arabic manuscripts are available online on the website of the BINA Digital library.

transcript for this dataset. The aim is to obtain polyvalent analysis models for this written tradition. Therefore, two manuscripts belong to the historical genre (*'ilm al-tārīḫ*), whereas the third has to do with inheritance law (*fiqh al-farā'iḍ*). The small number of manuscripts is justified by the necessity to quickly achieve HTR models. The pages are not sequential, and the pages have been randomly selected to cover all the variations of a same copyist and the different layouts within a single manuscript.

2.3   Selected Manuscripts

MS. ARA. 1977: The manuscript MS.ARA.1977[5] consists in a compilation of 249 pages: the most part of which (p. 1–201) is a historical treatise entitled *al-Ǧumān fī muḫtaṣar aḫbār al-zamān* written by the Andalusian historian Abū 'Abd Allāh Muḥammad b. 'Alī b. Muḥammad al-Šuṭaybī (d. 963/1556), disciple of the great Maliki jurist Aḥmad Zarrūq (d. 899/1493). The second 38-page long text (p. 205–243) deals with the customs and practices relating to the prophet Muḥammad; as for the third text (p. 247–249), it is a recollection of the words of a scholar al-Ḥasan b. Mas'ūd al-Yūsī (d. 1102/1691), native from the North-West of the Moroccan Middle Atlas. It deals with the mission that the prophet Muḥammad would have entrusted to the Berber tribes to conquer the Maghreb. The annotation and transcription have been achieved on the first and main part of the compilation, that has been copied by Muḥammad b. Mubārak al-Barāšī around 1259/1843[6]. Compiled on paper (305 × 210 mm.), the manuscript pages contain 31 lines, with the exception of the last three which contain 27 each, and the pages 67–68, 202–204 and 245–246 that are left blank. While the main text is written in black ink, here and there the copyist has used red (e.g. limited to section titles) and green inks (e.g. to indicate poetry verses). This manuscript features a series of marginalia: besides the catchwords at the bottom of the page, numerous corrections and notes are displayed along the text. The characteristics of the ink and the handwriting lead us to assume that they are made by the copyist himself. The same can be assumed for the manuscript MS.ARA.609 (see *infra*).

MS. ARA. 609: The manuscript MS.ARA.609[7] consists in a treatise in verse on arithmetic, on inheritances and wills. Written by the Maliki jurist 'Abd al-Raḥmān al-Aḫdarī (d. 953/1546 or 983/1575) around 946/1540, the poem and its commentary are about the science of successions (*'ilm al-farā'iḍ*) and the arithmetic knowledge it required. Abū Zayd 'Abd al-Raḥmān b. Muḥammad al-Aḫdarī, one of the great names of the Maliki school of the 10th/16th century, was born in 919/1513 near Biskra.

---

[5] مجموع – MS.ARA.1977, Collections patrimoniales numérisées de la BULAC.

[6] Muḥammad b. Mubārak al-Barāšī is also the copyist of the second text. There is no mention for the third text: the paleographical characteristics of the pages lead us to assume that it is the work of another hand.

[7] شرح الدرة البيضاء – MS.ARA.609, Collections patrimoniales numérisées de la BULAC.

He is the author of numerous didactic poems, often along with their commentaries, in several scholarly fields of study (logic, arithmetic, rhetoric, law). Compiled on paper (210 × 175 mm.), the copy of the manuscript was completed in 1146/1734 by Abū l-Qāsim b. Muḥammad b. Abū l-Qāsim al-Duraydī. The manuscript has 202 pages – 100 written folios, the folios 1 and 2 have been left blank – each page contains 25 lines. The main text is written in black ink, however the copyist has used red ink on several occasions (e.g. tables, numbers or poetry verses). This manuscript holds numerous numbers, fractions and tables throughout the text. The numbers are written in Indo-Arabic numerals. Like in the MS.ARA.1977, catchwords, additions, corrections and glosses are displayed on the lateral, top and bottom margins.

MS. ARA. 417: The manuscript[8], dated from 1292/1875, was copied on the manuscript n°1061 of the National Library of Algeria. It narrates the history of Beys of Oran in the 13th century: the *Tārīḫ Bāyāt Wahrān*, written by Ḥasān Ḥūǧah, secretary of Ḥasān Bey (1817–1831). It distinguishes itself from the two previous manuscripts by its length and its layout: it consists of 48 folios with pages of 12 lines, each of them with less than 10 words per line. The manuscript, very well written, is in black ink, even though the copyist uses red ink sporadically (e.g. chapters headings or some separators). In the lateral margins, another hand, which looks identical to the note of cataloguing in the first page, has added the names of the beys in Arabic and some dates. The same hand seems to have added some vocalizations marks written on some folios and some corrections with a blue ink.

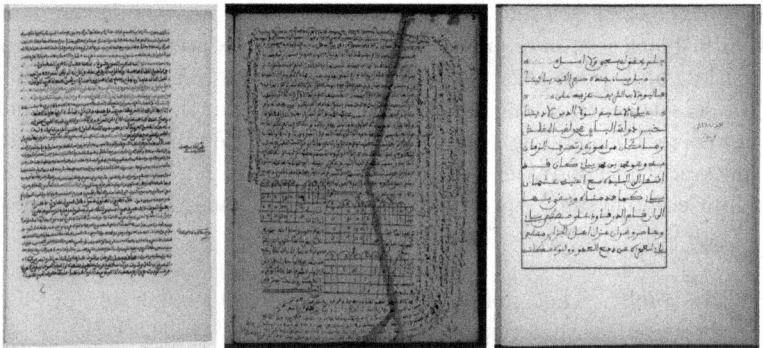

Fig. 1. MS.ARA.1977 (p. 42), MS.ARA.609 (p. 124) and MS.ARA.417 (f. 12v)

3   Ground-Truth Content and Creation

Each image is associated with a pageXML format file describing the entire ground truth and the associated metadata. The annotations have been realized automatically on Calfa Vision platform, then manually checked over the course of a collaborative hackathon.

---

[8] See bibliographic record on CALAMES.

The overall set of constitutive parts of a manuscript page is annotated. We offer for each image: (i) a semantic annotation of the regions, (ii) an annotation of baselines (polylines), (iii) a polygons framing every line associated to a baseline and (iv) the transcription. In figures, the dataset is comprised of 300 images, 676 text-regions, 7,540 lines and 483,725 characters (v1.1).

3.1 Structure Description

Text-Region: The layout analysis consisted in identifying all the text-regions. We have defined 5 classes: text (300), marginalia (171), catchword (102), table (53) and numbering (50). The classes are not uniformly distributed but constitute clearly identifiable items. To prevent an overwhelming variety and ambiguity of classes, the 5 defined classes can incorporate some regions that would be otherwise traditionally separated. Thus, the titles, often written in color, are not separated but included in the text class. Likewise, all content in the margins (everything outside from the main text-region) is encompassed in the class marginalia, with the exception of the catchwords. Table and numbering respectively refer to the tables located inside and outside the main text-region and to all fractions within the text (see Fig. 2).

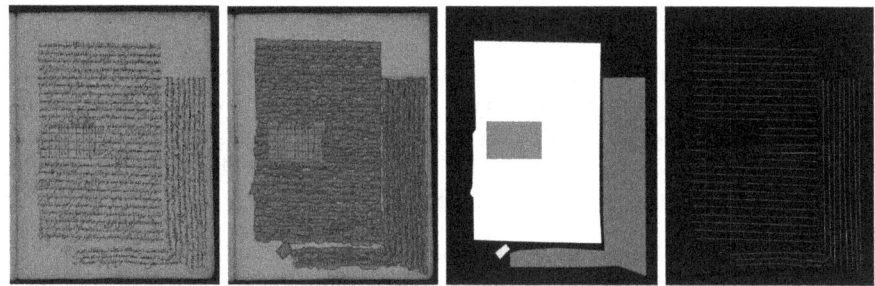

Fig. 2. Semantic classification of text-regions and baselines

Baseline: We adopted the annotation by baseline, suited for the Arabic scripts and interoperable with other datasets of the state of the art. Each segment of a sentence has its own baseline. In the event of overlap, as displayed by Fig. 2 for the marginal note, there is no continuity in the baseline. Besides, the outline of the latter follows the text line actually present in the manuscript and not a theoretical line that would link two segments of a single sentence (for instance, in the case of a line break for versification or due to a table – see Fig. 2). The reading order can be managed afterwards in post-processing. The manuscripts present numerous curved lines (see Fig. 2), in particular at the end of sentences and in the marginal notes. In this case, the baseline follows the same scheme that for BADAM [10], following a theoretical rotation point to match with the line curvature. Markers of verses or other signs for aesthetic purposes have not been taken into consideration (see Fig. 3, no 3). Lastly, in the event of characters composed of strokes expanding beyond the body of the character (e.g. the letter *nūn* in Fig. 3, no 1) and located at the end of a sentence, the baseline has been extended to cover the entire shape of the character, even in the absence of the theoretic writing line.

Polygons: Each line of text is extracted with a surrounding polygon, drawn with an adaptive seamcarve implemented on the annotation platform [14]. Polygons have been manually proofread to integrate all the constitutive strokes of a given character, including ascenders and descenders, as well as the associated diacritics (see Fig. 3, no 4). There remain overlaps between the polygons of lines (see Fig. 3, no 2), but the HTR results have demonstrated that these overlaps have very few impact on HTR predictions (see *infra* Sect. 4.2).

Fig. 3. Special cases for baselines and polygons annotation (MS.ARA.609 and MS.ARA.1977)

3.2   Specifications for Transcription

A common framework for the text input was defined to preserve the uniformity of the transcriptions of each participant. The aim is to achieve a HTR producing predictions as close as possible to the original text, and thus to offer a complete transcription and allow for a big panel of editorial choices. The dataset comprises 54 classes, whose detail we give in Table 2.

Table 2. Letters distribution in RASAM dataset (v1.1)

space	88,674	ت	12,386	ص	3,957	fatḥa	371	3	16		
ا	66,996	ف	12,064	خ	3,880	sukūn	357	4	12		
ل	48,324	د	9,508	ذ	3,500	šadda	344	6	10		
م	26,608	ق	9,219	ئ	2,924	ؤ	121	9	10		
و	25,606	لا	8,818	ض	2,675	#	100	5	9		
ن	22,378	س	8,793	ط	2,271	آ	91	7	8		
ي	22,105	ة	6,658	ز	2,096	fatḥat^{an}	78	8	7		
ه	19,213	ح	6,522	ء	1,696	kasra	44	0	3		
ر	16,559	ک	6,418	غ	1,278	ḍamma	31	ḍammat^{an}	2		
ب	15,956	ج	5,343	ظ	720	2	18	!	2		
ع	13,970	ث	4,312	ى	648	1	16				

We have notably realized transcriptions that restore the spaces in Arabic, even when there are visually no discernable spaces in the manuscript. In view of the great variety of character morphologies in the Maghrebi manuscripts scripts, we favored the word-based approach instead of the character-based approach where the word separation is managed in post-processing [5].

Table 3. Spelling conventions (examples from MS.ARA.609 and MS.ARA.1977)

	Example
Confusion between *ḍād* and *ẓā'*	 تحفض where the copist should have written تحفظ
tā' marbūṭa in final position	 الميت / المية
haḏā/haḏihi	 هاذه / هاذا
In case of an erased character	 Although we can guess that it may be: كالرجل, the transcription was addressing what can be actually identified, here: كالرج

In order to remain as close as possible to the text, the transcription follows the spellings habits of the copyist, even when they depart from the norm of standard Arabic. Hence, the frequent confusion, in particular in the MS.ARA.609, between *ḍād* and *ẓā*, and between *ṣād* and *ṭā'* have been retained. For instance, for the transcription of *ṭā'*, that could be spelled as *tā' marbūṭa*, or the other way around, the misspell was kept. Furthermore, the demonstratives *hāḏā, hāḏihi*, and in some instances *ḏālika*, that are spelled in modern Arabic with a defective form or with a dagger *alif*, are often spelled with their archaic form with a medial *alif*: we maintain in our transcriptions the spelling of this *alif* (see Table 3).

Table 4. Some realizations of the *hamza* in MS.ARA.609 and MS.ARA.1977

ان – *hamza* in أن or إن is not present	فان	انه	ان
Alif madda has been transcribed as it was done		اءدم where we would have written آدم	
When the *hamza* was not supported, we respected the way it was done		اسرائيل and not اسراءيل	

The punctuation, when present, has not been transcribed. In the event of vocalization signs or *šadda* (◌ّ), the participants were free to transcribe or not, the priority was on the characters and words. When not understood, the character was to be replaced by the sign #. When impossible to read because of an alteration of the manuscript, nothing was to be transcribed (see Table 3).

Two particular cases caught the attention of the transcribers and were much debated: the *hamza* (ء) and its different forms and the diacritic signs of the *yā'*. The main rule adopted for the *hamza* was the following: if there is no *hamza* do not add one and transcribe the *hamza* as it is written in other cases (see Table 4).

However, in the MS.ARA.609 and in the MS.ARA.417, the copyists used a singular *hamza* shape but consistent. In the MS.ARA.609, the *hamza* was written as a full *sukūn*. In these cases, we have considered that it was the way the copyist realized the letter (see Table 5).

Table 5. Specific realizations of the *hamza* in MS.ARA.609

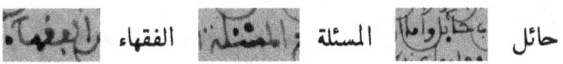

In the cases where the *hamza* was written below the line, with or without the diacritics of the ي, the *hamza* was drawn on *alif maqṣūra*. The issue of transcription of the ي was raised for the MS.ARA.1977 manuscript in particular, in which the copyist only wrote diacritics in rare instances. It was decided not to correct and to transcribe as close as possible to the text with only three exceptions: for the preposition في and the relative pronouns الذي / التي which, in the Maghrebi script, sometimes form some kind of glyphs (see Table 6).

Table 6. Exceptions regarding *yā'* and its diacritics

	MS.ARA.1977	MS.ARA.609	MS.ARA.417
في			
التي			
الذي			

4 Evaluation of the Crowdsourcing Campaign and HTR Models

The annotations have been realized with the Calfa Vision platform[9] [14], which incorporates – besides the online collaborative work on an image in real time – models for layout analysis and HTR prediction. These models are automatically assessed and fine-tuned, according to the corrections given by the contributor to the project, in order to speed up the checking task for the next images.

[9] https://vision.calfa.fr.

4.1   General Considerations and Implementation Protocol

The crowdsourcing campaign gathered 14 participants from January to April 2021, divided into three projects. The contributors were paired: one annotated and the other checked. Three main tasks were set to annotate each image.

1. Annotation and check of the layout analysis. A text-region (polygon) and a baseline (polyline) detection is automatically carried out beforehand. A first team is entrusted with the verification of the predictions and when necessary with the correction of the polygons and polylines shapes. The specifications to follow are defined in part Sect. 3.1.
2. Transcription of the text. Once the layout verified, the page is manually transcribed according to the specifications described in part Sect. 3.2. Some pre-annotations are realized in a second phase, once the HTR models are sufficiently precise (see Table 9).
3. Extraction of the lines with a surrounding polygon for each transcribed line.

After each task, the images are re-assigned, to enable cross-check and to smooth the annotation habits. When all three tasks are completed, a comprehensive verification is carried out by a team of administrators.

4.2   Benefits of Fine-Tuning and Transfer Learning for a Under-Resourced Language

The layout analysis models, provided by Calfa Vision for a project of handwritten documents annotation, are trained with an extensive dataset of various handwritten documents both medievals and recents [14]. A first assessment has been realized on BADAM with 0.9132% precision and 0.8575% recall [14].

Layout Analysis and Baseline Models: After each proofreading of predictions, models are evaluated. The automatic re-training has been processed with a batch of 50 verified images from the three manuscripts. Evaluation is carried out on the remaining images to annotate. The batch of 50 was compiled to encompass a wide range of assessed layouts.

Table 7. Fine-tuning of Calfa Vision models for baseline prediction (v1.0)

Model	Precision (%)	Recall (%)	F1-score (%)
Default	0.8886	0.9522	0.9193
Model 1 (Default + batch 1)	0.9627	0.9720	0.9673
Model 2 (Model 1 + batch 2)	0.9650	0.9716	0.9683
Model 3 (Model 2 + batch 3)	0.9680	0.9756	0.9718
Model 4 (Model 3 + batch 4)	0.9762	0.9694	0.9728
Model 5 (Model 4 + batch 5)	0.9769	0.9700	0.9734

We use the metric of the cBAD competition [8] and implemented on Calfa Vision [14]. From the first fine-tuning, we notice a significant increase in the model ability to correctly predict baselines on the dataset images. We also notice a steady improvement for the precision and the F1-score. The already high recall has little variation throughout the fine-tuning, with a slight dip when the batch is predominantly comprised of very curved lines. The results displayed in Table 7 constitute a baseline for the HTR ability to identify the lines of text in common Arabic Maghrebi scripts manuscripts. The very high score achieved by the first model demonstrates a strong ability to rapidly reach quality fine-tuning for a under-resourced language (see Fig. 4).

At the region level, we evaluate the relevance with an Intersection over Union (IoU) metric. The default model is already convincing to identify the area of the main text, but without distinction between the main text, catchwords and the marginal notes. Table and numbering are not considered (Table 8).

Table 8. Fine-tuning of Calfa Vision models for text-region prediction (v1.0)

Model	Average IoU (%)				
	T	M	C	Tab	N
Default	0.9780	–	–	–	–
Model 1 (Default + batch 1)	0.9673	0.2177	0.3221	0.0866	0.0095
Model 2 (Model 1 + batch 2)	0.9751	0.3809	0.4068	0.1823	0.0213
Model 3 (Model 2 + batch 3)	0.9720	0.3617	0.6197	0.1900	0.1285
Model 4 (Model 3 + batch 4)	0.9680	0.5528	0.7772	0.2737	0.1826
Model 5 (Model 4 + batch 5)	0.9685	0.6268	0.8853	0.2813	0.1219

The fine-tuning triggers mechanically a decrease in the score of main text identification, but also the rapid inclusion of the other classes (see Fig. 4). Concerning the text-regions, the latest model achieves an accuracy of 0.8534% on average. The margins and the catchwords are sometimes very close to the main text Sect. 1 which makes it difficult to distinguish from the main text. As for the tables and numbering, their very low distribution and unequal division in batches result in lower scores. The relevance of the numbering class may also be questioned based on the outcomes[10]. We achieve similar result with the creation from scratch of a model with the whole dataset. We nevertheless notice a quick integration of the new text-regions in the models (see Fig. 4).

At the polygon of lines level, with the same metric, we measure a global relevance of 94%, no matter the curve of the line. The main difficulty encountered concerns the diacritics, which sometimes are not encompassed by the polygon and must be manually corrected. An adjustment of the seam carve has occurred as soon as the batch 20 to better manage the line height. The polygons verification constitutes nevertheless the most time-consuming task. For the first proofreading task

---

[10] Numbering class is not kept in the v1.1 of the dataset, for which we notice a 9% gain in average for identification of catchword and table classes.

of layout, baseline and polygon predictions, the time saved amounts to 75%. On average, the full process (predictions and proofreading) takes 7 min for an image with the default model, bringing down to 4 min for an image from the first model. The time required is down to 3.2 min for the model 3, then to 2.5 min for the last model. At this point, the proofreading can be confined to the curved lines of the marginal notes. Results are summarized on Fig. 4.

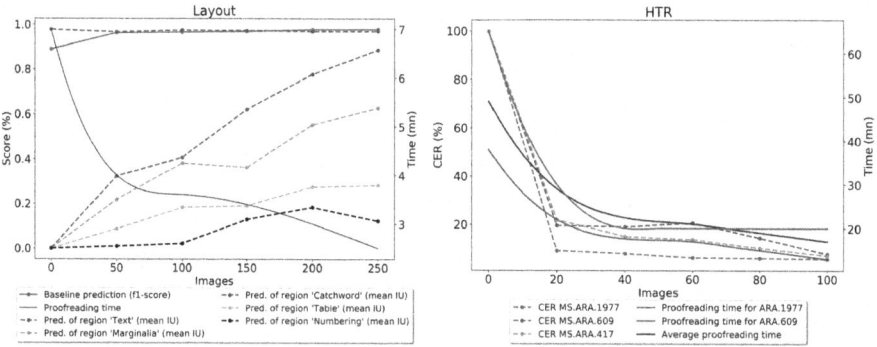

Fig. 4. Evolution of fine-tuning and effects on proofreading time for layout analysis and transcription (v1.0)

HTR Models: Two types of models have been created and evaluated with the default architecture proposed by Calfa Vision. The first are HTR models specific to a project, thus specialized on a manuscript to accompany the transcription. Here, we have chosen batches of 20 corrected images.

(i) Models Dedicated to a Project: Each manuscript has its own difficulties and a specific number of lines (see *supra* Sect. 2.3). We have measured the learning ability of HTR models on each manuscript to go along with the transcription (see Table 9). The models are trained incrementally as new transcriptions are checked and evaluated on the following folios of the project.

Table 9. Evolution of the CER for dedicated HTR models (v1.0)

Model	CER (%)		
	MS.ARA.1977	MS.ARA.609	MS.ARA.417
Model 1 (batch 1)	9.17	19.69	21.96
Model 2 (Model 1 + batch 2)	7.99	19.07	15.03
Model 3 (Model 2 + batch 3)	6.28	20.68	13.85
Model 4 (Model 3 + batch 4)	6.08	14.46	10.32
Final (Model 4 + batch 5)	5.71	7.80	7.10

The MS.ARA.609 and MS.ARA.417 have fewer text lines, thus the CER stays high until the batch 4^{11}. However, we observe a significant gain in the annotation upon application of the first model, with an average transcription time cut from 49 min for an unassisted transcription, to 29 min as of the first batch and 21 min for the last model: hence an average gain of 42% (see Fig. 4). In detail, the gain is 56% for ARA.1977 (with extrema of 1H15 unassisted and 20 min with a model) and 45% for ARA.609 (with extrema of 45 min unassisted and 13 min with a model). Word separation, HTR classical issue, is accurate at 80,4% for each specialized model.

Most character-level prediction errors seem to be more about the characters in the initial or final position in the word. Among the most frequent errors, the final *nūn* can be confused with the *rā'* or the *zāy*, as well as the *dāl* and the *rā'* or the *ḍād* and the *ḥā'*. Furthermore, we observe difficulties in identifying hyphenations between words, leading to misidentification of words. It should also be noted that when several characters with superscript or subscript diacritics follow each other (e.g. a sequence *tā'*, *nūn*, *šīn* or a sequence *bā'*, *yā'*, *fā'*), the prediction of this sequence of letters is frequently incorrect and random, but adding more context with mixed models show a significant improvement of predictions in these cases (see Fig. 5 and Table 10).

Table 10. Example of predictions on MS.ARA.1977 and MS.ARA.609

Pred	الناس بالزكام فشملت مروا ونيسابور والدى وهمدان رحلوان وجميع بلاد العراق وكادت ان تحمليهم بالوت
GT	الناس بالزكام فشملت مروا ونيسابور والرى وهمدان وحلوان وجميع بلاد العراق وكادت ان تحمليهم بالوت

Pred	لها اثنان ونصف تبع بها المدين ويفعل مثل ذلك سهام كل من الاختينة
GT	لها اثنان ونصف تبع بها المدين ويفعل مثل ذلك بسهام كل من الاختين

(ii) HTR models for Arabic Maghrebi Scripts: We have also measured the relevance of transfer learning in-between manuscripts. Some transcription campaigns have in deed progressed more quickly than others and to re-purpose a specialized model for another manuscript is proving beneficial. Results of transfer learning are described in the Fig. 5. For each model, we have used 80% of data for training and 20% for testing. Four mixed models have been evaluated.

The confusion matrix highlights the big discrepancies between the three manuscripts, and no specialized model can achieve a CER below 20% on the other manuscripts. Manuscripts display a wide variety of text density, loop shapes and diacritics management that could affect transfer benefits. These limitations also lead to very different shapes of polygons. But in contrast, we

---

[11] With better polygons (dataset v1.1), the CER decreases more quickly (16.6 for batch 1, then 15.87, 13.67, 11.52, and finally 6.67 for the last batch).

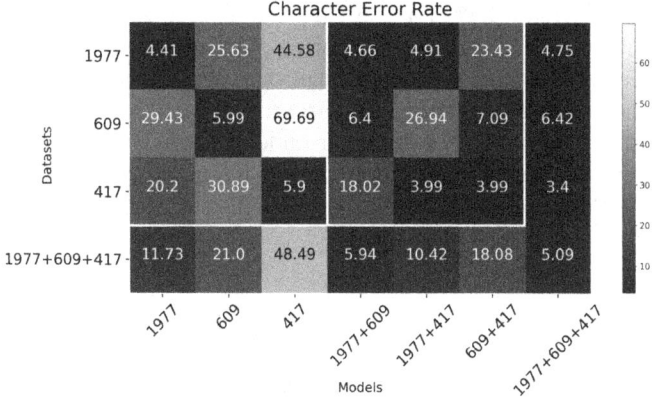

Fig. 5. Impact of transfer learning on CER (v1.0)

observe a much higher convergence of mixed models. The MS.ARA.1977 and MS.ARA.417 manuscripts benefit more from this transfer than MS.ARA.609 which presents specific difficulties. CER of MS.ARA.417 is below 4% with mixed models. If we observe no real gain for CER on each manuscript, the transfer favors greater robustness for word separation with a gain of 8.54% on average, that is consistent with the word-based approach. Experiments also show an ability to correct manual transcriptions misprint. The complete model demonstrates great versatility for these scripts with an average CER of 4.8% Moreover, first experiments on v1.1 tend to indicate that a precise polygonization of lines, with all diacritics included, is necessary with few data (see footnote 11). At the word-level however, gain remains marginal with a larger dataset.

In practice, the editorial choice to manually transcribe the majority of the dataset was made to limit possible typos. For an annotation project, this annotation approach with mixed models is deemed effective and will be implemented in future work, transfer showing a good and fast specialization with a slight fine-tuning.

5   Conclusion

We are presenting a new dataset comprising 300 annotated pages of Arabic Maghrebi script manuscripts of the BULAC. The chosen manuscripts display various layouts, sometimes very complex, with diverse deteriorations. The selected scripts encompass a representative panel of the handwritten production in Maghrebi scripts, in order to foster the emergence of robust HTR systems for these writings. Our work takes part in the commitment of the French scientific community towards the studies of Maghreb and for the promotion of the Maghrebi archives and manuscripts. Though the norms of transcription may be subject to evolution, our evaluations attest already a good recognition of these scripts, with a CER of 4.8% for the three manuscripts on average. The Arabic scripts and Arabic Maghrebi scripts in particular raise several difficulties for their layout processing and their recognition. We demonstrate that a crowd-sourcing approach incorporating automatic fine-tuning and transfer learning is a successful strategy for data creation for under-resourced languages. It achieves similar results to those of the state of the art for manuscripts in Latin scripts. Future work will focus on evaluating the versatility of this dataset and the HTR capabilities for Maghrebi Arabic scripts.

References

1. Abdelhaleem, A., Droby, A., Asi, A., Kassis, M., Asam, R.A., El-sanaa, J.: WAHD: a database for writer identification of Arabic historical documents. In: 2017 1st International Workshop on Arabic Script Analysis and Recognition, pp. 64–68 (2017)
2. Adam, K., Baig, A., Al-Maadeed, S., Bouridane, A., El-Menshawy, S.: KERTAS: dataset for automatic dating of ancient Arabic manuscripts. Int. J. Doc. Anal. Recogn. (IJDAR) **21**(4), 283–290 (2018). https://doi.org/10.1007/s10032-018-0312-3
3. Ben Azzouza, N.: Les corans de l'occident musulman médiéval?: état des recherches et nouvelles perspectives. Perspectives **2**, 104–130 (2017)
4. Bongianino, U.: The origins and developments of Maghribī rounds scripts, Arabic Paleography in the Islamic West (4th/10th-6th/12th centuries). Ph.D. thesis, University of Oxford (2017)
5. Camps, J.B., Vidal-Gorène, C., Vernet, M.: Handling heavily abbreviated manuscripts: HTR engines vs text normalisation approaches (2021). Accepted for IWCP workshop of ICDAR 2021
6. Clausner, C., Antonacopoulos, A., Mcgregor, N., Wilson-Nunn, D.: ICFHR 2018 competition on recognition of historical Arabic scientific manuscripts - RASM2018. In: 2018 16th International Conference on Frontiers in Handwriting Recognition (ICFHR), pp. 471–476 (2018)
7. Clérice, T.: Evaluating deep learning methods for word segmentation of Scripta Continua texts in old French and Latin. J. Data Min. Digit. Humanit. 2020 (2020). https://jdmdh.episciences.org/6264
8. Diem, M., Kleber, F., Sablatnig, R., Gatos, B.: cBAD: ICDAR2019 competition on baseline detection. In: 2019 International Conference on Document Analysis and Recognition (ICDAR), pp. 1494–1498 (2019)
9. Kassis, M., Abdalhaleem, A., Droby, A., Alaasam, R., El-Sana, J.: VML-HD: the historical Arabic documents dataset for recognition systems. In: 2017 1st International Workshop on Arabic Script Analysis and Recognition, pp. 11–14 (2017)
10. Kiessling, B., Ezra, D.S.B., Miller, M.T.: BADAM: a public dataset for baseline detection in Arabic-script manuscripts. In: Proceedings of the 5th International Workshop on Historical Document Imaging and Processing. HIP 2019, pp. 13–18. Association for Computing Machinery (2019)
11. Milo, T., Martínez, A.G.: A new strategy for Arabic OCR: archigraphemes, letter blocks, script grammar, and shape synthesis. In: Proceedings of the 3rd International Conference on Digital Access to Textual Cultural Heritage. DATeCH2019, pp. 93–96. Association for Computing Machinery, New York (2019)

12. Pantke, W., Dennhardt, M., Fecker, D., Märgner, V., Fingscheidt, T.: An historical handwritten Arabic dataset for segmentation-free word spotting - HADARA80P. In: 2014 14th International Conference on Frontiers in Handwriting Recognition, pp. 15–20 (2014)

13. Van Den Boogert, N.: Some notes on Maghribi script. Manuscripts Middle East **4**, 30–43 (1989)

14. Vidal-Gorène, C., Dupin, B., Decours-Perez, A., Riccioli, T.: A modular and automated annotation platform for handwritings: evaluation on under-resourced languages (2021). Accepted for ICDAR 2021 Main Conference

Towards Boosting the Accuracy of Non-latin Scene Text Recognition

Sanjana Gunna$^{(\boxtimes)}$ ⓘ, Rohit Saluja ⓘ, and C. V. Jawahar ⓘ

Centre for Vision Information Technology, International Institute of Information Technology, Hyderabad 500032, India
{sanjana.gunna,rohit.saluja}@research.iiit.ac.in, jawahar@iiit.ac.in
https://github.com/firesans/NonLatinPhotoOCR

Abstract. Scene-text recognition is remarkably better in Latin languages than the non-Latin languages due to several factors like multiple fonts, simplistic vocabulary statistics, updated data generation tools, and writing systems. This paper examines the possible reasons for low accuracy by comparing English datasets with non-Latin languages. We compare various features like the size (width and height) of the word images and word length statistics. Over the last decade, generating synthetic datasets with powerful deep learning techniques has tremendously improved scene-text recognition. Several controlled experiments are performed on English, by varying the number of (i) fonts to create the synthetic data and (ii) created word images. We discover that these factors are critical for the scene-text recognition systems. The English synthetic datasets utilize over 1400 fonts while Arabic and other non-Latin datasets utilize less than 100 fonts for data generation. Since some of these languages are a part of different regions, we garner additional fonts through a region-based search to improve the scene-text recognition models in Arabic and Devanagari. We improve the Word Recognition Rates (WRRs) on Arabic MLT-17 and MLT-19 datasets by 24.54% and 2.32% compared to previous works or baselines. We achieve WRR gains of 7.88% and 3.72% for IIIT-ILST and MLT-19 Devanagari datasets.

Keywords: Scene-text recognition · Photo OCR · Multilingual OCR · Arabic OCR · Synthetic data · Generative adversarial network

1 Introduction

The task of scene-text recognition involves reading the text from natural images. It finds applications in aiding the visually impaired, extracting information for map services and geographical information systems by mining data from the street-view-like images [2]. The overall pipeline for scene-text recognition involves a text detection stage followed by a text recognition stage. Predicting the bounding boxes around word images is called text detection [6]. The next step involves recognizing text from the cropped text images obtained from the labeled or the predicted bounding boxes [12]. In this work, we focus on improving text recognition in non-Latin languages. Multilingual text recognition has witnessed notable

© Springer Nature Switzerland AG 2021
E. H. Barney Smith and U. Pal (Eds.): ICDAR 2021 Workshops, LNCS 12916, pp. 282–293, 2021.
https://doi.org/10.1007/978-3-030-86198-8_20

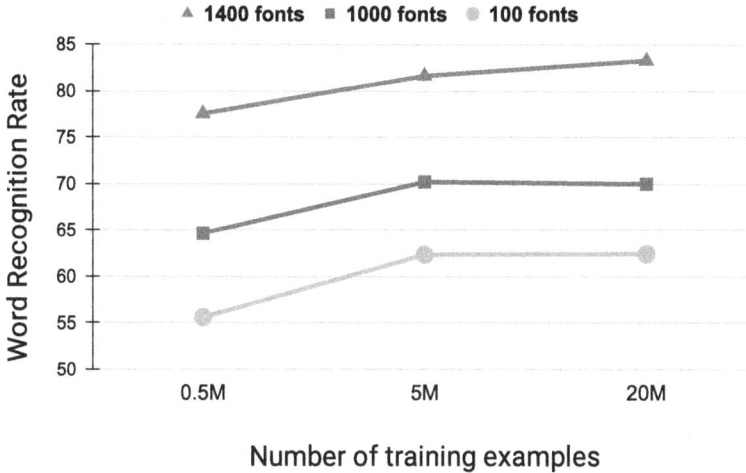

Fig. 1. Comparing STAR-Net's performance on IIIT5K [13] dataset when trained on synthetic data created using a varying number of fonts and training samples.

growth due to the impact of globalization leading to international and intercultural communication. Like English, the recognition algorithms proposed for Latin datasets have not successfully recorded similar accuracies on non-Latin datasets. Reading text from non-Latin images is challenging due to the distinct variation in the scripts used, writing systems, scarcity of data, and fonts. In Fig. 1, we illustrate the analysis of Word Recognition Rates (WRR) on the IIIT5K English dataset [13] by varying the number of training samples and fonts used in the synthetic data. The training performed on STAR-Net [11] proves extending the number of fonts leads to better WRR gains than increasing training data. We incorporate the new fonts found using region-based online search to generate synthetic data in Arabic and Devanagari. The motivation behind this work is described in Sect. 3. The methodology to train the deep neural network on the Arabic and Devanagari datasets is detailed in Sect. 4. The results and conclusions from this study are presented in Sect. 5 and 6, respectively. The contributions of this work are as follows:

1. We study the two parameters for synthetic datasets crucial to the performance of the reading models on the IIIT5K English dataset; i) the number of training examples and ii) the number of diverse fonts[1].

---

[1] We also investigated other reasons for low recognition rates in non-Latin languages, like comparing the size of word images of Latin and non-Latin real datasets but could not find any significant variations (or exciting differences). Moreover, we observe very high word recognition rates (> 90%) when we tested our non-Latin models on the held-out synthetic datasets, which shows that learning to read the non-Latin glyphs is trivial for the existing deep models. Refer https://github.com/firesans/STRforIndicLanguages for more details.

Table 1. Latin and non-Latin scene-text recognition datasets.

Language	Datasets
Multilingual	IIIT-ILST-17 ($3K$ words, 3 languages), MLT-17 ($18K$ scenes, 9 languages), MLT-19 ($20K$ scenes, 10 languages) OCR-on-the-go-19 (1000 scenes, 3 languages), CATALIST-21 (2322 scenes, 3 languages)
Arabic	ARASTEC-15 (260 signboards, hoardings, advertisements), MLT-17,19
Chinese	RCTW-17 ($12K$ scenes), ReCTS-25K-19 ($25K$ signboards), CTW-19 ($32K$ scenes), RRC-LSVT-19 ($450K$ scenes), MLT-17,19
Korean	KAIST-11 ($2.4K$ signboards, book covers, characters), MLT-17,19
Japanese	DOST-16 ($32K$ images), MLT-17,19
English	SVT-10 (350 scenes), SVT-P-13 (238 scenes, 639 words), IIIT5K-12 ($5K$ words), IC11 (485 scenes, 1564 words), IC13 (462 scenes), IC15 (1500 scenes), COCO-Text-16 ($63.7K$ scenes), CUTE80-14 (80 scenes), Total-Text-19 (2201 scenes), MLT-17,19

2. We share 55 additional fonts in Arabic, and 97 new fonts in Devanagari, which we found using a region-wise online search. These fonts were not used in the previous scene text recognition works.
3. We apply our learnings to improve the state-of-the-art results of two non-Latin languages, Arabic, and Devanagari.

2   Related Work

Recently, there has been an increasing interest in scene-text recognition for a few widely spoken non-Latin languages around the globe, such as Arabic, Chinese, Devanagari, Japanese, Korean. Multi-lingual datasets have been introduced to tackle such languages due to their unique characteristics. As shown in Table 1, Mathew et al. [12] release the IIIT-ILST Dataset containing around $1K$ images each three non-Latin languages. The MLT dataset from the ICDAR'17 RRC contains images from Arabic, Bangla, Chinese, English, French, German, Italian, Japanese, and Korean [15]. The ICDAR'19 RRC builds MLT-19 on top of MLT-17 to containing text from Arabic, Bangla, Chinese, English, French, German, Italian, Japanese, Korean, and Devanagari [14]. Recent OCR-on-the-go and CATALIST[2] datasets include around 1000 and 2322 annotated videos in Marathi, Hindi, and English [19]. Arabic scene-text recognition datasets involve ARASTEC and MLT-17,19 [26]. Chinese datasets cover RCTW, ReCTS-25k, CTW, and RRC-LSVT from ICDAR'19 Robust Reading Competition (RRC) [23,24,31,33]. Korean and Japanese scene-text recognition datasets involve KAIST and DOST [7,9]. Different English datasets are listed in the last row of Table 1 [3,10,13–17,20,27,28,30].

---

[2] https://catalist-2021.github.io/.

Various models have been proposed for the task of scene-text recognition. Wang et al. [29] present an object recognition module that achieves competitive performance by training on ground truth lexicons without any explicit text detection stage. Shi et al. [21] propose a Convolutional Recurrent Neural Network (CRNN) architecture. It achieves remarkable performances in both lexicon-free and lexicon-based scene-text recognition tasks as is used by Mathew et al. [12] for three non-Latin languages. Liu et al. [11] introduce Spatial Attention Residue Network (STAR-Net) with Spatial Transformer-based Attention Mechanism, which handles image distortions. Shi et al. [22] propose a segmentation-free Attention-based method for Text Recognition (ASTER). Mathew et al. [12] achieves the Word Recognition Rates (WRRs) of 42.9%, 57.2%, and 73.4% on $1K$ real images in Hindi, Telugu, and Malayalam, respectively. Bušta et al. [2] propose a CNN (and CTC) based method for text localization, script identification, and text recognition and is tested on 11 languages (including Arabic) of MLT-17 dataset. The WRRs are above 65% for Latin and Hangul and are below 47% for the remaining languages (46.2% for Arabic). Therefore, we aim to improve non-Latin recognition models.

3   Motivation and Datasets

This section explains the motivation behind our work. Here we also describe the datasets used for experiments on non-Latin scene text recognition.

Motivation: To study the effect of fonts and training examples on scene-text recognition performance, we randomly sample 100 and 1000 fonts from the set of over 1400 English fonts from previous works [5,8]. For 1400 fonts, we use the datasets available from earlier photo OCR works on synthetic dataset generation [5,8]. For 100 and 1000 fonts, we generate synthetic images by following a simplified methodology proposed by Mathew et al. [12]. Therefore, we create three different synthetic datasets. Moreover, we simultaneously experiment by varying the number of training samples from $0.5M$ to $5M$ to $20M$ samples. By changing the two parameters, we train our model (refer Sect. 4) on the above synthetic datasets and test them on the IIIT5K dataset. We observe that the Word Recognition Rate (WRR) of the dataset with around $20M$ samples and over 1400 fonts achieves state-of-the-art accuracy on the IIIT5K dataset [11]. As shown in Fig. 1, the WRR of the model trained on $5M$ samples generated using over 1400 fonts is very close to the recorded WRR ($20M$ samples). Moreover, models trained on 1400 fonts outperform the models trained on 1000 and 100 fonts by a margin of 10% because of improved (font) diversity and better but complex dataset generation methods. Also, in Fig. 1, as we increase the number of fonts from 1000 to 1400, the WRR gap between the models trained on $5M$ and $20M$ samples moderately improves (from 0% to around 2%). Finally, this analysis highlights the importance of increasing the fonts in synthetic dataset generation and ultimately improving the scene-text recognition models.

Table 2. Synthetic Data Statistics. μ, σ represent mean, standard deviation.

Language	# Images	μ, σ word length	# Fonts
English	17.5M	5.12, 2.99	>1400
Arabic	5M	6.39, 2.26	140
Devanagari	5M	8.73, 3.10	194

Fig. 2. Synthetic word images in Arabic and Devanagari.

Datasets: As shown in Table 2, we generate over $17M$ word images in English, $5M$ word images each in Arabic, and Devanagari, using the tools provided by Mathew et al. [12]. We use 140 and 194 fonts for Arabic and Devanagari, respectively. Previous works use 97 fonts and 85 fonts for these languages [1,12]. Since the two languages are spoken in different regions, we found 55 additional fonts in Arabic and 97 new fonts in Devanagari using the region-wise online search.[3] We use the additional fonts obtained by region-wise online search, which we will share with this work. As we will see in Sect. 5, we also perform some of our experiments with these fonts. Sample images of our synthetic data are shown in Fig. 2. As shown in Table 2, English has the lowest average word length among the languages mentioned, while Arabic and Devanagari have comparable average word lengths. Please note that we use over 1400 fonts for English, whereas the number of diverse fonts available for the non-Latin languages is relatively low. We run our models on Arabic and Devnagari test sets from MLT-17, IIIT-ILST, and MLT-19 datasets[4]. The results are summarized in Sect. 5.

---

[3] Additional fonts we found using region-based online search are available at: www. sanskritdocuments.org/, www.tinyurl.com/n84kspbx, www.tinyurl.com/7uz2fknu, www.ctan.org/tex-archive/fonts/shobhika?lang=en, www.hindi-fonts.com/, www. fontsc.com/font/tag/arabic, more fonts are shared on https://github.com/firesans/ NonLatinPhotoOCR.

[4] We could not obtain the ARASTEC dataset we discussed in the previous section.

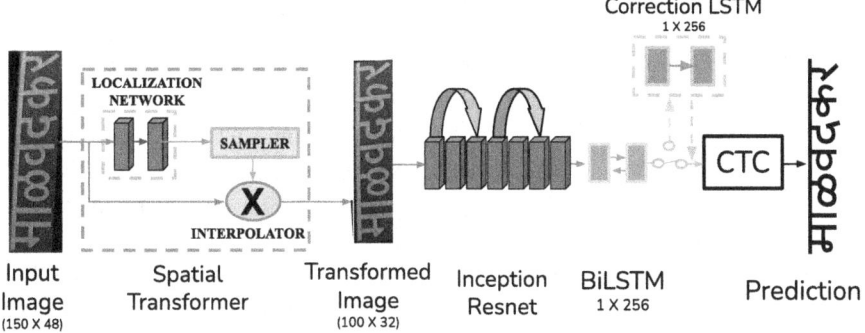

Fig. 3. Model used to train on non-Latin datasets.

4 Underlying Model

We now describe the model we train for our experiments. We use STAR-Net because of its capacity to handle different image distortions [11]. It has a Spatial Transformer network, a Residue Feature Extractor, and a Connectionist Temporal Classification (CTC) layer. As shown in Fig. 3, the first component consists of a spatial attention mechanism achieved via a CNN-based localisation network that helps predict affine transformation parameters to handle image distortions. The second component consists of a Convolutional Neural Network (CNN) and a Recurrent Neural Network (RNN). The CNN is inception-resnet architecture, which helps in extracting robust image features [25]. The last component provides the non-parameterized supervision for text alignment. The overall end-to-end trainable model consists of 26 convolutional layers [11].

The input to spatial transformer module is of resolution 150×48. The spatial transformer outputs the image of size 100×32 for the next stage (Residue Feature Extractor). We train all our models on $5M$ synthetic word images as discussed in the previous section. We use the batch size of 32 and the ADADELTA optimizer for our experiments [32]. We train each model for 10 epochs and test on Arabic and Devanagari word images from IIIT-ILST, MLT-17, and MLT-19 datasets. Only for the Arabic MLT-17 dataset, we fine-tune our models on training images and test them on validation images to fairly compare with Bušta et al. [1]. For Devanagari, we present the additional results on the IIIT-ILST dataset by fine-tuning our best model on the MLT-19 dataset. We fine-tune all the layers of our model for the two settings mentioned above. To further improve our models, we add an LSTM layer of size 1×256 to the STAR-Net model, pre-trained on synthetic data. The additional layer corrects the model's bias towards the synthetic datasets, and hence we call it correction LSTM. We plug-in the correction LSTM before the CTC layer, as shown in Fig. 3 (top-right). After attaching the LSTM layer, we fine-tune the complete network on the real datasets.

Table 3. Results of our experiments on real datasets. FT means fine-tuned.

Language	Dataset	# Images	Model	CRR	WRR
Arabic	MLT-17	951	Bušta et al. [2]	75.00	46.20
			STAR-Net (85 Fonts) FT	88.48	66.38
			STAR-Net (140 Fonts) FT	89.17	68.51
			STAR-Net (140 Fonts) FT with Correction LSTM	**90.19**	**70.74**
Devanagari	IIIT-ILST	1150	Mathew et al. [12]	75.60	42.90
			STAR-Net (97 Fonts)	77.44	43.38
			STAR-Net (194 Fonts)	77.65	44.27
			STAR-Net (194 Fonts) FT on MLT-19 data	79.45	50.02
			STAR-Net (194 Fonts) FT with Correction LSTM	**80.45**	**50.78**
Arabic	MLT-19	4501	STAR-Net (85 Fonts)	71.15	40.05
			STAR-Net (140 Fonts)	**75.26**	**42.37**
Devanagari	MLT-19	3766	STAR-Net (97 Fonts)	84.60	60.83
			STAR-Net (194 Fonts)	**85.87**	**64.55**

5   Results

Table 3 depicts the performance of our experiments on the real datasets. For the Arabic MLT-17 dataset and Devanagari IIIT-ILST dataset, we achieve recognition rates better than Bušta et al. [1] and Mathew et al. [12]. With STAR-Net model trained on <100 fonts (refer Sect. 3), we achieve 13.48% and 20.18% gains in Character Recognition Rate (CRR) and Word Recognition Rate (WRR) for Arabic, and 1.84% and 0.48% improvements for Devanagari over the previous works (compare rows 1, 2 and 5, 6 in the last column of Table 3). The CRR and WRR further improve by training the models on the same amount of training data synthesized with $>=$ 140 fonts (rows 3 and 7 in the last column of Table 3). By fine-tuning the Devanagari model on the MLT-19 dataset, the CRR and WRR gains raise to 3.85% and 7.12%. By adding the correction LSTM layer to the best models, we achieve the highest CRR and WRR gains of 15.19% and 24.54% for Arabic, and 5.25% and 7.88% for Devanagari, over the previous works. The final results for the two datasets discussed above can be seen in rows 3 and 7 of the last column of Table 3.

As shown in Table 3, for the MLT-19 Arabic dataset, the model trained on $5M$ samples generated using 85 fonts achieve the CRR of 71.15% and WRR of 40.05%. Increasing the number of diverse fonts to 140 gives a CRR gain of 4.11% and a WRR gain of 2.32%. For the MLT-19 Devanagari dataset, the model trained on $5M$ samples generated using 97 fonts achieves the CRR of 84.60% and WRR of 60.83%. Increasing the number of fonts to 194 gives a CRR gain of

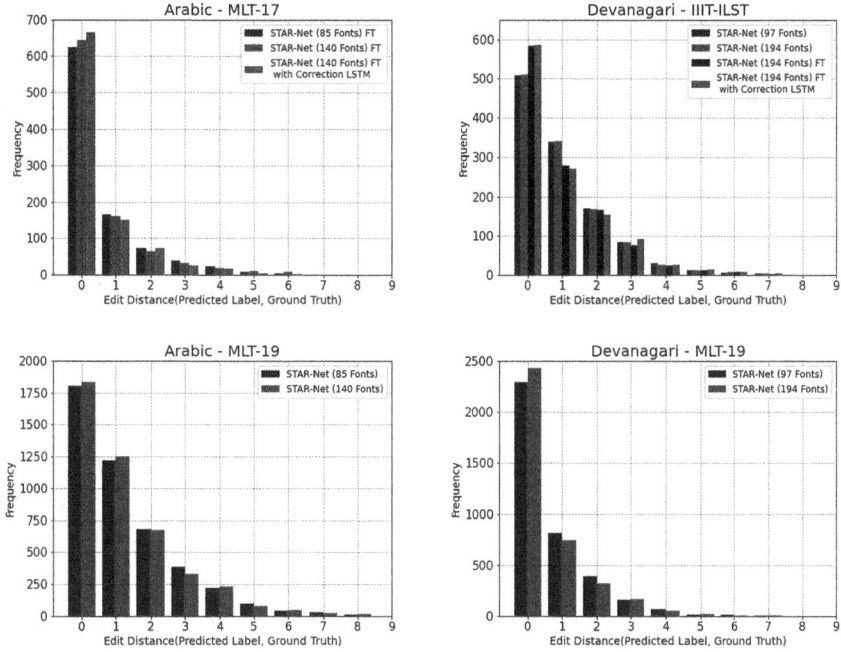

Fig. 4. Histogram of correct words ($x = 0$) and words with x errors ($x > 0$). FT represents the models fine-tuned on real datasets.

1.27% and a WRR gain of 3.72%. It is also interesting to note that the WRR of our models on MLT-17 Arabic and MLT-19 Devanagari datasets are very close to the WRR of the English model trained on $5M$ samples generated using 100 fonts (refer to the yellow curve in Fig. 1). It supports our claim that the number of fonts used to create the synthetic dataset plays a crucial role in improving the photo OCR models in different languages.

To present the overall improvements by utilizing extra fonts and correction LSTM at a higher level, we examine the histograms of edit distance between the pairs of predicted and corresponding ground truth words in Fig. 4. Such histograms are used in one of the previous works on OCR error corrections [18]. The bars at the edit distance of 0 represent the words correctly predicted by the models. The subsequent bars at edit distance $n > 0$ represent the number of words with x erroneous characters. As it can be seen in Fig. 4, overall, with the increase in the number of fonts and subsequently with correction LSTM, i) the number of correct words ($x = 0$) increase for each dataset, and ii) the number of incorrect words ($x > 0$) reduces for many values of x for the different datasets. We observe few exceptions in each histogram where the frequency of incorrect words is higher for the best model than others, e.g., at edit distance of 2 for the Arabic MLT-17 dataset. The differences (or exceptions) show that the recognitions by different models complement each other.

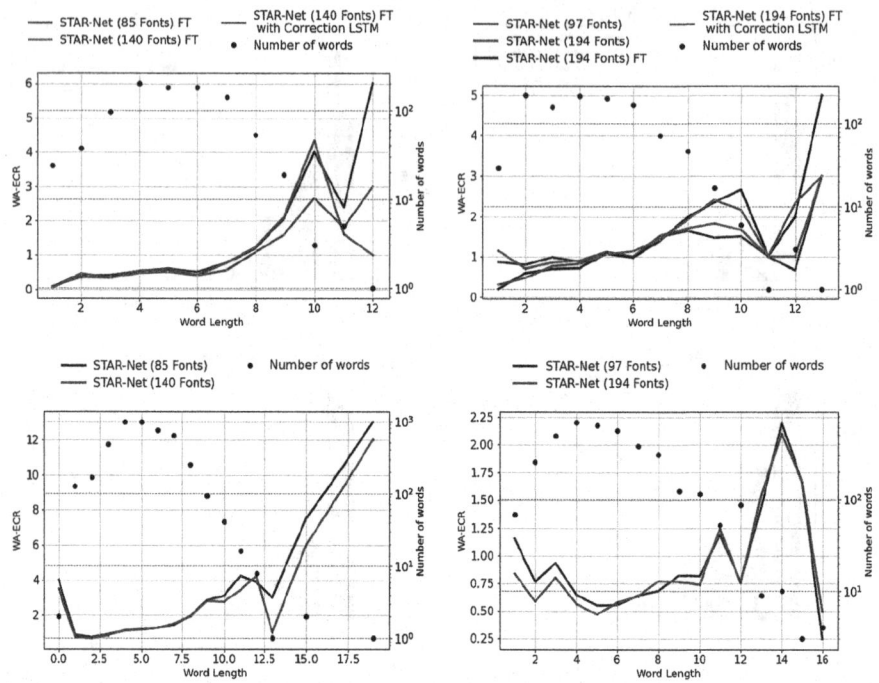

Fig. 5. Clockwise from top-left: WA-ECR of our models tested on MLT-17 Arabic, IIIT-ILST Devanagari, MLT-19 Devanagari, and MLT-19 Arabic datasets.

Another exciting way to compare the output of different OCR systems is Word-Averaged Erroneous Character Rate (WA-ECR), as proposed by Agam et al. [4]. The WA-ECR is the ratio of i) the number of erroneous characters in the set of all l-length ground truth words (e_l), and ii) the number of l-length ground truth words (n_l) in the test set. As shown in the red dots and the right y-axis of the plots in Fig. 5, the frequency of words generally reduces with an increase in word length after $x = 4$. Therefore, the denominator term tends to decrease the WA-ECR for short-length words. Moreover, as the word length increases, it becomes difficult for the OCR model to predict all the characters correctly. Naturally, the WA-ECR tends to increase with the increase in word length for an OCR system. In Fig. 5, we observe that our models trained on >=140 fonts (blue curves) are having lower WA-ECR across different word lengths as compared to the ones trained on <100 fonts (orange curves). For the IIIT-ILST dataset, the model, trained on 194 fonts, performs poorly on the long words ($x > 8$ in the top-right plot of Fig. 5), and the correction LSTM further enhances this effect. On the contrary, we observe that the Correction LSTM reduces WA-ECR for the MLT-17 Arabic dataset for word lengths in the range [6, 11] (compare green and blue curves in the top-left plot). Interestingly, the WA-ECR of some of our models drops after word-length of 10 and 14 for the MLT-19 Arabic and MLT-19

Fig. 6. Real word images in Arabic (top) and Devanagari (bottom), Below the images: predictions from i) baseline model trained on <100 fonts, ii) model trained on ≥ 140 fonts. Green & red represent correct predictions and errors. (Color figure online)

Devanagari datasets (see blue curve in the top-left plot and the two curves in the bottom-right plot of Fig. 5).

In Fig. 6, we present the qualitative results of our models. The green and red colors represent the predictions and errors. As shown, the models trained on over 140 fonts perform better than the models trained on <100 fonts. Overall, the experiments support our claim that the diversity in fonts used to generate synthetic datasets is crucial for improving the existing non-Latin scene-text recognition systems.

6   Conclusion

We carried out a series of controlled experiments in English to highlight the importance of font diversity and the number of synthetic examples in improving the scene-text recognition accuracy. We augmented the font set of two non-Latin scripts, Arabic and Devanagari, with new fonts obtained by region-based online search. We generated $5M$ synthetic images in two languages. Our experiments show improvements over the previous works and baselines trained on lesser fonts. We further improve our results by introducing the correction LSTM into the models to reduce the bias towards the synthetic data. Finally, we affirm that more fonts are required to improve the existing non-Latin systems. For future work in this area, we plan to employ human designers or Generative Adversarial Networks (GAN) based font generators to boost the accuracy of non-Latin scene-text recognition.

References

1. Bušta, M., Neumann, L., Matas, J.: Deep TextSpotter: an end-to-end trainable scene text localization and recognition framework. In: ICCV (2017)
2. Bušta, M., Patel, Y., Matas, J.: E2E-MLT - an unconstrained end-to-end method for multi-language scene text. In: Carneiro, G., You, S. (eds.) ACCV 2018. LNCS, vol. 11367, pp. 127–143. Springer, Cham (2019). https://doi.org/10.1007/978-3-030-21074-8_11
3. Chng, C.K., Chan, C.S.: Total-text: a comprehensive dataset for scene text detection and recognition. In: 2017 14th IAPR International Conference on Document Analysis and Recognition (ICDAR), vol. 1, pp. 935–942 (2017)
4. Dwivedi, A., Saluja, R., Kiran Sarvadevabhatla, R.: An OCR for classical Indic documents containing arbitrarily long words. In: The IEEE/CVF Conference on Computer Vision and Pattern Recognition (CVPR) Workshops, June 2020
5. Gupta, A., Vedaldi, A., Zisserman, A.: Synthetic data for text localisation in natural images. In: Proceedings of the IEEE Conference on Computer Vision and Pattern Recognition, pp. 2315–2324 (2016)
6. Huang, Z., Zhong, Z., Sun, L., Huo, Q.: Mask R-CNN with pyramid attention network for scene text detection. In: WACV, pp. 764–772. IEEE (2019)
7. Iwamura, M., Matsuda, T., Morimoto, N., Sato, H., Ikeda, Y., Kise, K.: Downtown Osaka scene text dataset. In: Hua, G., Jégou, H. (eds.) ECCV 2016. LNCS, vol. 9913, pp. 440–455. Springer, Cham (2016). https://doi.org/10.1007/978-3-319-46604-0_32
8. Jaderberg, M., Simonyan, K., Vedaldi, A., Zisserman, A.: Synthetic data and artificial neural networks for natural scene text recognition. In: Workshop on Deep Learning, NIPS (2014)
9. Jung, J., Lee, S., Cho, M.S., Kim, J.H.: Touch TT: scene text extractor using touchscreen interface. ETRI J. **33**(1), 78–88 (2011)
10. Karatzas, D., et al.: ICDAR 2013 robust reading competition. ICDAR 2013, pp. 1484–1493. IEEE Computer Society, USA (2013)
11. Liu, W., Chen, C., Wong, K.Y.K., Su, Z., Han, J.: STAR-Net: a SpaTial attention residue network for scene text recognition. In: BMVC, vol. 2 (2016)
12. Mathew, M., Jain, M., Jawahar, C.: Benchmarking scene text recognition in Devanagari, Telugu and Malayalam. In: ICDAR, vol. 7, pp. 42–46. IEEE (2017)
13. Mishra, A., Alahari, K., Jawahar, C.V.: Scene text recognition using higher order language priors. In: BMVC (2012)
14. Nayef, N., et al.: ICDAR2019 robust reading challenge on multi-lingual scene text detection and recognition-RRC-MLT-2019. In: 2019 International Conference on Document Analysis and Recognition (ICDAR), pp. 1582–1587. IEEE (2019)
15. Nayef, N., et al.: Robust reading challenge on multi-lingual scene text detection and script identification - RRC-MLT. In: 14th ICDAR, vol. 1, pp. 1454–1459. IEEE (2017)
16. Phan, T., Shivakumara, P., Tian, S., Tan, C.: Recognizing text with perspective distortion in natural scenes. In: 2013 IEEE International Conference on Computer Vision, pp. 569–576 (2013)
17. Risnumawan, A., Shivakumara, P., Chan, C.S., Tan, C.L.: A robust arbitrary text detection system for natural scene images. Expert Syst. Appl. **41**, 8027–8048 (2014)
18. Saluja, R., Adiga, D., Chaudhuri, P., Ramakrishnan, G., Carman, M.: Error detection and corrections in Indic OCR using LSTMs. In: 2017 14th IAPR International Conference on Document Analysis and Recognition (ICDAR), vol. 1, pp. 17–22. IEEE (2017)

19. Saluja, R., Maheshwari, A., Ramakrishnan, G., Chaudhuri, P., Carman, M.: OCR On-the-Go: robust end-to-end systems for reading license plates and street signs. In: 15th IAPR International Conference on Document Analysis and Recognition (ICDAR), pp. 154–159. IEEE (2019)
20. Shahab, A., Shafait, F., Dengel, A.: ICDAR 2011 robust reading competition challenge 2: reading text in scene images. In: 2011 International Conference on Document Analysis and Recognition, pp. 1491–1496 (2011)
21. Shi, B., Bai, X., Yao, C.: An end-to-end trainable neural network for image-based sequence recognition and its application to scene text recognition. IEEE Trans. Pattern Anal. Mach. Intell. **39**(11), 2298–2304 (2016)
22. Shi, B., Yang, M., Wang, X., Lyu, P., Yao, C., Bai, X.: ASTER: an attentional scene text recognizer with flexible rectification. IEEE Trans. Pattern Anal. Mach. Intell. **41**(9), 2035–2048 (2018)
23. Shi, B., et al.: ICDAR2017 competition on reading Chinese text in the wild (RCTW-17). In: 14th ICDAR, vol. 1, pp. 1429–1434. IEEE (2017)
24. Sun, Y., et al.: ICDAR 2019 competition on large-scale street view text with partial labeling - RRC-LSVT. In: 2019 International Conference on Document Analysis and Recognition (ICDAR), pp. 1557–1562. IEEE (2019)
25. Szegedy, C., Ioffe, S., Vanhoucke, V., Alemi, A.: Inception-v4, Inception-Resnet and the impact of residual connections on learning. In: Proceedings of the AAAI Conference on Artificial Intelligence, vol. 31 (2017)
26. Tounsi, M., Moalla, I., Alimi, A.M., Lebouregois, F.: Arabic characters recognition in natural scenes using sparse coding for feature representations. In: 13th ICDAR, pp. 1036–1040. IEEE (2015)
27. Veit, A., Matera, T., Neumann, L., Matas, J., Belongie, S.: COCO-Text: dataset and benchmark for text detection and recognition in natural images. arXiv preprint arXiv:1601.07140 (2016)
28. Wang, K., Babenko, B., Belongie, S.: End-to-end scene text recognition. In: 2011 International Conference on Computer Vision, pp. 1457–1464 (2011)
29. Wang, K., Babenko, B., Belongie, S.: End-to-end scene text recognition. In: ICCV, pp. 1457–1464. IEEE (2011)
30. Wang, K., Belongie, S.: Word spotting in the wild. In: Daniilidis, K., Maragos, P., Paragios, N. (eds.) ECCV 2010. LNCS, vol. 6311, pp. 591–604. Springer, Heidelberg (2010). https://doi.org/10.1007/978-3-642-15549-9_43
31. Yuan, T., Zhu, Z., Xu, K., Li, C., Mu, T., Hu, S.: A large Chinese text dataset in the wild. J. Comput. Sci. Technol. **34**(3), 509–521 (2019). https://doi.org/10.1007/s11390-019-1923-y
32. Zeiler, M.: Adadelta: An adaptive learning rate method 1212, December 2012
33. Zhang, R., et al.: ICDAR 2019 robust reading challenge on reading Chinese text on signboard. In: 2019 International Conference on Document Analysis and Recognition (ICDAR), pp. 1577–1581. IEEE (2019)

Aolah Databases for New Arabic Online Handwriting Recognition Algorithm

Samia Heshmat[⊠] and Mohamed Abdelnafea

Faculty of Engineering, Aswan University, Aswan 81542, Egypt
`samia.heshmat@aswu.edu.eg`

Abstract. Developing an online handwriting recognition system for Arabic script used in pen-based devices plays an important role in making these devices available and usable for Arabic society. This paper is carried out for Arabic script to overcome the difficulties presented in the Arabic language in cursive, overlapping, handwriting variability, different writing styles, delayed strokes, and other challenges. An algorithm for recognizing Arabic strokes written by hand is proposed; since there are some troubles in distinguishing the written stroke for similar characters. The uniqueness of the recommended algorithm is dealing with every stroke in the character separately. Furthermore, in the current research, two novel databases for Arabic characters and Arabic characters' strokes are generated. The two databases are presented, one for Arabic characters by different writers for the 28 Arabic characters, the other database is extracted from the previous database by taking only the Arabic character strokes. The algorithm used for data collection is distinguished by the ability to deal with each stroke in the written characters separately. The code acts as a simulation of a stylus pen and a touch screen. Stroke capturing is achieved by collecting data points along the path of an input device (stylus pen or mouse) same time those characters are written.

Keywords: Arabic online database · Data collection and preprocessing · Machine learning · Handwriting recognition · Artificial Intelligence

1 Introduction

Arabic script is an alphabet written from right to left which contains two types of symbols for writing words: letters and diacritics. Letters consist of two parts: letter form and letter mark. The letter form is an essential component in each letter with a total of 19 letter forms. The letter marks may be dots, short Kaf, or Hamza letter mark. Hamza is used for both the letter form and the letter mark, which appears with other letter forms. The Madda letter mark is a Hamza variant; Fig. 1 indicates the Arabic script. Diacritics, the second symbol in writing Arabic words which is not essential in writing like the main letter. Three types of diacritics are there: Vowel which are Fatha ó, Damma ó, Kasra ọ, or Sukun ọmeans no vowel, Nunation which is a doubled version of their corresponding short vowels are two Fathas, two Dammas, two Kasras, and Shadda which is a consonant doubling diacritic. Figure 2 shows types of Arabic diacritics [1–3].

E. H. Barney Smith and U. Pal (Eds.): ICDAR 2021 Workshops, LNCS 12916, pp. 294–311, 2021.
https://doi.org/10.1007/978-3-030-86198-8_21

There are many challenges do exist when the Arabic script is written by hand and that is due to its unique nature, it is cursive and overlapping occurs between Arabic letters, each character has more than one shape, and other challenges [4]. To build a recognition system for Arabic handwriting words researchers need a real and substantial database [5, 6]; hence, this work is a contribution in producing an Arabic database to help researchers in this field and to overcome challenges existed in previous databases. The databases are developed based on an algorithm that uses stroke capturing to facilitate recognition of Arabic characters [7]. The proposed databases (AOLAH) are typical formats of online handwritten data which is a sequence of coordinate points of the moving pen point. Connected parts of the pen trace, in which the pen point is touching the writing surface, are called strokes [8–10].

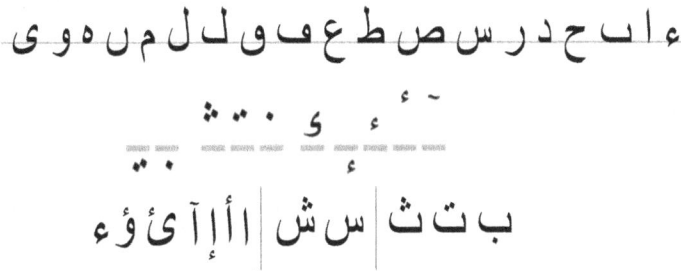

Fig. 1. Arabic script.

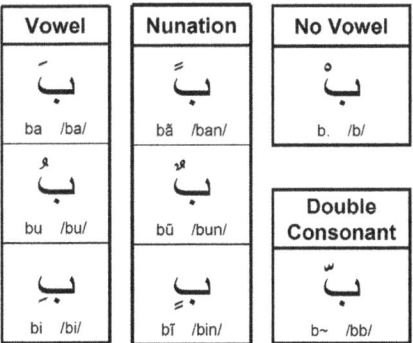

Fig. 2. Arabic diacritics types.

The remainder of this paper is organized as follows: a discussion about existing Arabic databases is shown in Sect. 2. Section 3 presents the proposed AOLAH databases for Arabic online handwriting letters and strokes. While Sects. 4 and 5 describe the proposed Arabic online handwriting recognition algorithm with showing the optimum proposed recognition model Conclusions are given in Sect. 6.

2 Present Databases for Handwritten Arabic Letters

This section describes the main databases used in online Arabic handwriting recognition researches. Table 1 shows a summary of these databases.

2.1 LMCA (2008) [11, 12]

The On/Off (LMCA) dual Arabic handwriting database; this abbreviation is from the French sentence which is Lettres, Mots et Chiffres Arabe. This database contains 30,000 digits, 100,000 Arabic letters and 500 Arabic words; there were 55 participants invited to contribute. This database was developed by REGIM laboratory which is abbreviated for REsearch Group on Intelligent Machines. Both on/off line handwritten characters and words were considered. LMCA database is limited to a small set of words, and the letters are collected separately which means not segmented from cursive text.

2.2 OHASD (2010) [13]

This database is considered as first online Arabic sentence database handwritten on tablet PC. The final version of this dataset is composed of 154 paragraphs, selected from public daily news, written by 48 writers, having a total of 3,825 words and 19,467 characters, after excluding erratic/illegible handwritings. This database has a limited lexicon, limited data, and a limited number of writers.

Table 1. Main present online databases.

Database	Chars	Words	Writers	Main Drawbacks
LMCA (2008)	100000	500	55	– Limited to a small set of words – Letters collected separately
OHASD (2010)	19467	3825	48	– Limited lexicon – Limited data – Limited number of writers
ADAB (2011)	174690	33164	166	– Isolated word samples "not a natural Arabic online handwriting" – No segmentation of words into letters
ALTEC (2014)	106433	152680	1001	– Data collected not in a natural Arabic online handwriting way
QHW (2014)	42800	12000	200	– Closed vocabulary database.- Limited number of words
Online-KHATT (2018)	801421	80931	623	– No dealing with characters on the base of its strokes

2.3   ADAB (2011) [14, 15]

This database was developed by the institut fuer Nachrichtentechnik and the research group on intelligent machines (REGIM). It contains online samples of 937 Tunisian city names that consist of 33,164 Arabic words which are 174,690 characters written by approximately 166 writers. It is used in competitions. The data are available in isolated word samples which are not a natural Arabic online handwriting, and no segmentation of the words into letters is provided.

2.4   ALTEC (2014) [16]

This database is produced by the Arabic language technology center (ALTEC) for online Arabic text with a large lexicon. It consists of 152,680 samples of 39,945 unique words, including 325,477 samples of 14,740 unique parts of a word, the database is collected from approximately 1,000 writers where samples are complete sentences that include digits and punctuation marks and the collected data is available on sentence, word and character levels. The main drawback of this database is that the data are collected by using a device digitally captures and stores everything written or drawn with ink on ordinary paper.

2.5   QHW (2014) [17]

The Quranic handwritten words database is the most commonly used words in the holy Quran. Handwritten words were chosen as the most common words repeated in the holy Quran. The initial version of QHW database includes 120 handwritten words and divided equally into two sets written by 200 writers in total. The QHW database contains 12,000 sample including more than 42,800 characters and 23,300 sub words. This database is a closed vocabulary database and has samples of a limited number of words.

2.6   Online-KHATT (2018) [18]

The Online-KHATT database contains more than 80,000 Arabic words written by 623 writers with approximation 801,421 characters using a source text that covers several domains to ensure a wide range of topics. Online-KHATT database may be considered as the largest Arabic online text database in terms of the number of lines written with electronic pens using natural Arabic text; however it ignored dealing with characters on the base of its strokes.

3   Proposed AOLAH Databases for Arabic Online Handwriting Letters and Strokes

Due to the drawbacks presented in the previous databases there is an essential need for databases overcomes those drawbacks. This work tries to seed a seed in this field. The proposed Arabic online handwriting recognition algorithm that is used in collecting databases mainly depends on the idea of collecting strokes as a separate unit as the stroke is the first base of any word. To do the process of stroke capturing we had developed an algorithm that was written by MATLAB. This algorithm provides a GUI to display the collected data from pen movements, theses pen movements were simulated by mouse where pen down is simulated by mouse left click, pen movement is simulated by holding the mouse left click while writing, and pen up is simulated by releasing mouse left click. The input pen movements are collected as a sequence of points and further are stored in a text file. The text file storage is required to retain original pen movements that are required at later stages in recognition beginning with preprocessing [19, 20]. Furthermore, those text files may also be used to verify the input stroke shape by the help of any application that may visualize data like Microsoft excel. Figure 3 indicates the graphical user interface of the developed application to collect the databases with the Arabic character zha which is written in three strokes and the screenshot of the data stored in the text file for this character is shown in Fig. 4, where the beginning and end of each stroke is clarified in the table.

The Proposed AOLAH databases are contributions from Faculty of Engineering, Aswan University to help researchers in the field of online handwriting recognition to build a powerful system to recognize Arabic handwritten script. AOLAH stands for Aswan On-Line Arabic Handwritten where Aswan is a small beautiful city located at the south of Egypt. Word On-Line in database's name means that the databases are collected the same time as they are written. While, Arabic word is used because these databases are just collected for Arabic characters; and Handwritten word since these databases are written by the natural human hand.

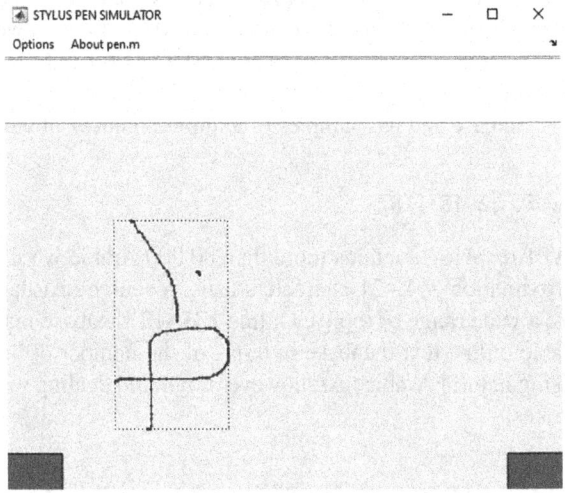

Fig. 3. GUI for the proposed data collection.

	A	B	C
1	x-co	y-co	stroke
99	0.20625	0.30392157	1
100	0.20446429	0.30392157	1
101	0.20089286	0.30112045	1
102	0.20089286	0.29831933	1
103	0.19732143	0.29831933	1
104	0.19553571	0.29551821	1
133	0.30446429	0.52521008	2
134	0.30625	0.51680672	2
135	0.30625	0.5140056	2
136	0.30625	0.50840336	2
137	0.30803571	0.50560224	2
138	0.30803571	0.50280112	2
149	0.34375	0.5952381	3
150	0.34553571	0.5952381	3
151	0.34732143	0.5952381	3
152	0.34732143	0.59243697	3
153	0.34910714	0.59243697	3
154	0.34910714	0.58963585	3
155	0.34910714	0.58683473	3

Fig. 4. Sample of data collected in csv file.

In order to collect data, we had used a help from volunteers students of Faculty of Engineering, Aswan University with ages from 18 to 20 years old.

To facilitate the procedure of collecting data to the volunteers we had prepared a collecting form with all steps needed to be done by students and we did not mention any constraints on the writing style. The indications include creating a folder for each volunteer and writing the 28 characters of Arabic script using the GUI. A total of 97 volunteers were participated All these files are reviewed to guarantee the accepted files for the database. A total of 2,520 files are accepted from the 97 volunteers, representing 90 files for each character after excluding unaccepted files. A second database is extracted from the previous accepted database by extracting strokes from characters. 17 strokes are separated from 28 characters and a database of 1,710 files representing strokes was created, strokes shapes selected with their IDs are shown in Table 2. We have demanded from Aswan University, that we had used their resources to collect the databases, to make these databases available for free.

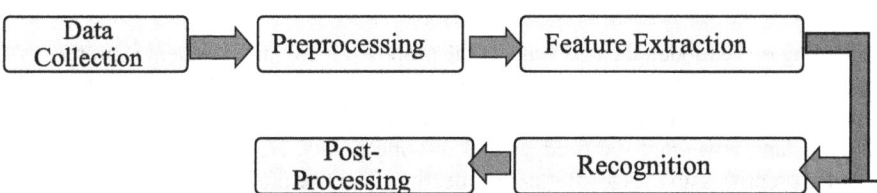

Fig. 5. Stages of online recognition system.

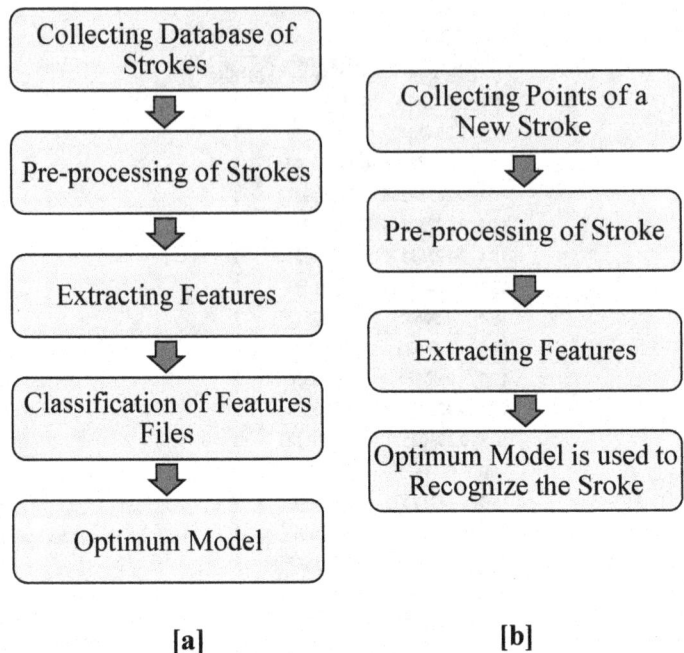

Fig. 6. System used in verifying the databases **[a]** Training and validation phase, **[b]** Testing phase.

In order to verify our collected databases, we should use them in building a recognition system. Most of the online recognition systems follow typical structure of pattern recognition systems; which basically consist of five major stages, data collection, preprocessing, feature extraction, recognition, and postprocessing as illustrated in Fig. 5 [21–24]. Other researchers claim that the recognition system typically comprises of two stages, training and test stages. In the training stage, data are refined, their remarkable features are extracted, similar symbols are merged (clustering) and their features' representatives are stored as training samples, while during the test stage matching takes place for identifying similar features with test features in classification. This recognition system will be used in verifying our databases and its block diagram is shown in Fig. 6 [25].

4   Proposed Arabic Online Handwriting Recognition Algorithm

4.1   Preprocessing Stage

According to the model, the first stage in the proposed recognition system is preprocessing. The preprocessing algorithm is needed to remove variations present in the stroke captured by tablet or smart phone. These variations are mainly present in the form of size, slant, unwanted sharp edges and missing points, etc., so there is a persistent need for preprocessing stage after data collection. The five preprocessing phases in proposed algorithm are used in sequential order after the process of data collection; which

Table 2. Arabic characters strokes with IDs.

Stroke ID	Stroke Shape	Stroke ID	Stroke Shape
10		11	
12		13	
14		15	
16		17	
18		19	
20		21	
22		23	
24		25	
26			

are as following: resizing and centering, interpolating missing points, smoothing, slant correction and resampling of points [26–28].

4.1.1 Resizing and Centering

Resizing and centering phase of stroke is a necessary process that should be performed in order to recognize the stroke. This can be done by assuming a certain frame with a fixed size then moving the stroke to the assumed center point of the frame.

4.1.2 Interpolation

The interpolation phase is used since the stroke may have been written with high speed, so that missing points in the stroke will be found. These missing points can be calculated using various interpolation techniques such as Bezier and B-Spline. We have opted piecewise Bezier interpolation in our procedure because it helps to interpolate points among fixed number of points. In piecewise interpolation technique, a set of consecutive four points is considered for obtaining the Bezier curve. The next set of four points gives the next Bezier curve [29]. The pseudocode of interpolation phase is shown in Fig. 7.

4.1.3 Smoothing

Flickers do exist in handwriting because of individual handwriting style and the hardware used. These flickers can be removed by modifying each point of the list with mean value of k-neighbors and the angle subtended at position from each end, this phase is called smoothing phase.

4.1.4 Slant Correction

Slant correction is required to correct the shape of input handwritten character as most of the writers handwriting is bend to left or right directions. Slant correction for a stroke becomes complex as no baseline can be assumed. In case of single stroke, no bottom-line marks can be made. As such the chain code estimation method by Yimei [30] has been applied for slant correction in Arabic strokes.

4.1.5 Resampling

Due to variations in writing speed, the acquired points are not distributed evenly along the stroke trajectory. Resampling is used to get a sequence of points which is almost equidistant. Besides the removal of variations, this step is essentially because it reduces the number of points in a stroke to a certain value. After resampling, the data is signifi-cantly reduced and the irregularly placed data points that create jitter on the trajectory of the stroke are removed. This makes the resampling step very useful in noise elimination as well as data reduction. In this phase new data points are calculated on the basis of the original points of list. After this phase, only 64 equidistant points will be present in the stroke, those 64 points is of great importance in the next step in recognition system, feature extraction. Figure 8 clarifies the five phases of preprocessing after data collection.

> Create an empty list L for storing the points generated from the Bézier function.

> Repeat the following steps for each stroke k, until $k \leq t$:

  ▪ Calculate m as the total number of points in the current stroke k.

  ▪ If $m \geq 4$ then call *Bezier* function for all points in the current stroke, *Bezier* $(P_i, P_{i+1}, P_{i+2}, P_{i+3})$

    1. u is a variable such that $u = 0 \leq u \leq 1$.

    2. Set $u = 0.1$ and $du = 0.1$.

    3. Repeat steps 4 and 5 until $u \leq 1$.

    4. Calculate x coordinate of new point as

    $P_{ix}*(1-u)^3+P_{(i+1)x}*3*u*(1-u)^2+P_{(i+2)x}*3*u^2*(1-u)+P_{(i+3)x}*u^3$,

    and calculate y coordinate of new point as

    $P_{iy}*(1-u)^3+P_{(i+1)y}*3*u*(1-u)^2+P_{(i+2)y}*3*u^2*(1-u)+P_{(i+3)y}*u^3$,

    5. Set $u = u + du$.

    6. Return

    else set $k = k + 1$

    endif.

  ▪ Update list L.

  ▪ Set $k = k + 1$.

> Exit.

Fig. 7. Algorithm for interpolation.

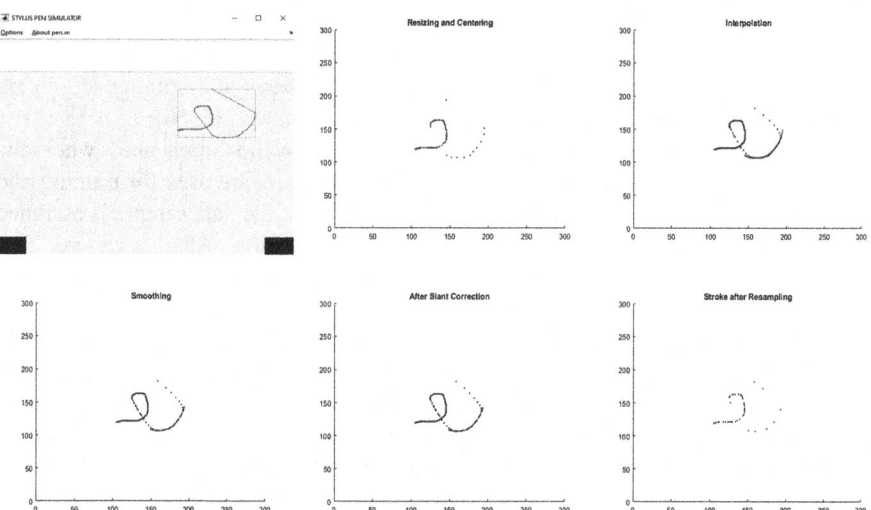

Fig. 8. Phases of preprocessing of an Arabic stroke.

4.2 Feature Extraction Stage

Feature extraction stage is one of the important stages in online handwritten character recognition, and selection of a feature extraction technique is an important task as efficiency of any online handwriting recognition system highly relies on the features which are considered as input to a classifier. There is no standard strategy for extracting features. Features that provide good results for one script may not provide good results for other scripts [31–33].

In the present study, we have presented two different techniques for feature extraction, one by just rearranging the preprocessed points without applying any transformation as shown in Fig. 9, while the second feature extraction technique is by applying Two-Dimensional Discrete Fourier Transform (2D-DFT) on the rearranged preprocessed points of the input stroke, Eq. 1 [34].

$$F[k, l] = \frac{1}{MN} \sum_{x=0}^{M-1} \sum_{y=0}^{N-1} f[x, y] . e^{-2\pi j(\frac{kx}{M} + \frac{ly}{N})} \qquad (1)$$

To reduce operations and computations we had used Fast Fourier Transform (FFT) instead of DFT, and after applying 2D-DFT, we got complex numbers as output. We had used experiments for both real part coefficients and imaginary part coefficients of these complex numbers as features and stored in a file, called feature file and this feature file is taken as input to the classifier.

4.3 Classification Stage

In machine learning and statistics, classification is the problem of identifying to which of a set of categories a new observation belongs, on the basis of a training set of data containing observations whose category membership is known [35, 36].

To evaluate the model after classification k-fold cross-validation is used, where the training data is divided into k parts; out of k parts, k-1 parts are used for training and remaining one part is used for testing. Each observation in the data sample is assigned to an individual group and stays in that group for the duration of the procedure. This means that each sample is given the opportunity to be used in the hold out set one time and used to train the model k-1 times [37, 38].

MATLAB Classification Learner application was used to train models to classify data, where we had used this application to perform automated training to search for the best classification model type, including decision trees, support vector machines, nearest neighbors, and ensemble classification. We had performed supervised machine learning by supplying a known set of input data which is our collected database and known responses to the data which is character stroke IDs. We had used the data to train a model that generates predictions for the response to new data [39–41].

	A	B	C	D	E	F	G	H	I	J	K
1	classid	xco1	yco1	xco2	yco2	xco3	yco3	xco4	yco4	xco5	yco5
2	160	194.2566	197.8709	194.3349	193.0007	195.517	189.4979	196.1959	184.5228	196.735	181.2176
3	160	204.678	223.6304	205.0713	219.3905	205.0533	214.5828	204.3333	207.9599	205.7814	201.4311
4	160	245.8452	218.7444	245.3712	211.7992	244.6667	201.476	244.0073	191.8145	243.5615	185.2822
5	160	271.1003	224.8184	260.2923	224.4483	250.5162	222.3773	239.6207	221.0323	228.3875	217.9118
6	160	212.0594	214.2792	211.7758	202.5776	212.0454	191.4635	210.5891	181.2874	209.6894	171.5721
7	160	217.597	209.9663	218.4831	204.4701	218.4662	196.445	216.3683	184.4777	215.1312	177.8791
8	160	260.122	211.2423	258.6051	205.1441	257.0363	199.1994	255.3848	188.6124	253.5537	179.1168
9	160	217.753	193.4647	217.6509	189.5962	217.3096	185.9756	216.5246	177.6499	215.8707	170.8649
10	160	224.9528	207.6866	223.3925	204.2178	221.7404	200.3087	220.5309	197.2542	219.0149	193.5981
11	160	209.9013	237.23	212.4574	225.9361	214.5051	213.9099	214.9484	201.4566	213.4717	190.2772
12	160	241.5044	166.2987	241.0336	160.6771	240.15	154.1877	235.7312	149.7727	230.1003	148.2379
13	160	240.3043	206.506	239.7096	198.7374	239.1969	192.0392	238.6307	184.6429	240.4392	179.4106
14	160	238.6908	194.8249	238.7121	190.2977	238.6048	186.4346	237.904	181.9654	237.061	177.988
15	160	222.0278	162.8082	221.3993	158.494	221.44	152.3369	221.3073	146.3086	220.2255	139.7526
16	160	239.1667	181.9846	238.7334	176.9549	239.3661	172.2394	239.2994	167.9104	239.9177	163.203
17	160	199.9671	180.1946	200.6212	173.8268	200.5255	167.4599	200.4499	162.4255	200.3464	155.5439
18	160	251.0856	218.7313	249.6819	209.4364	248.8716	203.1746	244.1546	196.4194	237.9107	190.9283
19	160	274.4553	199.9624	272.6341	188.3626	270.4776	172.9332	265.1188	164.9285	254.2905	162.2349
20	160	218.4996	192.005	218.1801	182.8867	217.9419	176.0865	217.1033	169.6563	213.5312	166.2689
21	160	183.6688	187.8642	183.6101	184.4836	183.5167	179.1036	183.432	174.2229	183.3613	170.8332
22											

Fig. 9. Preprocessed points rearranged in a single row.

Seven experiments were held to find the optimum accuracy, training time, and prediction speed:

– Without Applying FFT.
– Real part coefficients that was obtained after applying FFT is used as features and the feature file is taken as input to MATLAB classification learner app.
– Imaginary Part Coefficients as features.
– Real Part Coefficients normalized to 15.
– Imaginary Part Coefficients normalized to 15.
– Real Part Coefficients normalized to 100.
– Imaginary Part Coefficients normalized to 100.

5 Optimum Proposed Recognition Model

Here we had held a comparison between all experiments that were achieved in to decide which model we will use in our recognition system. The comparison was held in terms of accuracy and prediction speed because they are the parameters that are needed in our recognition system, training time is not so important because the training is done just one time and is not needed then in recognition. First, a comparison with the six experiments that had applied FFT will be held as they are common in applying the same transformation on the preprocessed points, then the best of those will be parts of the next comparison against the remaining experiment. The first comparison indicates that for all experiments tree classifiers give high prediction speeds with lower accuracies, SVM classifiers give lower accuracies with medium prediction speeds, KNN classifiers give low accuracies with medium prediction speed and Ensemble classifiers give higher accuracies with medium or low prediction speeds.

The best results were achieved almost from experiment 2 "using real part coefficients of FFT", also it is obvious that the best classifier learner among all classifiers of experiment 2 is SVM classifiers and Ensemble classifiers, Ensemble (Subspace KNN) classifier gives the highest accuracy (75.6%) but the prediction speed is so low (360 obs/sec), however Quadratic SVM gives a near accuracy of (74.4%) but with a better prediction speed of (1900 obs/sec). The other comparison that was held between experiment 2 using real part coefficients of FFT and experiment 1 without applying FFT is shown in Table 3. This comparison indicates that experiment 2 has better prediction speeds for almost all the classifiers, however experiment 1 gives more better accuracy. The highest accuracy from experiment 1 is for the Quadratic SVM classifier (86.4%) with a prediction speed of (1600 obs/sec). Notice that if we use another PC device in our experiments, the accuracy of models will still the same but the prediction speed and training time will differ according to the PC specifications. For example, when we used a PC with AMD A8–3870 CPU 3.00 GHz and 12.0 GB installed memory (RAM), the Quadratic SVM classifier gave an accuracy of 86.6% with a prediction speed 530 obs/sec. According to the previous comparisons it is clear that the optimum recognition model is the model from experiment 1 Quadratic SVM classifier with the accuracy (86.4%) in our recognition model.

Table 3. Comparison between FFT based feature extraction and without applying FFT.

Classification learner	Experiment 1		Experiment 2	
	Accuracy	Prediction speed obs/sec	Accuracy	Prediction speed obs/sec
Tree "Fine Tree"	72.1%	~13000	60.0%	~35000
Tree "Medium Tree"	52.4%	~12000	40.2%	~40000
Tree "Coarse Tree"	25.6%	~13000	19.4%	~34000
SVM "Linear SVM"	84.5%	~2400	71.6%	~3500
SVM "Quadratic SVM"	**86.4%**	**~1600**	74.4%	~1900
SVM "Cubic SVM"	85.8%	~1600	73.8%	~1800
SVM "Fine Gaussian SVM"	63.5%	~740	50.8%	~850

(*continued*)

Table 3. (*continued*)

Classification learner	Experiment 1		Experiment 2	
	Accuracy	Prediction speed obs/sec	Accuracy	Prediction speed obs/sec
SVM "Medium Gaussian SVM"	84.7%	~1300	73.8%	~1300
SVM "Coarse Gaussian SVM"	73.4%	~1200	46.3%	~1100
KNN "Fine KNN"	81.8%	~3700	60.1%	~4500
KNN "Medium KNN"	79.5%	~4000	52.6%	~4600
KNN "Coarse KNN"	61.6%	~3300	32.9%	~4000
KNN "Cosine KNN"	80.2%	~3600	62.5%	~4100
KNN "Cubic KNN"	79.5%	~160	59.4%	~160
KNN "Weighted KNN"	81.7%	~4000	55.1%	~4300
Ensemble "Boosted Trees"	69.9%	~6700	62.6%	~7600
Ensemble "Bagged Trees"	82.2%	~4700	71.5%	~5300
Ensemble "Subspace Discriminant"	80.4%	~2000	69.5%	~2000
Ensemble "Subspace KNN"	82.2%	~350	75.6%	~360
Ensemble "RUSBoosted Trees"	52.4%	~7500	36.4%	~8100

5.1 Testing of the Optimum Model

After creating classification models interactively in Classification Learner, we can export our optimum model to the workspace or make a standalone application. We can then use the produced application to make predictions using new data. The application will follow the stages that were used in the training phase by collecting data using the GUI, preprocessing points collected and outputs a total of 64 points, extracting features by just rearrange the preprocessed points to a single row with 128 features representing the entered stroke, then with the help of the trained model structure the app will predict ID of the entered stroke. Figure 10 shows the recognition of an Arabic handwritten cursive word (محمد) pronounced "Mohamad" after writing it with online handwriting letter by letter and predicts its characters by the proposed recognition model. Word Mohamad in

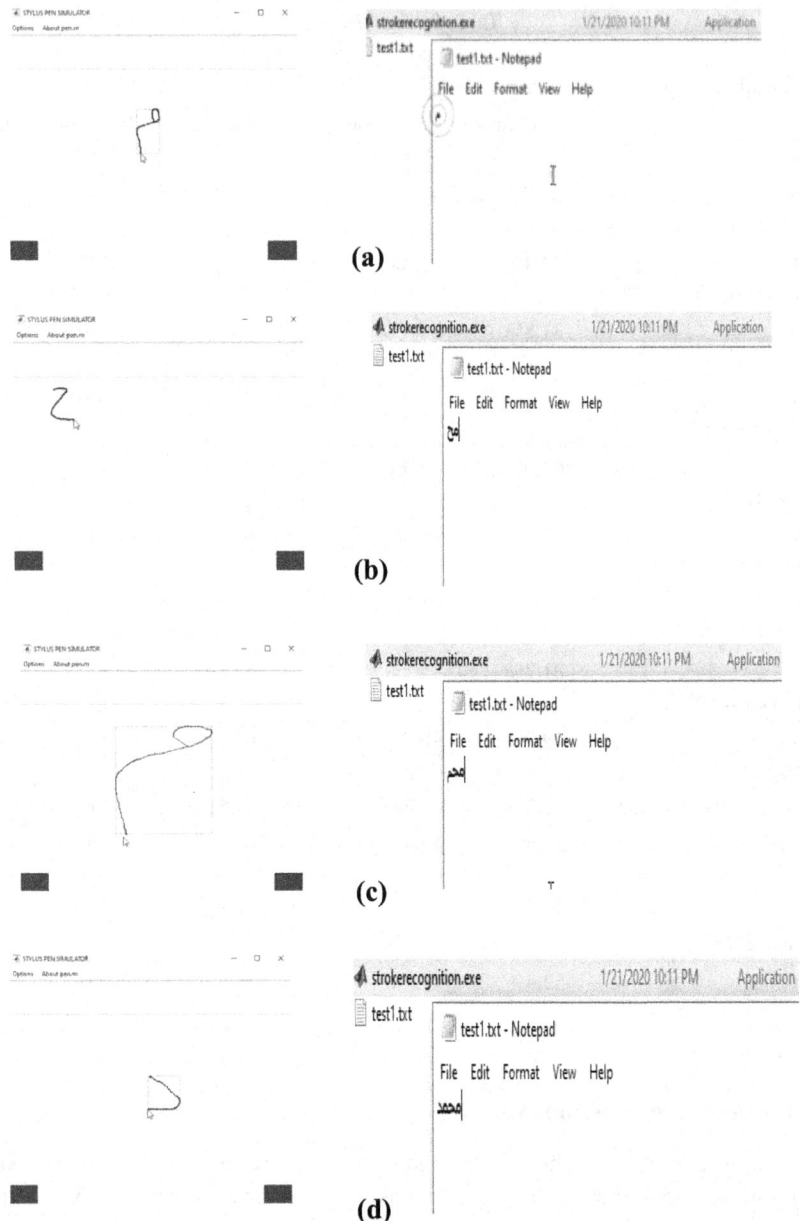

Fig. 10. The recognition of Arabic handwriting cursive word (محمد) "Mohamad" by the proposed model: From (a) to (d) at left-side GUI used to enter a stroke, at right-side predicted character by the proposed recognition model and written in text file

Arabic consists of four letters. In the stylus pen simulator, an Arabic stroke is written by hand (م) as shown in left-side in Fig. 10(a); the output of the code using Quadratic SVM model was ID22 with a prediction time of 0.896426 s. We export the output ID to a text file by first convert it to an Arabic character with identical shape. Therefore, the stroke ID22 will be converted to "meem" character (م) as shown in right-side in Fig. 10(a). After that, by repeated the same sequence for the three remaining letters of word Mohamed shown from Fig. 10(b) to Fig. 10(d). These indicate the prediction of the three Arabic strokes "hha" with ID13, "meem" with ID22, and "dal" with ID14 respectively.

6 Conclusion

This paper presented novel Arabic handwritten characters and strokes databases. These databases are focused only on Arabic handwritten characters with Naskh style. A lot of work is needed from researchers to supply Arabic society with this kind of strokes databases; Ruqaa,

Thuluth, Diwani are some styles of Arabic language that are needed to be part of the future databases. Furthermore, collecting databases for shapes of Arabic characters depending on their locations in the word is quite needed. Moreover, databases of diacritics will be also of great importance for more advanced character recognition. More volunteers from different ages are needed to make a powerful database.

Our study was based on mentioned machine learning technique using supervised learning as the database collected was with known stroke IDs and that is the cause of using classification, each stroke was given an ID and database was collected according to these IDs. The workflow for recognition was by collecting data, preprocess the data, derive features using preprocessed data, train models using features derived, iterate to find the best model, and then integrate the optimum-trained model into the recognition system.

Acknowledgments. The authors thank all participants' contribution to (AOLAH) databases formulation. They sincerely appreciated Dr. Omar Abdel-Reheem and Eng. Fatma Gamal, from Aswan faculty of engineering, Aswan University, for their help to facilitate the data collecting process.

References

1. Habash, N.Y.: Introduction to Arabic Natural Language Processing. Morgan & Claypool, San Rafael (2010)
2. AlMuallim, H., Yamaguchi, S.: A method of recognition of Arabic cursive handwriting. IEEE Trans. Pattern Anal. Mach. Intell. **PAMI-9**(5), 715–722 (1987)
3. Elbaati, A., Kherallah, M., Ennaji, A., Alimi, A.M.: Temporal order recovery of the scanned handwriting. In: 2009 10th International Conference on Document Analysis and Recognition, Barcelona, Spain, 26–29 July 2009
4. Alimi, A.: A neuro-fuzzy approach to recognize Arabic handwritten characters. In: Proceedings of International Conference on Neural Networks (ICNN 1997), Houston, TX, USA, 12–12 June 1997

5. Abuzaraida, M.A., Zeki, A.M., Zeki, A.M.: Recognition techniques for online Arabic handwriting recognition systems. In: 2012 International Conference on Advanced Computer Science Applications and Technologies (ACSAT), Kuala Lumpur, Malaysia (2012)

6. Al-Helali, B.M., Mahmoud, S.A.: Arabic online handwriting recognition (AOHR): a survey. ACM Comput. Surv. **50**(3), 1–35 (2017)

7. AbdElNafea, M., Heshmat, S.: Efficient preprocessing algorithm for online handwritten Arabic strokes. In: 2019 International Conference on Innovative Trends in Computer Engineering (ITCE), Aswan, Egypt, 2–4 February 2019

8. Sharma, A.: Online Handwritten Gurmukhi Character Recognition "thesis". Patiala, Punjab, India: School of Mathematics and Computer Applications, Thapar University, February 2009

9. Harouni, M., Mohamad, D., Rasouli, A.: Deductive method for recognition of on-line handwritten Persian/Arabic characters. In: 2010 The 2nd International Conference on Computer and Automation Engineering (ICCAE), Singapore, Singapore, 26–28 February 2010

10. Haraty, R., Ghaddar, C.: Arabic text recognition. Int. Arab J. Inf. Technol. **1**, 156–163 (2004)

11. Kherallah, M., Elbaati, A., El Abed, H., Alimi, A.M.: The On/Off (LMCA) Dual Arabic Handwriting Database. In: REGIM: Research Group on Intelligent Machines, University of Sfax (2008)

12. Kherallah, M., Tagougui, N., Alimi, A.M., El Abed, H., Margner, V.: Online Arabic handwriting recognition competition. In: 2011 International Conference on Document Analysis and Recognition, Beijing, China (2011)

13. Elanwar, R.I.M., Rashwan, M.A., Mashali, S.A.: OHASD: the first on-line Arabic sentence database handwritten on tablet PC. Int. J. Comput. Inf. Eng. **4**(12), 1907–1912 (2010)

14. El Abed, H., Kherallah, M., Märgner, V., Alimi, A.M.: On-line Arabic handwriting recognition competition 'ADAB database and participating systems.' Int. J. Doc. Anal. Recogn. (IJDAR) **14**, 15–23 (2011)

15. Azeem, S.A., Ahmed, H.: Recognition of segmented online Arabic handwritten characters of the ADAB database. In: 2011 10th International Conference on Machine Learning and Applications and Workshops, Honolulu, HI, USA, 18–21 December 2011

16. Abdelaziz, I., Abdou, S.: AltecOnDB: a large-vocabulary arabic online handwriting recognition database. arXiv, 24 December 2014

17. Abuzaraida, M.A., Zeki, A.M., Zeki, A.M.: Online database of Quranic handwritten words. J. Theor. Appl. Inf. Technol. **62**(2), 485–492 (2014)

18. Mahmoud, S.A., Luqman, H., Al-Helali, B.M., BinMakhashen, G., Parvez, M.T.: Online-KHATT: an open-vocabulary database for Arabic online-text processing. Open Cybern. Syst. J. **12**(1), 42–59 (2018)

19. Mahajan, L., Kulkarni, G.A.: Digital pen for handwritten digit and gesture recognition using trajectory recognition algorithm based on triaxial accelerometer. IOSR J. Electron. Commun. Eng. (IOSR-JECE) **10**(1), 24–31 (2015)

20. Nakkach, H., Hichri, S., Haboubi, S., Amiri, H.: A segmentation-free approach to strokes extraction from online isolated Arabic handwritten character. In: 2016 2nd International Conference on Advanced Technologies for Signal and Image Processing (ATSIP), Monastir, Tunisia, 21–23 March 2016

21. Plamondon, R., Srihari, S.N.: On-line and off-line handwriting recognition: a comprehensive survey. IEEE Trans. Pattern Anal. Mach. Intell. **22**(1), 63–84 (2000)

22. Sharma, A., Kumar, R., Sharma, R.K.: Online handwritten Gurmukhi Character recognition using elastic matching. In: 2008 Congress on Image and Signal Processing, Sanya, Hainan, China, 27–30 May 2008

23. Priya, A., Mishra, S., Raj, S., Mandal, S., Datta, S.: Online and offline character recognition: a survey. In: 2016 International Conference on Communication and Signal Processing (ICCSP), Melmaruvathur, India, 6–8 April 2016

24. Mezghani, N., Mitiche, A., Cheriet, M.: On-line recognition of handwritten Arabic characters using a Kohonen neural network. In: Proceedings Eighth International Workshop on Frontiers in Handwriting Recognition, Niagara on the Lake, Ontario, Canada, 6–8 August 2002

25. Santosh, K.C., Nattee, C.: A comprehensive survey on on-line handwriting recognition technology and its real application to the Nepalese natural handwriting. Kathmandu University J. Sci. Eng. Technol. **5**(1), 31–55 (2009)

26. Abuzaraida, M.A., Zeki, A.M., Zeki, A.M.: Problems of writing on digital surfaces in online handwriting recognition systems. In: 2013 5th International Conference on Information and Communication Technology for the Muslim World (ICT4M), Rabat, Morocco, 26–27 March 2013

27. El-Wakil, M.S., Shoukry, A.A.: On-line recognition of handwritten isolated Arabic characters. Pattern Recogn. **22**(2), 97–105 (1989)

28. Al-Emami, S., Usher, M.: On-line recognition of handwritten Arabic characters. IEEE Trans. Pattern Anal. Mach. Intell. **12**(7), 704–710 (1990)

29. Mortenson, M.E.: Mathematics for Computer Graphics Applications . Industrial Press Inc., South Norwalk (1999)

30. Ding, Y., Kimura, F., Miyake, Y., Shridhar, M.: Accuracy improvement of slant estimation for handwritten words. In: Proceedings 15th International Conference on Pattern Recognition. ICPR-2000, Barcelona, Spain, 3–7 September 2000

31. Ramzi, A., Zahary, A.: Online Arabic handwritten character recognition using online-offline feature extraction and back-propagation neural network. In: 2014 1st International Conference on Advanced Technologies for Signal and Image Processing (ATSIP), Sousse, Tunisia, 17–19 March 2014

32. Al-Habian, G., Assaleh, K.: Online Arabic handwriting recognition using continuous gaussian mixture HMMS. In: 2007 International Conference on Intelligent and Advanced Systems, Kuala Lumpur, Malaysia, 25–28 November 2007

33. Boubaker, H., El Baati, A., Kherallah, M., Alimi, A.M., Elabed, H.: Online Arabic handwriting modeling system based on the graphemes segmentation. In: 2010 20th International Conference on Pattern Recognition, Istanbul, Turkey, 23–26 August 2010

34. Gonzalez, R.C., Woods, R.E.: Digital Image Processing, 2nd edn. Prentice Hall, Upper Saddle River (2002)

35. White, R.L.: Methods for Classification, 16 August 1996. http://first.astro.columbia.edu/rick/SCMA/node2.html

36. W. Contributors, Training, Validation, and Test Sets, Wikipedia, The Free Encyclopedia, 14 December 2019. https://en.wikipedia.org/w/index.php?title=Training,_validation,_and_test_sets&oldid=930696087

37. Brownlee, J.: A Gentle Introduction to k-fold Cross-Validation, 23 May 2018. https://machinelearningmastery.com/k-fold-cross-validation/

38. Shalev-Shwartz, S., Ben-David, S.: Understanding Machine Learning: From Theory to Algorithms . Cambridge University Press, New York (2014)

39. MATHWORKS. Classification Learner App (2019). https://www.mathworks.com/help/stats/classificationlearner-app.html?s_tid=srchtitle

40. MATHWORKS. Choose Classifier Options (2019). https://www.mathworks.com/help/stats/choose-a-classifier.html

41. MathWorks. Machine Learning in MATLAB (2019). https://www.mathworks.com/help/stats/machine-learning-in-matlab.html

Line Segmentation of Individual Demographic Data from Arabic Handwritten Population Registers of Ottoman Empire

Yekta Said Can[(✉)] and M. Erdem Kabadayı

Department of History, Koc University, Rumelifeneri, Sarıyer Rumeli Feneri Yolu, 34450 Sarıyer/İstanbul, Turkey
ycan@ku.edu.tr

Abstract. Recently, more and more studies have applied state-of-the-art algorithms for extracting information from handwritten historical documents. Line segmentation is a vital stage in the HTR systems; it directly affects the character segmentation stage, which affects the recognition success. In this study, we first applied deep learning-based layout analysis techniques to detect individuals in the first Ottoman population register series collected between the 1840s and 1860s. Then, we used a star path planning algorithm-based line segmentation to the demographic information of these detected individuals in these registers. We achieved encouraging results from the selected regions, which could be used to recognize the text in these registers.

Keywords: Line segmentation · Convolutional neural networks · Page segmentation · Arabic document processing · Projection profiles · A* path planning

1 Introduction

Documents are often employed in daily life. The necessity of analyzing them and understand their content is perpetual. In the past, processing the manuscripts was performed manually because of the lack of comprehensive and high-quality digitized datasets where an automatic method could be employed. Because of the rarity of high-quality digital scanning solutions and devices with high-storage capability, transforming and saving manuscript images from paper form to digital

This work was supported by the European Research Council (ERC) project: "Industrialisation and Urban Growth from the mid-nineteenth century Ottoman Empire to Contemporary Turkey in a Comparative Perspective, 1850–2000" under the European Union's Horizon 2020 research and innovation program Grant Agreement No. 679097, acronym UrbanOccupationsOETR. M. Erdem Kabadayı is the principal investigator of UrbanOccupationsOETR.

© Springer Nature Switzerland AG 2021
E. H. Barney Smith and U. Pal (Eds.): ICDAR 2021 Workshops, LNCS 12916, pp. 312–321, 2021.
https://doi.org/10.1007/978-3-030-86198-8_22

form was difficult. Recently, this job has become more evident due to dramatic progress in digital scanning and storage solutions [1].

Nowadays, there are many digitized historical Arabic documents in the national libraries and archives around the world, thanks to the abovementioned advances in technology. Investigating data manually is a costly and challenging job. Therefore, an automatic method is needed to employ these documents rapidly. Processing historical manuscripts is an up-to-date research topic that has seen dramatic growth recently. However, historical Arabic document processing is a difficult research issue. The reasons could be listed as the complex nature of Arabic script compared to other scripts, and ancient documents, which are subject to degradation [2].

When recognizing handwritten documents, segmenting the document images into their primary objects such as words and text lines is a highly complex research issue in Arabic manuscripts because of the various problems faced in both word and text line segmentation processes. These main difficulties in the segmentation process of Arabic lines are overlapping words, very close neighboring text lines, and over the same text line or on the page between lines, the variances angle of the skew [3].

In this study, we developed software that automatically segments pages and detects lines in these objects from the population registered in Ottoman populated places. The data used is obtained from the first population registers of the Ottoman Empire that were conducted in the 1840s. The coverage of these registers is the entire Ottoman Empire in the mid-nineteenth century, which contained the regions of around two dozen successor countries of today in Southeast Europe and the Middle East. For this study, we focus on two areas: Nicaea and Manisa in western Anatolia in Turkey. We first applied a CNN-based page segmentation method to obtain demographic data of individuals by using the models developed in our previous studies [4,5]. After that, we used a star path planning algorithm-based line segmentation to the demographic information of these detected individuals in these registers. Detecting lines would be helpful for developing an automatic information retrieval system for these registers.

The structure of the paper is designed as follows. In Sect. 2, the related work for line segmentation studies will be examined. We describe the structure of the population registers in Sect. 3. Our method for page segmentation and line detection is described in Sect. 4. Experimental results and a discussion are presented in Sect. 5. We present the conclusion and future works of the study in Sect. 6.

2   Related Works

In the Arabic manuscript processing literature, a wide variety of segmentation techniques were reported. In [6], the authors started by removing outlier elements by using a threshold; then, the letters linked to two lines at the half distance are recognized and horizontally segmented. For detecting the lines, a rectangular neighborhood on a current element is centered and rises to contain particular conditions. The distance filtered elements to the corresponding lines are then allocated.

In another study, the authors developed a new algorithm for detecting lines from Arabic handwritten documents by mentioning the problems of multi-touching and overlapping characters [7]. The unsupervised method depends on the analysis of a block covering. The authors first analyze a statistical block that computes the specific number into vertical strips of manuscript decomposition. Then, they employed the fuzzy C means method, which accomplishes fuzzy-based line detection. Lastly, they assigned the blocks to their corresponding lines. They achieved 95% accuracy for detecting lines in Arabic manuscripts in their dataset.

In [8], the researchers applied morphological dilatation and projection profile techniques. In order to estimate the skew of the line, they used horizontal projection profiles. In every zone for smearing, they employed the slope, using dilation with adaptive structuring element to do the changes according to the zone, the slope, and the size. The big blobs are identified in the second stage with a recursive function that searches the cut point.

In [9], the projection profile method is again employed after joining cut characters and eliminating small elements to define the point of division within the horizontal projection profile; the curve of Fourier fitting is employed. The contour is employed for segmenting the baseline of the connected component, which permits defining the cut point between different neighbor lines. The nearest line is approximated by the curve of a polynomial that fits in the baselines of the pixels. As it could be seen from the literature, several techniques were applied to segment lines in Arabic manuscripts. Due to its convenience and wide usage, we selected a projection profile-based technique for our problem.

3   Dataset Description

The population registers resulted from an unprecedented governmental procedure, which aimed to record every male subject of the empire, irrespective of age, ethnic or religious affiliation, or military or financial status. They intended to have universal coverage for the male population. Government officials created these manuscripts without using hand-drawn or printed tables. Therefore, a predetermined page layout did not exist. Page structure can change in different districts, and structural variations occurred depending on the clerk in different registers. This research focused on Nicaea and Manisa district registers, with code names NFS.d. 1452, 1454, and NFS.d. 2865, 2866, 2867, respectively, available at the Turkish Presidency State Archives of the Republic of Turkey, Department of Ottoman Archives, in jpeg format, upon request. We aimed to implement a method for recognizing text of similar registers from different Empire regions collected between the 1840s and the 1860s.

These registers contain comprehensive demographic data on male members of the households, i.e., names, family relations, ages, and occupations. Females in the households were not recorded. The registers are provided for research purposes at the Ottoman State Archives in Turkey, as recently as 2011. Their total number is around 11,000. Until now, they have not been subject to any systematic study. Only individual registers were investigated in a piecemeal fashion. The size of the digital images of registers is 2210 3000 pixels. A sample register page is demonstrated in Fig. 1.

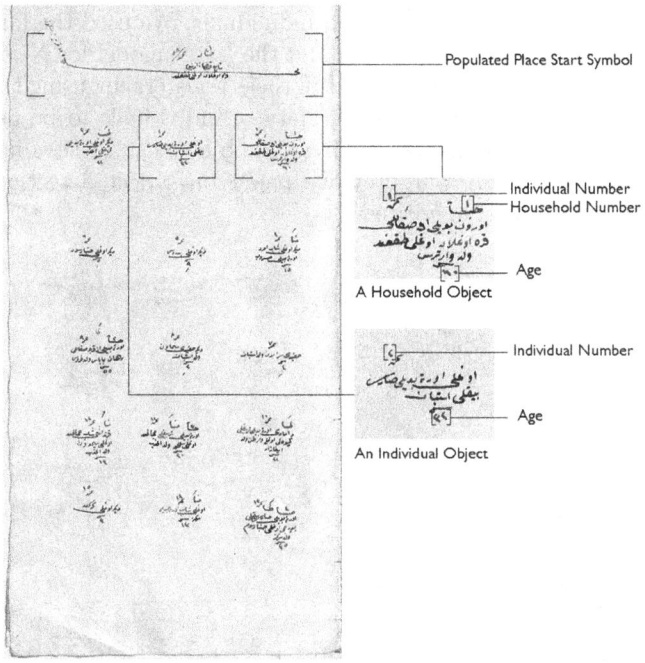

Fig. 1. The layout of a sample register page [10].

4 Automatic Object Detection and Line Segmentation Method

In this section, we will describe our object detection and line segmentation methods separately. We first detected objects that contain demographic data of individuals and applied line segmentation to them.

4.1 CNN-based Object Detection Method

First, we developed a deep learning algorithm for detecting individuals in the population registers in our previous studies. We created a manually labeled dataset by using several registers and train CNN models by using the dhSegment tool [11]. In the CNN-based dhSegment toolbox, paths use pretrained weights from well-known architectures such as Unet and Resnet50, where the system learns high-level features. They improve robustness and generalization. With the pretrained weights in the network, the training time and the number of parameters in the CNN architecture were reduced considerably [11].

We trained different models for different types of layouts. The first model was trained for registers with tightly placed individuals. The second model was trained for registers with loosely placed individuals. We used the former model for Manisa registers used in this study and the latter model for Nicea registers. After we detected the individual objects in these registers, by using the pixelwise locations, we cropped the demographic data of individuals to be used for line detection algorithms. The detected individual objects can be seen in Fig. 2. For more detailed information, our previous paper on object detection for these population registers could be visited [5].

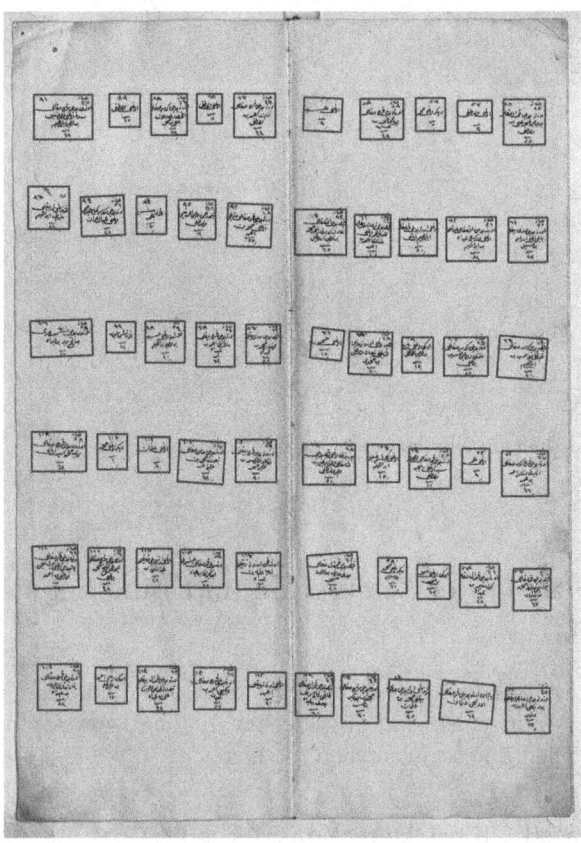

Fig. 2. Detected individuals are bounded with green boxes. They include demographic data of individuals. Since the pixelwise positions are detected, they are cropped for providing input to the line detection method.

4.2   Line Segmentation Method

One of the widely used methods for finding the line-height of a document is by examining its horizontal projection profile. Horizontal projection profile (HPP) is the array of sum or rows of a two-dimensional image. Wherever there are more white areas, we can observe more peaks. They give an idea of where should the segmentation between two lines can be applied. A sample HPP is provided in Fig. 4.

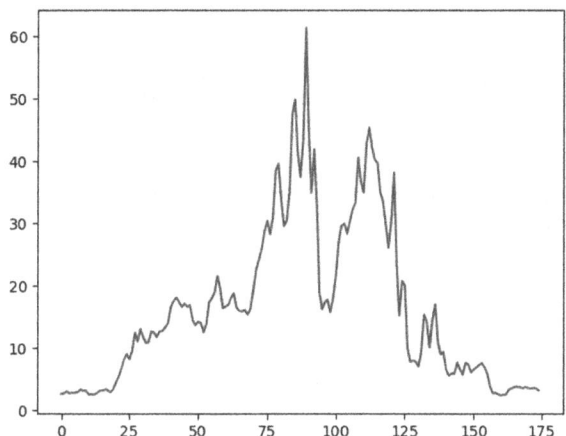

Fig. 3. In the white space regions, there are peaks in the figure. This information can be employed to determine the regions for the separation line.

We determined a threshold and extracted the peak regions. Then, the divider parameter value is employed to threshold the peak values from non-peak values. We applied a star path planning algorithm (A*) to these profiles to segment all the lines [12]. A sample output of our system can be seen in Fig. 5.

Fig. 4. We will run the path planning algorithm for line segmentation from the black areas.

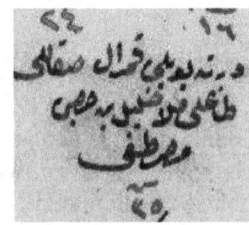 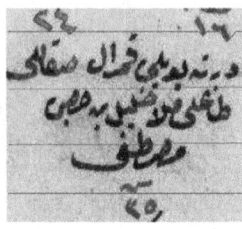

Fig. 5. A sample cropped individual demographic data (in the left side) and segmented lines are shown (in the right side).

5   Experimental Results and Discussion

In this section, we first provided our individual object detection results in two different population registers, namely Nicea and Manisa. After that, we presented line detection accuracies in these regions' registers.

5.1   Individual Object Detection Results

We tested the performance of our system on Nicea (over 8000 individuals) and Manisa (over 7000 individuals) separately. We employed two different pretrained architectures, namely Unet and Resnet50 and presented results by using both of them. In order to count individuals in the registers, we defined a high-level metric that can be calculated by dividing the predicted count errors over the ground truth count. We named this metric as Individual Counting Error (ICE).

$$ICE = || \frac{Predicted\,Individual\,Count - Ground - Truth\,Individual\,Count}{Ground - Truth\,Individual\,Count} ||$$

(1)

We used a loosely placed model for Nicea and a tightly placed model for Manisa since their structures are more suitable for these models. The results are presented in Table 1. As can be seen from Table 1, the counting errors are around %1. Using a pretrained Unet architecture reduces the error percentages. We achieved slightly better error rates for Manisa registers which could result from the layout structure of the objects.

Table 1. The individual counting error results are presented.

	ICE (%)
Nicea - Unet	0.984
Nicea - Resnet50	1.145
Manisa - Unet	0.3836
Manisa - Resnet50	0.6536

5.2   Line Detection Results

We evaluated the performance of our line detection system by choosing 100 objects randomly from both registers. We used again a high-level metric which is the line detection accuracy. It could be calculated by dividing the number of correctly detected lines by the total number of lines. We computed the results of both registers separately. The line detection accuracies are 80.30% for Manisa and 92.06% for Nicea registers. This could be explained by the fact that lines inside the objects are more intertwined in Manisa registers than Nicea registers. Furthermore, the lines are angular in Manisa registers and it makes the detection of lines more challenging. Several line detection examples from both registers are provided in Fig. 6.

Fig. 6. Line detection samples with the original objects. In the left side, objects from Nicea and in the right side, objects from Manisa registers are shown. The errors are explained with possible reasons.

6   Conclusion

In this study, we first developed a CNN-based layout analysis technique to detect individual objects in the first Ottoman population register series. Then, we implemented A star path planning algorithm-based line segmentation to the demographic information of these detected individual objects. We focused on

Nicaea and Manisa district registers. We detected objects that contain demographic information of individuals with less than 1% error by using the CNN-based segmentation algorithm. The line detection results for using both registers from Manisa and Nicea regions are presented. We achieved promising results, especially for Nicea registers. We detected lines in these registers with accuracy over 92%. We also showed issues for detecting lines such as angular and intertwined lines. Detecting lines would be helpful for developing an automatic information retrieval system for these registers. We plan to add word and letter detection systems as future works and develop a recognition system for these registers. It will reveal important demographic information from a wide area of the 19th century.

References

1. Saabni, R.M., El-Sana, J.A.: Keywords image retrieval in historical handwritten Arabic documents. J. Electron. Imaging **22**(1), 013016 (2013)
2. Khedher, M.I., Jmila, H., El-Yacoubi, M.A.: Automatic processing of historical Arabic documents: a comprehensive survey. Pattern Recogn. **100**, 107–144 (2020)
3. Ali, A.A.A., Suresha, M.: Efficient algorithms for text lines and words segmentation for recognition of Arabic handwritten script. In: Shetty, N.R., Patnaik, L.M., Nagaraj, H.C., Hamsavath, P.N., Nalini, N. (eds.) Emerging Research in Computing, Information, Communication and Applications. AISC, vol. 882, pp. 387–401. Springer, Singapore (2019). https://doi.org/10.1007/978-981-13-5953-8_32
4. Can, Y.S., Kabadayı, M.E.: Automatic CNN-based Arabic numeral spotting and handwritten digit recognition by using deep transfer learning in Ottoman population registers. Appl. Sci. **10**(16), 5430 (2020)
5. Can, Y.S., Kabadayı, M.E.: CNN-based page segmentation and object classification for counting population in Ottoman archival documentation. J. Imaging **6**(5), 32 (2020)
6. Khandelwal, A., Choudhury, P., Sarkar, R., Basu, S., Nasipuri, M., Das, N.: Text line segmentation for unconstrained handwritten document images using neighborhood connected component analysis. In: Chaudhury, S., Mitra, S., Murthy, C.A., Sastry, P.S., Pal, S.K. (eds.) PReMI 2009. LNCS, vol. 5909, pp. 369–374. Springer, Heidelberg (2009). https://doi.org/10.1007/978-3-642-11164-8_60
7. Boussellaa, W., Zahour, A., Elabed, H., Benabdelhafid, A., Alimi, A.M.: Unsupervised block covering analysis for text-line segmentation of Arabic ancient handwritten document images. In: 2010 20th International Conference on Pattern Recognition, pp. 1929–1932. IEEE (2010)
8. Khayyat, M., Lam, L., Suen, C.Y., Yin, F., Liu, C.L.: Arabic handwritten text line extraction by applying an adaptive mask to morphological dilation. In: 2012 10th IAPR International Workshop on Document Analysis Systems, pp. 100–104. IEEE (2012)
9. Adiguzel, H., Sahin, E., Duygulu, P.: A hybrid for line segmentation in handwritten documents. In: 2012 International Conference on Frontiers in Handwriting Recognition, pp. 503–508. IEEE (2012)
10. Can, Y.S., Kabadayı, M.E.: Curation of historical Arabic handwritten digit datasets from Ottoman population registers: a deep transfer learning case study. In: 2020 IEEE International Conference on Big Data (Big Data), pp. 1853–1860 (2020)

11. Oliveira, S.A., Seguin, B., Kaplan, F.: dhSegment: a generic deep-learning approach for document segmentation. In: 2018 16th International Conference on Frontiers in Handwriting Recognition (ICFHR), pp. 7–12. IEEE (2018)
12. Krishnan, M.: A* path planning line segmentation algorithm (2020). https:// github.com/muthuspark/line-segmentation-handwritten-doc

Improving Handwritten Arabic Text Recognition Using an Adaptive Data-Augmentation Algorithm

Mohamed Eltay[1], Abdelmalek Zidouri[1(✉)], Irfan Ahmad[2], and Yousef Elarian[3]

[1] Electrical Engineering Department, Interdisciplinary Research Center for Intelligent Secure Systems, King Fahd University of Petroleum and Minerals, Dhahran, Saudi Arabia
{g201410680,malek}@kfupm.edu.sa

[2] Information and Computer Science Department, Interdisciplinary Research Center for Intelligent Secure Systems, King Fahd University of Petroleum and Minerals, Dhahran, Saudi Arabia
irfan.ahmad@kfupm.edu.sa

[3] Cambrian College, Sudbury, ON, Canada
yousef.elarian@cambriancollege.ca

Abstract. Deep learning has increased the performance of classification and object detection, but it generally requires large amounts of labeled data for training. In this paper, we introduce a new data augmentation algorithm that promotes diversity between classes, representing the characters of the Arabic script, and can balance samples between different classes. This algorithm gives each word in the lexicon a weight. The weight of a word is based on the occurrence probabilities of the characters constituting the word. Minority classes are given higher weight as compared to the classes frequently occurring in the text. The data augmentation technique was evaluated on a handwritten word recognition task using the publicly available IFN/ENIT and AHDB datasets. We see significant improvement in results by employing our data augmentation technique, and we achieve state-of-the-art results on both datasets.

Keywords: Handwriting recognition · Deep Learning Neural Network · Data augmentation · Recurrent Neural Network · Connectionist temporal classification

1 Introduction and Related Works

One of the most researched pattern recognition problems is handwriting recognition. The problem has been studied extensively over many decades, and it has progressed from the recognition of isolated characters to complex cursive scripts. Recognition of handwritten text is a very important requirement in many applications around us such as identifying names on cheques, identifying written addresses on the mail, and many other applications. Because of writer-specific preferences in drawing character shapes (allographs) and joining various characters, handwritten text is considered more difficult to recognize than printed text.

© Springer Nature Switzerland AG 2021
E. H. Barney Smith and U. Pal (Eds.): ICDAR 2021 Workshops, LNCS 12916, pp. 322–335, 2021.
https://doi.org/10.1007/978-3-030-86198-8_23

A handwriting recognition system converts handwritten text into machine-readable text, and it can work with scanned or camera-based images (offline recognition) as well as writing captured directly on a digitizing device (online recognition). Offline handwriting recognition can be accomplished in two ways. The first approach is the segmentation approach [1], in which words are broken down into smaller pieces (e.g., characters, graphemes, allographs), and the features of each of these pieces are extracted and recognized. Despite its effectiveness, this method is entirely dependent on the selection of appropriate features, which is a difficult task in cursive writing. Second, there is the holistic approach, in which the entire word is processed without prior decomposition [2]. In this case, the feature vector is derived from the entire word. In a limited vocabulary handwriting database, holistic approaches have proven to be more successful [3–5]. Segmentation approaches, on the other hand, are used when the database's vocabulary is very large [6–8].

Traditionally, sequence features were chosen, fed into classification algorithms, and predictions were made to classify sequence data. It takes a significant amount of time and expertise to define input features suitable for machine learning algorithms. We need to identify distinguishing features for each character. Deep Learning Neural Networks (DLNNs) have simplified this process by eliminating the need for feature selection, which can be difficult in some cases. Instead of being treated as two separate problems, using raw data allows the visual and sequential aspects of handwriting recognition to be learned together. This type of 'end-to-end' training is often advantageous for machine learning algorithms because it gives them more flexibility in adapting to the task [9]. DLNNs have been shown to be superior to HMMs for sequence labeling tasks such as speech and online handwriting recognition in several studies and researches [10, 11]. One reason for this could be that DLNNs are discriminatively trained, whereas HMMs are generative. When a large set of data is available, discriminative methods tend to produce better results in pattern recognition tasks [12].

Recognition of Arabic handwriting has been investigated in several studies as the writing system for Arabic, Urdu, Persian, and other languages. DLNNs have been recently used for offline Arabic handwriting recognition, with high recognition rates reported in recent contributions [13]. Graves et al. [14] have presented one of the first RNN-based systems for handwritten Arabic text recognition. The presented system combined multidimensional LSTM with connectionist temporal classification and a hierarchical layer structure. On the IFN/ENIT database, the authors achieved a word accuracy rate of 93.37%.

Elleuch et al. [15] have demonstrated the use of a Convolutional Deep Belief Network (CDBN) to recognize Arabic words. On the IFN/ENIT Database, the authors achieved an accuracy rate of 83.7%. Maalej et al. [16] have proposed a new offline Arabic handwriting recognition system based on a specific RNN known as the MDLSTM, on which they have proposed using the dropout technique in various positions such as before, after, or inside the MDLSTM layers. This regularization technique has the benefits of protecting their system from overfitting and lowering the error recognition rate. They have conducted experiments using the well-known IFN/ENIT Database.

Elarian et al. [17] have presented an Arabic handwriting synthesis system. Extended-Glyph's connection and Synthetic-Extension's connection were two concatenation models used to synthesize Arabic words from segmented characters. The proposed system was used to synthesize handwriting from a dataset and inject it into a larger dataset. Regardless of previous work, utilizing deep models on Arabic handwriting recognition is still rare when compared to other languages due to the difficulty of dealing with cursive handwriting [18]. The main contribution of this paper is the introduction of our new adaptive data augmentation algorithm, which can generate a large amount of data from a small amount of minority labeled data. The algorithm was evaluated on the IFN/ENIT [19] and AHDB [20] database, and promising results are reported. We test our work using DLNN handwriting recognition systems that use raw image features for training.

The rest of this paper is structured as follows: Sect. 2 discusses the characteristics and challenges of Arabic handwriting. Section 3 goes over DLNN recognition systems in depth. Section 4 presents the technical details of our data-augmentation algorithm. Section 5 presents the experimental findings and discussions. Finally, Sect. 6 concludes the paper.

2   Challenges and Characteristics of Arabic Handwriting

Arabic is ranked among the top six languages in the world [21]. It is also widely used throughout the Muslim world as the language of the Qur'an, Islam's holy book. It is a member of the Semitic language family, which also includes Hebrew and Amharic [21]. Arabic is a cursive language, which means that most Arabic letters are connected when written. Arabic has 28 letters that are written from right to left. Depending on where it is in the word, each character has two or four distinct shapes. Table 1 shows Arabic alphabets and the shape of each character based on its position in the word. We can see that a number of the characters are drawn similarly but differ due to the presence or absence of diacritics.

In general, recognizing Arabic handwritten text is a difficult task. Below, we describe some of the most challenging structural characteristics of the characters that any designer can face when implementing a handwriting recognizer:

1. Each Arabic word has at least one associated segment (sub-word), and each one has at least one character that can be overlapping by different characters or diacritics. Also, different characters can be optionally consolidated vertically to form a ligature [37]. (Fig. 1a).
2. Some words have characters that are interlaced. (Fig. 1b).
3. Each writer has a distinct handwriting style. (Fig. 1c).
4. Some Arabic characters contain diacritics (a diacritic might be set above or beneath the body of the character). (Fig. 1d).

Table 1. The Arabic alphabets and the shapes for each letter based on its Position (Printed).

Name	Isolated	Final	Medial	Initial
'Alif	ا	ـا		
Bā'	ب	ـب	ـبـ	بـ
Tā'	ت	ـت	ـتـ	تـ
Thā'	ث	ـث	ـثـ	ثـ
Jīm	ج	ـج	ـجـ	جـ
Ḥā'	ح	ـح	ـحـ	حـ
Khā'	خ	ـخ	ـخـ	خـ
Dāl	د	ـد		
Dhāl	ذ	ـذ		
Rā'	ر	ـر		
Zāy	ز	ـز		
Sīn	س	ـس	ـسـ	سـ
Shīn	ش	ـش	ـشـ	شـ
Ṣād	ص	ـص	ـصـ	صـ
Ḍād	ض	ـض	ـضـ	ضـ
Ṭā'	ط	ـط	ـطـ	طـ
Ẓā'	ظ	ـظ	ـظـ	ظـ
Ayn	ع	ـع	ـعـ	عـ
Ghayn	غ	ـغ	ـغـ	غـ
Fā'	ف	ـف	ـفـ	فـ
Qāf	ق	ـق	ـقـ	قـ
Kāf	ك	ـك	ـكـ	كـ
Lām	ل	ـل	ـلـ	لـ
Mīm	م	ـم	ـمـ	مـ
Nūn	ن	ـن	ـنـ	نـ
Hā'	ه	ـه	ـهـ	هـ
Wāw	و	ـو		
Yā'	ي	ـي	ـيـ	يـ

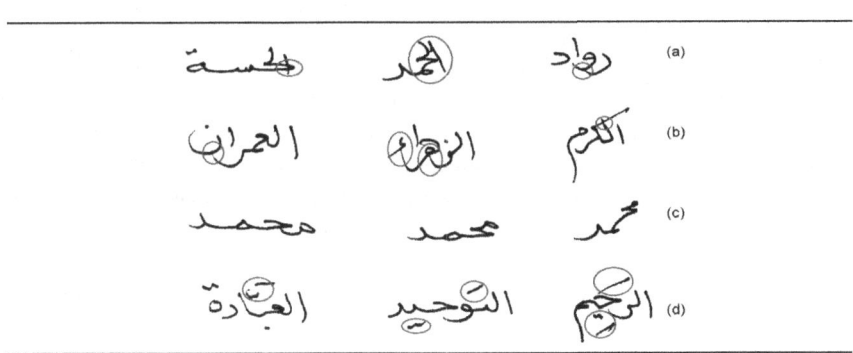

Fig. 1. Arabic handwriting recognition's challenges and complexities.

3 Deep Neural Networks for Handwritten Text Recognition

Because of recent progress on deep neural networks, the paradigm has moved from the traditional classification pipeline (including pre-processing, feature extraction, and classification) to the end-to-end trainable systems over the last few years. Deep learning algorithms perform better as the number of data increases, as opposed to traditional machine learning algorithms, which reach saturation almost immediately as the number of data increases [22]. Several deep learning algorithms have been reported in the literature such as Convolutional Neural Networks (CNNs) and Recurrent Neural Networks (RNNs). The most commonly used networks for image classification are CNNs. It's a neurobiology-inspired multilayer neural network model with convolutional layers and fully connected layers. There may be subsampling steps between these two types of layers. As a result, using CNN is very popular in the field of image processing, and it has now become a standard in the field [23].

On the other hand, RNNs are a type of neural network that was created to model the long-term dependencies that exist between data samples. The recurrent neural network avoids this problem by incorporating the notion of time, whereas the ordinary neural network does not respect the temporal order of the input data. RNNs, like other neural network architectures, have a hidden state. However, unlike other models, the RNN updates its hidden state after processing each time step in the input. This ensures that the input sequence's temporal structure is preserved. Modifying the RNN architecture, most notably producing the Long Short-Term Memory (LSTM) which is a successful and popular model. The (LSTM) [24] is applied to time-sorted data. In some cases, knowing what decisions were made in the past is necessary to make the best decision at time t. Besides the LSTM overcomes the vanishing gradients issue in RNN. In addition to the hidden state, the LSTM carries a cell state that preserves long-term information. The LSTM solves the problem by having a shortcut path to transmit the gradients back. This unique architecture involves three gates. The forget gate decides the contents of the cell state that are relevant to the problem and discards everything else. The input gate's function is to ensure that relevant information in the input is stored in the cell state. Finally, the output gate adds the relevant input to the cell state and passes it on to the next time step. This architecture ensures that the gradients are propagated back into the earlier time steps in the network.

Various deep learning algorithms have been used but we have just mentioned those algorithms that are very important for this article, and are widely used in the field of handwritten system recognition because certain approaches to the problem of handwritten Arabic words have yet to be implemented. In this work, the LSTM network was utilized to recognize handwritten Arabic words. That investigation also suggested that if a large amount of labeled training data is used to train the deep neural network (DNN), it can provide high accuracy.

Labeling each point of the input sequence is required by the traditional DNN learning method. Graves et al. [25] have presented the Connectionist Temporal Classification (CTC) as a loss function to meet this requirement. By interpreting the network output as a distributed probability over all possible label sequences on the given input sequence, CTC allows for the previously mentioned direct alignment between the input variables and the target label. The last layer in the network is N + 1 outputs where N is the total number of labels. These outputs specify the probability that each label is observed or not observed at a given time. The use of probabilities of a single character allows for a set of probabilities that can match possible outputs to an input. This can be described by Eq. (1).

$$p(\pi|x) = \prod_{t=a}^{T} y_{\pi_t}^t, \forall \pi \tag{1}$$

where:

x: The input sequence.

π: A possible output sequence.

$y_{\pi_t}^t$: The probability of a given label from π at time t.

A special β operator is defined for the correspondence between previous steps π and the final output sequence l through the removal of blank labels. As more than one sequence of π can be a single final sequence, the probability of a certain final sequence of l will be calculated as follows:

$$p(l|x) = \sum_{\beta(\pi)=l} p(\pi|x) \tag{2}$$

The maximum probability $p(l|x)$. is considered to be the final output for each given input x and sequence l. For handwriting recognition, a word picture is considered a time sequence, in which, time is modeled as one unit of time along the width of the picture with a one-pixel slice. Nonetheless, CTC still requires an enormous amount of training data with weak labels.

4 Adaptive Data Augmentation

Data Augmentation refers to the process of generating data from the training corpus [26]. This procedure can improve a model's robustness and prevent it from overfitting. Standard augmentation techniques, such as rotation, position shifting, zoom, shear, and so on, do not affect the distribution of labels in the original dataset. That is, if we have imbalanced data, the data will remain imbalanced after data augmentation. A neural network trained on an unbalanced database will bias itsecion toward the most prevalent class in the set [27]. When the training data has a class imbalance, classifiers tend to overclassify the majority due to the higher prior probability. As a result, minority classes are misclassified more frequently than majority classes.

The issue of class imbalance based on the difference in character frequencies is not specific to any language or dataset. It is a common observation for texts in natural languages. For example, characters 'e', 't', and 'a' are the most frequent characters whereas characters 'z', 'q', and 'x' are the least frequent characters in the English language [28]. Similarly, for Arabic, based on text analysis of an Arabic corpus containing more than five million characters [29], it was found that not only are the characters unevenly distributed but also the distribution is highly skewed. The two most common characters, i.e., Alif (‍ا) and Lām (‍ل), constituted almost a quarter of all the texts. On the other hand, the five least occurring characters (out of the total of 28 characters) constitute only around 2% of the entire text. The frequency of character Zā' (‍ظ) was almost 70 times lower than the frequency of character Alif (‍ا).

In Arabic handwriting recognition, if we consider the distribution of character shapes rather than characters, the distribution becomes even more skewed. There will be very few occurrences of some character shapes from low-frequency characters, so adequately training them will be a challenge. For this reason, we present here a novel data augmentation algorithm that can balance the distribution of character samples in a dataset. To begin, we must determine the size of the database lexicon (L) as well as the number of modeling units or classes (n). Consider using the IFN/ENIT database [19] and Arabic characters as modeling units [6]; we will end up with 45 models (28 basic letters, 10 Arabic digits, and 7 ligatures) [19, 20, 30]. So, the total number of modeling units $n =$

45. Then, let p_n be the probability of each modeling unit (class) in the training data sets, we can then define the term W_i as:

$$W_i = \frac{\sum_1^N \frac{1}{p_n}}{N} \quad i = \{1, .., L\} \tag{3}$$

where N is the total number of characters in the word i. After we calculate the term W_i for all the words in the lexicon, we normalize it by dividing each of its values by the sum of all W_i.

$$W_{i,normalized} = \frac{W_i}{\sum W_i} \quad i = \{1, \ldots, L\} \tag{4}$$

Let B be the total number of new augmented word images that we want to add to the original data. Then we can use the normalized entropy to decide the Number of Word Images (*NWI*) that we should augment for each word in the lexicon to balance the database.

$$NWI = W_{i,normalized} \times B \tag{5}$$

Because we augment new data based on the inverse average probability of occurrence of each class, we eventually balance the database. Words made up of infrequent characters will be augmented by a greater number than words made up of frequently occurring characters.

5   Experimentation and Results

In the next subsections, we present the IFN/ENIT database, AHDB database, experiment configurations, results of the evaluation, and the comparison with the state-of-the-art systems.

5.1   Handwritten Arabic Text Databases

A handwritten Arabic text database is a fundamental resource for handwritten Arabic text recognition. The availability of a database enables the results of the various systems developed to be compared objectively. The two popular databases IFN/ENIT [19] and AHDB [20] will be used as a benchmark in this paper. The IFN/ENIT-database contains material for training and testing Arabic handwriting recognition software. The database is consisting of seven sets 'a' to 'f' and 's' where each set contains word images of Tunisian cities and villages. There are more than 27,000 binary word images written by 411 writers. A ground truth file for each word in the database has been compiled. This file contains information about the word such as the position of the words baseline, and information on the individual used characters in the word. The lexicon size is 937-word names, with one or more words for each name. Figure 2 shows samples from the IFN/ENIT database.

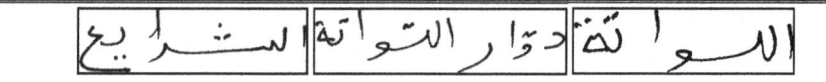

Fig. 2. Samples from the IFN/ENIT database [19].

The Arabic handwritten database (AHDB) includes approximately 10,081 handwritten Arabic words representing numbers and quantities used in cheques, as well as the most popular words in Arabic writing. The lexicon size is 96 words, 67 of which were handwritten words corresponding to numbers that can be used in handwritten cheque writing. The other 29 words were the most popular Arabic words. Figure 3 shows some sample images from the AHDB database.

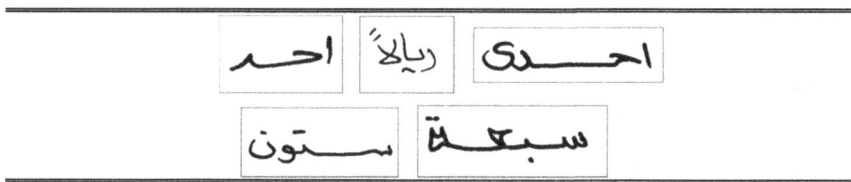

Fig. 3. Samples from the AHDB database [20].

5.2 Applying Adaptive Data Augmentation Algorithm

Table 2 gives information about the number of occurrences of each character in the IFN/ENIT database along with its percentage probability p_n. We can see that there is a significant difference in the appearance of some characters in the database. E.g., the letter Alif (ا) is repeated more than 36 thousand times, compared to the letter Za (ظ), which we can barely see in the database. The frequency distribution of the IFN/ENIT

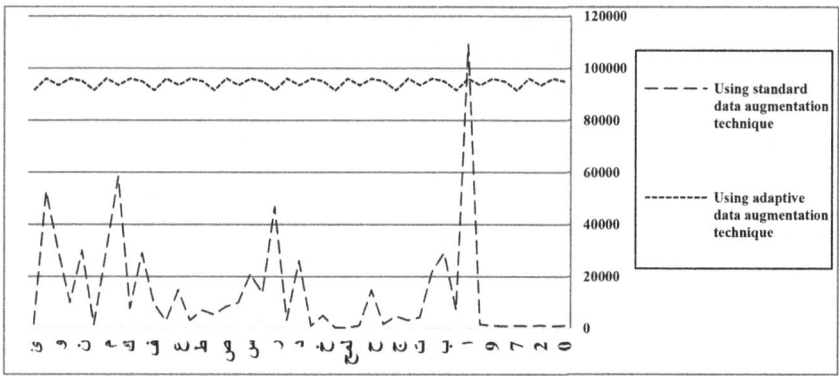

Fig. 4. Proposed data augmentation vs. standard data augmentation on the IFN/ENIT database.

characters after using our adaptive data augmentation algorithm versus the standard data augmentation technique is shown in Fig. 4. We can see that standard data augmentation methods do not affect class distribution, whereas our method attempts to balance classes by increasing the number of minority classes.

Table 2. IFN/ENIT database [19] statistics.

Character	No. of Occurrences	$\%p_n$	Character	No. of Occurrences	$\%p_n$
0	342	0.162	د	8657	4.106
1	279	0.132	ذ	1056	0.500
2	384	0.182	ر	15621	7.410
6	311	0.147	ز	4482	2.126
7	354	0.167	س	7012	3.326
8	284	0.134	ش	3270	1.551
9	341	0.161	ص	2764	1.311
ء	520	0.246	ض	1752	0.831
ا	36394	17.265	ط	2310	1.095
ى	2357	1.118	ظ	1029	0.488
ب	9718	4.610	ع	4902	2.325
ة	7259	3.443	غ	926	0.439
ت	1402	0.665	ف	3180	1.508
ث	1018	0.482	ق	9660	4.582
ج	1564	0.741	ك	2536	1.203
جـ	539	0.255	ل	19475	9.238
ح	4896	2.322	م	9938	4.714
خ	365	0.173	لم	458	0.217
جـ	64	0.030	ن	10021	4.753
حـ	100	0.047	ه	3318	1.574
خـ	1640	0.778	و	10040	4.762
لح	310	0.147	ي	17591	8.345

The AHDB database, on the other hand, contains approximately 10,000 handwritten Arabic words and approximately 31,200 characters. Unlike the IFN/ENIT database, this database lacks many Arabic characters because it was created specifically to identify the most popular Arabic words and words used in cheque writing. Figure 5 shows the frequency distribution of the AHDB database's characters after using our adaptive data augmentation algorithm compared to the conventional data augmentation technique on the.

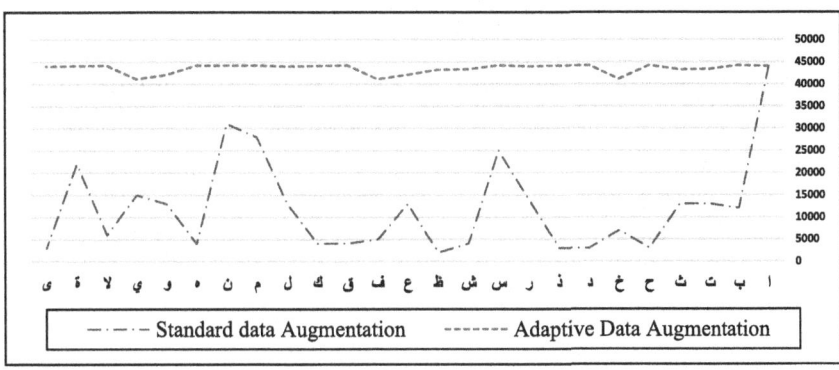

Fig. 5. Adaptive data augmentation vs. standard data augmentation on the AHDB database.

5.3 Results of Arabic Handwriting Recognition

In this section, we shall present the results of Arabic handwriting recognition after applying our augmentation algorithm to the database. Experimental models were implemented by using the Recurrent Neural Network Library (RNNLIB) [31]. This library has different deep learning architectures e.g. (LSTM, BiLSTM, and RNN). It also includes the ability to use Connectionist Temporal Classification (CTC), which allows a given system to transcribe unsegmented sequence data.

Our network has 200 extended LSTM memory blocks in each of the forward and backward layers. Each memory block has one memory cell, an input gate, an output gate, a forget gate, and three peephole connections. The cell input and output activation functions are *tanh* while the gate activation function is logistic sigmoid. A decaying learning rate has been used in all the CTC systems. The initial learning rate was set to 0.01 and was reduced gradually to 0.0001.

The IFN/ENIT database contains the transcriptions for the word image at the character-shape level [12] and we have created similar transcriptions for the AHDB database. We ended up with a total of 157 models in our recognition system. Table 3 shows the accuracy rate of the designed system when augmenting data using standard methods (rotation, shearing, etc.) and the accuracy rate when using augmented data by our algorithm on the IFN/ENIT and AHDB databases. The results show that the WAR% is improved when using our adaptive data augmentation techniques.

Table 3. Summary of the word accuracy rates on the IFN/ENIT and AHDB databases.

Database	Train-test configuration	Word Accuracy Rates (WAR) %	
		Standard data augmentation	Adaptive data augmentation
IFN/ENIT	abc-d	96.22%	98.99%
	bcd-a	93.17%	96.91%

(continued)

Table 3. (*continued*)

Database	Train-test configuration	Word Accuracy Rates (WAR) %	
		Standard data augmentation	Adaptive data augmentation
	abcd-e	90.14%	95.05%
	abcde-f	87.94%	93.57%
AHDB	70% training	95.99%	98.10%
	20% training		
	10% training		

5.4 Comparison with Other State-Of-The-Art Systems

To demonstrate the performance of the proposed model, we here compare the results of recent state-of-the-art systems evaluated on the IFN/ENIT and AHDB database to our model. Table 4 presents a comparison of the results from the state-of-the-art systems evaluated on the IFN/ENIT and AHDB databases. In comparison to other state-of-the-art systems, our systems' overall results are among the best. To the best of our knowledge, our systems outperform the best-reported systems in the literature on the IFN/ENIT database's evaluation set f. Recently, the best recognition result on the database's evaluation set e was obtained by Ghanim et al. [32] because the authors have constructed clusters support that represented the database as a big search tree model which helps to attain a reduced complexity in matching each test image with a cluster. On the other hand, Hassan et al. [34] had the highest accuracy on the AHDB database because they have used a Supported Vector Machine (SVM) method evaluated with only 2072 images for training and 868 images for testing, whereas our system was evaluated with 7057 images for training and 2017 images for testing. The SVM typically performs better on small datasets [36] which explains why their system had a slightly higher accuracy rate than ours.

Table 4. Comparison with other state-of-the-art systems.

Author	Word Accuracy Rate (WAR) %		
	The IFN/ENIT Database		
	Train-Test Configuration		
	abc-d	*abcd-e*	*abcde-f*
Ahmad et al. [8]	97.71%	94.76%	93.32%
Graves et al. [14]	–	–	93.37%
Elleuch et al. [15]	83.70%	–	–

(continued)

Table 4. (*continued*)

Author	Word Accuracy Rate (WAR) %		
	The IFN/ENIT Database		
	Train-Test Configuration		
	abc-d	*abcd-e*	*abcde-f*
Ghanim et al. [32]	99.00%	95.60%	—
Abandah et al. [33]	98.96%	93.46%	92.46%
Present work	**98.99%**	**95.05%**	**93.57%**
Author	Word Accuracy Rate (WAR) %		
	The AHDB Database		
Hassan et al. [34]	99.08%		
Zafar et al. [35]	97.80%		
Present work	**98.10%**		

6 Conclusion

We have proposed an adaptive data-augmentation algorithm in this paper to deal with the problem of imbalanced classes. It has been demonstrated that this method can effectively solve the data imbalance problem. Our method is designed in such a way that the minority classes are expanded by calculating each class's average probability. The results show that when applied to the IFN/ENIT and AHDB databases, the presented algorithm can balance the distribution of minority classes. The system presented in this paper is one of the state-of-the-art systems available. As future work, we are considering a two-stage data augmentation method that can further improve the current work. In the first stage, we will be using the presented algorithm here and in the second stage, a words' synthesizer will be implemented to generate unique handwritten image words which will increase the diversity of classes.

Acknowledgment. This research was supported by the King Fahd University of Petroleum and Minerals (KFUPM).

References

1. Khobragade, R., Koli, N., Lanjewar, V.: Challenges in recognition of online and off-line compound handwritten characters: a review. In: Zhang, Y.-D., Mandal, J.K., So-In, C., Thakur, N.V. (eds.) Smart Trends in Computing and Communications. SIST, vol. 165, pp. 375–383. Springer, Singapore (2020). https://doi.org/10.1007/978-981-15-0077-0_38
2. Sulaiman, A., Omar, K., Nasrudin, M.F.: Two streams deep neural network for handwriting word recognition. Multimedia Tools Appl. **80**(4), 5473–5494 (2020). https://doi.org/10.1007/s11042-020-09923-1

3. Kumar, P., Sharma, A.: Segmentation-free writer identification based on a convolutional neural network. Comput. Electrical Eng. **85**, 106707 (2020)
4. Vásquez, J.L., Ravelo-García, A.G., Alonso, J.B., Dutta, M.K., Travieso, C.M.: Writer identification approach by holistic graphometric features using off-line handwritten words. Neural Comput. Appl. **32**(20), 15733–15746 (2018). https://doi.org/10.1007/s00521-018-3461-x
5. Nashwan, F., Rashwan, M.A., Al-Barhamtoshy, H.M., Abdou, S.M., Moussa, A.M.: A holistic technique for an Arabic OCR system. J. Imaging **4**(1), 6 (2018)
6. Eltay, M., Zidouri, A., Ahmad, I.: Exploring deep learning approaches to recognize handwritten Arabic texts. IEEE Access **8**, 89882–89898 (2020)
7. Rabi, M., Amrouch, M., Mahani, Z.: Cursive Arabic handwriting recognition system without explicit segmentation based on hidden Markov models. J. Data Mining Digital Hum. 2018
8. Ahmad, I., Fink, G.A.: Handwritten Arabic text recognition using multi-stage sub-core-shape HMMs. Int. J. Doc. Anal. Recogn. (IJDAR) **22**(3), 329–349 (2019). https://doi.org/10.1007/s10032-019-00339-8
9. Shrestha, A., Mahmood, A.: Review of deep learning algorithms and architectures. IEEE Access **7**, 53040–53065 (2019)
10. Graves, A., Mohamed, A.R., Hinton, G.: Speech recognition with deep recurrent neural networks. In: 2013 IEEE International Conference on Acoustics, Speech and Signal Processing, pp. 6645–6649, May 2013
11. Maalej, R., Kherallah, M.: Improving the DBLSTM for online Arabic handwriting recognition. Multimedia Tools Appl. **79**(25), 17969–17990 (2020)
12. Ahmad, M., Protasov, S., Khan, A.M., Hussain, R., Khattak, A.M., Khan, W.A.: Fuzziness-based active learning framework to enhance hyperspectral image classification performance for discriminative and generative classifiers. PloS One **13**(1), e0188996 (2018)
13. Elnagar, A., Al-Debsi, R., Einea, O.: Arabic text classification using deep learning models. Inf. Process. Manag. **57**(1), 102121 (2020)
14. Graves, A.: Offline Arabic handwriting recognition with multidimensional recurrent neural networks. In: Märgner V., El Abed H. (eds) Guide to OCR for Arabic Scripts, pp. 297–313. Springer, London (2012). https://doi.org/10.1007/978-1-4471-4072-6_12
15. Elleuch, M., Tagougui, N., Kherallah, M.: Deep learning for feature extraction of Arabic handwritten script. In: International Conference on Computer Analysis of Images and Patterns, pp. 371–382. Springer, Cham (2015)
16. Maalej, R., Kherallah, M.: Improving MDLSTM for offline Arabic handwriting recognition using dropout at different positions. In: Villa, A.E.P., Masulli, P., Pons Rivero, A.J. (eds.) ICANN 2016. LNCS, vol. 9887, pp. 431–438. Springer, Cham (2016). https://doi.org/10.1007/978-3-319-44781-0_51
17. Elarian, Y., Ahmad, I., Awaida, S., Al-Khatib, W.G., Zidouri, A.: An Arabic handwriting synthesis system. Pattern Recogn. **48**(3), 849–861 (2015)
18. Korichi, A., Slatnia, S., Aiadi, O., Tagougui, N., Kherallah, M.: Arabic handwriting recognition: between handcrafted methods and deep learning techniques. In: 2020 21st International Arab Conference on Information Technology (ACIT), pp. 1–6, November 2020
19. Pechwitz, M., Maddouri, S.S., Märgner, V., Ellouze, N., Amiri, H.: IFN/ENIT-database of handwritten Arabic words. In: Proceedings of CIFED, vol. 2, pp. 127–136, October 2002
20. Al-Ma'adeed, S., Elliman, D., Higgins, C.A.: A database for Arabic handwritten text recognition research. In: Proceedings Eighth International Workshop on Frontiers in Handwriting Recognition, pp. 485–489, August 2002
21. Ethnologies: Languages of the World, 22nd edn. SIL International. Released on February 21st 2019
22. Zhang, Q., Yang, L.T., Chen, Z., Li, P.: A survey on deep learning for big data. Inf. Fusion **42**, 146–157 (2018)

23. Wu, C., Fan, W., He, Y., Sun, J., Naoi, S.: Handwritten character recognition by alternately trained relaxation convolutional neural network. In: 2014 14th International Conference on Frontiers in Handwriting Recognition, pp. 291–29, September2014

24. Schmidhuber, J., Hochreiter, S.: Long short-term memory. Neural. Computer **9**(8), 1735–1780 (1997)

25. Graves, A., Fernández, S., Gomez, F., Schmidhuber, J.: Connectionist temporal classification: labelling unsegmented sequence data with recurrent neural networks. In: Proceedings of the 23rd International Conference on Machine Learning, pp. 369–376, June 2006

26. Elarian, Y., Abdel-Aal, R., Ahmad, I., Parvez, M.T., Zidouri, A.: Handwriting synthesis: classifications and techniques. Int. J. Doc. Anal. Recogn. (IJDAR) **17**(4), 455–469 (2014). https://doi.org/10.1007/s10032-014-0231-x

27. Marceau, L., Qiu, L., Vandewiele, N., Charton, E.: A comparison of Deep Learning performances with other machine learning algorithms on credit scoring unbalanced data. arXiv preprint arXiv:1907.12363 (2019)

28. English letter frequency : https://en.wikipedia.org/wiki/Letter_frequency. Accessed 19 May 2021

29. Intellaren: A study of Arabic letter frequency analysis (2016). Accessed 19 May 2021. http://www.intellaren.com/articles/en/a-study-of-arabic-letter-frequency-analysis

30. Ahmad, I., Fink, G.A.: Class-based contextual modeling for handwritten Arabic text recognition. In: 2016 15th International Conference on Frontiers in Handwriting Recognition (ICFHR), pp. 554–559, October 2016

31. Graves, A.: RNNLIB: a recurrent neural network library for sequence learning problems (2016). http://sourceforge.net/projects/rnnl

32. Ghanim, T.M., Khalil, M.I., Abbas, H.M.: Comparative study on deep convolution neural networks DCNN-based offline Arabic handwriting recognition. IEEE Access **8**, 95465–95482 (2020)

33. Abandah, G.A., Jamour, F.T., Qaralleh, E.A.: Recognizing handwritten Arabic words using grapheme segmentation and recurrent neural networks. Int. J. Doc. Anal. Recogn. (IJDAR) **17**(3), 275–291 (2014). https://doi.org/10.1007/s10032-014-0218-7

34. Hassan, A.K.A., Mahdi, B.S., Mohammed, A.A.: Arabic handwriting word recognition based on scale-invariant feature transform and support vector machine. Iraqi J. Sci., 381–387 (2019)

35. Zafar, A., Iqbal, A.: Machine reading of Arabic manuscripts using KNN and SVM classifiers. In: 2020 7th International Conference on Computing for Sustainable Global Development (INDIACom), pp. 83–87, March 2020

36. Liu, P., Choo, K.-K., Wang, L., Huang, F.: SVM or deep learning? A comparative study on remote sensing image classification. Soft. Comput. **21**(23), 7053–7065 (2016). https://doi.org/10.1007/s00500-016-2247-2

37. Elarian, Y., Ahmad, I., Awaida, S., Al-Khatib, W., Zidouri, A.: Arabic ligatures: analysis and application in text recognition. In: 2015 13th International Conference on Document Analysis and Recognition (ICDAR), pp. 896–900. IEEE, August 2015

High Performance Urdu and Arabic Video Text Recognition Using Convolutional Recurrent Neural Networks

Abdul Rehman[1]([✉]), Adnan Ul-Hasan[2], and Faisal Shafait[1,2]

[1] School of Electrical Engineering and Computer Science,
National University of Sciences and Technology (NUST), Islamabad, Pakistan
{arehman.bese17seecs,faisal.shafait}@seecs.edu.pk
[2] Deep Learning Laboratory, National Center of Artificial Intelligence,
Lahore, Pakistan
adnan.ulhassan@seecs.edu.pk

Abstract. Text extraction from videos is an emerging research field in the document analysis community. We propose a simple Convolutional Recurrent Neural Network to perform text recognition on both Arabic and Urdu scripts. We use a large variety of data augmentation techniques to generalize the model and prevent over-fitting. We also use a slightly improved loss function that helps the model converge faster. Using the proposed method we achieved 99.73% CRR, 88.37% WRR and 89.92% LRR on the Urdu Ticker Text dataset and 96.82% CRR, 90.41% WRR and 76.78% LRR on the AcTiVComp20 dataset. The proposed method has significantly outperformed Google Vision API on both of the datasets.

Keywords: Urdu · Arabic · Video text recognition · CRNN

1 Introduction

Text recognition in its pure form - recognizing text from scanned documents - has been transformed in modern times. Smart phones have introduced a new capturing mechanism with additional challenges like page warps and view translations. Moreover, scene text recognition is a relatively new domain in text recognition that deals with text recognition in natural scenes and images. The text may occur in any shape, style or orientation, thereby increasing the text recognition challenge multi folds. An extension of scene text recognition is the emerging field of video text recognition. A video frame has additional challenges such as frame rate estimation, unique text determination, etc. in addition to the challenges of natural scenes.

Latin-based scripts receive most of the attention in both academia and industry. Remarkable results in Latin text recognition for both printed as well as handwritten text have been achieved. There has been very little or non-existent research carried out for Arabic scripts, including Arabic, Urdu, Persian, etc.

© Springer Nature Switzerland AG 2021
E. H. Barney Smith and U. Pal (Eds.): ICDAR 2021 Workshops, LNCS 12916, pp. 336–352, 2021.
https://doi.org/10.1007/978-3-030-86198-8_24

Arabic and the derived scripts pose several additional challenges for reliable recognition such as joined character, contextual shape change, diagonal writing flow (in case of Urdu Nastaleeq script) are some of these challenges.

The text recognition in video sequences has several important applications in the real world. News classification (sports, entertainment, current affairs, etc.) and coverage, Analytic dashboards highlighting the insights gained from entity extraction and news monitoring (identification of banned content) are some of the widely used scenarios of video text recognition. News channels use these applications to not only monitor their own transmission but also to keep an eye on their competition.

In contrast to text recognition from a scanned document, video text recognition poses several complex challenges. We need to extract individual frames from the video stream and apply text detection and recognition methods to get the text content. In video streams, the transition from one frame to another can cause some distortion that can be troublesome for text detection and recognition. We face this issue in NEWS channels. Static tickers fade in and out of the frame and in-between these transition models do not perform well. Distortion in scrolling tickers can cause a problem in text recognition. In video streams, we also need to extract unique instances of the text. We do so by keeping the track of position and duration of a text shown in the video stream. Text is video streams might also be in a different font, size or style. So in general, video text recognition is more challenging than documents.

The Urdu language is the national language of Pakistan. It is an Indo-Aryan language and shares similar phonology and syntax with the Hindi language; however, it is written in Arbic-like script. It has more than 170 million speakers. Urdu is mostly written in Nastaleeq script. Nastaleeq script was developed during the Persian region in the 14th and 15th centuries. The Urdu language has 37 unique alphabets. Alphabets are not written individually but are joined together to form ligatures.

The Arabic language is a universal language and the official language of 25 countries. It has over 300 million speakers. There are different writing styles for Arabic scripts. But it is mostly written in Naskh script. It first emerged in the 1st to 4th centuries CE. Similar to Urdu language, the Arabic alphabets are not written individually but are joined together to form ligatures.

In this paper, we focus on video text recognition in Arabic and Urdu that applies to natural images, videos and live streams. We pose the video text recognition as a sequence recognition problem. The problem can be described as converting an input image to a sequence of characters that represent text written in that image. It is also known as Image-based sequence recognition. Convolutional Neural Networks are excellent tools for extracting visual features from images. These visual features may represent different structures and shapes. In our case, these could be parts of letters, numbers or special characters. But these visual features alone are not enough to recognize the text written inside the image. We need to translate the sequence of visual features into character probability scores. Recurrent Neural Networks (RNNs) are excellent in sequence-to-sequence translation. So using CNNs as visual feature extractors and RNNs as sequence-to-sequence translation we can effectively model our text recognition technique. We trained an end-to-end Convolutional Recurrent Neural Network (CRNN) to recognize Arabic and Urdu video text.

We test our proposed solution on two different datasets. The first dataset is the our own developed NUST-Urdu Ticker Text (NUST-UTT) dataset. It contains images of static and scrolling ticker text from different Urdu channels. The second dataset is the AcTiVComp20 dataset [1]. It also consists of cropped ticker text from different Arabic channels. We calculated Character Recognition Rate (CRR), Word Recognition Rate (WRR) and Line Recognition Rate (LRR) on each dataset, with and without spaces. We achieved CRR, WRR and LRR of 99.73%, 88.37% and 89.92% on the Ticker Text dataset and 96.82%, 90.41% and 76.78% on the AcTiVComp20 dataset.

The remaining of the paper is organized as follow. The Sect. 2 provides an overview of related work in the field of video text recognition, Sect. 3 provides the details of the proposed approach, Sect. 4 describes the dataset and outline the performed experiments. Section 5 details the results obtained and Sect. 6 concludes the paper with future directions.

2   Related Work

Text recognition is an old and well-researched field. But it is mainly researched in English or Chinese [15]. The field is wide-open for other languages. One of the main hurdles is the availability of data. But it is possible to generate printed text synthetically using different techniques. Researchers have tried different approaches for text recognition. These methods can be divided into segmentation-based techniques and non-segmentation-based techniques.

Older methods use a segmentation-based approach. These methods focus on isolated character or word recognition and recognizing them individually, which is possible in English and Chinese [11,20]. This sounds logical but it may become difficult when the characters are overlapping especially in handwriting recognition. Also, it is not easy to isolate each character in the languages Urdu as well as Arabic text. Another approach is to isolate ligatures instead of individual characters [3,4]. This approach works well for Urdu and Arabic text [13]. But this requires a lot of data. The technique becomes ineffective when the ligature count becomes very large. To classify each ligature there should be enough data samples in the dataset. In the NUST-UTT dataset alone there are 5,450 unique ligatures. Some other techniques also try to achieve word-level recognition but suffer the same problem [12]. The total number of classes grows so much, it becomes difficult to classify. Techniques that perform an isolated character, ligature, or word recognition are not effective for Urdu and Arabic scripts.

Modern techniques use a non-segmentation-based approach. We use a neural network to recognize the whole sequence instead of recognizing individual characters or ligatures. Non-segmentation approach uses Hidden Markov Models and recognizes the entire sequence [6,10]. Convolutional Neural Networks (CNNs) are excellent at understanding visual features. Models like AlexNet and ResNet are exceptionally good at recognition objects [7,9]. But CNNs perform poorly in the sequence recognition tasks. RNNs are another family of Neural Networks which are pretty good at sequence-to-sequence tasks [16,19]. So, instead of recognizing isolated characters or words, researchers worked on sequence recognition. In sequence recognition, we recognize the entire line instead of recognizing individual parts. Recurrent Neural Networks are context-aware and can recognize patterns occurring in time series [19]. But visual features are troublesome for RNNs to learn. So a Convolutional Recurrent Neural architecture was proposed [14,18]. It uses a Convolutional Neural Network as a feature extractor and RNNs for sequence recognition. It takes advantage of both architectures and performs exceptionally well.

In ICDAR2017 Competition on Arabic Text Detection and Recognition in Multi-resolution Video Frames, the THDL-Rec technique used GRUs and achieved second best scores in evaluation [1,21]. GRU layers usually perform better than RNNs and LSTMs [5,8]. The most recent work is of OCR framework for detecting and recognizing Urdu in News channels [17]. It also uses a similar approach. It uses Convolutions Neural Networks for extracting visual features and Bidirectional LSTM layers for sequence recognition. We have further improved the method mentioned above by introducing a new model architecture, an improvement in loss function and an aggressive data augmentation for generalization.

3   Methodology

Since we only focus on Text Recognition in images, videos or live streams, we assume that we already have a Text Detection module to extract bounding boxes where the text is present. And we also need dataset containing those cropped images and their corresponding labels (in form of machine-readable text). In our case, we have NUST-UTT and the AcTiVComp20 dataset. We evaluate our method on both datasets.

To easily load datasets for training and inference we propose a formal method for storing them. This reduces CPU bottleneck and we do not need to write separate training and evaluation scripts for each dataset. We compute necessary hyper-parameters for each dataset during this step. This is further discussed in the Experimental Evaluation section (Sect. 4).

Once the dataset is ready, we train our model. In the training loop, we fetch images and labels from the dataset, perform data augmentation and preprocessing, feed it through the network, compute loss and back-propagate to adjust weights. We save the weights after each epoch if the validation loss improves. After training the model, we compute all five metrics (individual for each sample as well as collective) on the test set. We analyze the weaknesses of the model and adjust hyper-parameters to improve results.

3.1   Preprocessing

Urdu and Arabic Text is written from Right to Left. The data in computers (such as in arrays) are stored from Left to Right. So our labels (sequences of characters) are stored in reverse order. We either need to invert our label or the image itself to compute loss. We decided to invert the image (flip it on the x-axis). The color range of a grey-scale image is 0 to 255. So we normalize the image to have a range of -1 to $+1$. Normalizing input helps in gradient flow and the model converges faster. At this point, we have images that have fixed height but variable width. We need to have fixed height and width to create batches. So we pad zeros to images to make its width equal to max-width (computed in Data Preparation step). For labels, we simply convert each character to its corresponding index in characters plus 1 and pad the rest of the sequence with zeros (blank class) to make length equal to max label size.

3.2   Model Architecture

The network architecture is similar to CRNN model [14]. Original CRNN paper takes input image of height 32px while we input image of 64px. We also use fewer Convolution filters and fewer hidden units in RNN layers as compared to CRNN paper. Lastly, we use Bidirectional GRU layers instead of vanilla RNN layers. The architecture is further discussed below.

Our model has three parts. A CNN features an extractor, RNN layers for sequence recognition and a classification layer to classify feature vectors to characters. Initializing the model requires two hyper-parameters. Max Width W and Total Characters C. These two parameters differ from dataset to dataset and are computed during the Data Preparation step. It is important to notice Total Characters also includes a blank character which is used by the CTC Loss function. For NUST-UTT, dataset we used $W = 1300$ and $C = 91$. For AcTiVComp20 dataset, we used $W = 1600$ and $C = 83$.

CNN Feature Extractor. The main purpose of the CNN feature extractor is to extract visual features from the image. RNN layers can be directly applied to input images but RNN layers are not very good at extracting visual features. CNN feature makes it easier for RNNs to understand sequence. Most of the noise and augmentations are filtered out by these CNN layers. This part consists of Convolution and Max Pooling layers. Each Convolution layer is followed by a Batch Normalization layer and Leaky ReLU activation function. We also use Dropout layers for regularization. Max pooling operations reduce the dimensions of the features. We only apply max pooling on the x-axis twice, reducing the width by a factor of 4. Applying further max-pooling can increase receptive field on CNNs but the operation results in poor results as multiple characters start to overlap on a single feature vector. The max-pooling operations along with the final Convolution layer squash the height dimension. We simply squeeze the height dimension.

Table 1. CNN feature extractor architecture

Layer	Kernel size	Padding	Output shape	Parameters
Input			(1, 64, W)	0
Conv2d (16)	3 × 3	1 × 1	(16, 64, W)	160
BN (16) + LReLU			(16, 64, W)	32
MaxPool2d	2 × 2		(16, 32, W//2)	0
Conv2d (32)	3 × 3	1 × 1	(32, 32, W//2)	4,640
BN (32) + LReLU			(32, 32, W//2)	64
MaxPool2d	2 × 2		(32, 16, W//4)	0
Conv2d (48)	3 × 3	1 × 1	(48, 16, W//4)	13,872
BN (48) + LReLU			(48, 16, W//4)	96
Conv2d (64)	3 × 3	1 × 1	(64, 16, W//4)	27,712
BN (64) + LReLU			(64, 16, W//4)	128
MaxPool2d	2 × 1		(64, 8, W//4)	0
Dropout (0.2)			(64, 8, W//4)	0
Conv2d (96)	3 × 3	1 × 1	(96, 8, W//4)	55,392
BN (96) + LReLU			(96, 8, W//4)	192
Conv2d (128)	3 × 3	1 × 1	(128, 8, W//4)	110,720
BN (128) + LReLU			(128, 8, W//4)	256
MaxPool2d	2 × 1		(128, 4, W//4)	0
Dropout (0.2)			(128, 4, W//4)	0
Conv2d (256)	4 × 4		(256, 1, W//4 − 3)	524,544
BN (256) + LReLU			(256, 1, W//4 − 3)	512
Reshape			(256, W//4 − 3)	0
Total parameters				738,320

Table 2. RNN layers architecture

Layer	Output shape	Parameters
Input	(256, W//4 − 3)	0
Bidirectional-GRU (256)	(512, W//4 − 3)	789,504
BatchNorm (512) + LReLU	(512, W//4 − 3)	1024
Bidirectional-GRU (512)	(1024, W//4 − 3)	3,151,872
BatchNorm (1024) + LReLU	(1024, W//4 − 3)	2048
Total parameters		3,944,448

RNN Layers. RNN layers are used for sequence-to-sequence translation. In our case, we need to translate visual features into feature vectors that can be easily classified into characters. In short, we are performing sequence recognition. In this section, we use specifically two Bidirectional GRU layers, each one followed

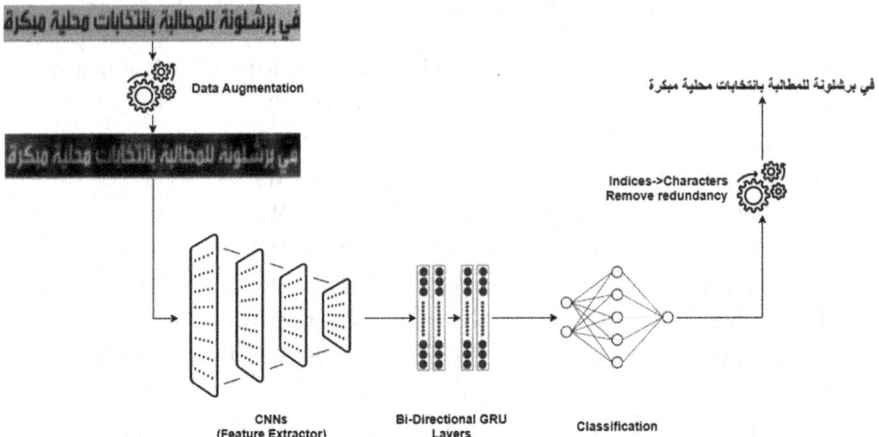

Fig. 1. A diagram summarizing the Model Architecture. The input sample is taken from AcTiVComp20 dataset [1]. The model consists of CNN Feature Extractor, two Bi-directional GRU layers and a Classification layer. Data augmentation is applied only during the training loop.

by a Batch Normalization layer and a Leaky ReLU activation function. We used GRU layers as these layers suffer least from gradient vanishing and exploding problems. GRU layers are faster and lightweight compared to LSTM layers.

Classification Layer. The classification layer consists of a single 1-d Convolution layer of kernel size of 1 followed by a LogSoftmax activation function. This layer classifies each feature vector of RNN Layer output into a character. The output channels of Convolution are equal to Total Characters C. Total number of parameters of the classification layer are $1025 * C$.

3.3 Model Training

To start training, we first load the target dataset containing train, validation and test sets. We initialize the model with default parameters suited for the target dataset. We initialize the AdamW optimizer to optimize the weights of the network.

The training loop takes a batch of images, passes them through the network and gets output activation of the model. After that, it computes loss between output activation and ground truths, and backpropagates through the network to compute gradients. The optimizer optimizes the model based on computed gradients. The same training loop is used to compute validation loss as well but the loss is not backpropagated. The validation loss is used to identify if the model is converging or not. Model checkpoint is saved only when validation loss decreases compared to the previous epoch. This makes sure we do not save a checkpoint that is overfitted on the dataset. Once the training is finished we take the best checkpoint and evaluate it on the test set. It is important to notice that we do not use the test set during the whole training process. It is only used to compute metrics once the whole training process is finished (Table 1).

We train the models in two steps. In the first step, we train the model with a learning rate of 0.0003. Model mostly learns in the first step. We train the model for nearly 100 epochs or until it stops converging. In the second step, we further fine-tune the model with a learning rate of 0.0001 for nearly 50 epochs. In this step, the optimizer tweaks the parameters to slightly improve accuracy (Fig. 1).

3.4 Loss Function

Similar to the CRNN model we also use Connectionist Temporal Classification Loss (CTC Loss) [14]. It is used to predict as well as align the output with the ground truth. Since we pad images during Pre-processing, we also need to ignore the output activations for the padded region of the image. Max Width is the maximum possible width an image can have in the dataset. Most of the images will not even reach close to the maximum possible width. The padded region will not contribute to the output of the model, making it difficult for the model to align text from such a large region. We can make this job easier for the model by simply ignoring the activations of the padded region. The shape of the activation region for any image with width W can be calculated using the formula (C, $W//4 - 3$) given in Table 2. So, We can only consider the first $W//4 - 3$ vectors.

We also need to consider receptive fields of CNNs and RNNs. To make the model fail-safe we add another 16 vectors after the $W//4 - 3$ vectors. We call these vectors CTC pad. So total number of output feature vectors we use for loss function are $W//4 - 3 + 16$. But the term may exceed the total output feature vectors returned by the model. So, the fail-safe formula is $min(Width//4 - 3 + 16, MaxWidth//4 - 3)$.

Figure 2 shows an input image and the output activations of the model. The input image is pre-processed. In the pre-processing step, we invert the image and pad it to make it fixed-sized. In the output, we can see scores of each class or character. The Image Activations classify the target character while the CTC pad section has the highest score for the blank class. The Padding Activations are random. These are ignored in both training and inference.

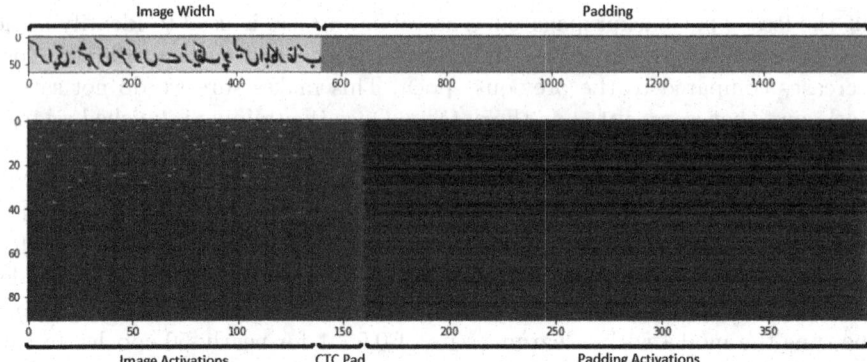

Fig. 2. The diagram shows the input image and the output activations of the model on the same image. The input can be divided into two parts, the image itself and the padding to make it fixed sized. The output activation maps show Image Activations, CTC Pad and Padding Activations. Image Activations are class scores of each character. CTC Pad consists of regions separating image activations from padding activations. Padding Activations are ignored in both training and inference.

4 Experimental Evaluation

In this section we will discuss about Data Preparation, Data Augmentation, Evaluation Metrics and Experiments performed on the datasets.

4.1 Datasets

We focus on two main dataset, AcTiVComp20 dataset and NUST-Urdu Ticker Text datasets. Both datasets contain ticker text data for News channels. Each dataset has its own set of alphabets. Both datasets have different characteristics which are mentioned below.

AcTiVComp20 Dataset. AcTiVComp20 dataset contains 7,943 training samples which have 36,593 words and 212,393 characters. It has total 83 unique characters. The dataset contains images and their corresponding XML files. XML files contain labels in both Arabic and Latin transcriptions. We only used Arabic transcription. The test set is given separately, so we only split the train set into train (90%) and validation (10%) sets while keeping the test set the same. We do not perform any data cleaning or correction for AcTiVComp20 dataset. It is important to notice that AcTiVComp20 dataset contains RGB images while our method operates at grey scale images. So we convert images to grey scale as mentioned in Algorithm 1.

NUST-Urdu Ticker Text (NUST-UTT) Dataset. The NUTS-Urdu Ticker Text (NUST-UTT) dataset[1] contains 19,437 data samples. But the dataset contains ligatures separated by spaces instead of words. It has total of 503,273 ligatures, 5,450 unique ligatures and 90 unique characters. We need to convert ligature into words. We cannot compute WRR without converting text to words. To solve this issue, we trained a simple 2 layer Bidirectional GRU model that predicts the positions where the spaces should be inserted in a space less text. We trained our model on a huge Urdu Corpus. Once the model was trained we corrected our ground truth with the trained model during Data Preparation step.

4.2   Data Preparation

The first step is to formalize the given dataset. In short, we convert it into an easy-to-use format. It reduces the CPU bottleneck for model training. We simply convert images to grey scale, resize all the images to a fixed height of 64px while keeping the image aspect ratio the same, store them in the "images" directory. Then we create three index files. Each one for train-set, validation-set and test-set. The index files contain a list of image-label pairs separated by a tab. We can quickly load any set by reading the index files. We also compute a character file which contains all unique characters that appear in the dataset. We also compute two hyper-parameters Max Image Width and Max Label Length which are later used in training process. We use the following algorithm to formalize the dataset.

The algorithm for each dataset is the same except for data loading and the data cleaning & data correction steps.

4.3   Data Augmentation

Data augmentation refers to the process of increasing data samples by transforming existing samples. It is very useful in image-based learning tasks. It prevents over-fitting. We use data augmentation during model training. Data augmentation is only performed on the train set. We have a total of 9 different augmentation functions. 4 of them are shape-based augmentation functions and the rest are color-based augmentations. Shape-based augmentations transform the image into different dimensions. So, after each shape-based augmentation, we resize the image to a height of 64px while keeping the aspect ratio the same.

We do not apply all the augmentations to every batch we fetch from the train set. We select K random augmentation functions for each image fetched from the train set. Applying all augmentation at the same time can distort the image to such an extent that it becomes unrecognizable. In other words, the distribution of the dataset becomes different. The resultant model does not perform well on the test-set. Applying all augmentation can also cause a CPU bottleneck because

---

[1] The Urdu Ticker Text Dataset will be available publicly at https://tukl.seecs.nust. edu.pk/downloads.html.

Data: List of image-label pairs
Result: Processed Dataset
data cleaning & correction;
compute maximum image width;
compute maximum label size;
save max image width and max label size;
create unique characters list;
foreach *image_path, label* ∈ *pairs* **do**
  | add unseen characters from label to characters list;
  | load image and convert it to grey scale;
  | resize image to height 64px while keeping aspect ratio same;
  | save image to new path and update path in pairs;
end
sort and save characters;
shuffle pairs;
if *test set given* **then**
  | split images into training and validation sets;
else
  | split images into training, validation and test sets;
end
foreach *set* ∈ {*trainingset, validationset, testset*} **do**
  | create index file;
  | **foreach** *image, label* ∈ *set* **do**
  | | append image path;
  | | append tab character;
  | | append image label ;
  | | append new line character;
  | **end**
  | close index file ;
end

Algorithm 1: Data preparation steps.

these augmentation functions are CPU-intensive tasks. After testing different values for K, we decided to use K = 3 for all datasets. Following is the list of all augmentation functions, their default hyper-parameters and the affect of each augmentation on the image given below.

Invert Colors. It is a color-based augmentation function that randomly inverts color. There is a 50% chance for image colors to be inverted.

Pad Image. It is a shape-based augmentation function that randomly pads each side of the image. It samples the number of pixels for each side (left, right, top and bottom) from a range of 0px to 20px and pads image accordingly.

Brightness and Contrast. It is a color-based augmentation function that randomly changes the brightness and contrast of the image. It samples α from a range of 0.75 to 1.25 and β from a range of -0.25 to 0.25. It computes new value x' given original value as x using formula $x' = x * \alpha + \beta$. The output value x' may exceed the pixel intensity range of 0 to 255. So we clip values that fall out of this pixel intensity range.

Cloudy Effect. It is a color-based augmentation that adds random soft noise to the image. It creates a small image containing grain noise. Pixel intensity values of grain noise image range from -255 to 255. The grain noise image has dimensions of (W//DOWN_FACTOR, H//DOWN_FACTOR). The variable DOWN_FACTOR is sampled from a range of 8 to 16. Another variable intensity i is sampled from a range of 0 to 1 that refers to the intensity at which the noise will be applied. To apply the effect, we resize the grain noise image to the dimensions of the input image using bilinear interpolation, multiply it with intensity i and add it to the input image. Similar to "Brightness & Contrast" augmentation, the output image pixel intensity values may exceed the of 0 to 255. We clip values that fall outside of this range.

Blur. It is a color-based augmentation that applies box blur of a random kernel size to the image. The size of kernel, height and width is sampled from a range of 1 to 5.

Squeeze. It is a shape-based augmentation function that resizes the image with a random ratio. X & Y ratios are sampled from a range of 0.9 and 1.1, and the image is resized to new ratio.

Degrade. It is a color-based augmentation function that down-samples input image by a random factor and up-samples to original dimensions. It keeps aspect ratio same when down-sampling the image. Down-sample factor is sampled from a range of 1 to 2.

Rotate. It is a shape-based augmentation function that rotates image by a random angle. The angle is sampled from a range of -0.5 to 0.5.

Stretch. It is a shape-based augmentation function that stretches or compresses random chunks of the image. We divide image from left to right in chunks of random width ranging from 64px to 128px. Stretch or compression is only applied on x-axis (width). These chunks are then concatenate to get output image (Figs. 3 and 4).

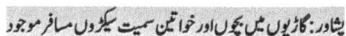

Fig. 3. Reference image to show different augmentations.

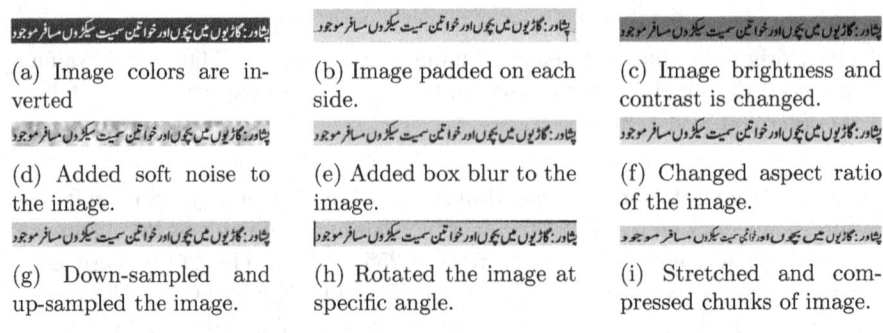

(a) Image colors are inverted

(b) Image padded on each side.

(c) Image brightness and contrast is changed.

(d) Added soft noise to the image.

(e) Added box blur to the image.

(f) Changed aspect ratio of the image.

(g) Down-sampled and up-sampled the image.

(h) Rotated the image at specific angle.

(i) Stretched and compressed chunks of image.

Fig. 4. Visualization of all augmentation functions respectively.

4.4   Evaluation Metrics

We trained and evaluated our model on both datasets. The evaluation is based on five different metrics. We compute character recognition space (CRR), word recognition rate (WRR) and line recognition rate (LRR) which are used to evaluate any text recognition model. We also compute CRR-WS and LRR-WS. WS referring to Without Spaces. In languages like Arabic or Urdu, some words can be understood without having a space between them. So space in such cases is optional. A space between two ligatures of a single word is optional as well. Word Recognition Rate (WRR) and Line Recognition Rate (LRR) are very sensitive on it. A single incorrectly predicted character can invalidate the entire line causing LRR to be zero. So also we compute CRR and LRR by ignoring spaces. We call them CRR-WS and LRR-WS.

Computing CRR and WRR without spaces are also important for the NUST-UTT dataset. The dataset originally contained ligatures separated by spaces. We fixed dataset by predicting spaces after words in a spaceless Urdu text using another model. The reconstructed dataset contains words separated by space. But the reconstruction is not perfect. So it is important to compute WRR-WS and LRR-WS metrics.

4.5   Hyper-parameter Selection

Before training the models, we performed a hyper-parameter search to look for the best parameters to train the model. We tested three optimizers with two different learning rates. We tested SGD, Adam and AdamW. For learning rate, we tested learning rates of 0.001 and 0.0003. Hyper-parameter search is a very time-consuming task, so we ran our simulations for 10 epochs and computed test metrics. This gives us a general idea about what parameters should we use.

4.6   Experiments Performed

We compared our model with Google Vision APIs as well [2]. Google Vision APIs provide Optical Character Recognition for both Urdu and Arabic language.

Urdu APIs are currently in experimental phase. We used Google Vision APIs to extract text from all the test images for both datasets. We compared the output of APIs with the ground-truth. We computed all five evaluation metrics for Google Vision APIs as well. We compared our model with the Google Vision APIs.

5   Results and Discussion

The hyper-parameter search gave us a general idea about parameters to choose for training. As Table 3 shows, the model optimized using AdamW optimizer with a learning rate of 0.0003 performed the best. SGD performed the worst. It could be because SGD takes much longer to optimize. SGD optimizer with a learning rate of 0.0003 did not learn anything in 10 epochs. After performing the hyper-parameter search we decided to use AdamW optimizer with a learning rate of 0.0003. We trained the models for 100 epochs and then further fine-tuned models with a learning rate of 0.0001 for 50 more epochs. Table 3 shows the final metrics of the model on each dataset.

Table 3. Results of the hyper-parameter search for Urdu Ticker Text dataset trained for 10 epochs. AdamW with a learning rate of 0.0003 performed the best.

Optimizer	Learning rate	CRR	WRR	LRR	CRR-WS	LRR-WS
AdamW	0.0003	**98.52%**	**88.37%**	**48.92%**	**99.73%**	**89.92%**
AdamW	0.001	98.39%	87.6%	46.4%	99.65%	86.57%
Adam	0.0003	98.48%	88.11%	48.66%	99.68%	87.91%
Adam	0.001	98.2%	86.01%	41.41%	99.65%	87.04%
SGD	0.0003	0%	0%	0%	0%	0%
SGD	0.001	42.16%	5.74%	0%	34.28%	0%

Once the training finished, we evaluated both models on their respective datasets. We also evaluate results for Google Vision APIs on both datasets as well. We compare each sample prediction with its ground turth and take mean of all individual metric values. The Table 4 shows the final results of our models compared with Google APIs. [2]

In Urdu Ticker Text Dataset, metrics without spaces are much better than normal ones. The LRR is only 48.92% while LRR-WS is 89.92%. Its shows that space inconsistencies in Urdu Ticker Text Dataset are huge as compared to the AcTiVComp20 dataset. The results show that OCR is difficult to do in the Arabic language as compared to Urdu. But it could be due to the incorrect labeling of digits, as mentioned in the Failure Cases section.

Finally, we analyze the failure cases of our model. We created a table containing ground truth and model predictions of all data samples in the test set.

Table 4. Final results on both datasets compared with Google Vision APIs.

Dataset	Method	CRR	WRR	LRR	CRR-WS	LRR-WS
NUST-UTT	Google Vision	93.71%	65.80%	8%	96.21%	40.07%
NUST-UTT	Ours	**98.66%**	**89.32%**	**52.05%**	**99.76%**	**90.84%**
AcTiVComp20	Google Vision	91.69%	76.62%	64.61%	92.95%	71.25%
AcTiVComp20	Ours	**96.82%**	**90.41%**	**76.78%**	**96.83%**	**79.38%**

We also computed all of the five metrics for individual data samples. We were able to identify major problems with our approach.

In AcTiVComp20 dataset, the model mostly fails to recognize numeric digits or special character. It is because the digits in Arabic transcription are stored in reverse order but do visually appear correct as shown in Fig. 5. This makes harder for RNN layers to recognize digits in AcTiVComp20 dataset. This issue can be solved by inverting sequence of digits of all numbers in the dataset or simply using Latin transcription instead.

Fig. 5. The diagram shows the problem we faced with Arabic transcription of AcTiV-Comp20 dataset [1]. The order in which the characters are stored were incorrect for digits in Arabic transcription. Visually the text below aligns with the image but it is not read correctly by computers. The arrows at the bottom show the order in which the label is actually stored.

In NUST-UTT dataset, the failure cases are mostly related to space inconsistencies. This can be observed from results. There is huge difference in normal and WS metrics. The LRR improved from 52.05% to 90.84% when the spaces were ignored.

6   Conclusion

In this paper, we proposed a technique to efficiently recognize Urdu and Arabic text from video. This includes the different types of augmentation functions, how to apply them, and how it helps regularize the model. We also explained the model architecture, the purpose of each part and its total parameters. Lastly, we explained the training and evaluation process for our model. In the training process, we explained how the CTC Loss function can be modified to achieve better results.

There is a lot of room for improvement. We only performed a hyper-parameter search for only 10 epochs. And the total search space was very limited. A better hyper-parameter search can help us improve the results. Results can be further improved by solving issues mentioned in the Failure Cases section. In the AcTiVComp20 dataset, we need to fix the order of digits for numbers. And for the NUST-UTT dataset, we need to fix the issue with space inconsistencies.

Acknowledgement. This work has been partially funded by the Higher Education Commission of Pakistan's grant for National Center of Artificial Intelligence (NCAI).

References

1. Competition on superimposed text detection and recognition in Arabic news video frames. https://diuf.unifr.ch/main/diva/AcTiVComp/index.html. Accessed 23 Feb 2021
2. Detect text in images — cloud vision API — google cloud. https://cloud.google.com/vision/docs/ocr. Accessed 26 May 2021
3. Ahmad, I., Wang, X., Li, R., Ahmed, M., Ullah, R.: Line and ligature segmentation of Urdu Nastaleeq text. IEEE Access **5**, 1–17 (2017)
4. Al-Wzwazy, H.: Handwritten digit recognition using convolutional neural networks. Int. J. Innovative Res. Comput. Commun. Eng. **4**, 1101–1106 (2016)
5. Bengio, Y., Simard, P., Frasconi, P.: Learning long-term dependencies with gradient descent is difficult. IEEE Trans. Neural Netw. Publ. IEEE Neural Netw. Council **5**, 157–66 (1994)
6. Graves, A., Fernández, S., Gomez, F., Schmidhuber, J.: Connectionist temporal classification: labelling unsegmented sequence data with recurrent neural networks, vol. 2006, pp. 369–376 (2006)
7. He, K., Zhang, X., Ren, S., Sun, J.: Deep residual learning for image recognition, vol. 7 (2015)
8. Informatik, F., Bengio, Y., Frasconi, P., Schmidhuber, J.: Gradient flow in recurrent nets: the difficulty of learning long-term dependencies. In: A Field Guide to Dynamical Recurrent Neural Networks (2003)
9. Krizhevsky, A., Sutskever, I., Hinton, G.: ImageNet classification with deep convolutional neural networks. In: Neural Information Processing Systems, vol. 25 (2012). https://doi.org/10.1145/3065386
10. Melnikoff, S.J., Quigley, S.F., Russell, M.J.: Implementing a hidden Markov Model speech recognition system in programmable logic. In: Brebner, G., Woods, R. (eds.) FPL 2001. LNCS, vol. 2147, pp. 81–90. Springer, Heidelberg (2001). https://doi.org/10.1007/3-540-44687-7_9
11. Mollah, A., Majumder, N., Basu, S., Nasipuri, M.: Design of an optical character recognition system for camera-based handheld devices. Int. J. Comput. Sci. Issues, vol. 8 (2011)
12. Rehman, A., Hussain, S.: Large scale font independent Urdu text recognition system (2020)

13. Sabbour, N., Shafait, F.: A segmentation free approach to Arabic and Urdu OCR. In: Proceedings of SPIE - The International Society for Optical Engineering, vol. 8658 (2013)

14. Shi, B., Bai, X., Yao, C.: An end-to-end trainable neural network for image-based sequence recognition and its application to scene text recognition. IEEE Trans. Pattern Anal. Mach. Intell. **39**(11), 2298–2304 (2016)

15. Ul-Hasan, A., Shafait, F., Breuel, T.: High-performance OCR for printed English and fraktur using lstm networks. In: Proceedings of the International Conference on Document Analysis and Recognition, ICDAR (2013)

16. Ul-Hasan, A., Ahmed, S., Rashid, S.F., Shafait, F., Breuel, T.: Offline printed Urdu Nastaleeq script recognition with bidirectional LSTM networks (2013)

17. Ur-Rehman, S., Tayyab, B., Naeem, M., Ul-Hasan, A., Shafait, F.: A multi-faceted OCR framework for artificial Urdu news ticker text recognition. In: 13th IAPR International Workshop on Document Analysis Systems, DAS 2018, Vienna, Austria, 24–27 April 2018, pp. 211–216. IEEE Computer Society (2018)

18. Xie, Z., Sun, Z., Jin, L., Feng, Z., Zhang, S.: Fully convolutional recurrent network for handwritten Chinese text recognition (2016)

19. Yanikoglu, B., Sandon, P.: Off-line cursive handwriting recognition using style parameters (1970)

20. Yuan, T.L., Zhu, Z., Xu, K., Li, C.J., Hu, S.M.: Chinese text in the wild (2018)

21. Zayene, O., Hennebert, J., Ingold, R., Essoukri Ben Amara, N.: ICDAR 2017 competition on Arabic text detection and recognition in multi-resolution video frames, pp. 1460–1465 (2017)

ASAR 2021 Online Arabic Writer Identification Competition

Thameur Dhieb[1,2(✉)] , Houcine Boubaker[2] , Sourour Njah[2,3] ,
Mounir Ben Ayed[2,4] , and Adel M. Alimi[2,5]

[1] University of Sousse, ISITCom, 4011 Sousse, Tunisia
[2] REsearch Groups in Intelligent Machines (REGIM Lab), National Engineering School of Sfax
(ENIS), University of Sfax, BP 1173, 3038 Sfax, Tunisia
{thameur.dhieb,houcine.boubaker,sourour.njah,mounir.benayed,
adel.alimi}@regim.usf.tn
[3] High Institute of Commerce of Sfax, University of Sfax, Sfax, Tunisia
[4] Computer Sciences and Communication Department, Faculty of Sciences of Sfax,
University of Sfax, Sfax, Tunisia
[5] Department of Electrical and Electronic Engineering Science, Faculty of Engineering and the
Built Environment, University of Johannesburg, Johannesburg, South Africa

Abstract. This paper describes the online Arabic writer identification competition held at ICDAR 2021 Workshop on Arabic and derived Script Analysis and Recognition (ASAR 2021, 4th edition). This first competition of online writer identification uses the ADAB database with online Arabic handwritten words which represent the Tunisian town names. Four systems are participating in the competition. The systems are evaluated on test data that are unknown to the participants. The systems are compared based on the writer identification rate. A detailed description of the systems and the results achieved are presented.

Keywords: Writer identification · Online handwriting · Arabic script

1 Introduction

The objective of writer identification is to retrieve the writer of handwritten documents through a list of writers known by the system [1]. The idea to automate writer identification came due to its variety of applications such as banking, access control and forensic document examination [2, 3]. However, there are several challenges to achieve effective writer identification systems including the inter-writer variability and the intra-writer variability. In fact, the psychological and neurophysiological characteristics of an individual may influence the writer's handwriting style [4].

Writer identification can be categorized into two different methods based on the presentation of the data to the system. The first method, offline, uses a scanned image of the handwritten text as input for the identification steps. The second method, online, uses data that comprises information about the pen movement over time [5].

© Springer Nature Switzerland AG 2021
E. H. Barney Smith and U. Pal (Eds.): ICDAR 2021 Workshops, LNCS 12916, pp. 353–365, 2021.
https://doi.org/10.1007/978-3-030-86198-8_25

The task of writer identification from online handwriting has attained research interest recently, with most of the advancements done especially in the last decade [6, 7]. But, the field of research on online writer identification from Arabic handwriting remains yet an under-exploited compared to Latin script [8]. Bulacu et al. say that the writer identification on Arabic script seems to be more difficult than on Western script [9].

Indeed, Arabic letters are used in many languages like Arabic, Malay, Farsi, Kurdish, Urdu and Pashto. However, the work on online writer identification from Arabic handwriting is complicated and has great difficulties due to different factors:

- Arabic script is naturally cursive and written from right to left [10].
- Letters' form is context-sensitive, can have one to four different shapes depending on their position in a word (i.e., initial, medial, final, or isolated).
- Presence of ligatures in Arabic script.
- Most Arabic letters contain dots beside the letter body like "خ"which consists of "ح"letter body and one dot above it.
- The diacritic signs like dumma ('), fatha (-) or chadda (ω) have a heavy presence in the Arabic language.

The lack of works to online writer identification from Arabic handwriting motivate us to organize the first online Arabic writer identification competition.

Our paper is written as follows: Sect. 2 describes the training and the test data extracted from the ADAB database. Section 3 presents detailed descriptions of the participating systems. Section 4 is devoted to results and discussion. Finally, the paper ends with some concluding remarks.

2   ADAB Database

ADAB Database (The Arabic handwriting DataBase) consists of more than 33000 Arabic words which are Tunisian town and village names contributed by 166 writers. This database is made in cooperation between Institut fuer Nachrichtentechnik (IfN) and Research Groups in Intelligent Machines, University of Sfax, Tunisia to advance the research on online Arabic handwritten text [11–15]. ADAB database is freely available for non-commercial research (https://ieee-dataport.org/open-access/adab-database).

In ADAB database, the writers do not contribute the same number of words. To give the equal chance to all writers, we selected a sub-database detailed as follows:

- **The training data** consists of 3000 online handwritten word documents which have been written by 60 different writers. Each writer has contributed 50 handwritten words, while the validation data contains 600 online handwritten words document of the 60 different writers. The data extracted from ADAB Database (An InkML file including trajectory information).
- **The test data** consists of 170 tests (each test contains 10 online handwritten word documents written by one of the 60 writers).

Figure 1 shows some examples of samples performed by different writers from ADAB database.

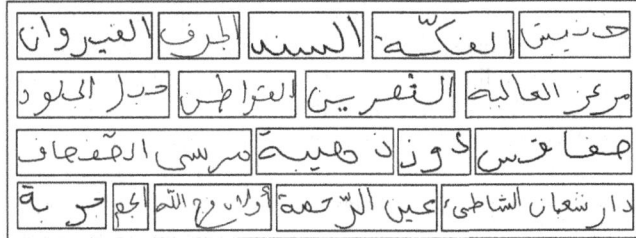

Fig. 1. Examples of Arabic words handwritten by different writers selected from ADAB Database

3 Participating Systems

The REsearch Groups in Intelligent Machines (REGIM Lab) at the National Engineering School of Sfax (ENIS), University of Sfax, Tunisia participated with four systems submitted by Thameur Dhieb, Houcine Boubaker, Sourour Njah, Mounir Ben Ayed and Adel M. Alimi which are:

1) Beta-Elliptic Model system (BEM system) [16].
2) Combination of Beta-Elliptic Model with Fuzzy Elementary Perceptual Codes system (BEM_FEPC system) [17].
3) Extended Beta-Elliptic Model system (EBEM system) [18].
4) Combination of Extended Beta-Elliptic Model with Fuzzy Elementary Perceptual Codes system (EBEM_FEPC system) [19].

Figure 2 provides an overview of their systems.

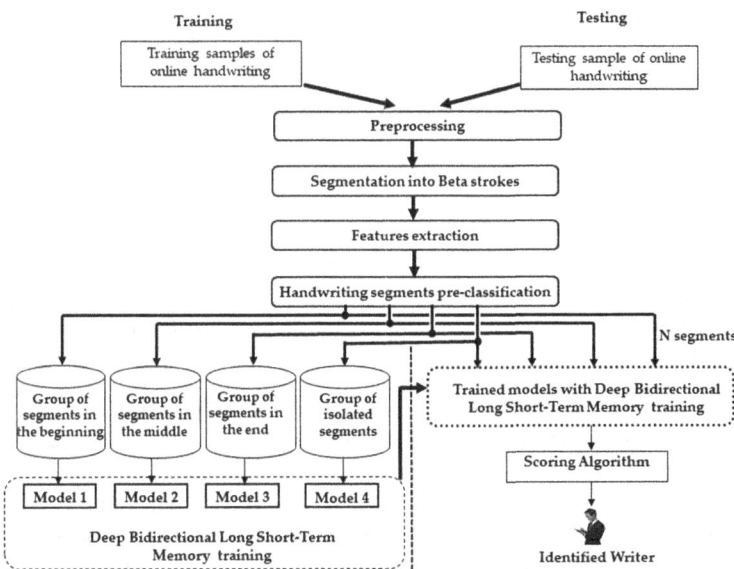

Fig. 2. The general architecture of the proposed systems

The systems have six steps: 1) Preprocessing, 2) segmentation of online handwriting samples into Beta strokes, 3) features extraction, 4) handwriting segments preclassification, 5) Deep Bidirectional Long Short-Term Memory classification and 6) identification process. The four systems differ only in step 2) segmentation of online handwriting samples into Beta strokes and step 3) features extraction.

- **Preprocessing:** For preprocessing step, a Chebyshev type II low pass filter [20] is used with a cut-off frequency of fcut = 12 Hz to eliminate the noise caused by spatial and temporal sampling.
- **Segmentation into Bea strokes:** To segment the online handwritings which are characterized by the sequence $(x(t), y(t))$ representing the trajectory coordinates in time, the curvilinear velocity profile Vσ(t) computed by the following equation:

$$V_\sigma(t) = \sqrt{\left(\frac{dx(t)}{dt}\right)^2 + \left(\frac{dy(t)}{dt}\right)^2} \tag{1}$$

Three types of points are detected, presented in Fig. 3, from the curvilinear velocity profile $V_\sigma(t)$ which are:

- The velocity local minimum representing the local minima of the curvilinear velocity $V_\sigma(t)$;
- The velocity local maximum representing the local maxima of the curvilinear velocity $V_\sigma(t)$;
- The double inflexion point representing the variation of speed.

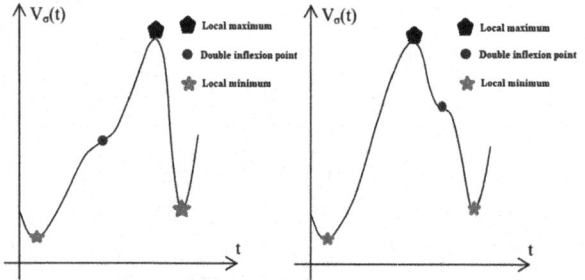

Fig. 3. Detection of the local extrema from the curvilinear velocity profile

After that, the curvilinear velocity profile is split into Beta strokes using the local extrema according to the system used in features extraction:

- For BEM system, the curvilinear velocity profile is split into Beta strokes limited between two local extrema (maximum, minimum, or double inflexion point);
- For EBEM system, the curvilinear velocity profile is split into Beta strokes limited between two successive local minimums or double inflexion points.

– **Features extraction:** The Beta-Elliptic Model (BEM), the Extended Beta-Elliptic Model (EBEM) and the Fuzzy Elementary Perceptual Codes (FEPC) will be presented as follow:

The Beta-Elliptic Model (BEM)

The Beta-Elliptic Model is used in several areas of research like the study of the effect of age on hand movement [21] and handwriting recognition [22–29]. The Beta-Elliptic Model is characterized by a description combining dynamic and static profiles. In the dynamic profile, each trajectory stroke corresponds to the generation of one Beta signal computed by the following equation:

$$pulse\ \beta(K, t, q, p, t_0, t_1) = \begin{cases} K.\left(\frac{t-t_0}{t_c-t_0}\right)^p.\left(\frac{t_1-t}{t_1-t_c}\right)^q & if\ t \in [t_0, t_1] \\ 0, & elsewhere \end{cases} \tag{2}$$

Where t_0 is the starting time of Beta function, t_1 is the ending times of Beta function, t_c is the instant when the Beta function reaches its maximum value K named Beta impulse amplitude, and p, q are intermediate parameters.

The generation of a velocity model is represented by the following expression:

$$V_\sigma(t) = \sum_{i=1}^{n} V_i(t - t_{0i}) \approx \sum_{i=1}^{n} pulse\ \beta_i(K_i, t, q_i, p_i, t_{0i}, t_{1i}) = V_r(t) \tag{3}$$

In the static profile, each trajectory stroke can be assimilated to an elliptic arc characterized by four parameters (a, b, θ, and θp).

The retained number of Beta strokes per segment in the Beta-Elliptic Model which maximizes the identification rate is $N = 4$. Thus, a vector of 10 features models each Beta stroke as detailed in Table 1.

Table 1. Features extraction from Beta-Elliptic Model

Feature	Parameter and formula	Signification
$f1$	$\Delta t = (t_1 - t_0)$	Beta impulse duration
$f2$	$RapT_c = \frac{t_c-t_0}{\Delta t}$	Beta impulse asymmetry report or culminating time
$f3$	p	Beta shape parameter of the neuromuscular impulse
$f4$	k	Beta impulse amplitude
$f5$	$\frac{k_i}{k_{i+1}}$	Successive Beta impulse amplitude report
$f6$	a	Ellipse major axis half-length
$f7$	b	Ellipse small axis half-length
$f8$	θ	Ellipse major axis inclination angle
$f9$	θ_p	Inclination angle of the tangents at the stroke endpoint M2
$f10$	POS_STROKE	Stroke position in the pseudo-word

The Extended Beta-Elliptic Model (EBEM)

The neuromuscular action is delimited in the interval time $[t_0, t_1]$ between two successive local minimums of velocity or double inflexion points. The curvilinear velocity inside this interval is divided into two components:

1) An impulsive component $V_{Imp}(t)$ in which all the implied articulations contribute as presented in (4):

$$V_{Imp}(t) = K \cdot \left(\frac{t - t_0}{t_C - t_0}\right)^p \cdot \left(\frac{t_1 - t}{t_1 - t_C}\right)^q \tag{4}$$

2) A continuous training component that allows the continuous passage from one segment of the layout to another with a not null velocity as calculated through the following formula:

$$V_{Tra}(t) = A \cdot \left[\frac{(t - t_0)^3}{3} - \frac{(t_1 - t_0) \cdot (t - t_0)^2}{2}\right] + V_i \tag{5}$$

Where:

$$A = -6 \cdot \frac{V_f - V_i}{(t_1 - t_0)^3} \tag{6}$$

Where:

- t_0 and t_1 are respectively the starting and the ending times of the Beta function.
- V_i and V_{fin} are respectively the velocity at the starting and the velocity at the ending time of the Beta function.

The reconstituted curvilinear velocity is the sum of its components as expressed in (7):

$$V_R(t) = V_{Imp}(t) + V_{Tra}(t) \tag{7}$$

The retained number of Beta strokes per segment in the Extended Beta-Elliptic Model which maximizes the identification rate is $N = 2$. Thus, a vector of 14 features models each Beta stroke as detailed in Table 2.

Table2. Features extraction from the Extended Beta-Elliptic Model

Feature	Parameter and formula	Signification
$f1$	$t_1 - t_0$	Beta impulse duration
$f2$	$\frac{t_c - t_0}{t_1 - t_0}$	Beta impulse asymmetry report or culminating time
$f3$	P	Beta shape parameter of the neuromuscular impulse
$f4$	K	Beta impulse amplitude
$f5$	V_i	Initial training velocity amplitude of the current stroke
$f6$	V_{fin}	Final training velocity amplitude of the current stroke
$f7$	$\frac{k_i}{training}$	Beta impulse amplitude report with respect to the medium value of the training component
$f8$	a_1	Major axis half-length of the ellipses supporting the first arc
$f9$	b_1	Half-length of the small axis of the ellipse including the first arc
$f10$	b_2	Half-length of the small axis of the ellipse including the second arc
$f11$	θ_{p1}	Inclination angle of the tangents at the stroke endpoint M1
$f12$	θ	Ellipse major axis inclination angle
$f13$	θ_{p2}	Inclination angle of the tangents at the stroke endpoint M3
$f14$	Stroke position	Stroke position in the pseudo-word

The Fuzzy Elementary Perceptual Codes (FEPC)

According to several theories of visual perception [30–33], each Beta stroke is assigned to one of the four types of Elementary Perceptual Codes (EPC) presented in Table 3 using the ellipse major axis inclination angle.

Table 3. Form of EPC

EPC	Form
EPC_1: Valley	—
EPC_2: Left oblique shaft	/
EPC_3: Shaft	│
EPC_4: Right oblique shaft	\

The trigonometric circle is split into 8 regions corresponding to the EPC as shown in Fig. 4.

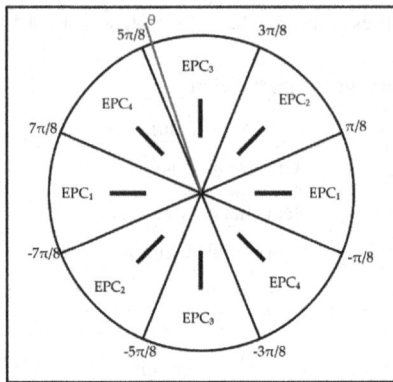

Fig. 4. Different EPC presented on the trigonometric circle

Yet, EPC suffers from vagueness and uncertainty due to diverse constraints like hand disorder. To solve this problem, the fuzzy logic theory is adopted to attribute a membership degree for each EPC. Consequently, a vector of 4 features is obtained for each Beta stroke as shown in Table 4.

Table 4. Features extraction generated by the Fuzzy Elementary Perceptual Codes

Feature	Parameter	Signification
$f1$	$FEPC_1$	Membership degree of EPC1
$f2$	$FEPC_2$	Membership degree of EPC2
$f3$	$FEPC_3$	Membership degree of EPC3
$f4$	$FEPC_4$	Membership degree of EPC4

– **Handwriting segments pre-classification:** This phase intends to pre-classify all handwriting segments into groups according to their position in the pseudo-words. The latter refers to a hand-drawn shape comprised between pen up and pen down. The trajectory length and shape for the same word may differ from one person to another depending on the style of handwriting. To take into consideration all the segments with the details of their position in the pseudo-word and to have more chance to determine the author's identity of a handwriting sample amongst a set of known writers, each segment is pre-classified by sliding on the pen trajectory to one of the four following groups:
  1) Group of segments in the beginning;
  2) Group of segments in the middle;
  3) Group of segments in the end;
  4) Group of isolated segments.
– **Deep Bidirectional Long Short-Term Memory classification:** Four models are created for the four groups which are model of segments in the beginning, model of

segments in the middle, model of segments in the end and model of isolated segments using Deep Bidirectional Long Short-Term Memory. For each of the four models, the input vectors $[S_1, S_2, ..., S_k]$ where S_k represents the features vector for segment k feed to the forward LSTM layer. Thus, a hidden state sequence of $[h_1, h_2, ..., h_k]$ is achieved. Meantime, the input vectors $[S_k, S_{k-1}, ..., S_2, S_1]$ feed to the backward LSTM layer, another hidden state sequence of $[h'_1, h'_2, .., h'_k]$ achieved. Hence, h_k and h'_k can be viewed respectively as summaries of the input sequence in forward and backward directions. After that, the outputs of the two LSTM networks are concatenated together and fed into the fully connected layer. Finally, a SoftMax layer is performed for final classification. The architecture of Deep Bidirectional Long Short-Term Memory is presented in Fig. 5.

– **Identification step:** A scoring algorithm to identify the writer is proposed. The decisions made by the different networks are aggregated and scores are given to each writer.

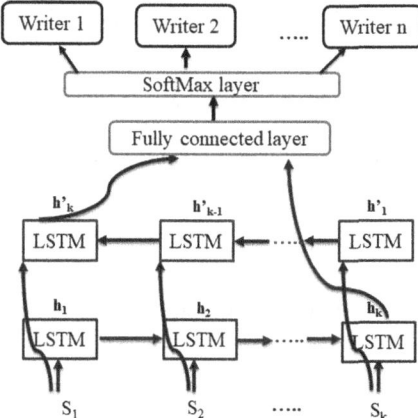

Fig. 5. Architecture of Deep Bidirectional Long Short-Term Memory model for online writer identification

Then the writers are ranked according to their obtained sum of output affectation rate T_{affect}. The writer with the maximum output sum obtained is the final answer as expressed in (8):

$$Iw = \underset{k \in \{1,...,Nw\}}{ArgMax} \left\{ \sum_{i=1}^{S} T_{affect}(X_i, k) \right\} \tag{8}$$

where:

– Iw: index number of the identified writer.
– k: index number of the kth suspect writer.
– Nw: number of suspect writers.

- S: number of segments recuperated from the tested handwriting data.
- T_{affect}: rate of affectation of the ith tested segment to the kth writer, as represented in Eq. (9):

$$T_{affect}(X_i, k) = \left[net_j(X_i)\right]_k \qquad (9)$$

where:

- Xi: the feature vector of the ith tested segment assigned to the jth group.
- net_j: the output of the jth trained model to which the feature vector X_i is assigned.

4   Results and Discussion

Participants provide the score of each test to each writer. The writer with the maximum score is the final answer. Results is evaluated using the writer identification rate in terms of Top1, Top2 and Top5. knowing that TopN indicates if the correct writer is found in the N first writers ranked results. The experimental results achieved are reported in Table 5.

Table 5. Experimental results

System	Writer identification rate (%)		
	Top 1	Top 2	Top 5
BEM system	95.29	97.65	98.82
BEM_FEPC system	96.47	97.65	99.41
EBEM system	95.88	98.24	100
EBEM_FEPC system	97.65	100	100

From experimental results, we observe that firstly, the features extracted by Beta-Elliptic Model (BEM system) and by the Extended Beta-Elliptic Model (EBEM system) are effective to discriminate the styles of handwriting and identify the writer. Secondly, we remark that the integration of the Fuzzy Elementary Perceptual Codes (FEPC) enhances the writer's identification rate as presented in Fig. 6. Finally, the combination of the Extended Beta-Elliptic Model and the Fuzzy Elementary Perceptual Codes (EBEM_FEPC system) has yielded the best result with an identification rate of 97.65% compared to the other systems.

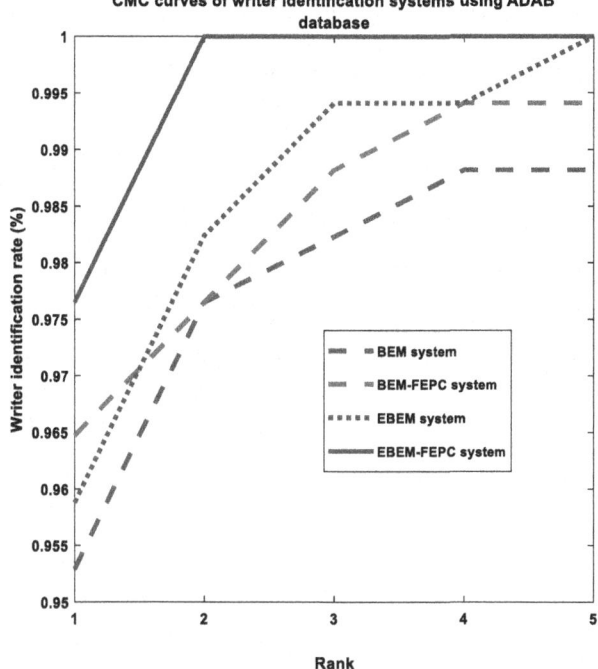

Fig. 6. CMC curves of writer identification systems using ADAB database

5   Conclusion

Four systems have participated in this ASAR 2021 Online Arabic Writer Identification Competition. All systems are based on RNNs with Deep Bidirectional Long Short-Term Memory model that show a very high writer identification rate. The EBEM_FEPC system using the combination of the Extended Beta-Elliptic Model and the Fuzzy Elementary Perceptual Codes is the winner of this first competition. This field remains a challenging task even though the latest improvements of wrier identification methods and systems are very promising.

Acknowledgments. The research leading to these results has received funding from the Ministry of Higher Education and Scientific Research of Tunisia under the grant agreement number LR11ES4.

References

1. Dhieb, T., Ouarda, W., Boubaker, H., Halima, M.B., Alimi, A.M.: Online Arabic writer identification based on Beta-Elliptic model. In: 2015 15th International Conference on Intelligent Systems Design and Applications (ISDA), pp. 74–79 (2015)
2. Dhieb, T., Ouarda, W., Boubaker, H., Alimi, A.M.: Deep neural network for online writer identification using Beta-Elliptic model. In: 2016 International Joint Conference on Neural Networks (IJCNN), pp. 1863–1870 (2016)

3. Dhieb, T., Ouarda, W., Boubaker, H., Alimi, A.M.: Beta-Elliptic model for writer identification from online Arabic handwriting. J. Inf. Assur. Secur. **11**, 263–272 (2016)
4. Dhieb, T., Njah, S., Boubaker, H., Ouarda, W., Ayed, M.B., Alimi, A.M.: An extended Beta-Elliptic model and fuzzy elementary perceptual codes for online multilingual writer identification using deep neural network. arXiv:1804.05661 (2018)
5. BabaAli, B.: Online writer identification using statistical modeling-based feature embedding. Soft Comput. **25**(14), 9639–9649 (2021). https://doi.org/10.1007/s00500-021-05729-x
6. Chen, Z., Yu, H.-X., Wu, A., Zheng, W.-S.: Letter-level online writer identification. Int. J. Comput. Vis. **129**(5), 1394–1409 (2021). https://doi.org/10.1007/s11263-020-01414-y
7. Venugopal, V., Sundaram, S.: Online writer identification system using adaptive sparse representation framework. IET Biometrics **9**, 126–133 (2020)
8. Awaida, S.M., Mahmoud, S.A.: State of the art in off-line writer identification of handwritten text and survey of writer identification of Arabic text. ERR **7**, 445–463 (2012)
9. Bulacu, M., Schomaker, L., Brink, A.: Text-independent writer identification and verification on offline Arabic handwriting. In: Ninth International Conference on Document Analysis and Recognition (ICDAR 2007), pp. 769–773 (2007)
10. Hassen, H., Al-Maadeed, S.: Arabic handwriting recognition using sequential minimal optimization. In: 2017 1st International Workshop on Arabic Script Analysis and Recognition (ASAR), pp. 79–84 (2017)
11. ElAbed, H., Märgner, V., Kherallah, M., Alimi, A.M.: ICDAR 2009 Online Arabic Handwriting Recognition Competition. In: 2009 10th International Conference on Document Analysis and Recognition, pp. 1388–1392 (2009)
12. ElAbed, H., Kherallah, M., Märgner, V., Alimi, A.M.: On-line Arabic handwriting recognition competition: ADAB database and participating systems. Int. J. Doc. Anal. Recogn. **14**, 15–23 (2011)
13. Kherallah, M., Tagougui, N., Alimi, A.M., Abed, H.E., Margner, V.: Online Arabic handwriting recognition competition. In: 2011 International Conference on Document Analysis and Recognition, pp. 1454–1458 (2011)
14. Boubaker, H., Elbaati, A., Tagougui, N., ElAbed, H., Kherallah, M., Alimi, A.M.: Online Arabic databases and applications. In: Märgner, V., El Abed, H. (eds.) Guide to OCR for Arabic Scripts, pp. 541–557. Springer, London (2012). https://doi.org/10.1007/978-1-4471-4072-6_22
15. Tagougui, N., Kherallah, M., Alimi, A.M.: Online Arabic handwriting recognition: a survey. Int. J. Doc. Anal. Recogn. (IJDAR) **3**, 209–226 (2013)
16. Dhieb, T., Boubaker, H., Ouarda, W., Ayed, M.B., Alimi, A.M.: Deep bidirectional long short-term memory for online Arabic writer identification based on Beta-Elliptic model. In: 2019 International Conference on Document Analysis and Recognition Workshops (ICDARW), pp. 35–40 (2019)
17. Dhieb, T., Njah, S., Boubaker, H., Ouarda, W., Ayed, M.B., Alimi, A.M.: An online writer identification system based on Beta-Elliptic model and fuzzy elementary perceptual codes. arXiv preprint arXiv:1804.05661 (2018)
18. Dhieb, T., Njah, S., Boubaker, H., Ouarda, W., Ben Ayed, M., Alimi, A.M.: Towards a novel biometric system for forensic document examination. Comput. Secur. **97**, 101973 (2020)
19. Dhieb, T., Boubaker, H., Ouarda, W., Njah, S., Ben Ayed, M., Alimi, A.M.: Deep bidirectional long short-term memory for online multilingual writer identification based on an extended Beta-Elliptic model and fuzzy elementary perceptual codes. Multimedia Tools Appl. **80**(9), 14075–14100 (2021). https://doi.org/10.1007/s11042-020-10412-8
20. Paarmann, L.D. (ed.): Chebyshev type II filters. In: Design and Analysis of Analog Filters: A Signal Processing Perspective, pp. 155–176. Springer, Boston (2001). https://doi.org/10.1007/0-306-48012-3_5

21. Dhieb, T., Rezzoug, N., Boubaker, H., Gorce, P., Alimi, A.M.: Effect of age on hand drawing movement kinematics. Comput. Methods Biomech. Biomed. Engin. **22**, S188–S190 (2019)
22. Boubaker, H., Chaabouni, A., Kherallah, M., Alimi, A.M., Abed, H.E.: Fuzzy segmentation and graphemes modeling for online Arabic handwriting recognition. In: 2010 12th International Conference on Frontiers in Handwriting Recognition, pp. 695–700 (2010)
23. Boubaker, H., Kherallah, M., Alimi, A.M.: Optimization of the beta–elliptic model features estimation. In: 16th International Conference of Graphonomix Society (IGS 2013), pp. 151–154. International Graphonomix Society (2013)
24. Hamdi, Y., Boubaker, H., Dhieb, T., Elbaati, A., Alimi, A.M.: Hybrid DBLSTM-SVM Based Beta-Elliptic-CNN models for online Arabic characters recognition. In: 2019 International Conference on Document Analysis and Recognition (ICDAR), pp. 545–550 (2019)
25. Akouaydi, H., et al.: Neural architecture based on fuzzy perceptual representation for online multilingual handwriting recognition. arXiv:1908.00634 (2019)
26. Rabhi, B., Elbaati, A., Hamdi, Y., Alimi, A.M.: Handwriting recognition based on temporal order restored by the end-to-end system. In: 2019 International Conference on Document Analysis and Recognition (ICDAR), pp. 1231–1236 (2019)
27. Akouaydi, H., Njah, S., Ouarda, W., Samet, A., Zaied, M., Alimi, A.M.: Convolutional neural networks for online arabic characters recognition with Beta-Elliptic knowledge domain. In: 2019 International Conference on Document Analysis and Recognition Workshops (ICDARW), pp. 41–46 (2019)
28. Rabhi, B., Elbaati, A., Boubaker, H., Hamdi, Y., Hussain, A., Alimi, A.: Temporal order and pen velocity recovery for character handwriting based on sequence-to-sequence with attention mode (2021). https://doi.org/10.36227/techrxiv.13902650.v1
29. Hamdi, Y., Boubaker, H., Alimi, A.M.: Data augmentation using geometric, frequency, and beta modeling approaches for improving multi-lingual online handwriting recognition. IJDAR (2021). https://doi.org/10.1007/s10032-021-00376-2
30. Njah, S., Bezine, H., Alimi, A.M.: Linguistic interpretation for on-line handwriting using PerTOHS theory. In: 16th International Graphonomics Society (IGS), pp. 175–178 (2013)
31. Njah, S., Ltaief, M., Bezine, H., Alimi, A.M.: The PerTOHS theory for on line handwriting segmentation (2012)
32. Njah, S., Bezine, H., Alimi, A.M.: On-line Arabic handwriting segmentation via perceptual codes: application to MAYASTROUN database. In: Eighth International Multi-Conference on Systems, Signals Devices, pp. 1–5 (2011)
33. Gordon, I.E.: Theories of Visual Perception. Psychology Press, London (2004)

ASAR 2021 Competition on Online Signal Restoration Using Arabic Handwriting Dhad Dataset

Besma Rabhi[1]([⊠]) (iD), Abdelkarim Elbaati[2] (iD), Tarek M. Hamdani[1] (iD), and Adel M. Alimi[1,3] (iD)

[1] University of Sfax, National Engineering School of Sfax, REGIM-Lab.: REsearch Groups in Intelligent Machines, LR11ES48, 3038 Sfax, Tunisia
{besma.rabhi,adel.alimi}@regim.usf.tn
[2] University of Monastir, Higher Institute of Applied Sciences and Technology of Mahdia, REGIM Lab.: REsearch Groups in Intelligent Machines, LR11ES48, 5121 Mahdia, Tunisia
[3] Department of Electrical and Electronic Engineering Science, Faculty of Engineering and the Built Environment, University of Johannesburg, Johannesburg, South Africa

Abstract. Stroke reconstruction from offline handwriting is an important research field. This article presents Online Signal Restoration (OSR) using Arabic Handwriting Dhad Dataset competition organized at ASAR 2021. The goal of this competition is to collect different systems and compare recent advances in online handwriting recovery. This competition has attracted 4 teams from Regim lab. The participating systems were evaluated on known data and tested on unknown dataset. The evaluation metrics are based on Root Mean Squared Error (RMSE), Euclidean Distance (ED) and visual comparison of the recovered velocity. This paper details the proposed competition by describing the used dataset, the participating systems and their effectiveness based on the evaluation metrics.

Keywords: Temporal order reconstruction · Pen velocity recovery · Dhad-database

1 Introduction

Recently, handwriting recovery research field have become increasingly important [1–3]. It has many applications such as a) signature verification [5], where the investigators convert an offline signature into its counterpart online which is passed to the online verification mode, b) offline handwriting recognition, where the temporal order script is reconstructed from handwriting images and then recognized based on online recognition system [3]. Temporal order reconstruction from the offline handwriting is a great challenge and a complicated task for researchers and developers, since dynamic information such as the temporal order of the pen trajectory can be useful for many systems, such as handwriting recognition system [9, 12, 16], writer identification [7, 8] and signature verification to help investigators detect falsified signatures and to simplify the document analyses task. Thus, the performance of online systems is more encouraging than offline

© Springer Nature Switzerland AG 2021
E. H. Barney Smith and U. Pal (Eds.): ICDAR 2021 Workshops, LNCS 12916, pp. 366–378, 2021.
https://doi.org/10.1007/978-3-030-86198-8_26

systems. An overview of these methods is presented in [3, 4, 6, 13–15, 17–19]. Online Signal Restoration (OSR) is a real challenge depending to many factors such as limited assumptions about start/end points, ambiguous zone and double trace segments which give rise to a difficult decision for obtaining the right trajectory. In addition, the recovered signal must be described with dynamic features such as: the temporal order and the pen velocity. The challenge of the proposed competition is to recover the online script of Arabic Handwriting character and digit, which is characterized by the pen velocity and the non-normalized number of points.

In the next section, the new Dhad-dataset and the competition task are described. Section 3 describes the participating systems. Section 4 presents the tests and analyses of the obtained results.

2 Dataset and Task Description

A perceivable Arabic handwritten database is an important resource for handwriting recovery context. The database Dhad was created by Abdelkarim Elbaati. It was constructed to advance the investigation in online handwriting recovery field and encourage the development of various frameworks. We present our comprehensive Arabic online and offline Handwritten database (Dhad) consisting of 389 Handwritten samples written by 40 writers from Tunisia. This dataset contains Arabic digits, isolated characters and words. The size is small for training deep learning models, but, participants were welcome to increase the size using data-augmentation of their choice [10]. The dataset Dhad contains 2 samples for each isolated Arabic digits, characters and word, presented as

Fig. 1. Dhad's collection tool

offline image of the writing and its corresponding online signal. This dataset is collected by an android application installed on Smartphone. The collection step begins when the writer clicks on enter bottom. Then, the application shows at the top of the interface the Arabic words, characters and digits as it shown in Fig. 1. The obtained samples will be automatically saved to each appropriated files as offline handwriting (.tif) and its counterpart online (.txt). Figure 2 and Fig. 3 show two examples of the digit <<8>> and the character <<Dha>>. Table 1 presents the number of training, testing and validation samples. The training database digits and characters will be sent to participants in offline and online forms. In general, the first step in handwriting reconstruction is the preprocessing step. This process include image normalization, knowing that all images have the same size (64 × 64). For the sake of simplicity, the online signal will be used without pen up/down, Fig. 5 shows the distribution of the recovered X and Y coordinates on the Arabic Dhad dataset relative to the number of iteration steps. Since we normalized the points to (64 * 64), it can be deduced that the coordinate values are limited in the range (64 * 64). Figure 4 shows an example of the offline and its counterpart online Arabic letter <<noun>>. In actual fact, the online script is a set of non-equidistant points stored over the writing process. Naturally, the pen velocity decreases at some zones such as the terminal points and the ambiguous curve. Those points are introduced as two-dimensional matrix with pen velocity feature.

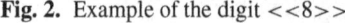

Fig. 2. Example of the digit <<8>>

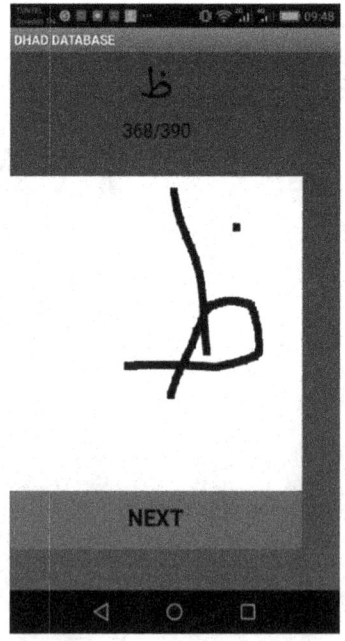

Fig. 3. Example of the Arabic letter <<dha>>

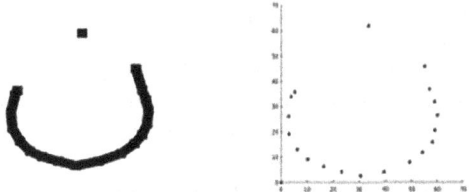

Fig. 4. Example of the offline and its corresponding online Arabic letter <<noun>>

Table 1. Dhad Dataset details.

Scripts	Training	Testing	Validation
Dhad digits	23340	3890	3890
Dhad characters	23340	3890	3890
Dhad words	23340	3890	3890

To the best of our knowledge, existing works have not recovered a signal character-ized by pen velocity and with adaptable number of points. However, in [4], the authors generated a significant script with trajectory order and velocity feature at the same time but with a fixed number of points (50 points). By way of contrast, participants are invited to recover a signal with trajectory order and velocity feature without fixed number of points.

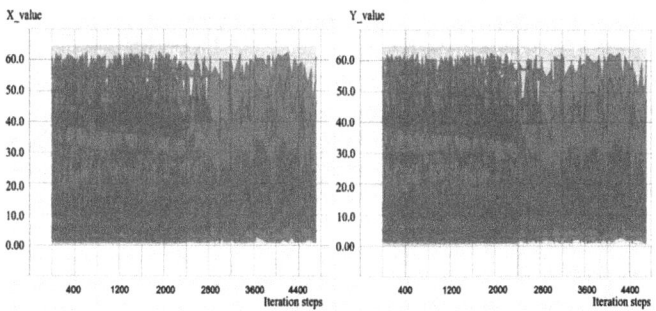

Fig. 5. Region distributions of the recovered X and Y coordinate values for the Arabic validation data.

3   Participating Methods

This section describes the architecture-based of different participating systems. In this competition, 4 teams participated.

3.1 RHM (REGIM-Heuristic-Method)

The RHM system was submitted by Abdelkarim Elbaati, member in REsearch Groups in Intelligent Machines at national school of engineers, university of sfax from Tunisia. RHM is a traditional system for Arabic handwriting recovery.

In fact, surrounding areas of online handwriting reconstruction could be divided into two main groups: local and global methods. The first search method is based on the selection of the smoothest trajectory at each ambiguous zone. The goal of the second method is to create a graph model of the skeleton images and find the optimal path based on a search technique. The RHM system has merged these two search methods together. It was based on the graph method to represent the input handwriting image as a set of segments. The Genetic Algorithm (GA) was used to select the smoothest path of segments. This system generates equidistant points.

3.2 RDL (REGIM-DL-LSTM)

The RDL system was submitted by Besma Rabhi, Abdelkarim Elbaati and Adel M. Alimi, from REsearch Groups in Intelligent Machines at national school of engineers, university of Sfax from Tunisia. This system is inspired from [11] which used the end-to-end model for handwriting recovery. Their proposal is based on an encoder-decoder with LSTM neural network. Figure 6 shows an overview of the participating system RDL. Here, a convolutional neural network (CNN) followed by LSTM is used as encoder module. It receives as input the offline Arabic handwriting and generates a set of points. The CNN is used to extract a set of features from the image. Table 2 depicts the proposed

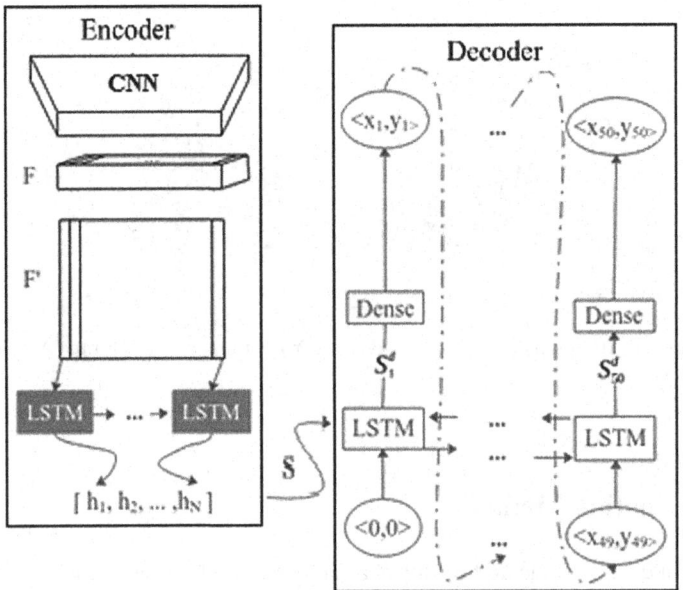

Fig. 6. Architecture of the system RDL.

CNN parameters configuration. The encoder processes the final feature map of CNN to produce a single hidden state of all the input. The BLSTM NN is used as the decoder and its state is initialized by the last encoder state. This system generates 50 equidistant points.

Table 2. CNN Configuration for six layers.

Systems	Parameters	Values for each layer
RDL	Conv. filters	$64 - 128 - 128 - 256 - 256 - 512$
	MaxPooling	$(2,2) - (2,1) - (2,1) - (2,1) - (2,1) - (2,1)$
	Dropout	$0 - 0.1 - 0.1 - 0.1 - 0.1 - 0.1$

3.3 RDV (REGIM-DL-VGG)

RDV was submitted by Besma Rabhi, Abdelkarim Elbaati and Adel M. Alimi, from REsearch Groups in Intelligent Machines at national school of engineers, university of Sfax from Tunisia. RDV addressed the use of the classical Seq2Seq for handwriting recovery process [3]. This system is similar to RDL but the encoder is based on VGG-16 and BLSTM rather than CNN-BLSTM. Figure 7 shows an overview of the participating system RDAG. The VGG-BLSTM receives the input image and produces a feature

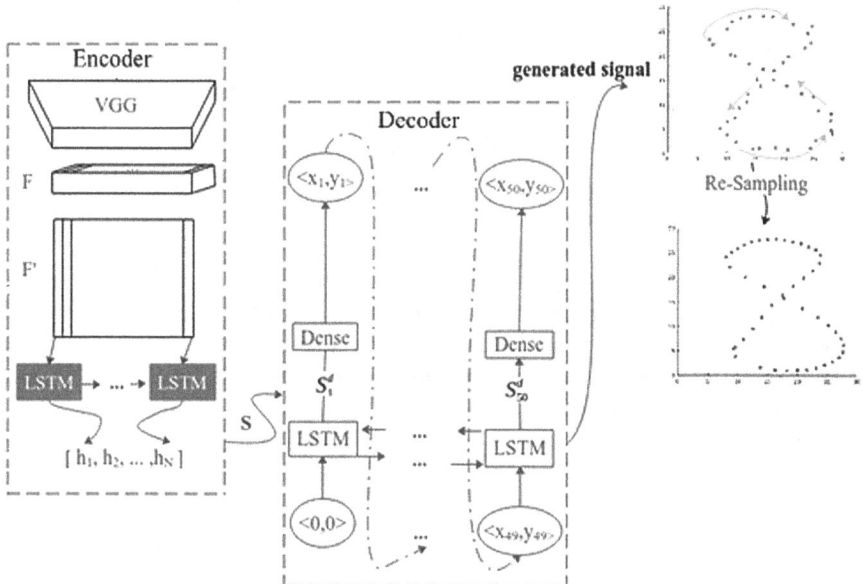

Fig. 7. Architecture of the system RDV.

map which represents the most important pixels that composes the offline handwriting image. Table 3 depicts the proposed VGG parameters configuration. Participants use 16 conv layers with kernel size of (3 * 3). The role of Max-Pooling is to select the most interesting feature of its previous convolution output. A Rectified Linear Unit (RELU) is used after each layer and a batch normalization BN is applied after the third layer and the last convolution one. This system produces 50 equidistant points. Here, the participants applied a post-processing step on the obtained script to add the velocity. Thus, their recovered signal is a set of 50 points characterized by the velocity.

Table 3. VGG Configuration for just eight layers.

Systems	Parameters	Values for each layer
RDV	Conv. filters	$64 - 64 - 128 - 128 - 256 - 256 - 256 - 512$
	MaxPooling	$(2,2) - (2,1) - (2,1) - (2,1) - (2,1) - (2,1) - $No-No
	Dropout	$0 - 0 - 0 - 0.1 - 0.1 - 0.1 - 0.1 - 0.1$

3.4 RDAG (REGIM-DL-Attention-GRU)

This system was submitted by Besma Rabhi, Abdelkarim Elbaati and Adel M. Alimi, from REsearch Groups in Intelligent Machines at national school of engineers, university of Sfax from Tunisia. RDAG consists in using the CNN-BGRU as encoder and the Attention-BGRU as decoder [4]. This system is intended to reconstruct an online script with velocity and with non-normalized point number. Figure 8 shows an overview of the participating system RDAG.

The first step is to fed the normalized images (64 * 64) into the CNN which will generate the feature maps. These features are extracted from left to right. Each feature vector is a rectangular region. Table 4 represents the proposed CNN parameters configuration The second step is to build the encoder with BGRU model that receives the CNN features and produce a fixed-length vector. The last step is the generation of point coordinates via the decoder BGRU. The architecture of both encoder and decoder models is based on three layers with 512 units for each one. The participants propose to add the attention model between the encoder and the decoder model. At each step, the decoder with attention layer generates a probability distribution of points characterized by velocity feature and non-normalized points. Table 5 presents the Encoder/Decoder configurations of different participating systems.

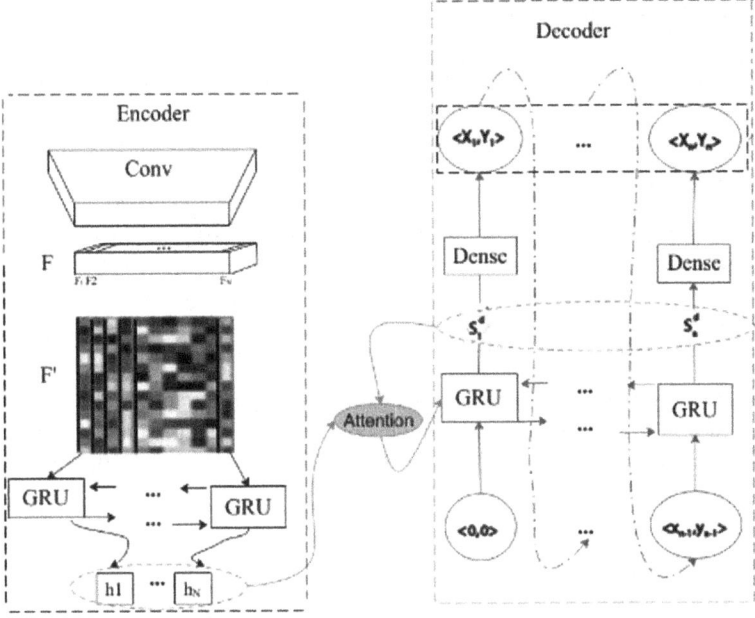

Fig. 8. Architecture of the system RDAG.

Table 4. CNN Configuration for eight layers.

Systems	Parameters	Values for each layer
RDAG	Conv. Filters	64 – 128 – 256 –256 – 256 – 256 – 256 – 512
	MaxPooling	(2,2) – (2,1) – No -(2,1) -(2,1) -No-(2,1)-No
	Dropout	0 – 0 – 0 – 0.1 – 0.1 – 0.1 – 0.1 – 0.1

Table 5. Encoder/Decoder configurations of participating systems.

Parameters	RDL	RDV	RDAG
Hidden units	256	256	512
Layer	3	3	3
Learning rate	0.00001	0.0001	0.0001
Batchsize	128	32	32
Gradient	Adam	Adam	Adam

4   Evaluation and Results

For the evaluation step, the metrics consist of using the Root Mean Squared Error (RMSE) and the Euclidean Distance (ED).

4.1   Root Mean Square Error

RMSE measures the difference between the ground truth script and the recovered signal according to the following equation:

$$RMSE = \frac{1}{2LC}\left(\sum_{i=1}^{ns}\sum_{t=1}^{li}\left(x_t - x'_t\right)^2 + \sum_{i=1}^{n}\sum_{t=1}^{li}\left(y_t - y'_t\right)^2\right) \qquad (1)$$

Where l_i is the number of point coordinates in each sample and ns is the number of samples. x_t and y_t represent the coordinates of the ground truth signal and x'_t, y'_t are the reconstructed coordinates. LC notices the total length of letters. Different results are depicted on Fig. 9.

4.2   Euclidean Distance

ED calculates the difference between the ground truth signal and the recovered one via a warping path. Hence, the best system is the one that have the lowest values.

It is calculated according to the following formula:

$$ED = \sqrt{\left(x_q - x'_p\right)^2 = \left(y_q - y'_p\right)^2} \qquad (2)$$

Where q and p vary between 1 and LC. Figure 10 shows the Euclidean distance of the participating systems.

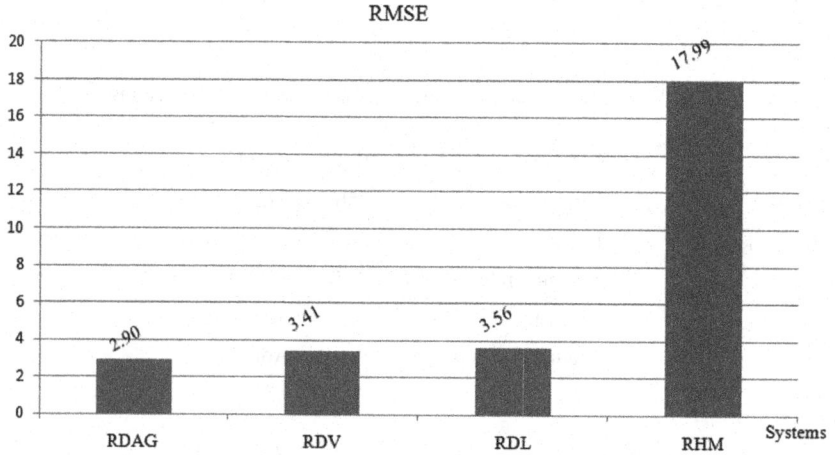

Fig. 9. RMSE values of the participating systems

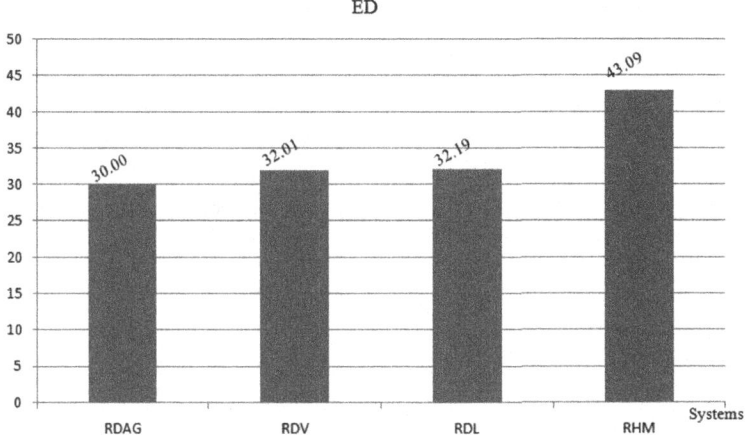

Fig. 10. ED values of the participating systems

4.3 Visual Comparison of the Recovered Velocity

We compare the velocity prediction of the participating systems using a visual graphic on the Arabic letter <<sad>> from Dhad dataset.

Figure 11 (a)–(d) shows the trajectory recovery and Fig. 11 (e)–(h) presents the velocity curves of RDAG, RDV and (RDL/RHM). As exposed in these figures, the Arabic character <<sad>> is well reconstructed by different systems, but the velocity is lost in some cases (RDL/RHM).

RDV system recovered the letter and generated an online signal with 50 points with velocity.

RDAG system recovered a trajectory characterized by a flexible curvature as human writing because the attention layer can focus on the detailed curves.

4.4 Analysis and Discussion

Figure 9 and Fig. 10 show the RMSE and the ED measures for all participating methods, respectively. From these figures we can see the following:

(1) The system RHM achieves a low margin performance compared to all deep systems. In addition, deep learning systems (RDL, RDV and RDAG) can be handled efficiently. The drawback of RHM system that used the heuristic method is the high computational time caused by the complexity of the search techniques. Furthermore, the systems RHM and RDL remains below both participant systems RDV and RDAG because the pen acceleration is lost.
(2) RDV performs better than both RDL and RHM, thanks to the VGG-BLSTM which outperforms CNN-LSTM for handwriting recovery process. By comparing between the RDV and RDAG with velocity, we affirm that the pen velocity is efficient for

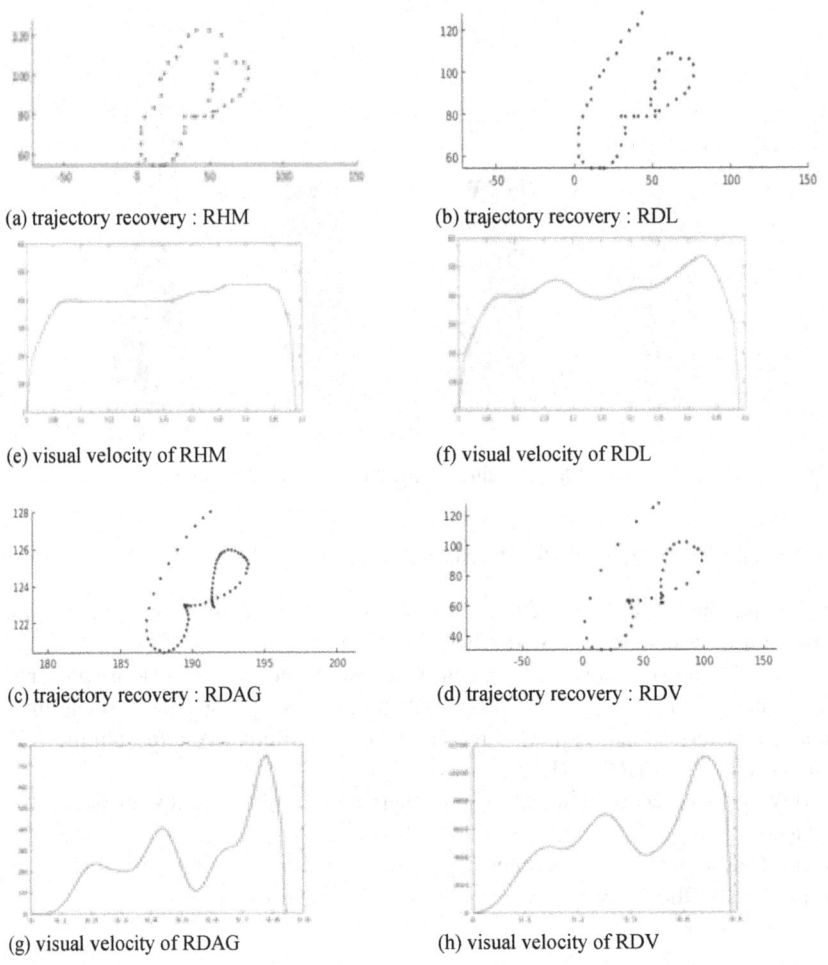

(a) trajectory recovery : RHM

(b) trajectory recovery : RDL

(e) visual velocity of RHM

(f) visual velocity of RDL

(c) trajectory recovery : RDAG

(d) trajectory recovery : RDV

(g) visual velocity of RDAG

(h) visual velocity of RDV

Fig. 11. A comparison of the reconstructed velocity of participating systems.

handwriting recovery. The obtained results demonstrate that RDAG with the attention model and BGRU enhances the performance of the handwriting reconstruction process. This indicates that BGRU could be a better choice for the temporal information process compared to LSTM.

(3) RDAG achieves lower RMSE than all systems. This demonstrates that the attention model can improve the effectiveness of handwriting recovery. This system is trained by signal with pen velocity, thus, the pen velocity recovery improves its performance. This system achieves best results over all participating systems and over all metrics.

5   Conclusion

In order to encourage, to renew research in this context, and to compare the performance of different advances on OSR systems, we propose the ASAR2021 Competition on Arabic Handwriting character and digit. The results of this competition will be reported and presented in specific session at ASAR 2021.

The online signal is used without pen up/down. It is described by the velocity feature and also the number of points is not specified. This online process can appear the challenge, thus, all participants must recover a signal with trajectory order, velocity feature and non-normalized number of points.

Acknowledgements. The research leading to these results has received funding from the Ministry of Higher Education and Scientific Research of Tunisia under the grant agreement number LR11ES4.

References

1. Diaz, M., Crispo, G., Parziale, A., Marcelli, A., Ferrer, M.A.: Writing order recovery in complex and long static handwriting. Int. J. Interactive Multimedia Artif. Intell. (2021)
2. Nguyen, H.T., Nakamura, T., Nguyen, C.T., Nakagawa, M.: Online trajectory recovery from offline handwritten japanese kanji characters of multiple strokes. In: 25th International Conference on Pattern Recognition (ICPR). IEEE (2020)
3. Rabhi, B., Elbaati, A., Hamdi, Y., Alimi, A.M.: Handwriting recognition based on temporal order restored by the end-to-end system. In: International Conference on Document Analysis and Recognition (ICDAR), pp. 1231–1236 (2019). https://doi.org/10.1109/ICDAR.2019. 00199
4. Rabhi, B., Elbaati, A., Boubaker, H., Hamdi, Y., Hussain, A., Alimi, A.M.: Temporal Order and pen velocity recovery for character handwriting based on sequence-to-sequence with attention mode. In: TechRxiv. Preprint. https://doi.org/10.36227/techrxiv.13902650.v1 (2021)
5. Hassaïne, A., Al Maadeed, S., Bouridane, A.: ICDAR 2013 competition on handwriting stroke recovery from offline data. In: 12th International Conference on Document Analysis and Recognition, pp. 1412–1416 (2013). https://doi.org/10.1109/ICDAR.2013.285
6. Elbaati, A., Kherallah, M., Ennaji, A., Alimi, A.M.: Temporal order recovery of the scanned handwriting. In: 10th International Conference on Document Analysis and Recognition, pp. 1116–1120 (2009). https://doi.org/10.1109/ICDAR.2009.266
7. Dhieb, T., Njah, S., Boubaker, H., Ouarda, W., Ayed, M.B., Alimi, A.M.: Towards a novel biometric system for forensic document examination. Comput. Secur. **97**, 101973 (2020)
8. Dhieb, T., Rezzoug, N., Boubaker, H., Gorce, P., Alimi, A.M.: Effect of age on hand drawing movement kinematics. Comput. Meth. Biomechan. Biomed. Eng. **22**(sup1), S188–S190 (2019)
9. Hamdi., Y., Boubaker, H., Dhieb, T., Elbaati., A, Alimi, A.M.: Hybrid DBLSTM-SVM based beta-elliptic CNN models for online Arabic characters recognition. In: International Conference on Document Analysis and Recognition (ICDAR), pp. 803–808 (2019)
10. Hamdi, Y., Boubaker, H., Alimi, A.M.: Data augmentation using geometric, frequency, and beta modeling approaches for improving multi-lingual online handwriting recognition. IJDAR (2021)

11. Bhunia, A.K., et al.: Handwriting trajectory recovery using end-to-end deep encoder-decoder network. In: 24th International Conference on Pattern Recognition (ICPR), pp. 3639–3644. IEEE (2018)

12. Akouaydi, H., Njah, S., Wael O., Anis, S., Mourad, Z., Alimi, A.M.: Convolutional neural networks for online arabic characters recognition with beta-elliptic knowledge domain. In: ICDARW, pp. 1–6 (2019)

13. Rabhi, B., Dhahri, H., Alimi, A.M., Alturki, F.A.: Grey wolf optimizer for training elman neural network. In: Abraham, A., Haqiq, A., Alimi, A.M., Mezzour, G., Rokbani, N., Muda, A.K. (eds.) HIS 2016. AISC, vol. 552, pp. 380–390. Springer, Cham (2017). https://doi.org/10.1007/978-3-319-52941-7_38

14. Dinh, M., Yang, H.J., Lee, G.S., Kim, S.H., Do, L.N.: Recovery of drawing order from multi-stroke English handwritten images based on graph models and ambiguous zone analysis. Expert Syst. Appl. **64**, 352–364 (2016)

15. V. A. Kha, H. H. Kha, M. Blumenstein, "Extraction of Dynamic Trajectory on Multi-Stroke Static Handwriting Images Using Loop Analysis and Skeletal Graph Model," REV Journal on Electronics and Communications, 6(1–2), (2016).

16. Elbaati, A., Boubaker, H., Kherallah, M., Ennaji, A., El Abed, H., Alimi, A.M.: Arabic handwriting recognition using restored stroke chronology. In: Document Analysis and Recognition ICDAR 2009. 10th International Conference, pp. 411–415 (2009)

17. ElBaati, A., Alimi, A.M., Charfi, M., Ennaji, A.: Recovery of temporal information from off-line arabic handwritten. In: AICCSA, pp. 127-vii, January 2005

18. Akouaydi, H., et al.: Neural architecture based on fuzzy perceptual representation for online multilingual handwriting recognition, pp.1–14, arXiv preprint arXiv:1908.00634 (2019)

19. Akouaydi, H., Njah, S., Alimi, A.M.: Android application for handwriting segmentation using PerTOHS theory. In: Ninth International Conference on Machine Vision, ICMV, pp.1–5 (2016)

ASAR 2021 Competition on Online Arabic Character Recognition: ACRC

Yahia Hamdi[1]([✉]) [iD], Houcine Boubaker[1] [iD], Tarek M. Hamdani[1] [iD],
and Adel M. Alimi[1,2] [iD]

[1] REGIMLab.: Research Groups in Intelligent Machines, University of Sfax, National Engineering School of Sfax, LR11ES48, 3038 Sfax, Tunisia
{yahia.hamdi,houcine.boubaker,tarek.hamdani,
adel.alimi}@regim.usf.tn
[2] Department of Electrical and Electronic Engineering Science, Faculty of Engineering and the Built Environment, University of Johannesburg, Johannesburg, South Africa

Abstract. The online Arabic handwriting recognition task always presents a challenge due to the existence of some complexity and variability of its writing style. In this paper, we describe an online Arabic handwritten Character recognition competition (ACRC) held at ASAR 2021. The aim is to evaluate the limits of Arabic character recognition systems on collected LMCA database (with and without noising). Four systems are participating in this competition which were tested on an unknown test dataset to all participants. Two metrics; notably Character Error Rate (CER) and speed are used to compare the systems. The achieved results in ACRC 2021 demonstrate the high potential of competitive participating systems which are based on deep learning methods.

Keywords: ACRC 2021 · Online handwriting · LMCA_Database · Character recognition · Systems evaluation · Deep learning

1 Introduction

On-line handwritten character recognition has been considered a very active scope of research because of its high popularity for recognition scenarios [1, 7, 25] and the variety of open challenges that are still being explored nowadays. Indeed, the performance of on-line handwriting recognition systems has been improved lately due to assorted factors such as the evolution of acquisition technology with the appearance of touch screens mobile devices allowing the capturing of any *ink* format through the finger or stylus, and the extensive usage of deep learning technology [30, 31] in abundant different fields of handwriting recognition [9, 11, 14, 18, 24], signature verification [5, 6], handwriting recovery [16, 17] and emotion recognition [21], etc.

Compared to Latin and English scripts, a lot of studies has remained to be done on online Arabic handwriting recognition task. Indeed, the performance of recognition systems depends generally on the used architecture as well as the validation database. A few of publicly available databases in the case of Arabic script. Therefore, the approaches and

© Springer Nature Switzerland AG 2021
E. H. Barney Smith and U. Pal (Eds.): ICDAR 2021 Workshops, LNCS 12916, pp. 379–389, 2021.
https://doi.org/10.1007/978-3-030-86198-8_27

benchmarks to evaluate the advance of the community in online handwriting recognition topic are more than ever needed.

Since 2005, more than three competitions on online Arabic words have been organized in the different editions of ICDAR (one in 2005 [20], 2009 [12], and 2011 [13]) but none of them were related to online Arabic characters.

In this paper, we present the experimental results of the ASAR 2021 competition on On-Line Arabic Handwriting Character Recognition (ACRC 2021). The goal of ACRC 2021 is to validate the effectiveness of Arabic character recognition systems on available LMCA database.

The rest of this paper is organized as follows. In Sect. 2 the LMCA-database is presented in some detail. Section 3 describes the competition setup. The description of the participating systems is introduced in Sect. 4. Section 5 provides the different tests and discusses the results performed with the different systems. Finally, the paper ends with some concluding remarks.

2 LMCA Database

The LMCA (abbreviation of the French sentence "Lettres, Mots et Chiffres Arabe,") database is taken into consideration in this competition in order to evaluate the advance of the research works and the development of online Arabic handwritten character recognition systems.

This database was constructed in our laboratory REGIM (REsearch Group on Intelligent Machines) [12]. It contains 100.000 Arabic characters, 30.000 digits, and 500 Arabic words in both online and offline formats collected by 55 different writers. Different categories of writers, aged from 8 to 66 years old, contributed to the development of this database; Two-thirds were male, of whom 90% were right-handed. Data collection from this database was done using a digital Wacom UltraPad A4 tablet with a 7-spatial resolution of 200 dpi and a sampling rate of 100 points/seconds, were stored using the UNIPEN format. For the online forms, the information (x, y, z) related to the coordinates (x, y) and pressure (z) of the handwriting trajectory is stored, whereas, for offline forms, the image of its trajectory is also saved.

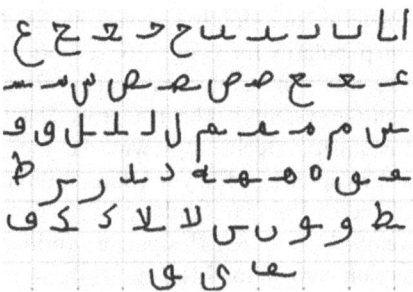

Fig. 1. The 56 shapes of Arabic characters

For this competition, we have employed only the dataset for Arabic characters to appraise the limits of online handwriting recognition systems. As shown in Fig. 1, the total number of characters' class existing in this dataset is 56 which represents the main shapes of Arabic characters without considering the diacritics. The used databases are divided into distinct sets, a training set of 70% and a test set of 30%.

Ain 'ع'   Seen 'س'   waw 'و'

Fig. 2. Some of noised Arabic characters of LMCA dataset.

Furthermore, we have applied a white noise on test set to evaluate the robustness of participant systems. Some example of noised Arabic characters of LMCA dataset are shown in Fig. 2.

3 ACRC 2021: Competition Setup

The competition aims to promote innovation in Online Arabic character recognition field as well as to provide objective and fair comparisons among methods using available LMCA database. The principal tasks, evaluation metric, and experimental protocol of ACRC 2021 are described in the following:

3.1 Tasks

The aim of ACRC 2021 is to validate and compare the effectiveness of the Arabic character recognition systems on available LMCA database. Consequently, three tasks are considered in the competition:

– **Task 1**: Analysis of participant systems using collected LMCA database.
– **Task 2**: Analysis of participant systems using noised LMCA test database.
– **Task 3**: Evaluate the relative speed of systems.

3.2 Evaluation Criteria

The evaluation metric employed in this competition is the popular Character Error Rate (CER) which is used in most online handwriting recognition studies in the literature. This competition proceeds a ranking based on points (3 points: for the first system, 2: second, 1: third). We have evaluated each task separately. Indeed, according to the points awarded for the three tasks, we have announced the winner of the ACRC 2021.

3.3 Experimental Protocol

Two principle steps are considered in ACRC 2021:

- *Development*: the aim of this step is to provide the participants with data necessary to train online character recognition systems. In addition, participants can employ other databases or adopt data augmentation methods [32, 33] to train their systems. In our case, we have used some geometric techniques such as baseline inclination angle, magnitude ratio, etc., and frequency methods [9] to increase the training set by generating more samples.

    In order to permit participants to test the performance of their trained systems, we divide the LMCA database into training and validation datasets. The training dataset used for this competition consists of 13,715 samples, whereas 2,150 observations for evaluation.

- *Evaluation*: The final validation of ACRC 2021 is performed by providing the participant with the test datasets (with and without noise) without specifying the Ground-truth labels. In this step, the participants are allowed to submit the scores (CER) obtained by the different recognition systems for each task considered in the competition.

4   Description of Participated Systems

In this section, we introduce a brief description of the participating systems to the present competition. The systems description was provided by the authors and edited by the contest organizers.

4.1 REGIM-GS-DBLSTM

The *REGIM-GS-DBLSTM* [8] system is submitted by Yahia Hamdi, Houcine Boubaker and Adel M. Alimi from REGIM-Lab (REsearch Groups in Intelligent Machines Regim) at National School of Engineers, University of Sfax, Tunisia. This system is based on graphemes segmentation (GS) strategy [3] and deep BLSTM (DBLSTM) recurrent neural network as shown in Fig. 3.

The developed system proceeds by a preprocessing module using handwriting normalization size and low-pass filtering Chebyshev type II to eliminate the acquisition system's noise. The second step consists of detecting the baseline [4] by considering concordance between the alignment of the handwriting trajectory points and their tangent directions. Thereafter, the handwritten trajectory is decomposed in continuous parts called graphemes, delimited by the ligature bottoms points neighboring the baseline. Then, a set of pertinent parameters combining dynamic and static features are extracted by using enhanced Beta-elliptic model [2], Fourier descriptors parameters [29] for trajectory shape modeling, and other normalized features for grapheme dimensions modeling, positions relative to the baseline and the assignment diacritics codes. The constructed features vectors are then studied by DBLSTM for character recognition module.

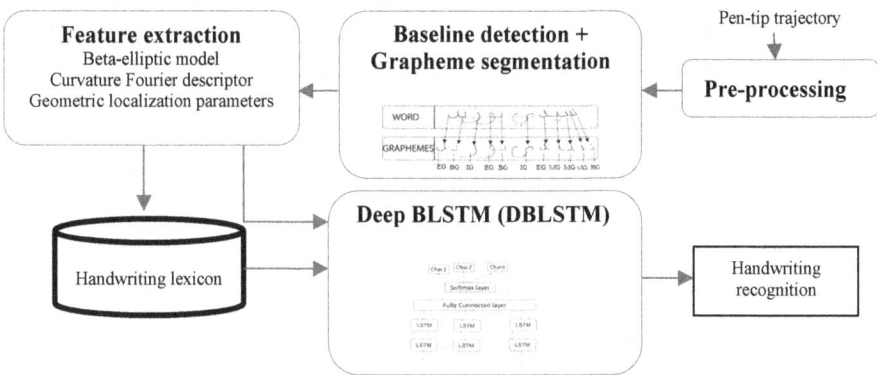

Fig. 3. REGIM-GS-DBLSTM architecture.

4.2 REGIM-DBLSTM-SVM

The second system submitted by Yahia Hamdi, Houcine Boubaker and Adel M. Alimi [7]. It based on hybrid Beta-Elliptic model (BEM) and convolutional neural network (CNN) for handwritten feature extraction, and DBLSTM with SVM classifiers (see Fig. 4). Firstly, the online trajectory is denoised and normalized by a preprocessing module. Secondly, two types of features are extracted from the online trajectory: the online Beta-Elliptic parameters and offline CNN generic features after transforming the online trajectory to image. The online features are studied using DBLSTM whereas the offline on SVM classifier. The last step combines the two modules in order to increase the discrimination power of the overall system.

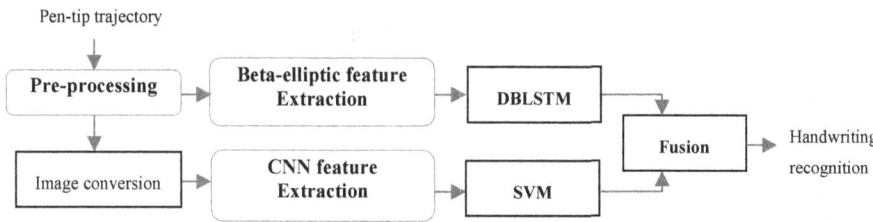

Fig. 4. REGIM-DBLSTM-SVM architecture.

4.3 REGIM-EPC-LSTM

The *REGIM-EPC-LSTM* [24] is submitted by Hanen Akouaydi, Houcine Boubaker, Sourour Njah, Mourad Zaied and Adel M. Alimi from REGIM-Lab.

This system is based on Perceptual codes detection and LSTM recurrent neural network as depicted in Fig. 5. The developed system proceeds by a preprocessing module such as interpolation, de-hooking, and removing noise to eliminate the imperfections caused by the acquisition system. The second step consists of detecting elliptic strokes and classifying them into Elementary Perceptual Code (EPC) based on Beta-Elliptic parameters. Then, the authors use the EPC handwriting representation with LSTM for LMCA characters recognition.

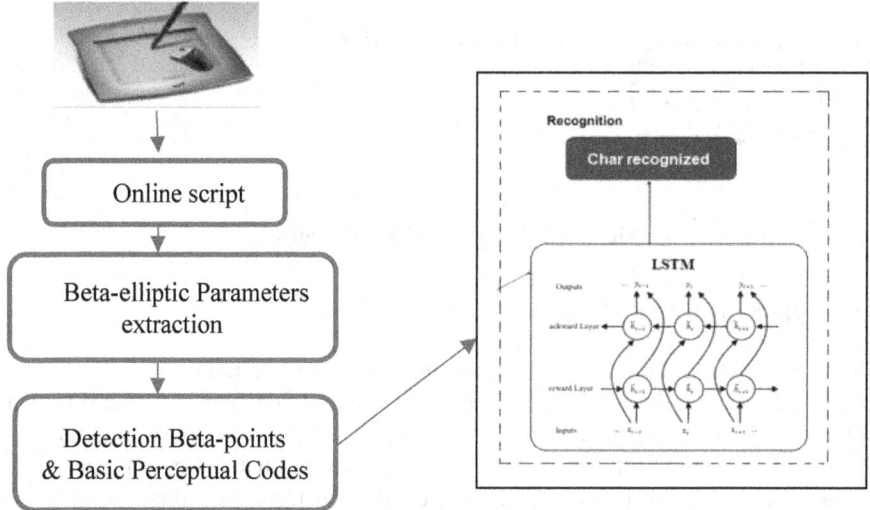

Fig. 5. REGIM-EPC-LSTM architecture.

4.4 RDAG (REGIM-DL-Attention-GRU)

RDAG [15] is submitted by Besma Rabhi, Abdelkarim Elbaati, Yahia Hamdi, and Adel M. Alimi. It consists in using the CNN-BGRU as encoder and the Attention-BGRU as decoder as presented in Fig. 6. This system is intended to reconstruct an online script with velocity and with non-normalized point number. The first step is to fed the normalized images (64*64) into the CNN which will generate the feature maps. These features are extracted from left to right. Each feature vector is a rectangular region. The second step is to build the encoder with BGRU model that receives the CNN features and produce a fixed-length vector. The last step is the generation of point coordinates via the decoder BGRU. The architecture of both encoder and decoder models is based on three layers with 512 units for each. The participants propose to add the attention model between the encoder and the decoder model. At each step, the decoder with attention layer generates a probability distribution of points characterized by velocity feature and non-normalized points. After obtaining the online signal from its counterpart offline one, participants use the online recognition system described in *REGIM-GS-DBLSTM* which is based on grapheme segmentation and DBLSTM models.

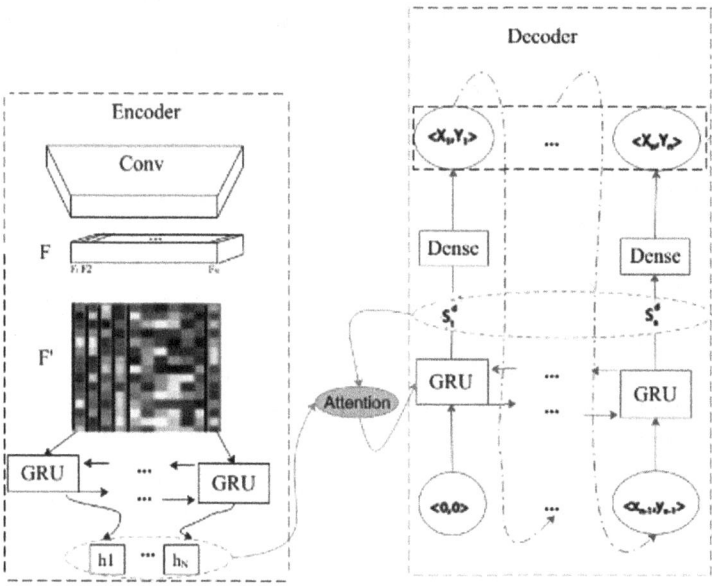

Fig. 6. REGIM-DBLSTM-SVM architecture.

5   Experimental Results

In this section, we present the evaluation results of the competition using LMCA database acquired for ACRC 2021. For that, we have employed the unknown test dataset to validate the performance of participating systems based on CER metric and speed test criteria.

Table 1. Evaluation results of ACRC 2021 using LMCA dataset. For each particular task, we include the points obtained by each system depending on the ranking points (first: 3, second: 2, and third: 1).

Systems	Task 1		Task 2		Task 3	
	Points	CER	Points	CER	Points	Time (s)
REGIM-GS-DBLSTM	1	1.43%	0	2.85%	3	250.13
REGIM-DBLSTM-SVM	3	**0.89%**	3	**1.30%**	1	270
REGIM-EPC-LSTM	0	2.50%	1	2.75%	2	255.23
RDAG	2	1.20%	2	2.01%	0	300

5.1   Evaluation Results

The evaluation of the four participating systems is performed by computing the CER on test set (with and without noising), and speed test criteria. Table 1 shows the results

Table 2. Overall ranking ACRC 2021.

Systems	Total points
REGIM-GS-DBLSTM	4
REGIM-DBLSTM-SVM	**7**
REGIM-EPC-LSTM	3
RDAG	4

achieved by different participating systems in each task including the ranking based on assigned points. Three systems are developed under Microsoft Windows environment (*REGIM-GS-DBLSTM, REGIM-DBLSTM-SVM*, and *REGIM-EPC-LSTM*) and one under Linux (*RDAG*).

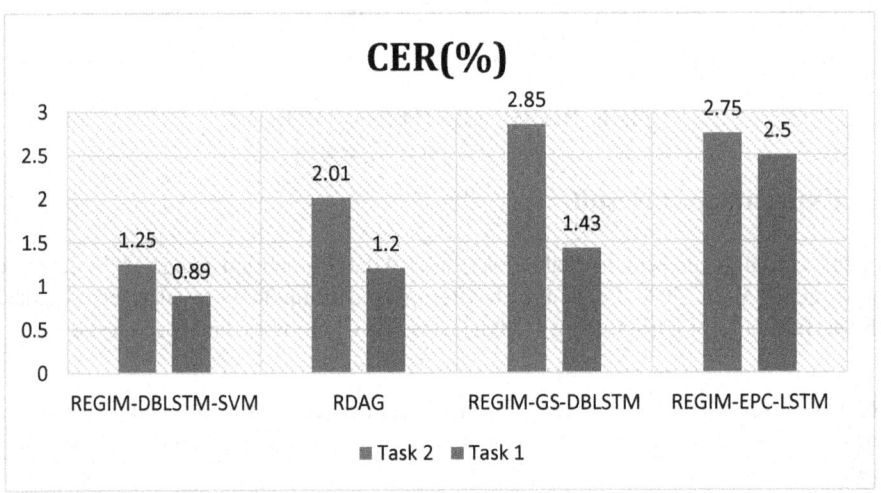

Fig. 7. CER of participating systems using LMCA dataset.

In terms of recognition rate, we can see from Fig. 7 that the lowest CER is achieved by REGIM-DBLSTM-SVM of 0.89% followed by RDAG system of 1.2% and 1.43% by REGIM-GS-DBLSTM. Also, the CER is increased by 0.36% and 0.25% for REGIM-GS-DBLSTM and *REGIM-EPC-LSTM* respectively using the noised test data set. This slight increase of CER demonstrates the robustness of the two later systems for adding noise compared to REGIM-GS-DBLSTM and RDAG systems.

In addition, we observe a substantial difference in speed between systems. The fastest system is about 250.13 times for *REGIM-GS-DBLSTM* system followed by *REGIM-EPC-LSTM* comparing with the other systems.

5.2 Discussion

We can see from Table 2 that REGIM-DBLSTM-SVM is the winner of ACRC 2021 which achieves the best ranking points of 7 followed by RDAG and *REGIM-GS-DBLSTM by 4 and REGIM-EPC-LSTM* by 3 points. It can be explained by the combination of DBLSTM and SVM classifier, and hybrid CNN-BEM features.

Also, it is important to clarify the competitive result achieved by the three other systems such as REGIM-GS-DBLSTM and RDAG which use the same features extraction model and DBLSTM classifier which provided very good results. The same happens with the REGIM-EPC-LSTM system based on perceptual code and LSTM network which shown its robustness for adding noise.

These results justify the superiority of deep learning methods for the online handwriting character recognition field, as commented in previous studies [27, 28].

6 Conclusion

This paper introduced the experimental results of the ASAR 2021 Competition on On-Line Arabic handwriting character recognition (ACRC 2021). The goal of ACRC 2021 is to evaluate the recognizing systems on online Arabic characters through the available LMCA database. Two evaluation criteria, namely Character Error Rate (CER) and speed, are employed to analyze the performance of participant systems.

Experimental results testify to the effectiveness of the participant systems as most of them have shown a low error rate and a very high speed. These results prove on the one hand the high potential systems based on deep learning methods and the challenging conditions of successful AHCRC 2021 on the other hand. In fact, the winner of ACRC 2021 has been REGIM-DBLSTM-SVM that proposed the combination of DBLSTM and SVM classifier based on CNN and BEM features extraction models.

Despite the limited number of participating systems in this competition and based on mentioned results, we have announced their success as the first competition on online Arabic character, and the 4th competition after ICDAR 2005, ICDAR 2007, ICDAR 2011 which are organized on On/Offline Arabic words.

ACRC 2021 will be established as an ongoing competition, where researchers can easily compare their systems against the state of the art in an open common platform using public databases such as LMCA, and standard experimental protocols.

Acknowledgments. The research leading to these results has received funding from the Ministry of Higher Education and Scientific R search of Tunisia under the grant agreement number LR11ES4.

References

1. Altwaijry, N., Al-Turaiki, I.: Arabic handwriting recognition system using convolutional neural network. Neural Comput. Appl. **33**(7), 2249–2261 (2020). https://doi.org/10.1007/s00521-020-05070-8

2. Boubaker, H., Chaabouni, A., Tagougui, N., Kherallah, M., Alimi, A.M.: Handwriting and hand drawing velocity modeling by superposing beta impulses and continuous training component. Int. J. Comput. Sci. Issues (UCS), pp S7–63 (2013)

3. Boubaker, H., Tagougui, N., ElAbed, H., Kherallah, M., Alimi, A.M.: Graphemes segmentation for Arabic on-line handwriting modelling. J. Inf. Process. Syst. (JIPS) 10(4), 503–522 (2014)

4. Boubaker, H., Chaabouni, A., El-Abed, H., Alimi, A.M.: GLoBD: geometric and learned logic algorithm for straight or curved handwriting baseline detection. Int. Arab J. Inf. Technol. 15(1) (2018)

5. Dhieb, T., Njah, S., Boubaker, H., Ouarda, W., Ayed, M.B., Alimi, A.M.: Towards a novel biometric system for forensic document examination. Comput. Secur. 97, 101973 (2020)

6. Dhieb, T., Rezzoug, N., Boubaker, H., Gorce, P., Alimi, A.M.: Effect of age on hand drawing movement kinematics. Comput. Meth. Biomechan. Biomed. Eng. 22(sup1), S188–S190 (2019)

7. Hamdi, Y., Boubaker, H., Dhieb, T., Elbaati, A., Alimi, A.: Hybrid DBLSTM-SVM based eeta-elliptic-CNN models for online Arabic characters recognition. In: International Conference on Document Analysis and Recognition (ICDAR), pp. 803–808 (2019)

8. Hamdi, Y., Boubaker, H., Alimi, A.M.: Online arabic handwriting recognition using graphemes segmentation and deep learning recurrent neural networks. In: Hassanien, A.E., Darwish, A., Abd El-Kader, S.M., Alboaneen, D.A. (eds.) Enabling Machine Learning Applications in Data Science. AIS, pp. 281–297. Springer, Singapore (2021). https://doi.org/10.1007/978-981-33-6129-4_20

9. Hamdi, Y., Boubaker, H., Alimi, A.M.: Data augmentation using geometric, frequency, and beta modeling approaches for improving multi-lingual online handwriting recognition. IJDAR (2021)

10. Hamdi, Y., Chaabouni1, A., Boubaker, H., Alimi, A.M.: Off-lexicon online Arabic hanwriting recognition using neural network. In: Proceedings of the SPIE 10341, Ninth International Conference on Machine Vision (ICMV 2016), 103410G, 17 March 2017. https://doi.org/10.1117/12.2268650

11. Hamdi, Y., Chaabouni, A., Boubaker, H., Alimi, A.M.: Hybrid neural network and genetic algorithm for off-lexicon online Arabic handwriting recognition. In: Abraham, A., Haqiq, A., Alimi, A.M., Mezzour, G., Rokbani, N., Muda, A.K. (eds.) HIS 2016. AISC, vol. 552, pp. 431–441. Springer, Cham (2017). https://doi.org/10.1007/978-3-319-52941-7_43

12. Kherallah, M., Elbaati, A., ElAbed, H., Alimi, A.M.: The On/Off (LMCA) dual arabic handwriting database. In: International Conference on Frontiers in Handwriting Recognition (2008)

13. Kherallah, M., Tagougui, N., Alimi, A.M., Abed, H.E., Margner, V.: Online Arabic handwriting recognition competition. In: 2011 International Conference on Document Analysis and Recognition, pp. 1454–1458 (2011). https://doi.org/10.1109/ICDAR.2011.289

14. Maalej, R., Kherallah, M.: Improving the DBLSTM for on-line Arabic handwriting recognition. Multimedia Tools Appl. 79(25–26), 17969–17990 (2020). https://doi.org/10.1007/s11042-020-08740-w

15. Rabhi, B., Elbaati, A., Hamdi, Y., Alimi, A.M.: Handwriting recognition based on temporal order restored by the end-to-end system. In: 2019 International Conference on Document Analysis and Recognition (ICDAR), pp. 1231–1236 (2019). https://doi.org/10.1109/ICDAR.2019.00199.

16. Rabhi, B., Dhahri, H., Alimi,, A.M., Alturki, F.A.: Grey Wolf Optimizer for Training Elman Neural Network. In: Abraham, A., Haqiq,, A., Alimi, A., Mezzour, G., Rokbani,, N., Muda,, A. (eds.) Proceedings of the 16th International Conference on Hybrid Intelligent Systems (HIS 2016), HIS 2016. Advances in Intelligent Systems and Computing, vol 552, pp. 380—390. Springer, Cham (2017). https://doi.org/10.1007/978-3-319-52941-7_38

17. Rabhi, B., Elbaati, A., Boubaker, H., Hamdi, Y., Hussain, A., Alimi, A.: Temporal order and pen velocity recovery for character handwriting based on sequence-to-sequence with attention mode. TechRxiv (2021)
18. Rubén, T., et al.: ICDAR 2021 Competition on On-Line Signature Verification (2021)
19. Volker, M., Mario, P., Abed, H.E.: Eighth International Conference on Document Analysis and Recognition (ICDAR 2005), 29 August–1 September 2005, Seoul, Korea. IEEE Computer Society (2005). ISBN 0–7695–2420–6
20. Abed,H.E., Märgner, V., Kherallah, M., Alimi, A.M.: ICDAR 2009 online arabic handwriting recognition competition. In: 2009 10th International Conference on Document Analysis and Recognition, pp. 1388–1392 (2009). https://doi.org/10.1109/ICDAR.2009.284
21. Wilson-Nunn, D., Lyons, T., Papavasiliou, A., Ni, H.: A path signature approach to online arabic handwriting recognition. In: International Workshop on Arabic and Derived Script Analysis and Recognition (ASAR), pp. 135–139 (2018)
22. Yongqiang, Y., Xiangwei, Z., Bin, H., Yuang, Z., Xinchun, C.: EEG emotion recognition using fusion model of graph convolutional neural networks and LSTM. Appl. Soft Comput. **100**, 06954 (2021)
23. Chen, Z., Yin, F., Zhang, X.-Y., Yang, Q., Liu, C.: Multilingual handwritten text recognition via multi-task learning of recurrent neural networks. Pattern Recogn. **108**, 107555 (2020)
24. Akouaydi, H., Njah, S., Wael, O., Anis, S., Mourad, Z., Alimi, A.M.: Convolutional neural networks for online arabic characters recognition with beta-elliptic knowledge domain. In: ICDARW, pp. 1–6 (2019)
25. Hanen, A., Sourour, N., Alimi, A.M.: Android Application for handwriting segmentation using PerTOHS theory. In: Ninth International Conference on Machine Vision, ICMV, pp.1–5 (2016)
26. Najiba. T., Kherallah, M.: Recognizing online Arabic handwritten characters using a deep architecture. In: Proceedings of the SPIE 10341, Ninth International Conference on Machine Vision, 17 March 2017
27. Mezghani, N., Mitiche, A., Cheriet, M.: Bayes classification of online Arabic characters by gibbs modeling of class conditional densities. IEEE Trans. Pattern Anal. Mach. Intell. **30**(7), 1121–1131 (2008). https://doi.org/10.1109/TPAMI.2007.70753
28. Elleuch, M., Zouari, R., Kherallah, M.: Feature extractor based deep method to enhance online arabic handwritten recognition system. In: Villa, A.E.P., Masulli, P., Pons Rivero, A.J. (eds.) ICANN 2016. LNCS, vol. 9887, pp. 136–144. Springer, Cham (2016). https://doi.org/10.1007/978-3-319-44781-0_17
29. Persoon, E., Fu, K.S.: Shape discrimination using Fourier descriptors. J. IEEE Trans. Pattern Anal. Mach. Intell. 388–397 (1986)
30. Sun, L., Su, T., Liu, C., Wang, R.: Deep LSTM networks for online Chinese handwriting recognition. In: 2016 15th International Conference in Frontiers in Handwriting Recognition (ICFHR), pp. 271–276 (2016). https://doi.org/10.1109/ICFHR.2016.0059
31. Bhateja, V., Coello, C.A., Satapathy, S.C., Pattnaik, P.K. (eds.): Intelligent Engineering Informatics. AISC, vol. 695. Springer, Singapore (2018). https://doi.org/10.1007/978-981-10-7566-7
32. Shen, X., and Messina, R.: "A method of synthesizing handwritten Chinese images for data augmentation. In: The 15th International Conference on Frontiers in Handwriting Recognition (ICFHR), pp. 114–119 (2016)
33. Shorten, C., Khoshgoftaar, T.M.: A survey on image data augmentation for deep learning. J. Big Data **6**, 60 (2019)

ASAR 2021 Competition on Online Arabic Word Recognition

Hanen Akouaydi[1]([✉]) [ID], Houcine Boubaker[1] [ID], Sourour Njah[1],
Mourad Zaied[2] [ID], and Adel M. Alimi[1,3] [ID]

[1] REGIM-Lab.: REsearch Groups in Intelligent Machines, National Engineering
School of Sfax (ENIS), University of Sfax, BP 1173, 3038 Sfax, Tunisia
{hanen.akouaydi,houcine.boubaker,sourour.njah,adel.alimi}@regim.usf.tn
[2] Research Team in Intelligent Machines, University of Gabes,
National School of Engineers of Gabes (ENIG), BP 6072, Sidi rzig, Gabes, Tunisia
[3] Department of Electrical and Electronic Engineering Science,
Faculty of Engineering and the Built Environment, University of Johannesburg,
Johannesburg, South Africa

Abstract. This paper presents the Online Arabic words recognition
competition (OAWRC 2021) held at ASAR 2021 workshop in conjuction
with ICDAR 2021. This competition uses the ADAB database, which is
a database of Arabic handwritten words. Three groups with four systems are participating in the competition this year. A brief description
of participating systems and the experimental results is doing. The systems were tested on unknown set to the participants of 100 labels. In
particular, we wish to investigate and compare participating methods
that can be robust against competition challenges. The main goal of this
competition is to encourage the research in Arabic script and the development of Arabic handwritten systems. Within good recognition rate,
the competition has two new challenges: working with small training set
and being robust against noise. The evaluation process of participating
systems adapts various metrics such as the recognition rate, the speed,
etc. All systems produce good results. The winner and his runner-up are
very close while they use different methods.

Keywords: Online Arabic handwritten words · Recognition ·
Competition

1 Introduction

Handwriting recognition is a arduous task to teach machine how to identify
script. Recognition of handwritten words can be classified into two different
approaches. The first approach, handcrafted approach or traditional approach

Supported by REGIM-Lab.: REsearch Groups in Intelligent Machines.

E. H. Barney Smith and U. Pal (Eds.): ICDAR 2021 Workshops, LNCS 12916, pp. 390–401, 2021.
https://doi.org/10.1007/978-3-030-86198-8_28

based on segmentation and extraction features [7–9,11,19–22,25] and the second approach is the deep learning approach [31,32] related to time-sequence interpretation [12,13][?]. The field of online handwriting recognition is still a challenging task through the latest improvements of recognition tools. In fact, various researches focus on this field that deal with different scripts, for instance, Arabic [7,30], Chinese [27], devanagari and bengali scripts [28,29], mathematic expressions [30], etc. It also exist multi-script and multi-language systems for online handwriting recognition like those presented in [19] and commercial systems Google [26] and My Script. Research methods still weak in terms of performance comparing to commercial ones. Online Arabic Handwritten words is still ripe for research compared to Latin script. The series of competition for Arabic handwriting recognition systems has shown a positive effect for the improvement of recognition systems [1,2]. Arabic Word recognition is considered as an challenged research topic in pattern recognition and is essential for understanding Arabic script. In fact, Arabic script, written from right to left, is a cursive script written both printed and handwritten versions. It includes strokes and dots above and under letters, with the connectivity of characters comparing to Latin scripts. There are 28 letters and the shape of those letters change depending on their position in the word: preceded and/or followed by other letters or isolated. The goal of this competition is to evaluate the performance of participating methods. For the current OAWRC 2021 competition, two new challenges are presented: noise and dealing with small dataset. The benchmark ADAB is described in details in Sect. 2. This paper is organized as follows: First, in Sect. 2 presents ADAB database. Section 3 illustrates methods of participants while Sect. 4 describes the evaluation protocols, the results of this competition and reports the challenges of our OWRC 2021 competition. Finally, conclusions are drawn in Sect. 5.

2   ADAB Database

2.1   Presentation

The ADAB database (The Arabic handwriting Data Base) is a cooperation between the Institut fuer Nachrichtentechnik (IfN) and Research Groups in Intelligent Machines, University of Sfax, Tunisia. It contains 937 Tunisian town/village names of text written that aims to advance the research in Arabic field and to raise the development of Arabic handwritten systems. It is an online/offline database [4–6]. Figure 1 shows some samples from the ADAB dataset. The ADAB-database composes of 3 sets. Table 1 reports some informations about ADAB like the number of files, characters, words and writers for each set 1 to 3. This latter composes of more than 33,000 Arabic words handwritten by 170 different writers. The ADAB dataset is useful in the context of online Arabic handwriting word recognition, but the numbers of words for the various writers are not equals, it currently encompasses labels with a limited number of samples. As we mention working with small training set. The training data used consists of 811 online handwritten words of 100 classes extracted from ADAB

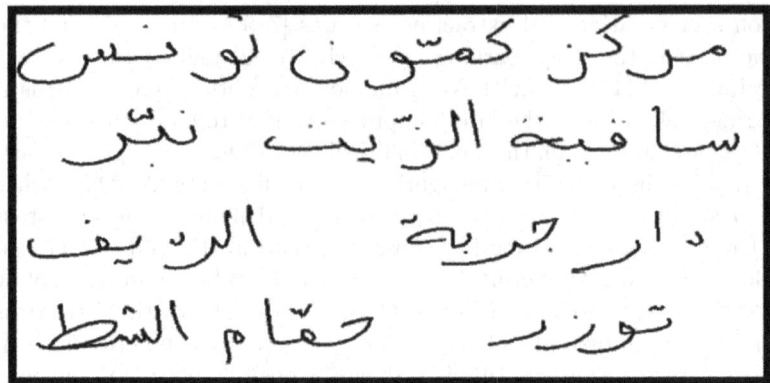

Fig. 1. Some samples from the ADAB dataset.

Table 1. Characteristics of ADAB-datasets 1, 2, and 3

Set	Files	Words	Characters	Writers
1	5037	7670	40500	56
2	5090	7851	41515	37
3	5031	7730	40544	39
Sum	15158	23251	122559	132

Database (An inkml file including trajectory information). The test data used for this competition consists of 200 words. The competition aims to promote innovation in Online Arabic Words Recognition, as well as to provide objective and fair comparisons among methods. The ranking of participants is based on word recognition rate with respect to test samples. Participants will be given a limited time to submit their results after the competition started. The OAWRC 2021 challenges are continuous including effort. The OWRC 2021 competition is the first edition will organize in the context of ASAR 2021 and will take place in another future events.

3 Participating Systems

In this section, we present the systems submitted to the competition. A brief description has been provided by the system's authors and edited by the competition organizers.

3.1 REGIM-FPC-Segmentation

REGIM-EPC-Segmentation is submitted by Hanen Akouaydi, Houcine Boubaker, Sourour Njah, Mourad Zaied and Adel M. Alimi belonging to REGIM-Lab (REsearch Groups in Intelligent Machines Regim). This system is based on fuzzy perceptual codes (FPC) and deep recurrent neural networks. The developed system proceeds by a preprocessing module to eliminate the acquisition imperfections. The second step consists of detecting elliptic strokes and classify them into FPC. Then, we use constructed features vector with LSTM for adab online words recognition [15]. The main assumption of our proposed system consists in the fact that handwriting is a group of visual codes grouped together so as to get a shape, a character or a digit. The principal perceptual codes to write are: $-$, $/$, $|$,n. [15,16].

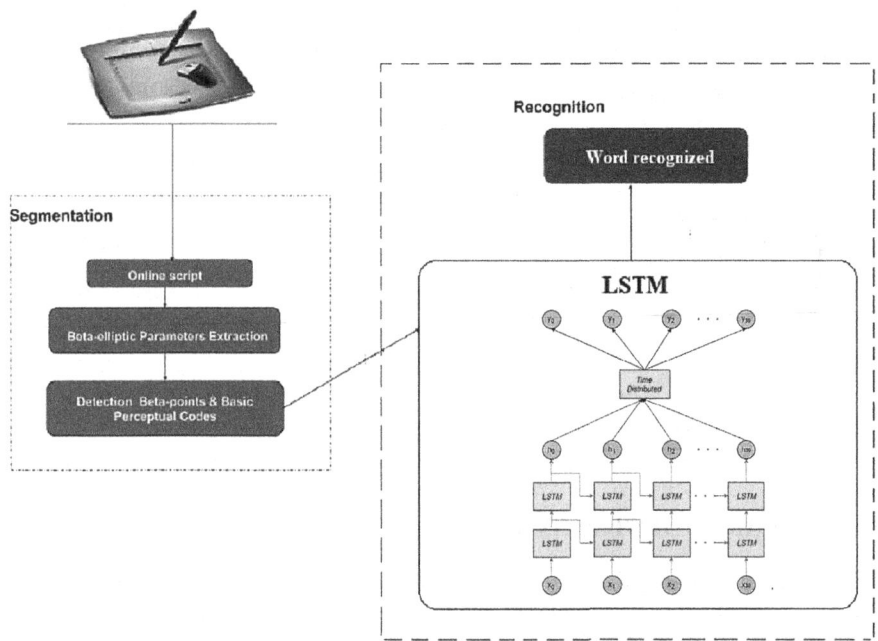

Fig. 2. Architecture of REGIM GS Segmentation system.

3.2 REGIM-GS-Segmentation

REGIM-GS-Segmentation system of Yahia Hamdi, Houcine Boubaker and Adel M. Alimi from REGIM-Lab (REsearch Groups in Intelligent Machines Regim) at National School of Engineers, University of Sfax Tunisia. This system is based on graphemes segmentation (GS) strategy and deep recurrent neural network. It is presented in Fig. 3. The developed system proceeds by a preprocessing module using handwriting normalization size and low-pass filtering Chebyshev type II to eliminate the acquisition system's noise. The second step consists of detecting the baseline by considering concordance between the alignment of the handwriting trajectory points and their tangent directions. Thereafter, the handwritten trajectory is decomposed in continuous part called graphemes delimited by the

ligature bottoms points neighboring the baseline. Then, a set of pertinent parameters combining dynamic and static features are extracted by using Beta-elliptic
model, Fourier descriptors parameters for trajectory shape modeling, and other
normalized features for grapheme dimensions modeling, positions relative to the
baseline and the assignment diacritics codes. The constructed features vectors
are classified with deep BLSTM/LSTM for word recognition module [10]. This
system proposes two architectures one with BLSTM and another with BLSTM.

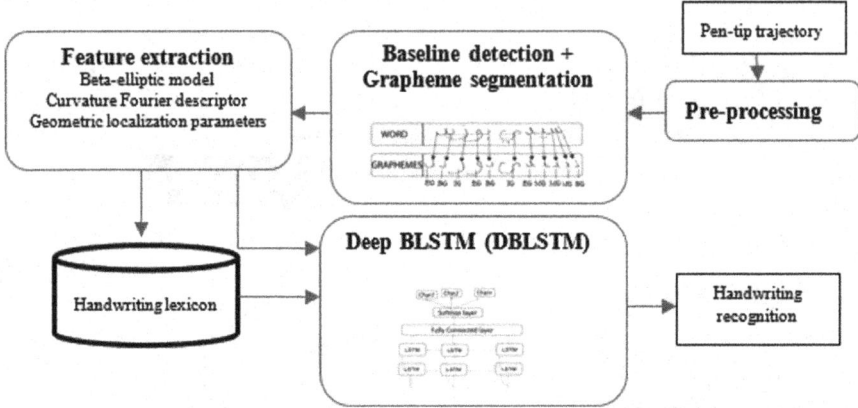

Fig. 3. Architecture of REGIM GS Segmentation system.

3.3   REGIM-DBLSTM-SVM

REGIM-DBLSTM-SVM system of Yahia Hamdi, Houcine Boubaker, Besma
Rabhi, Wael Ouarda and Adel M. Alimi belonging to REGIM-Lab (REsearch
Groups in Intelligent Machines Regim) at National School of Engineers, University of Sfax Tunisia. The proposed system composes of multi-stage architecture
of deep learning networks associated with effective feature vectors that integrate dynamic and visual parameters. It segments acquired script into Segments
of Online Handwriting Trajectories (SOHTs). Then, it extracts two types of
feature: Beta-Elliptic Model (BEM) and Convolutional Neural Network (CNN)
from each SOHT in order to fuzzy classify them into k sub-groups by using
DBLSTM neural networks. Finally, it combines the trained models by using
SVM engine. This work is a combination of hybrid online and line models based
on DBLSTM and SVM for word recognition [18].

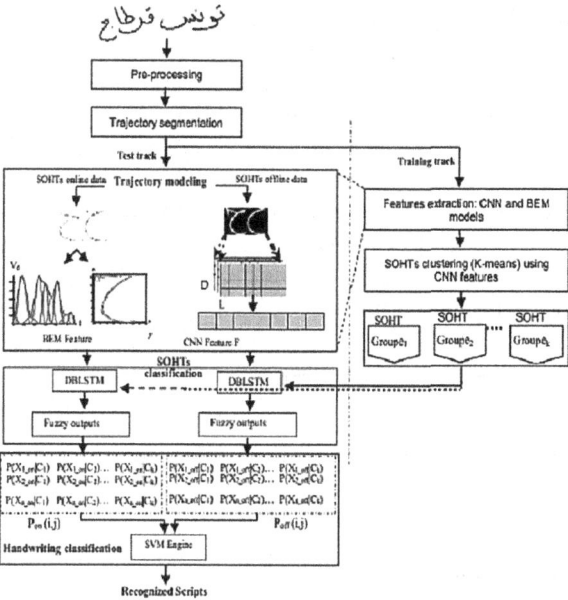

Fig. 4. Architecture of REGIM DBLSTM SVM.

4 Tests and Results

In this section, 2 different online Arabic handwriting recognition systems are evaluated in two steps. In a first step we used a new set unknown to all participants. Then in the next step, we compare the speed performance of the 2 systems.

4.1 General Remarks

The participants sent us running versions of their recognition systems trained on the provided train database. All systems are developed under Microsoft Windows environment. An unknown set with 200 Arabic words was used to test the functionality of the systems. Each system passed this test.

4.2 Challenge 1

Working with small training set is our first challenge. All participating systems need to apply data augmentation due you to the few numbers of training samples. Some of systems [10,18] introduce new techniques of data augmentation that will be illustrated in this section.

4.2.1 Classical Data Augmentation

This deformation technology is to apply shape variation and to generate numerous online training data. This transformation aims to extend the data-set by applying affine transformations including scaling, rotations, and translations. Stroke jiggling is also used to generate local distortions to enrich our data with local diversity. Those techniques of data augmentation generate more samples and also to prevent over-fitting, and to make the learning more uniform. In fact, system **REGIM-FPC-Segmentation** uses this kind of data augmentation. In fact, it also applies the geometric transformations such talicity angle, baseline inclination angle as presented in Fig. 5.

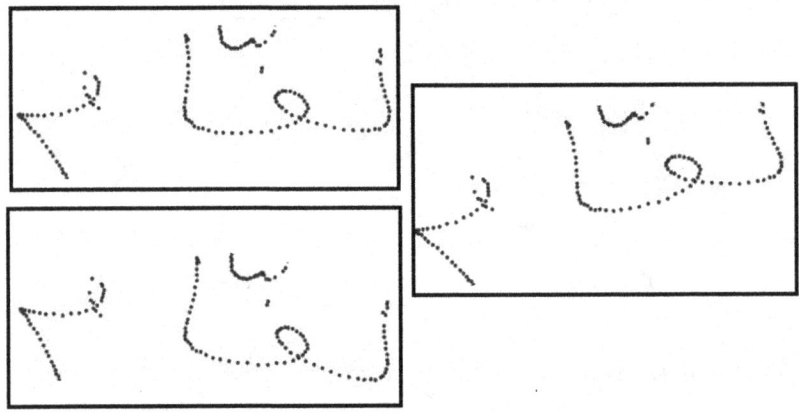

Fig. 5. Generation using geometric transformation.

4.2.2 New Approach of Data Augmentation

In this competition, system 2 & system 3 apply new data augmentation methods that generate more data to improve the performance of recognition systems due to our small training dataset. Various data augmentation techniques are employed. They use a strategy that attenuates or amplifies the trajectory with high harmonics to generate different handwriting styles. Their new technique uses the beta-elliptic model to extract a combined static and dynamic representation of the handwritten trajectory that undergoes a random change by the introduction of white noise to produce new geometric parameters in order to generate more scripts, as illustrated in the figure Fig. 6 above [14]. They also classical techniques of data augmentation. In fact, Table 2 illustrates how those deformation technologies raise the number of samples of training set. They use 4 methods to generate more data.

4.3 Challenge 2

Our second challenge is having a good recognition rate. The important results of our tests are illustrated in Table 3. The best result is marked in bold font.

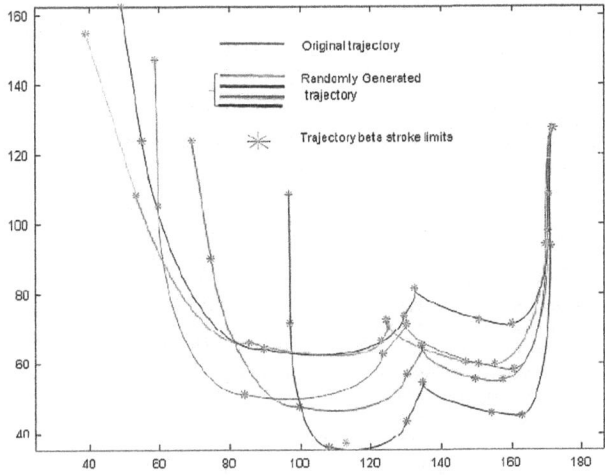

Fig. 6. Generation using beta-elliptic model by modifying geometric parameters.

Table 2. Characteristics of ADAB-datasets 1, 2, and 3

Original-data	576,7 (70% of the original dataset)
ADAB-geometric	2270,8 (576,7 × 3 + 576,7)
ADAB-frequency	2270,8(576,7 × 2 +576,7)
ADAB-beta-elliptic	2270,8
ADAB-hybrid	(2270,8 × 3 + 2270,8 × 2)

We had three submissions with four different architectures. The two first participants achieve a very good performance, with an error rate lower than 2%. It is interesting to point out that they use different approaches, one based on kinematic and geometric features,

Table 3. Recognition results in %

Systems	Features	Recognition rate
REGIM-FPC-LSTM	Fuzzy PC	96,50%
REGIM-GS-LSTM	Graphemes segmentation	96,65%
REGIM-GS-DBLSTM	Graphemes segmentation	**98,73%**
REGIM-DBLSTM-SVM	BEM+CNN	98,72%

4.3.1 Main Test

The most important test to compare the different systems by using the new test set. Interesting results are illustrated Table 2.

- Four systems have more than 96% recognition rate on unknown set, the variation of best system between top 1, top 5 and top10 is also tested.
- The best system has a recognition rate of almost 2% higher than the second-best system.

4.3.2 Speed Tests

The average processing time per word of the test, time features extraction and response, is presented in Table 4. Remarkable difference in speed can be observed between the 4 systems. Fastest system is more than 380 times faster than the slowest one. In fact, the last system take more time in response because it extracts online features by Beta-elliptic model and offline features with CNN and classify them with LSTM. The winner system **REGIM-GS-BLSTM** in recognition rate take more time because it uses beta-elliptic and Fourier descriptors parameters. The **REGIM-FPC-LSTM** is the faster because it extract only perceptual beta-elliptic features.

Table 4. Speed test

Systems	Time (ms)
REGIM-FPC-LSTM	**430**
REGIM-GS-LSTM	580
REGIM-GS-BLSTM	650
REGIM-BLSTM-SVM	780

4.4 Challenge 3

In order to study the robustness of the participating recognizer systems, a randomly noise signal is applied in our test database. Noise in handwriting can be caused by the support of writing, the person who write that produces trembled script. In our competition OWRC 2021, we use white noise or gaussian noise with differents values in order to produce noisy samples. Table 5 demonstrates that the Word Error Rate (WER) of all systems are not too affected and resist facing noise. All systems with Beta-elliptic features as a matter of fact yield best accuracy and even REGIM-BLSTM-SVM that uses CNN. The Beta-elliptic method [23,24,33–39] can improve the online script and eliminate, showing its robustness with noise. Therefore, all participating systems pass on its preprocessing step can improve the quality of online script via removing noise and applying sampling. Beta-elliptic model can eliminate trembling from script. Table 5 illustrates the Word Error Rate (WER) of participating systems with differents values of noise. Actually, REGIM-GS-DBLSTM resists with noise more than other participating systems.

Table 5. Recognition rate ADAB database with noise.

Systems	WER	WER Noise1 (0.2)	WER Noise1 (0.6)
REGIM-FPC-LSTM	3,5	3,8	3,9
REGIM-GS-LSTM	3,35	3	3,7
REGIM-GS-DBLSTM	**1,27**	2	**2,8**
REGIM-DBLSTM-SVM	1,28	2,2	2.9

5 Conclusion

Competition results show that Arabic handwriting recognition systems made a remarkable progress. Participating systems show a very high accuracy and also a very high speed. Used features of the systems cannot be presented in this short paper. The system**REGIM-GS-DBLSTM** is the winner of this competition that have the WER even with noise and **REGIM-FPC-LSTM** is the system with the shortest average processing time. The challenges of OAWRC 2021 are continuous. The OWRC 2021 competition is the first edition will organize in the context of ASAR 2021 and will maintain in another future events.

Acknowledgments. The research leading to these results has received funding from the Ministry of Higher Education and Scientific Research of Tunisia under the grant agreement number LR11ES48.

References

1. Märgner, V., Abed, H.: ICDAR 2007 - Arabic Handwriting Recognition Competition. In: ICDAR, pp. 1–5 (2007)
2. Abed, H., Märgner, V., Kherallah, M., Alimi, A.M.: ICDAR 2009-online Arabic handwriting recognition competition. In: ICDAR, pp. 1–5 (2009)
3. Abed, H., Märgner, V.: ICDAR 2011 - Arabic handwriting recognition competition. In: International Conference on Document Analysis and Recognition, pp. 1444–1448 (2011)
4. Boubaker, H., Elbaat, A., Tagougui, N., Kherallah, M., Abed, H., Alimi, A.M.: Online Arabic Databases and Applications, pp. 541–557 (2012)
5. Kherallah, M., Tagougui, N., Alimi, A.M., El Abed, H., Margner, V.: Online Arabic handwriting recognition competition. In: International Conference on Document Analysis and Recognition, pp. 1454–1458 (2011)
6. Abed, H., Kherallah, M., Märgner, V., Alimi, A.M.: Online Arabic handwriting recognition competition ADAB database and participating systems. In: International Conference on Document Analysis and Recognition, pp. 1–5, July 2010
7. Hamdi, Y., Boubaker, H., Dhieb, T., Elbaati, A., Alimi, A.M.: Hybrid DBLSTM-SVM based beta-elliptic-CNN models for online Arabic characters recognition. In: ICDAR, pp. 545–550 (2019)
8. Akouaydi, H., Njah, S., Ouarda, W., Samet, A., Zaied, M., Alimi, A.M.: Convolutional neural networks for online Arabic characters recognition with beta-elliptic knowledge domain. In: ICDARW, pp. 1–6 (2019)

9. Akouaydi, H., Njah, S., Alimi, A.M.: Android Application for handwriting segmentation using PerTOHS theory. In: Ninth International Conference on Machine Vision, ICMV, pp. 1–5 (2016)

10. Hamdi, Y., Boubaker, H., Alimi, A.M.: Online Arabic handwriting recognition using graphemes segmentation and deep learning recurrent neural networks. In: Hassanien, A.E., Darwish, A., Abd El-Kader, S.M., Alboaneen, D.A. (eds.) Enabling Machine Learning Applications in Data Science. AIS, pp. 281–297. Springer, Singapore (2021). https://doi.org/10.1007/978-981-33-6129-4_20

11. Hamdi, Y., Chaabouni, A., Boubaker, H., Alimi, A.M.: OffLexicon online Arabic handwriting recognition using neural network. Proc. SPIE **10341**, 1–5 (2017)

12. Rabhi, B., Elbaati, A., Hamdi, Y., Alimi, A.M.: Handwriting recognition based on temporal order restored by the end-to-end system. In: ICDAR, pp. 1231–1236 (2019)

13. Graves, A.: Generating Sequences with Recurrent Neural Networks. arXiv: 1308.0850, pp. 1–5 (2014)

14. Hamdi, Y., Boubaker, H., Alimi, A.M.: Data augmentation using geometric, frequency, and beta modeling approaches for improving multi-lingual online handwriting recognition. Int. J. Doc. Anal. Recogn. (IJDAR), 1–17 (2021). https://doi.org/10.1007/s10032-021-00376-2

15. Akouaydi, H., et al.: Neural architecture based on fuzzy perceptual representation for online multilingual handwriting recognition. arXiv preprint arXiv:1908.00634, pp. 1–14 (2019)

16. Pinales, J.R.: Reconnaissance hors-ligne de lcriture cursive par lutilisation de modles perceptifs et neuronaux. University of Paris, Ph.D. (2002)

17. Malaviya, A., Peters, L., Camposano, P.: A fuzzy online handwriting recognition system: FOHRES. In: Proceedings of the International Conference on Fuzzy Theory and Technology, pp. 1–15 (1993)

18. Hamdi, Y., Boubaker, H., Rabhi, B., Ouarda, W., Alimi, A.: Hybrid architecture based on RNN-SVM for multilingual handwriting recognition using Beta-elliptic and CNN models. TechRxiv, pp. 1–21 (2021)

19. Carbune, PV., et al.: Fast multi-language LSTM-based online handwriting recognition. Int. J. Doc. Anal. Recogn. (IJDAR) **2**, 89–102 (2020)

20. Akouaydi, H., Abdelhedi, S., Njah, S., Zaied, M., Alimi, A.M.: Decision trees based on perceptual codes for on-line Arabic character recognition. In: 1st International Workshop on Arabic Script Analysis and Recognition, pp. 1–6 (2017)

21. Zitouni, R., Bezine, H., Arous, N.: Online handwritten Arabic scripts recognition, using stroke-based class labeling scheme. Int. J. Comput. Intell. Syst. **14**, 187–198 (2020)

22. Tagougui, N., Kherallah, M.: Recognizing online Arabic handwritten characters using a deep architecture. In: Ninth International Conference on Machine Vision (ICMV 2016), vol. 10341, pp. 107–111. International Society for Optics and Photonics, SPIE (2017)

23. Alimi, A.M.: An evolutionary neuro-fuzzy approach to recognize on line Arabic handwriting. In: 4th International Conference Document Analysis and Recognition (ICDAR 1997), pp 382–386 (1997)

24. Bezine, H., Alimi, A.M., Sherkat, N.: Generation and analysis of handwriting script with the beta-elliptic model. In: Ninth International Workshop on Frontiers in Handwriting Recognition, pp. 515–520 (2004)

25. Eraqi, H.M., Azeem, S.A.: An on-line Arabic handwriting recognition system Based on a new on-line graphemes segmentation technique. In: International Conference on Document Analysis and Recognition, pp. 409–413 (2011)

26. Keysers, D., Deselaers, T., Rowley, H.A., Wang, L.L., Carbune, V.: Multilanguage online handwriting recognition. IEEE Trans. Pattern Anal. Mach. Intell. **39**, 1180–1194 (2017)
27. Zhang, X.-Y., Bengio, Y., Liu, C.-L.: Online and offline handwritten Chinese character recognition: a comprehensive study and new benchmark. Pattern Recogn. **61**, 348–360 (2017)
28. Kubatur, S., Sid-Ahmed, M., Ahmadi, M.: A neural network approach to online Devanagari handwritten character recognition. In: 2012 International Conference on High Performance Computing Simulation (HPCS), pp. 209–214 (2012)
29. Ghosh, R., Vamshi, C., Kumar, P.: RNN based online handwritten word recognition in Devanagari and Bengali scripts using horizontal zoning. Pattern Recogn. **92**, 203–218 (2019)
30. Pranoto, Y.M., Setyati, E., Pramana, E., Kristian, Y., Budiman, R.: Real time handwriting recognition for mathematic expressions using Hidden Markov Model. In: International Seminar on Intelligent Technology and Its Applications (ISITIA), pp. 1–6 (2016)
31. Hamdi, Y., Chaabouni, A., Boubaker, H., Alimi, A.M.: Hybrid neural network and genetic algorithm for off-lexicon online Arabic handwriting recognition. In: Abraham, A., Haqiq, A., Alimi, A.M., Mezzour, G., Rokbani, N., Muda, A.K. (eds.) HIS 2016. AISC, vol. 552, pp. 431–441. Springer, Cham (2017). https://doi.org/10.1007/978-3-319-52941-7_43
32. Rabhi, B., Dhahri, H., Alimi, A.M., Alturki, F.A.: Grey wolf optimizer for training Elman neural network. In: Abraham, A., Haqiq, A., Alimi, A.M., Mezzour, G., Rokbani, N., Muda, A.K. (eds.) HIS 2016. AISC, vol. 552, pp. 380–390. Springer, Cham (2017). https://doi.org/10.1007/978-3-319-52941-7_38
33. Rabhi, B., Elbaati, A., Boubaker, H., Hamdi, Y., Hussain, A., Alimi, A.M.: Temporal order and pen velocity recovery for character handwriting based on sequence-to-sequence with attention mode. TechRxiv. Preprint, pp. 1–10 (2021). https://doi.org/10.36227/techrxiv.13902650.v1
34. Njah, S., Bezine, H., Alimi, A.M.: A fuzzy genetic system for segmentation of on-line handwriting: Application to ADAB database, pp. 95–102 (2011)
35. Dhieb, T., Ouarda, W., Boubaker, H., Hlima, M.B., Alimi, A.M.: Online Arabic writer identification based on beta-elliptic model, pp. 74–79 (2015)
36. Dhieb, T., Njah, S., Ouarda, W., Boubaker, H., Ayed, M.B., Alimi, A.M.: An online writer identification system based on beta-elliptic model and fuzzy elementary perceptual codes, p. 12 (2018)
37. Njah, S., Bezine, H., Alimi, A.M.: A new approach for the extraction of handwriting perceptual codes using fuzzy logic, pp. 302–307 (2008)
38. Dhieb, T., Njah, S., Ouarda, W., Boubaker, H., Ayed, M.B., Alimi, A.M.: Towards a novel biometric system for forensic document examination. Comput. Secur. **97**, 1–17 (2020)
39. Dhieb, T., Rezzoug, N., Boubaker, H., Gorce, P., Alim, A.M.: Effect of age on hand drawing movement kinematics. Comput. Methods Biomech. Biomed. Eng. **22**, 188–190 (2019)
40. Akouaydi, H., Abdelhedi, S., Njah, S., Zaied, M., Alimi, A.M.: Decision trees based on perceptual codes for online Arabic character recognition, pp. 153–157 (2017)

ICDAR 2021 Workshop on Computational Document Forensics (IWCDF)

IWCDF 2021 Preface

We are glad to welcome you to the proceedings of the 3rd edition of the International Workshop on Computational Document Forensics (IWCDF 2021), which built on the success of the two previous editions held in Kyoto, Japan (IWCDF 2017) and Sydney, Australia (IWCDF 2019).

IWCDF 2021 is an event that aims at addressing theoretical and practical works related to computational forensics applied to documents and creating a space for discussions between people working on these issues in different areas such as document and speech processing, digital security, biometry, and forensic sciences.

Everywhere around the world, industries and government processes are being more and more digitized. Document management systems and digital safe-boxes are particularly concerned with these questions, since documents generally remain the basis of many decisions for transactions, contracts, communication..., and documents also remain the proofs for many legal issues. As a consequence, it becomes absolutely essential to develop computational forensic science applied to documents and to create the conditions for protecting documents, for confirming their authenticity or detecting forgeries.

Like in previous years, the workshop aimed at building a forum for in-depth discussion of the work taking place in both academia and industry, documenting the advances in the related fields and creating mutual collaboration on related areas.

The workshop received six submissions, and each of them was reviewed carefully. The Program Committee took into account the relevance of the papers to the workshop, the technical merit, the potential impact, and the originality and novelty. From these submissions, five papers were selected.

We hope you will enjoy the workshop and we are looking forward to welcome you to IWCDF 2021.

Nicolas Sidère
Imran Ahmed Siddiqi
Jean-Marc Ogier
Chawki Djeddi
Haikal El Abed
Xunfeng Lin

Organization

Workshop Chairs

Nicolas Sidère   La Rochelle Université, France
Imran Ahmed Siddiqi   Bahria University, Pakistan
Jean-Marc Ogier   La Rochelle Université, France
Chawki Djeddi   Larbi Tebessi University, Algeria
Haikal El Abed   Technische Universitaet Braunschweig,
                                 Germany
Xunfeng Lin   Deakin University, Australia

Recognition of Laser-Printed Characters Based on Creation of New Laser-Printed Characters Datasets

Takeshi Furukawa[✉][iD]

Forensic Science Laboratory, Ibaraki Prefectural Police Headquarters, Mito, Japan
tfurukawa@ieee.org

Abstract. We report a new method for the recognition of laser-printed characters. We create new datasets to confirm the accuracy of our methods. When creating our datasets, printing counts are controlled, and printed character type is standardized. We purchase eight brand new manufactures laser printers and printed characters on one thousand paper sheets at each manufacture brand. We chose pages 1, 2, 299, 300, 499, 500, 799, 800, 999, and 1000 from one thousand printed sheets. One hundred printed characters on each paper sheet were captured with a high-resolution CCD camera attached with a microscope and are arranged to datasets. We extracted contours of the characters' images of the datasets and traced x and y coordinates on the contours. Each of the coordinate was considered as a wave and was analyzed with eighth scale wavelet decomposition. We recognize between the eight brands of laser printers based on the decomposition result. In the learning phase, two hundred numbers of 'a' and forty four numbers 'e' characters printed on each sheet was leaned while in the testing phase remains of eight hundred numbers of 'a' and forty hundred 'e' characters printed on the same sheets are tested. The result shows the error rate (ER) of recognition is 2.78% using the subspace-based method. In addition, the ER of recognition is 5.67% using a support vector machine and 2.56% using a convolutional neural network. As a result, we confirm that our method is able to recognize manufacturers and brands of the laser-printed characters using our datasets.

Keywords: Laser printers · Wavelet decomposition · Subspace-based method · SVM · CNN

1 Introduction

Laser printers are widely used in offices because of their high printed speed and low cost. Recently they are also used at home because they are now available in a compact size and are falling prices. In crime scenes, stalkers often print threaten

Supported by Japan society for the promotion of science (JSPS) KAKENHI Grant Number 20H01165.

E. H. Barney Smith and U. Pal (Eds.): ICDAR 2021 Workshops, LNCS 12916, pp. 407–421, 2021.
https://doi.org/10.1007/978-3-030-86198-8_29

letters by using a laser printer. Police officers ask forensic document examiners to specify manufacturers and laser printers' brands to confirm a suspect crime. A conventional method for the recognition of laser printers is chemical analysis from toners. The chemical analysis is classified into two methods such as destructive and non-destructive. In chemical analysis, destructive methods are usually used; meanwhile, non-destructive methods keep criminal evidences. To recognize manufacturers and brands of laser-printed characters, we have already developed a non-destructive method [12,13]. The method we developed uses magnified images of laser-printed characters captured with a high-resolution flatbed scanner, and in feature extraction phases, we used wavelet decomposition. In the recognition phase, we used subspace-based methods. A conceptual diagram we used is shown in Fig. 1. Details of our proposed method are described in Sect. 2. Although the method we proposed showed significant results of recognition of laser-printed characters properties, the printed counters were not able to be controlled systematically, i.e.; printers used in the experiments were working on daily routine, so that range between printed counters were wide and printed character type and image were different each other. As a result, we have to plan a new experiment that is able to be controlled printed counters so that we purchase eight manufactures and brands new laser printers. We print samples by using the new printers and confirm the reliability of our proposed method. We also used SVM (Libsvm) and CNN (ResNet-50) as the same as the subspace-based methods. The details of the procedures are described in the following section. The main contribution of the paper we propose is followed.

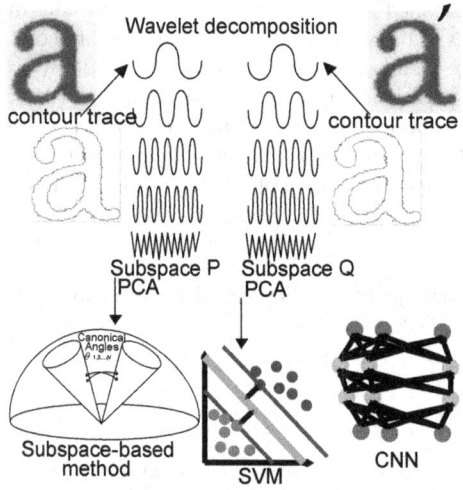

Fig. 1. The conceptual diagram of the proposed method. Taken from [Furukawa2015]

Table 1. Related work

Author	Features	Printers	Kinds	Resolution(dpi)	Recognition	Accuracy(%)	Year
Mikkikineni	variance / entropy GLCM	laser	10	2400	5-NN classifier	over 50	2005
Lampert	edge roughness	laser between inkjet	laser 8 ink 5	3200	RBF-SVM	94.8	2006
Gupta	GLCM	laser and inkjet	laser 2 ink 2	microscope 125×	only quantified	not calculated	2007
Wang	ratio between stroke and whole	laser laser	10 printouts 1000	professional scanner	fuzzy synthetic	84.2	2007
Schulze	edge roughness	laser between inkjet	laser printouts 49 ink printouts 14	400	SVM Muti -layer NN	over 80	2008
Schulze	DCT	laser between inkjet photocopies	laser printouts 49 ink printouts 14 photo printouts 46	200- 300- 400 -800	SVM	99.08 (800dpi) 92.92 (400dpi) 80.85 (200dpi)	2009
Gebhard	edge roughness	laser and inkjet	laser 13 ink 7 printouts 200	400	k-NN Grubb's test	1.16 1.00 is perfect	2013
Furukawa	edge profiles wavelet decomposition number of zero-crossing point	laser	3 printouts 30	5400	SQRT	82.0	2013
Elkasraw	subtracting image	laser and inkjet	laser 13 ink 7	400	SVM	76.75	2014
Furukawa	edge profiles wavelet decomposition all ten scales profiles	laser	10 printouts 200	5400	subspace -based method	76.36 17.27 (ERR)	2015
Wang	Fourier-Laplacian coefficients of a text line	laser	50 2500 images	400	SVM	90.1 Wang dataset 90.0 DKFI dataset	2017

1. We create new datasets controlled print counts and printed type characters for computer and forensic science researchers.
2. We confirm our proposed method to recognize printer manufactures and brands using our new datasets.

2   Related Work

We study related works in the following fields; Document forgery detection, forensic document examination using machine learning, and wavelet.

2.1   Document Forgery Detection

Recognition of laser-printed characters was studied for detecting altered or forged official documents such as identification cards, receipts, and invoices scanned at a high resolution of 3200 dpi [15,18,24,25] as shown in Table 1. Lampert et al. proposed printing technique classification from sharp edges on printed characters. They extracted the following qualities to form a feature vector: Line Edge Roughness, Area Difference, Correlation Coefficient, Texture and classified using support vector machine with Gaussian Kernel (RBF-SVM). They tested on a dataset consisted of 26 printouts of 8 laser and 5 inkjet printouts. This resulted in a classification accuracy of 94.8% [18]. Gupta proposed for detecting a fraudulent document and fixing it to color laser printers or color inkjet printers. They used three indicators: the variance of the intensity of the image, the total number of unique color count, and gray level co-occurrence matrix uniformity (GLCM) [15]. German Research Center for Artificial Intelligence (DFKI) researchers Schulze et al. gathered 49 different laser printed samples captured with image scanners. The method they proposed was able to detect altered parts of documents. They used edge-based gray-level features, which showed edge roughness as indicators. They pointed out the cause of edge roughness was dependent on the noise which the printers produced. Their methods used relatively low-resolution images, i.e., 400 dpi, scanned with low-cost flatbed scanners. They used machine learning methods such as SVM and multilayers neural networks and obtained over 80% accuracy rate [25]. They also proposed a method to use a frequency domain approach for document printing technique recognition. Their experimental results demonstrated that using discrete cosine transformation (DCT) coefficients and machine learning, they were able to distinguish between inkjet and laser printed documents and also detect first-generation photocopies at low scan resolution [24]. Elkasrawi et al. also reported printer identification using supervised learning [6]. They proposed the method to automatically identify source printers using common resolution scanners (400 dpi). They used distinctive noise produced by printers. The classification accuracy they obtain using SVM was 76.75%. Gebhardt et al. identified document authentication using printing technique features and unsupervised anomaly detection [14]. They have created a dataset featuring 1200 document images from different domains (invoices, contracts, scientific papers) printed by 7 different inkjet and 13 laser printers called DFKI Printing Technique Dataset on the web [4]. They used edge roughness or degeneration features, i.e., standard deviation of pixel gray values along vertical edges of characters. In recognition, they used global k-NN, a nearest neighbor based approach, and obtained 1.16 accuracy (1.00 was perfect). Mikkilineni et al. described image texture analysis to identify laser printers used to print a document. They used features based on gray level co-occurrence features (GLCM) from documents scanned at 2400dpi and 5-nearest-neighbor (5NN) classifier in the recognition process. Percent correct classification at the output of the 5NN classifier for all features was over 55% [22]. Wang et al. proposed a method to identify the type of laser printers according to a printed character image [26]. Their method was that the character image was segmented into 16 equal-area quadrate sections. Specific area of the ratio between an area of stroke and an area of whole character image is the indicator. They used the correlation coefficient between them was calculated, and fuzzy synthetic identification was conducted. The maximum accuracy rate they obtained was 84.2%. Wang et al. [27] proposed Fourier-residual for printer identification based on texture and spatial features extracted from residual images. Their prosed method was able to obtain over 90% accuracy when using high-resolution samples.

2.2 Causes of Differences on Laser Printed Characters Qualities

Difference in Electrophotography Development Process. The methods recently used in printing are inkjet and laser. We focus on the laser printing method, especially monochrome printing. Forensic document examiners often observe differences among magnified images that are printed with laser printers. Several causes which derived from differences among manufactures and brands are considered [7]. DFKI researchers expressed the difference among the different manufactures and brands as noise [6,18,25]; however causes conducted noise was not described. Accordingly, we investigate the literature. According to Ito et al. [17], methods in the development process are two. One of them is a single component method, in which only toner is used in the developer process; another is the two-component method, in which both toner and carriers are used in the process. The former merit is to make printers compact; however, the demerit is printed images are so degraded. The later merit is stable of printed image qualities; however, demerit is machine construction is complicated. In our experiments, we collect only single component laser printers; meanwhile the previous our experiments both single and two components laser printers were used together. Accordingly, the previous experimental results depended on the difference of development methods, i.e., single or two components, so that the numbers of components are regulated only one in this experiment. We study the above-related work and find a feature that is able to distinguish printer properties. The most available feature that we find in the above previous studies is edge roughness in printed characters. In addition, we classify edge roughness from details to approximation at several scales. Conveniently, edges are considered profiles such as signal waves, so that a lot of analysis methods for signal processing such as fast Fourier transform (FFT) and discrete cosine transform (DCT) which was used by Schulze et al. as described before [24]. However, FFT and DCT have a fatal defect that is not able to detect multi-scale differences from details to approximation. We resolve the problem using wavelet decomposition. Details are described in Sect. 3. One of our studies' shortages is that there is no common datasets: controlled print counts and printed type characters.

Another research field such as face recognition, handwritten character recognition, has used common datasets. We have to create new common datasets for laser printed character recognition. The details of our new datasets are described in the following section.

3   Method

3.1   Creation of New Datasets

We create datasets to verify our methods. The datasets contain characters printed with laser printers, which are regularity controlled pages counters. We newly purchase new eight different manufactures and brands printers. The manufactures and the brands are indicated in Table 2 and Fig. 2. Those printers are low-cost brands of which prices under approximately 100$. We print the following English sentences with the printers; 'The quick brown fox jumps over the lazy dog' and 'THE QUICK BROWN FOX JUMPS OVER THE LAZY DOG.' Those sentences are sequentially printed on paper sheets at thirty-seven times. In addition, English characters 'a' are also printed in a header and a footer on the paper sheet. Each printer prints one thousands of paper sheets. We capture one hundred English characters 'a' on pages 1, 2, 299, 300, 499, 500, 799, 800, 999, 1000 of paper sheets using a CCD camera (ProgRes 3012 C14plus, JENOPTIK Laser, Optik System GmbH, Germany) which of image pixel sizes are 4080 × 3072. The camera is attached to a measuring microscope (MF-UD4020D, Mitutoyo, Japan). In addition 'e' in pages 1, 2, 999, 1000 are also captured. The magnified rate is approximately fifty. The one hundred images via one of the paper sheets are captured. The total number of characters is 100 × 10 pages × 8 brands so that 8000 characters of 'a' and 111 × 4 pages × 8 brands, 3552 characters of 'e' are obtained. One of the examples shows in Fig. 3.

3.2   Features

FFT and DCT. As described in the previous section, DFKI researchers found that the edge roughness of laser-printed characters was useful for recognizing brands and manufacturers of laser printers. Thanks to the finding, we consider contours of laser-printed characters as wave signals. The procedures are following. Firstly, as preprocessing, images of laser-printed characters are made binary, and the contours of laser-printed characters are detected using image processing. Next, the contours are traced along x and y axes directions so that two profiles like wave signals are obtained. Finally, the profiles are analyzed as signals. In signal analysis, FFT and DCT are often used. The base is the sine function in FFT while it is cosine in DCT. The base represents the characteristic of signal waves. Both methods have advantages; however, they also have disadvantages.

Table 2. New purchased the printers to create laser-printed characters datasets: CVLAB laser-printed characters datasets

	Manufacture	Brand
1	ELECOM	EPR-LS01W
2	Canon	LBP6030
3	NEC	PR-L5100
4	Brother	HL-L2300
5	KONICA MINOLTA	PagePro 1350W
6	FUJI XEROX	DocuPrint P250 dw
7	OKI	B801
8	Ricoh	IPSio SP3400LE

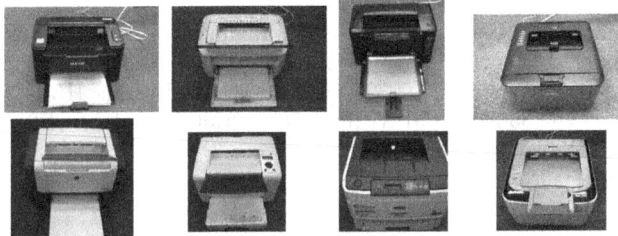

Fig. 2. Overview of laser printers used in the experiment. From Upper left ELECOM EPR- LS01W, Canon LBP6030, NEC PR-L5100, Brother HL-L2300. From lower left, KONICA MINOLTA PagePro 1350W, FUJI XEROX DocuPrint P250 dw PCI 6, OKI B801and Ricoh, IPSio SP3400LE.

For example, one of the advantages of FFT has the high ability to detect subtle changes in the case of using fixed window size, which is adequately fitted on the size of wave changes. One of the advantages of DCT is the ability of high rate data compression. On the contrary, one of the disadvantages in FFT has not flexible adjustments to changes in different scales, which is similar to the case when we need different scales to measure various objects' sizes. FFT is useful in a case when we have already known the size of changes. One of the disadvantages of DCT is in a case when a lot of factors are complicatedly combined. In other words, we are not able to find primary factors that are involved in multiple scales. With taking the advantages and disadvantages described before into considerations, we use wavelet decomposition instead of FFT and DCT when detecting feature extraction. In the following section, we describe the wavelet decomposition we use in our experiments.

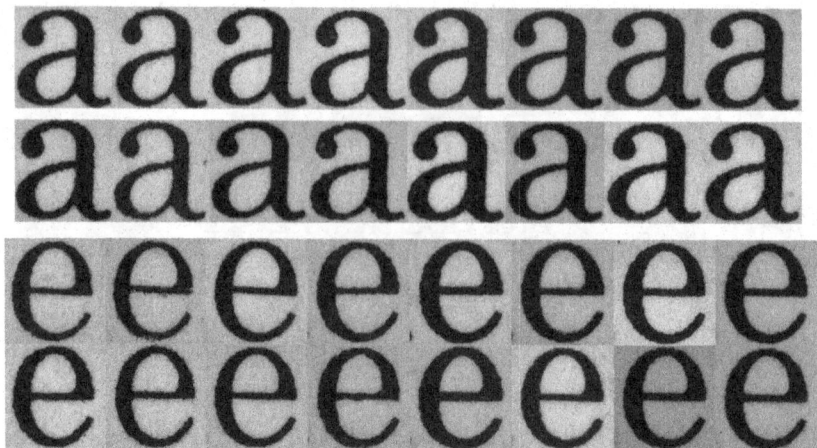

Fig. 3. An example of datasets contained eight manufactures and brands laser printed characters. Upper row is first print. Lower row is 1000 page print. From left ELE-COM EPR- LS01W, Canon LBP6030, NEC PR-L5100, Brother HL-L2300, KONICA MINOLTA PagePro 1350W, FUJI XEROX DocuPrint P250 dw PCI 6, OKI B801and Ricoh, IPSio SP3400L.

Multiscale Wavelet Decomposition. Wavelet has been widely used to analyze signals and images. One of the reasons is distinct from FFT is corresponds to multi resolutions [20], Mallat showed that multi-resolution representation effectively analyzed the information of the content of images and studied the properties of operators that estimated signals at given resolutions. FFT has a window of which size is fixed. Wavelet, meanwhile, has a window of which size is flexible from details to approximation. The wavelet bases are a lot of types such as Haar wavelet, Mexican hat, which are differentiated from gauss distribution profiles. The ability for corresponding multi-resolution is able to detect subtle changes that are added to waves and high frequency, such as noise detection. In addition, the bases in which the wavelet is used are mathematically defined. Daubechies described details of a theory of wavelet [3]. He provided a mathematical foundation to wavelet and had been one of the researchers who developed the wavelet theory. Mayer edited proceedings which described applications of wavelets [21]. This showed that researchers had widely applied wavelet to many fields such as acoustic and signal processing, image processing and fractals, turbulence. As described in previous section, the feature used conventional study is edge roughness derived from noise in which print mechanism involved. However, we add another indicator, i.e., the raster method described in Sect. 2. Our method is to detect contours on laser printed characters and to trace coordinate both along x and y axes separately to obtain a profile like a wave. The profile is decomposed to eighth scales sub-profiles using wavelet as shown in Fig. 4. The first profile indicates noise meanwhile the eighth profile indicates the raster method. The details are described in Sect. 3.

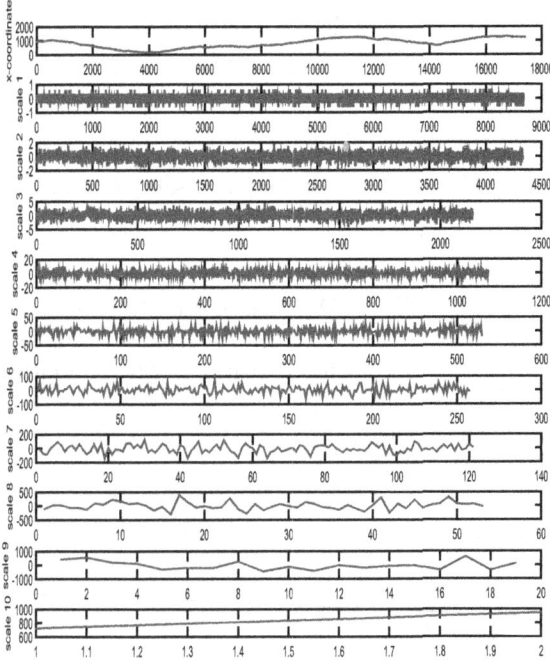

Fig. 4. An example of the results of the wavelet decomposition. The top column showed an original profile. The second column showed the first scale, second, ..., eighth. Taken from [Furukawa2015].

3.3   Recognition

Subspace-Based Methods. As shown in Fig. 1, we used subspace-methods in recognition phase. Subspace-based methods are widely used to be recognized 3D objects and human faces because of ease of extension for multi classes. There are several subspace-based methods. We review the flow of progress of the methods. The basic subspace-based method (SM) uses Karhunen-Lovève expansion [11] or principal component analysis (PCA) to reduce dimensions in input vectors and construct subspaces. Subspace-based methods have been gradually developed by ours and a lot of researchers. We describe the concept of the subspace-based method briefly. Angles between subspaces are indicators for the recognition of objects.

$$\cos^2 \theta = \frac{\sum_{i=1}^{N} (\mathbf{p} \cdot \Psi)^2}{\parallel \mathbf{p} \parallel^2} \tag{1}$$

As shown in Eq. 1, $(\mathbf{p} \cdot \Psi)^2$ denotes inner product between input vector $\mathbf{p}$ and i th orthogonal base vector in dictionary subspace-based. $\parallel \mathbf{p} \parallel$ denotes norm of vector $\mathbf{p}$.

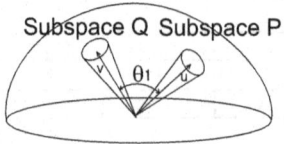

Fig. 5. Conceptual diagram of the mutual subspace method. Taken from [Furukawa2015].

Mutual Subspace Method (MSM). MSM is naturally extended from basic SM. SM makes a reference dictionary in a learning phase from feature vectors. In the test phase, a test feature vector is compared to reference subspace using canonical angles between a feature vector and a reference subspace as shown in Fig. 1. In a test phase, there are plural test feature vectors so that a test subspace i.e., a mutual subspace is constructed in showed Fig. 5. In the test phase, both subspaces are compared using the canonical angles between both subspaces are made [28]. We applied MSM to face recognition using a temporal image sequence.

$$
\cos^2 \theta_i = \max_{\substack{\mathbf{u}_i \perp \mathbf{u}_{j(=1,\ldots,i-1)} \\ \mathbf{v}_i \perp \mathbf{v}_{j(=1,\ldots,i-1)}}} \frac{(\mathbf{u}_i \cdot \mathbf{v}_i)^2}{\| \mathbf{u}_i \|^2 \| \mathbf{v}_i \|^2} \tag{2}
$$

As shown in Eq. 2 $\mathbf{u}_i$ denotes i th input vector subspace, $\mathbf{v}_i$ denotes i th dictionary subspace. In using SM, MSM similarities are defined as shown in Eq. 3.

$$
Similarity = \frac{1}{N} \sum_{i=1}^{N} \cos^2 \theta_i \tag{3}
$$

Constrained Mutual Subspace Method (CMSM). CMSM is a unique concept to the above subspace methods in terms of removing common area between subspaces. The subspace is projected to plane, which is removed common space [9].

Kernel Orthogonal Mutual Subspace Method (KOMSM). Although OMSM has high accuracy to discriminate classes, the accuracy decreases when relationships among classes have non-linear structures. For example, in a case of comparing the human face under various angles and illumination conditions, there are linear deformation such as length between eyes and non-linear deformation such as brightness of image pixels value cascade from the illuminated face angles. Subspaces that distribute non-linearly are difficult to distinguish from each other. To overcome the difficulty of non-linearly, kernel methods are often used such as support vector machine (SVM). When comparing features in subspaces that distribute non-linearly angles or lengths are not the same as linear spaces, features in the subspaces are mapped to corresponding features in multi-dimensions spaces Eq. 4.

$$\phi : \mathbf{x} \rightarrow \phi(\mathbf{x}) = (\phi_1(\mathbf{x}, \ldots, \phi_{d\phi}(\mathbf{x}))^\top \tag{4}$$

$$h(\mathbf{x} \cdot \mathbf{y}) = \exp(-\frac{\| \mathbf{x} - \mathbf{y} \|^2}{2\sigma^2}) \tag{5}$$

To project maps in non-linear spaces to non-linear subspaces, it is necessary to calculate the inner product between the map $\phi(\mathbf{x})$ and $\phi(\mathbf{y})$. It is difficult to calculate the inner product directly because there is data in high dimensions. Defining non-linear transform ϕ through kernel function, $h(\mathbf{x}, \mathbf{y})$, however, inner product, $(\phi(\mathbf{x}) \cdot \phi(\mathbf{y}))$ is able to calculate from original pattern vectors $\mathbf{x}$ and $\mathbf{y}$. This is called the kernel trick to deal with non-linear distribution of data. For example, there is a Gaussian which is used as the kernel function as shown in Eq. 5. Fukui et al. proposed the Kernel Orthogonal Mutual Subspace Method (KOMSM) to resolve the problems [10]. To adapt nonlinear data classes, KOMSM uses kernel trick so that nonlinear distribution is mapped to high dimension spaces to discriminate classes. KOMSM applied methods for face recognition.

Kernel Constrained Mutual Subspace Method (KCMSM). As OMSM is progress to KOMSM, CMSM is also progressing to KCMSM. Fukui et al. applied it to three-dimensional object recognition under multi-view images [8].

Support Vector Machine (SVM). Vapnik et al. proposed SVM [2] was implemented to input vectors that are non-linearly mapped to a very high dimensional feature space. In this feature space, a linear decision surface is constructed. Special properties of the decision surface ensure a high generalization ability of the learning machine. We applied SVM (Libsvm) [1] to recognize printed characters.

Convolutional Neural Network (CNN). LeCun et al. proposed CNN [19] was designed to recognize visual patterns directly from pixel images. They developed LeNet designed for handwritten and machine-printed character recognition. He et al. developed ResNet [16] which had more deep layers. They found the importance of residual error and use them. We applied ResNet-50 to printer identification using our proposed features. The following section shows the details of our experiments.

Table 3. Result of recognition

	Subspace-based					SVM	CNN
	SM	MSM	KMSM	KCMSM	KOMSM	LIBSVM	ResNet-50
a ER%	22.22	20.83	20.83	5.56	2.78	5.67	2.56
a EER%	6.94	15.28	14.09	4.17	1.49		
e ER%	33.88	2.50	2.50	2.50	2.50	5.83	1.64
e EER%	25.59	3.75	3.75	3.75	3.75		

4 Experiments

4.1 Data Acquirement

We conduct the following experiments. The eight manufactures and brands of laser printers are purchased. Table 2 shows the details of the printers. These printers have printed a total of one thousand paper sheets. The printing samples are able to be distinguishable due to be described by the brands, the manufacturers, printed counters and printed dates, times in header and footer of samples. Microsoft Word version 14 is used to describe the sentences in the samples. The written sentences are 'The quick brown fox jumps over the lazy dog. THE QUICK BROWN FOX JUMPS OVER THE LAZY DOG', both lower and upper cases letter sizes. The sentence is repeatedly printed 37 times. The font used is Century, 12 points. The paper used in the experiments are 'Premium white PPCKA4, Kyokuto'. The one hundred laser printed characters 'a' which are printed on page 1, 2, 299, 300, 499, 500, 799, 800, 999 1000, and 'e', which are printed on page 1. 2. 999, 1000 are captured with the CCD camera attached with microscope described Sect. 3. The characters images are made binary using Otsu's method [23]. After the preprocessing, contours on the binary characters are detected. The contours are traced along the x and y axes separately so that two profiles are obtained. The profiles are conducted using wavelet decomposition with eighth scales so that we obtain 'a' profiles, which are composed of 10 pages × 100 characters × 8 printers × 8 scales, total 64,000 and 'e' profiles which are composed of 4 pages × 111 characters × 8 printers × 8 scales, total 28,416. The base used in the experiments is the orthogonal wavelet, as be referred to in the study by Ding et al. [5].

4.2 Data Recognition

We recognize the profiles using subspace-based methods. In addition we also use SVM (Libsvm [1].) with 0.1 RBF kernel and CNN (ResNet-50 [16]). The kinds of subspace-based methods are SM, MSM, KMSM, KCMSM, and KOMSM. In using ResNet-50, batch size is 7, the epoch is 10. Firstly we recognize manufacturers and brands from printed characters in the ten pages and conduct the following experiment. We tested the laser-printed characters in which are printed the above ten pages. We used one hundred 'a' characters on each page so that the total numbers of characters are one thousand. In each method's learning phase,

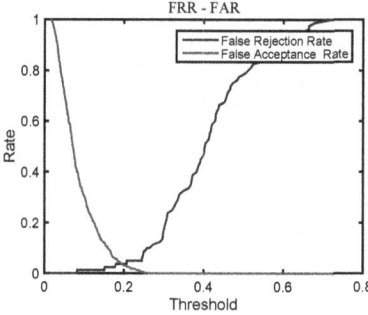

Fig. 6. FRR and FAR of the experiment: recognition using KOMSM.

200 out of 1000 laser printed characters images in ten pages are learned. After the learning phase, the test phase is started with the following procedure. Remains of 800 out of 1000 laser printed characters images in ten pages are tested. Out of 900 images, 50 images are extracted in ascending order and sliding 50 by 50. In addition, we used one hundred eleven 'e' characters on each page. In learning, 44 images out of 444 images are extracted in ascending order and sliding 40 by 40. Each set is calculated to construct a subspace and are tested on a subspace constructed in the learning phase. The number of dimensions is thirty in the test phase, and that of dimension is eight in test phases. The results are shown in the Table 3. In addition, the false rejection rate (FRR) and false acceptance rate (FAR) of the experiment using KOMSM is shown in Fig. 6.

5   Conclusion

5.1   Create New Laser-Printed Characters Dataset

We confirm our proposed method using the new datasets. The datasets are contained printed characters with the eight manufacturers laser printers. In addition, the printers are only used the experiments so that each printer started to print from 1 page to 1,000 pages. We make 8 brands and manufacturers × 1,000 pages; a total of 8,000 sheets of printed paper. In addition, we extract 10 pages and capture 100 'a' from 10 pages and 111 'e' characters from 4 pages so that we construct the datasets that contain 11552 characters. These datasets will be useful for forensic document examiners to confirm the differences among laser printers made from different manufacturers. In the future, we have a plan to open our datasets after we are adjusted the huge sizes of the datasets.

5.2   Confirm the Proposed Method Accuracy

We are able to confirm the accuracy of our method using our created datasets. The error rate was under 5%. Because the new printers we purchased have the same low-quality brands, our methods' performance is adequate results.

Although we use only 'a' and 'e', which are one of 26 kinds of English alphabet characters, we can obtain an adequate equal error rate. In the future, we are going to digitize remain sample images and extend our datasets. We plan our datasets called the computer vision laboratory (CVLAB) laser-printed datasets for public use for forensic science and computer science researchers. They are able to collaborate to help crime investigation or eradication of terrorism.

Acknowledgment. The author would like to thank Prof. Dr. Fukui of the University of Tsukuba for the discussion on machine learning. This work was supported by JSPS KAKENHI Grant Number 20H01165.

References

1. Chang, C.C., Lin, C.J.: LIBSVM: a library for support vector machines. ACM Trans. Intell. Syst. Technol. **2**, 1–27 (2011)
2. Cortes, C., Vapnik, V.: Support-vector networks. Mach. Learn. **20**, 273–297 (1995)
3. Daubechies, I.: Ten Lectures of Wavelets. Springer (1992)
4. Dengel, A.: DFKI printing technique dataset (2013). http://madm.dfki.de/downloads-ds-printing-technique
5. Ding, X., Chen, L., Wu, T.: Character independent font recognition on a single Chinese character. IEEE Trans. Pattern Anal. Mach. Intell. **29**, 195–204 (2007)
6. Elkasrawi, S., Shafait, F.: Printer identification using supervised learning for document forgery detection. In: 2014 11th IAPR Internal Workshop on Document Analysis Systems (DAS), Tours, France, pp. 146–150, April 2014
7. Ferreira, A., Navaro, L.C., Pinherio, G., Santos, J.A., Rocha, A.: Laser printer attribution: exploring new features and beyond. Forensic Sci. Int. **247**, 105–125 (2015)
8. Fukui, K., Stenger, B., Yamaguchi, O.: A framework for 3D object recognition using kernel orthogonal mutual subspace method. In: Proceedings of Asian Conference on Computer Vision (ACCV 2006), Hyderabad, India, pp. 315–324, January 2006
9. Fukui, K., Yamaguchi, O.: Face recognition using multi-viewpoint patterns for robot vision. In: Dario, P., Chatila, R. (eds.) Robotics Research. The Eleventh International Symposium. STAR, vol. 15, pp. 192–201. Springer, Heidelberg (2005). https://doi.org/10.1007/11008941_21
10. Fukui, K., Yamaguchi, O.: The kernel orthogonal mutual subspace method and its application to 3D object recognition. In: Yagi, Y., Kang, S.B., Kweon, I.S., Zha, H. (eds.) ACCV 2007. LNCS, vol. 4844, pp. 467–476. Springer, Heidelberg (2007). https://doi.org/10.1007/978-3-540-76390-1_46
11. Fukunaga, K., Koontz, W.L.G.: Application of the Karhunen-Loève expansion to feature selection and ordering. IEEE Trans. Comput. C **19**(4), 311–318 (1970)
12. Furukawa, T.: A new method for discriminating printers based on contours qualities of printed characters using wavelet decomposition. In: Proceedings of IEEE International Workshop on International Conference on Document Analysis and Recognition (ICDAR 2013), Washington, DC, USA, pp. 1147–1151, August 2013
13. Furukawa, T.: Subspace method with multi scale wavelet for recognition of printer property. In: Proceedings of IEEE International Workshop on International Conference on Document Analysis and Recognition (ICDAR 2015), Nancy, France, pp. 471–475, August 2015

14. Gebhardt, J., Goldstein, M., Shafait, F., Dengel, A.: Document authentication using printing technique features and unsupervised anomaly detection. In: Proceedings of IEEE International Workshop on International Conference on Document Analysis and Recognition (ICDAR 2013), Washington, DC, pp. 479–483, August 2013

15. Gupta, G., Saha, S.K., Chakraborty, S., Mazumdar, C.: Document frauds: identification and linking fake document to scanners and printers. In: Proceedings of the International Conference on Computing: Theory and Applications (ICCTA 2007), Kolkata, India, pp. 497–501, March 2007

16. He, K., Zhang, X., Ren, S., Sun, J.: Deep residual learning for image recognition. In: Proceedings of 2016 Conference on Computer Vision and Pattern Recognition (CVCR 2016), Las Vegas, USA, June 2016

17. Ito, T., Kawamoto, H., Okamoto, H.: Numerical analysis of image defects in single component magnetic development in electrophotography. Trans. Jpn. Soc. Mech. Eng. Ser. C 72(714), 418–425 (2006)

18. Lampert, C.H., Mei, L., Breuel, T.M.: Printing technique classification for document counterfeit detection. In: 2006 International Conference on Computational Intelligence and Security, Guangzhou, China, pp. 639–644, November 2006

19. LeCun, Y., Bottou, L., Bengio, Y., Haffner, P.: Gradient-based learning applied to document recognition. Proc. IEEE 86(11), 2278–2324 (1998)

20. Mallat, S.G.: A theory for multiresolution signal decomposition: the wavelet representation. IEEE Trans. Pattern Anal. Mach. Intell. 11(7), 674–693 (1989)

21. Meyer, Y.: Wavelets and Applications. Masson, Saint-Germain (1991)

22. Mikkilineni, A.K., Chiang, P.J., Ali, G.N., Chiu, G.T.C., Allebach, J.P.: Printer identification based on graylevel co-occurrence features for security and forensic applications. In: Proceedings of the SPIE, vol. 5681, San Jose, USA, pp. 430–440, March 2005

23. Otsu, N.: A threshold selection method from gray-level histograms. IEEE Trans. Syst. Man Cybern. 9(1), 62–66 (1979)

24. Schulze, C., Schreyer, M., Stahl, A., Breuel, T.: Using DCT features for printing technique and copy detection. In: Peterson, G., Shenoi, S. (eds.) DigitalForensics 2009. IAICT, vol. 306, pp. 95–106. Springer, Heidelberg (2009). https://doi.org/10.1007/978-3-642-04155-6_7

25. Schulze, C., Schreyer, M., Stahl, A., Breuel, T.M.: Evaluation of graylevel-features for printing technique classification in high-throughput document management systems. In: Srihari, S.N., Franke, K. (eds.) IWCF 2008. LNCS, vol. 5158, pp. 35–46. Springer, Heidelberg (2008). https://doi.org/10.1007/978-3-540-85303-9_4

26. Wang, N., Han, G.: Laser printer fuzzy identification based on correlative specific area of character image. In: 2007 International Conference on Computational Intelligence and Security (CIS 2007), Harbin, China, pp. 415–419, December 2007

27. Wang, Z., Shivakumara, P., Lu, T., Basavanna, M., Pal, U., Blumenstein, M.: Fourier-residual for printer identification. In: Proceedings of IEEE International Conference on Document Analysis and Recognition (ICDAR 2017), Kyoto, Japan, pp. 1114–1119, November 2017

28. Yamaguchi, O., Fukui, K.: Face recognition using temporal image sequence. In: Proceedings of Third IEEE International Conference on Automatic Face and Gesture Recognition, Nara, Japan, pp. 318–323, April 1998

CheckSim: A Reference-Based Identity Document Verification by Image Similarity Measure

Nabil Ghanmi[(✉)], Cyrine Nabli, and Ahmad-Montaser Awal

ARIADNEXT - Research department, Rennes, France
{nabil.ghanmi,cyrine.nabli,montaser.awal}@ariadnext.com

Abstract. This paper presents a generic deep-learning based framework for identity document verification, which we call *CheckSim*. We tackle the document verification problem as a similarity measure between two aligned images: a query document image and a predefined reference representing the document model. To this end, we explore the use of two particular architectures of Convolutional Neural Network (CNN) models that are well adapted to find similarity or relationship between two comparable inputs, namely siamese and triplet CNN. The models are trained using specific loss function, then used for extracting high-dimensional features from the query and the reference images. The distance between these two feature vectors is used to decide if the two images are similar or different based on a threshold optimized by minimizing a custom cost function. An experimental comparison of these two models as well as traditional approaches based on handcrafted features is conducted on a real-world data set of patches extracted from identity documents. The obtained results show that our approach achieves good results and outperforms handcrafted features based methods.

Keywords: Identity document verification · Siamese/Triplet Network · Similarity measure · Forgery detection

1 Introduction

Digital customer onboarding allows companies (banks, financial institutions or any other service provider) to acquire new customers without the need to receive them physically in their offices or stores. Before giving access to services or products (credit, internet subscription, online gambling, etc.) to a new customer, the service provider ask him to send or upload copies (scans or photos) of his identity documents for KYC (Know Your Customer) requirements. KYC refers to the process of clients' identity verification when indulging in business with companies.

It is obvious that this digitization has several advantages. In fact, it makes a tedious and complex process a fluid and immediate one. It also provides more efficient and faster document management, better customer monitoring and better

© Springer Nature Switzerland AG 2021
E. H. Barney Smith and U. Pal (Eds.): ICDAR 2021 Workshops, LNCS 12916, pp. 422–436, 2021.
https://doi.org/10.1007/978-3-030-86198-8_30

traceability. However, it creates new challenges in terms of document verification and fraud detection. Indeed, document fraud represents economic and security risks, ranging from petty crimes in small services to money laundering, illegal immigration and terrorism financing. The advancement of digital image processing software makes the generation of fraudulent documents easy and frequent nowadays. To fight against document forgery threats, advanced verification systems are today necessary more than ever.

Document verification aims to confirm the authenticity of a given document by carrying out a series of automatic and/or manual checks on two main levels:

- Content level checks the conformity of the physical document structure and ensures that the information contained in the document makes sense and goes together by checking some predefined business-rules.
- Visual level checks whether the document is visually similar to the authentic version of this document (presence of specific security patterns or texture, etc.).

In this paper, we address the document verification problem at the visual level. To our knowledge, most of the works dealing with forgery detection within document images are based on content analysis. In the other hand, the most of visual-based forgery detection researches treat non-textual images (natural scene, portrait, picture, etc.).

The current work is a part of a real-world application that includes a full identity document analysis chain, from image capture to information extraction and verification. The document type (french ID card, french resident permit, Italian driving licence, etc.) being identified during the first steps of the chain, we approach the visual verification by an image comparison problem that checks the similarity of the query document image to a reference image representing the document model.

The rest of this paper is organized as follows. In Sect. 2, we review the state-of-the-art approaches of image forgery detection and image similarity measures. The problem statement and the related challenges are then presented in Sect. 3. The proposed deep learning approach is presented in Sect. 4. Experiments and obtained results are detailed in Sect. 5, followed by the conclusion and some perspectives.

2   State of the Art

Image forgery detection has found, over these last years, a particular interest from the scientific community. The state-of-the art of this field and the related challenges have been summarized in a number of surveys [2,14]. Some pixel-based techniques are presented in [1] to detect copy-move and splicing alterations. Basically, these techniques rely on 3 main steps: image decomposition into overlapping blocks, feature extraction (such as SIFT, SURF, etc.) and feature sorting. In this context, L. Kang et al. [9] used Singular Value Decomposition (SVD) and

Discrete Cosine Transform (DCT) as feature extractors for respectively copy-move and splicing detection. These methods are very specific and not adapted to real-world noisy images. In [12], Popescu et al. explored Principal Component Analysis (PCA) for detecting duplicate image regions. The main limitation of this algorithm is that it is time consuming. Authors in [8] used Discrete Wavelet Transform (DWT) for copy-move forgery. Unlike SVD, the proposed algorithm works well on noisy images. In addition, it presents less computation time compared to other methods like PCA. In [10], a fast algorithm for finding tampered regions in JPEG images is proposed. It is based on the analysis of the double quantization effect using DCT. It gives good performance but it is not applicable on other image formats.

More recently, a two-stream deep neural network based on Faster R-CNN model is proposed [20]. The first stream uses the RGB image to find tampered regions by predicting bounding boxes using a Region Proposal Network. The second stream uses a noise feature map (extracted from the RGB image using a steganalysis rich model filter layer) to detect inconsistency between genuine and tampered regions. Then, spatial features from both RGB and noise streams are combined via bilinear pooling. Finally, these features are processed through a fully connected layer and a softmax layer to predict label for each region. The obtained results on NIST16 dataset show good overall performance on multi-class image manipulation detection such as splicing, removal, and copy-move.

Among works that are more related to the current problem and deal with document images, we can cite [3,5–7]. The main idea in [6] is to check identity document authenticity by comparing predefined regions to reference ones. For this end, the authors propose a new descriptor *Grid-3CD* to classify document as genuine or forged by computing the difference (using Canberra distance) or by one-class classification (using SVM). Tested on identity document data set, *Grid-3CD* combined with one-class SVM shows good performance and exceeds the best combination of common descriptors.

Two methods based on SIFT features are proposed in [7] to detect textual changes in document images. The first method works at word level. However, the second one is segmentation-free and tries to align text blocks using dense SIFT to overcome segmentation problems and handle graphical objects. An euclidean distance is then calculated between local SIFT feature vectors in the two images and a fixed threshold is used to decide if they are altered. In [5], a classification method using a SVM classifier based on Local Binary Pattern (LBP) is presented. This classification works at patch level. The LBP descriptor extracted from a given patch, combined with other descriptors extracted from neighboring regions are fed to a SVM to predict if the patch is forged or not.

An exhaustive evaluation of the state-of-the-art descriptors including handcrafted and CNN learned features, was presented in [3]. The obtained results show that, even if learned features globally outperform handcrafted ones, Histogram Of Gradient (HOG) still an efficient descriptor for industrial applications since it is less memory consuming.

When a reference image is available, verifying the integrity of a given image can be performed by comparing it to this reference. For such approaches, it would be important to have a good similarity measure which could help a lot to detect image forgeries.

Similarity measure is a fundamental task in computer vision that has been widely studied in several application such as image retrieval, object recognition, face matching, etc. To check the similarity between two images, a Faster R-CNN based framework is presented in [17]. This model aims to resolve the problem of forgery detection by finding differences between a given image and its reference. It has been tested on book cover images where a query image is compared to a reference one that have the same bar code. After alignment, the two images are concatenated as a 6-channel image which is fed to the R-CNN to spot the difference. This latter is considered as an object to detect in the concatenation image.

Three CNN models (siamese, pseudo-siamese and 2-channel) were explored in [19] to learn a similarity function that will be used for deciding if two patches are similar or not. For the siamese model, two copies of the same network, sharing the same parameters are used as feature extractor. The features computed from two input images are concatenated and sent to a fully connected network which model the similarity function. The pseudo-siamese model is similar to the first one but the parameters are not shared between the two branches, which allows to learn much more complicated functions. Differently from these two models, the 2-channel network combines the pair of images into a unique 2-channel image. Similar CNN architectures were also used for image retrieval in [11].

In the context of similar image ranking, Wang et al. [15] propose a triplet-based network architecture to characterize fine-grained image similarity relationships using a specific loss ranking function. A similar architecture is employed in [16] for aerial image similarity estimation. In [18], triplet network combined with the spatial pyramid pooling is presented. For training the model, the authors use an improved triplet loss function that help the model to learn the most efficient features such that an inter-class distance is always greater than the intra-class distance.

In conclusion, several methods have been proposed for forgery detection but few of them are dealing with identity documents. In this paper, we focus on this kind of documents and more precisely on the visual aspects (not on the textual content). To have an efficient system for identity document verification, it is necessary to use discriminant features and good similarity measure. In this paper we try to resolve this two issues simultaneously by learning a generic similarity function.

3   Problem Statement

An identity document has an invariable part composed of its background and fixed textual zones containing field labels (name, surname, address, etc.) and a variable part containing its holder information and eventually his photo. The variable part of an identity document is verified after recognizing its textual content (using an OCR) and crossing the data extracted from the various document zones (such as Machine Readable Zone (MRZ), electronic chip, visible fields, etc.) to check their consistency. For more advanced document verification, invariable background and patterns also should be verified to detect some kinds of falsification that do not violate the content consistency but affect the appearance invariable zones. We tackle the problem of document verification as an image comparison problem. Given an input image, the document is firstly localized and corrected based on a predefined reference. Then, the cropped document is compared to its reference while ignoring the variable parts. A set of predefined regions of interest (ROIs) are used in this comparison.

As the two images are aligned, the comparison may seem easy by pixel-wise difference for example or any other distance between common features extracted from the images. Nevertheless, many factors can cause large differences between pixels or computed features in similar images and thus make the comparison task more difficult.

In fact, we work on real-world document images that are captured in uncontrolled environment with different devices (flat scanners, smartphones having various settings, etc.). Thus, these images may suffer from many imperfections:

- Various lighting conditions, which often lead to too dark or too bright images or also images containing shaded areas, etc.
- Variable resolutions, which results on different pixels or features values even for two images of a similar scene.
- Blur which is generally caused by incorrect focal length (defocus blur) or by capture device and/or document movement (motion blur). It can alter the image content and thus lead to incomparable images.
- Glare which results from an overexposing of the captured document to the light of the acquisition device, or from reflective zones such as Optically Variable Device (OVD). It can alter (by occlusion) the image pixel values.

It is clear that, under such conditions, a direct pixel-wise image comparison is not suitable. It is also hard to craft resistant descriptors to achieve this task using descriptor distances. In fact, we need a system that should be both able to detect small difference between two images when one of them is intentionally altered (for fraud purposes) and tolerant to the differences induced by the image capture. That is why, we are turned to deep learning approaches as they:

- tend to resolve the problem end-to-end by learning high-level features directly from the images. The main advantage is that the learned features are more robust than handcrafted ones, especially in our case of application where designing discriminant features is non trivial;
- are able to absorb, thanks to the training process, the natural image variability induced by the capture;
- are able to learn differences between image "natural" dissimilarities (induced by the capture) and "intentional" dissimilarities (made by a human to falsify the document).

4   Proposed Method

We attempt to learn the image similarity function between two images directly from annotated pairs of pre-aligned images. Based on recent advances in neural network models, we choose to use siamese and triplet neural network to model the image similarity problem. This kind of models is well adapted for tasks that involve finding similarity or relationship between two comparable inputs such as paraphrases or images. It is composed of two or more instances of the same neural network that share weights. During the training step, parameters updating is mirrored across the subnetworks. Once trained, the model takes two images as input and outputs a distance between these two images. Then, the computed distance is compared to an optimized threshold to decide whether the images are similar or different. Our choice for siamese and triplet neural network was motivated by:

- Unlike traditional CNN where the learned features are dependent on training images, these models allow to learn generic similarity function that can be applied for any new images pairs (representing scenes that are never seen in the training). This eliminates the need to re-train the model when we want to apply it for comparing images of a completely different field.
- Siamese and triplets models are particularly efficient when we are facing a problem of limited data which is the case for the current application as we deal with identity documents that are sensitive and thus difficult to collect.

4.1   Model Architectures

Siamese Neural Network (SNN) is composed of two identical sub-networks (same configuration and same parameters). Each sub-network is a classical CNN composed of a stack of convolutional layers, rectification linear units, pooling layers, etc. To train the model, two images I_1 and I_2 are injected into the CNN networks to compute high-dimensional feature vectors f_1 and f_2. Then, the distance $d = norm(f_1 - f_2)$ between the two vectors is computed in the embedding space. This distance is used to calculate the contrastive loss as defined in the Eq. 1 (see Fig. 1). The network weights are finally updated via back propagation such as the loss function is optimized.

$$L_{contrastive} = 0.5 \times (Y \times d^2) + 0.5 \times ((1 - Y) \times (\max(0, m - d))^2) \quad (1)$$

Where Y is the ground truth of the image pair (I_1, I_2). It is equal to 1 if I_1 and I_2 are similar, 0 otherwise. m is a parameter, arbitrary set to 1 in this work.

Minimizing the contrastive loss (Eq. 1) aims at finding optimal feature vectors that minimize the distance between similar images and maximize it between different images. In fact:

- when two images are similar, *i.e.* $Y = 1$, $L_{contrastive} = 0.5 \times d^2$. Minimize the loss function amounts to minimize the distance d between the two similar images.

– when two images are different, *i.e.* $Y = 0$, $L_{contrastive} = 0.5 \times (\max(0, m - d))^2$. Minimize the loss function amounts to maximize the distance d between the different images (such that $d > m$).

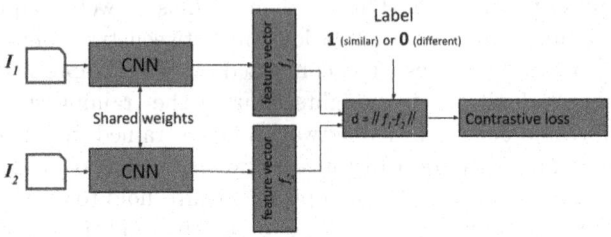

Fig. 1. Siamese model.

As already mentioned, the advantage of using this architecture is that we can generate sufficient number of pairs for training even if only a few number of images is available. In fact, from a dataset of N images, we can generate $N * (N - 1)/2$ pairs. For example, from only 100 images, we can generate 19900 pairs.

Triplet Neural Network can be seen as an extension of the siamese network. As its name suggests, it is made up of three identical branches which take as input three images representing an anchor sample a, a positive sample p and a negative sample n (see Fig. 2). A positive sample is an image similar to the anchor whereas a negative sample is different from the anchor. The triplet network training tends to optimize the neural network weights such that anchor samples are closer to positive samples than to negative samples by a margin m, *i.e.* $d(a, p) - m < d(a, n)$. For this end, the triplet loss function defined in Eq. 2 is used.

$$L_{triplet} = max(0, d(a, p) - d(a, n) + m) \qquad (2)$$

Compared to siamese network, the triplet model provides two main advantages:

– it optimizes both distances (to negative and positive samples) simultaneously;
– it allows generating much more data combinations for training as it uses images triplets.

Features Distance. Most of the existing systems use L^2-norm to compute the distance between feature vectors. In this work, we tested three types of distance: L^1-norm, L^2-norm, and cosine distance and we found that L^1-norm performs better than the two other distances.

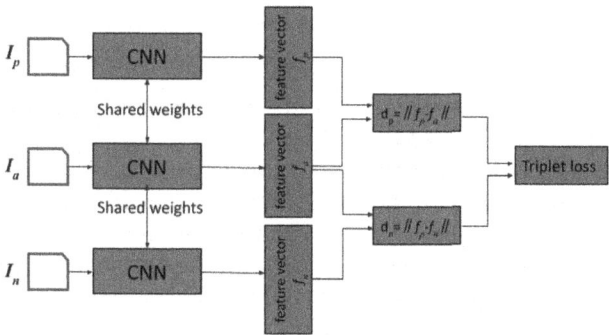

Fig. 2. Triplet model.

4.2   CNN Architecture Selection

Whether for the siamese or the triplet model, the feature extractor is a CNN. To build an efficient CNN architecture, we were inspired from the VGG-16 network [13] that has proven to be one of the most efficient vision model architectures till date. Its particularity is that it uses convolutional filters of small sizes (3×3) instead of large convolution filters (like in AlexNet 11×11 and $5\times$). It also uses the same maxpool layer with 2×2 kernel and a stride of 2, and follows a definite arrangement of convolution and max pool layers throughout the whole architecture. In order to optimize the CNN architecture for our task, we carried-out several experiments by changing the number of layers and filters while respecting the basic principle of VGG16. The final architecture we retained is illustrated on Fig. 3.

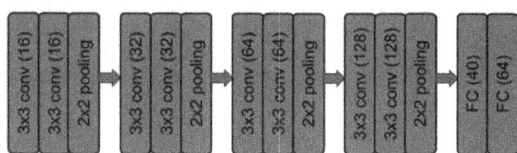

Fig. 3. The CNN architecture that we use as feature extractor.

4.3   A Custom Cost Function for Prediction

We remind that the trained models will be used to predict if a document image is forged or not by comparing a set of query patches to the reference ones. For each patch in the query, a feature vector f_q is computed and the distance $d = ||f_q - f_r||$ is calculated, where f_r is the feature vector calculated from the correspondent patch in the reference image. Based on an optimized threshold T_0, the distance d is used to decide if the query image is genuine or forged.

– if $d \leq T_0$ then the document image is considered as genuine.
– if $d > T_0$ then the document image is considered as forged.

Two kinds of errors are possible. A false rejection (FR) corresponds to a genuine document that is predicted as forged. On the other hand, a false acceptance (FA) correspond to a forged document that is predicted as genuine. Both errors are undesirable, but depending on the application, one may be more serious than the other. In the context of identity document verification, a FR have more serious consequences than a FA. Therefore, a FR error should be more penalized than a FA error. We define a global cost function C_{global} as follows:

$$C_{global} = \frac{3 \times C_{FR} + C_{FA}}{4} \tag{3}$$

Where C_{FR} and C_{FA} are the cost of FR and FA errors respectively. The value 3 of the FR penalty factor is fixed empirically.

This cost function (Eq. 3) will be used to find the optimum value T_0 of the decision threshold.

(a) A genuine FR_ID document.

(b) A genuine FR_RP document.

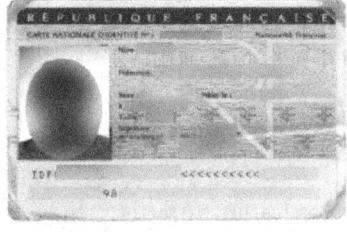

(c) A forged FR_ID document where the photo is replaced (its background is different from Fig. 4a).

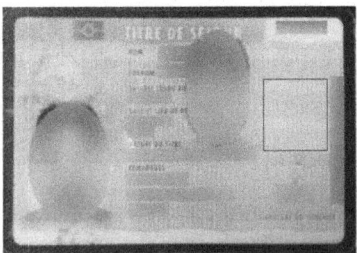

(d) A forged FR_RP document where the altered pattern is shown in red.

Fig. 4. Examples of genuine and forged documents with masked personal information.

5   Experiments and Results

The presented models are tested on two verification tasks *photoVerif* and *patternVerif*. These tasks concerns the verification of the following ROIs respectively:

- Photo background: check that the identity photo was not altered or replaced, by comparing the query photo background to the reference one.
- Security visual pattern: check the presence of security patterns and their conformity to those defined in the reference model as in [6].

Our data set is built from a real-life application. As in a real-world data set there are only very few forged documents, we have generated few hundreds of altered documents by randomly tampering visual pattern areas (using copy-paste of randomly selected areas within the document) and/or photo areas (by replacing original photos by another ones taken from the web). The majority of these alterations are generated automatically by the means of a specific tool developed for this purpose. Other alterations are manually made in order to imitate realistic document falsifications. Figure 4 illustrates some examples of genuine and forged documents taken from our data set. Four types of french identity documents are used for these experiments: identity card (FR_ID), passport (FR_PA), driving licence (FR_DL), and residence permit (FR_RP), as described in Table 1. All these documents contain identity photo but only FR_RP contains visual patterns and thus it is the only one concerned by the *patternVerif* task.

Table 1. Training and test data sets.

Type	#samples			
	Training and validation set		Test set	
	Genuine	Forged	Genuine	Forged
FR_ID	117	100	564	299
FR_PA	104	100	345	125
FR_DL	122	100	205	172
FR_RP	149	100	304	198

5.1   Training Data Preparation

Given an input image, the identity document is localized and aligned with the reference image as described in [4] (see Fig. 5a). Then, the photos as well as the security patterns are localized based on their positions in the reference. The corresponding patches are extracted to be compared to their counterparts in the reference image. Examples of the extracted patches are shown in Fig. 5b and 5c.

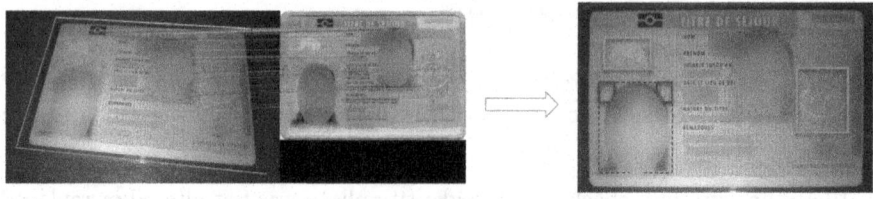

(a) Image alignment and patches extraction. On the left, document identification and localization from an input image based on keypoint matching as detailed on [4]. On the right, the query document image is aligned with the reference and the patches are extracted: in green the patches used for patternVerif and in red those used for photoVerif (top left and top right sub-regions which are not covered by the photo are used)

(b) Examples of patches used for patternVerif

(c) Examples of patches used for photoVerif

Fig. 5. Patches extraction.

Training Pairs and Triplets Construction. To select training samples, we adopted an offline random mining: the pairs/triplets are randomly created beforehand as follows. Once extracted, the patches are grouped based on similarity criterion, *i.e.* the patches corresponding to the same ROI are grouped together if they belong to genuine documents, let $\{P_i\}_i$ the set of these groups. All the altered patches (even if they are not similar) are grouped into the same class N.

Pairs Construction. Firstly, a random image is taken from a randomly selected group P_k while respecting an equitable distribution between the different groups. Then, alternately, a similar image (from the same group P_k) or a different image (from the groups $N \cup \{P_i\}_i \backslash P_k$) is randomly selected. This operation is repeated until obtaining a consequent data set. A number of 55000 pairs is experimentally fixed as being a good trade-off between efficient and rapid learning.

Triplet Construction. As for pairs generation, we firstly select randomly an anchor sample from a group P_k (which is itself randomly selected). Then, a positive sample (from the same group P_k) and a negative sample (from other groups) are randomly selected, until obtaining the desired number of triplets (also fixed to 55000).

Once pairs (or triplets) data set is constructed, we split it into two sets: training and validation.

5.2   Some Implementation Details

We train the proposed models using SGD with momentum. A minibatch of 20 samples and a fixed learning rate of 10^{-4} are used. We use a momentum of 0.9, a weight decay of 0.05 and a maximum epoch number of 10. All trainings were held on a GeForce GTX 1080 Ti.

5.3   Results and Interpretation

Evaluation Metric. To evaluate the proposed approach, the following error rates are used:

- False Rejection Rate (FRR) defined as the ratio of similar images that are predicted as different by the total number of similar image.
- False Acceptance Rate (FAR) defined as the ratio of different images that are predicted as similar by the total number of different image.

Results. During the training, the models are periodically evaluated (after each 500 iterations) in term of loss and accuracy. This evaluation is used to save, at each moment during the training, the model instance that has achieved the best accuracy. One can notice that siamese model converge more quickly than the triplet one. It achieves an accuracy of $98, 7\%$ (the blue point in Fig. 6a) in less than 50000 iterations and remains stable around an accuracy of about 97%. As for the triplet network, it converges gradually after more than 400000 iterations and achieves an accuracy of 99.2% (the blue point in Fig. 6b).

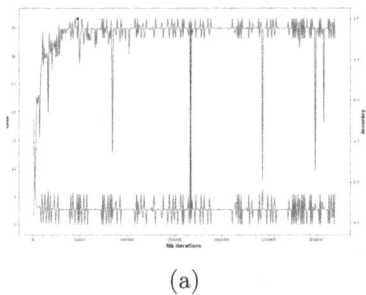
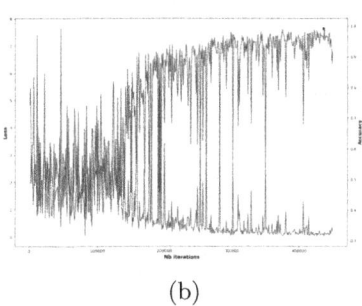

(a)   (b)

Fig. 6. Evolution of loss (in red) and accuracy (in green) during the training of the siamese network (6a) and the Ttiplet network (6b). (Color figure online)

Before using the trained model for prediction on the test set, the decision threshold is optimized as discussed in Sect. 4.3, using the validation set. To this end, the curve $C_{global} = f(T)$ is built (see Fig. 7). The optimum thresholds are 1.7 and 1.2 for the siamese and the triplet models respectively.

Using the optimized threshold, the trained models are evaluated on the test set at patch level as well as document level. A document is considered forged

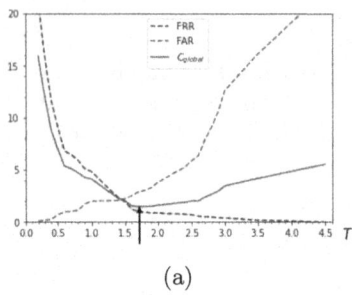
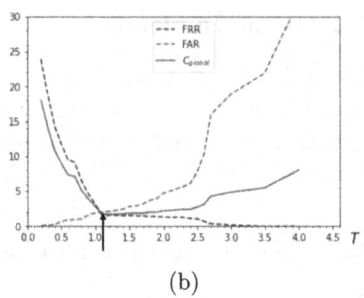

(a)	(b)

Fig. 7. Optimization of the threshold decision for siamese network (7a) and triplet network (7b) based on global cost minimization.

if the two photo patches (or the two visual patterns for the FR_RP document) are predicted as different from the reference. Table 2 summarizes the obtained results on the test set. One can notice that, at both levels, siamese network gives better FRR for most of the document types, leading to a weighted average FRR of 1.27 which is better than those of the triplet network 1.55. Whereas, the triplet network is better in terms of FAR with a weighted average of 2.27 vs. 2.52 for the siamese network. This can be explained by the nature of the loss functions used for both networks. As the triplet network optimizes the loss function using similar and different pairs simultaneously, it tends to have a good balance between similarity and dissimilarity detection. On the other hand, the SNN tends to better learn similarity and thus identifying non-similar pairs is harder.

Based on the obtained results, the siamese model is retained and integrated in our industrial solution.

Table 2. FRR and FAR on test set obtained by the siamese and the triplet network.

Type	Patch level				Document level			
	Siamese network		Triplet network		Siames network		Triplet network	
	FRR (%)	FAR (%)	FRR (%)	FAR (%)	FRR (%)	FAR (%)	FRR (%)	FAR (%)
FR_ID	1.95	3.68	2.04	3.34	1.24	2.68	1.60	2.34
FR_PA	0.72	3.20	1.01	2.4	0.58	2.4	0.87	1.60
FR_DL	1.46	2.30	1.71	2.61	0.98	1.16	1.46	1.74
FR_RP	2.8	4.29	2.46	3.78	2.30	3.54	2.30	3.03

To show the efficiency of the proposed models, we compare them to classical image comparison approaches based on usual features. To this end, we implement 3 of the most used descriptors in image matching and image retrieval fields. Table 3 shows that the proposed models significantly outperforms HOG, LBP, and SIFT based methods, in terms of both FRR and FAR.

Table 3. Comparison of the proposed models to classical approaches.

Methods	Patch level		Document level	
	FRR (%)	FAR (%)	FRR (%)	FAR (%)
Siamese model	**1.76**	3.40	**1.27**	2.52
Triplet model	1.84	**3.21**	1.55	**2.27**
HOG + euclidean dist	7.44	10.20	4.72	9.70
LBP + euclidean dist	8.78	15.05	6.84	13.22
SIFT matching	5.04	6.49	4.37	5.92

6 Conclusion

We proposed a reference-based approach that compares a set of predefined ROIs within the input image to their equivalents in the reference one. For this purpose, two different CNN-based architectures were explored. Known to be well adapted for comparing two images, it was experimentally shown that siamese and triplet models are robust to various image capture imperfections. The contrastive and triplet loss functions allow these networks to learn relevant high level features for image comparison giving better results than classical approaches based on handcrafted descriptors.

The main advantage of this work is that it presents generic algorithm for image comparison based on a learned generic similarity function. This latter can be applied for several other verification tasks on identity document, such as face matching, background verification, etc.

In order to improve our system, we are currently studying different combination strategies to benefit from the advantages of each model. Furthermore, we plan to improve the pairs/triplets sampling algorithm in order to use the data more efficiently by studying the online pairs/triplets sampling techniques.

References

1. Ansari, M.D., Ghrera, S., Tyagi, V.: Pixel-based image forgery detection: a review. IETE J. Educ. **55**, 40–46 (2014)
2. Birajdar, G.K., Mankar, V.H.: Digital image forgery detection using passive techniques: a survey. Digit. Investig. **10**(3), 226–245 (2013)
3. Centeno, A.B., Terrades, O.R., Canet, J.L., Morales, C.C.: Evaluation of texture descriptors for validation of counterfeit documents. In: 14th IAPR International Conference on Document Analysis and Recognition, ICDAR 2017, Kyoto, Japan, 9–15 November 2017, pp. 1237–1242. IEEE (2017)
4. Chiron, G., Ghanmi, N., Awal, A.M.: Id documents matching and localization with multi-hypothesis constraints. In: 25th International Conference on Pattern Recognition (ICPR), pp. 3644–3651 (2020)

5. Cruz, F., Sidere, N., Coustaty, M., D'Andecy, V.P., Ogier, J.-M.: Local binary patterns for document forgery detection. In: 14th IAPR International Conference on Document Analysis and Recognition, ICDAR 2017, Kyoto, Japan, 9–15 November 2017, pp. 1223–1228. IEEE (2017)

6. Ghanmi, N., Awal, A.M.: A new descriptor for pattern matching: Application to identity document verification. In: 2018 13th IAPR International Workshop on Document Analysis Systems (DAS), pp. 375–380 (2018)

7. Jain, R., Doermann, D.: VisualDiff: document image verification and change detection. In: 2013 12th International Conference on Document Analysis and Recognition, pp. 40–44 (2013)

8. Zhang, J., Feng, Z., Su, Y.: A new approach for detecting copy-move forgery in digital images. In: 2008 11th IEEE Singapore International Conference on Communication Systems, pp. 362–366 (2008)

9. Kang, L., Cheng, X.: Copy-move forgery detection in digital image. In: 2010 3rd International Congress on Image and Signal Processing, vol. 5, pp. 2419–2421 (2010)

10. Lin, Z., He, J., Tang, X., Tang, C.-K.: Fast, automatic and fine-grained tampered JPEG image detection via DCT coefficient analysis. Pattern Recognit. **42**(11), 2492–2501 (2009)

11. Melekhov, I., Kannala, J., Rahtu, E.: Siamese network features for image matching. In: 2016 23rd International Conference on Pattern Recognition (ICPR), pp. 378–383 (2016)

12. Popescu, A.C., Farid, H.: Exposing digital forgeries by detecting duplicated image regions (2020)

13. Simonyan, K., Zisserman, A.: Very deep convolutional networks for large-scale image recognition. arXiv preprint arXiv:1409.1556 (2014)

14. Stamm, M.C., Wu, M., Ray Liu, K.J.: Information forensics. An overview of the first decade. IEEE Access **1**, 167–200 (2013)

15. Valaitis, V., Marcinkevicius, V., Jurevicius, R.: Learning aerial image similarity using triplet networks. In: Sergeyev, Y.D., Kvasov, D.E. (eds.) NUMTA 2019. LNCS, vol. 11974, pp. 195–207. Springer, Cham (2020). https://doi.org/10.1007/978-3-030-40616-5_15

16. Wang, J., et al.: Learning fine-grained image similarity with deep ranking. In: 2014 IEEE Conference on Computer Vision and Pattern Recognition, CVPR 2014, Columbus, OH, USA, 23–28 June 2014, pp. 1386–1393. IEEE Computer Society (2014)

17. Wu, J., Ye, Y., Chen, Y., Weng, Z.: Spot the difference by object detection (2018)

18. Yuan, X., Liu, Q., Long, J., Lei, H., Wang, Y.: Deep image similarity measurement based on the improved triplet network with spatial pyramid pooling. Inf. **10**(4), 129 (2019)

19. Zagoruyko, S., Komodakis, N.: Learning to compare image patches via convolutional neural networks. In: Proceedings of the IEEE Conference on Computer Vision and Pattern Recognition (CVPR), June 2015

20. Zhou, P., Han, X., Morariu, V.I., Davis, L.S.: Learning rich features for image manipulation detection. CoRR, abs/1805.04953 (2018)

Crossing Number Features: From Biometrics to Printed Character Matching

Pauline Puteaux[1] and Iuliia Tkachenko[2(✉)]

[1] LIRMM, Université de Montpellier, CNRS, Montpellier, France
`pauline.puteaux@lirmm.fr`
[2] LIRIS, Université Lumière Lyon 2, CNRS, Lyon, France
`iuliia.tkachenko@liris.cnrs.fr`

Abstract. Nowadays, the security of both digital and hard-copy documents has become a real issue. As a solution, numerous integrity check approaches have been designed. The challenge lies in finding features which are robust to print-and-scan process. In this paper, we propose a new method of printed-and-scanned character matching based on the adaptation of biometrical features. After the binarization and the skeletonization of a character, feature points are extracted by computing crossing numbers. The feature point set can then be smoothed to make it more suitable for template matching. From various experimental results, we have shown that an accuracy of more than 95% is achieved for print-and-scan resolutions of 300 dpi and 600 dpi. We have also highlighted the feasibility of the proposed method in case of double print-and-scan operation. The comparison with a state-of-the-art method shows that the generalization of proposed matching method is possible while using different fonts.

Keywords: Printed document · Feature extraction · Print-and-scan process · Crossing numbers · Matching method

1 Introduction

Due to the broad availability of professional image editing tools, cheap scanning devices and advancements of high-quality printing technologies, there is a high need in fast, reliable and cost-efficient document authentication techniques. The actual pandemic situation forces the people and the administrative centers to use digital copies of hard-copy documents. The official hard-copy documents have specific security elements that can be efficiently used for document authentication as moiré patterns, holograms, or specific inks [24]. Nevertheless, the digital copies of such documents can be easily tampered using some image editing tools (like Photoshop or Gimp) [2] or using some novel deep learning approaches [18,27,28]. That is why, there is a high need of designing efficient and robust solutions for printed-and-scanned document integrity check.

© Springer Nature Switzerland AG 2021
E. H. Barney Smith and U. Pal (Eds.): ICDAR 2021 Workshops, LNCS 12916, pp. 437–450, 2021.
https://doi.org/10.1007/978-3-030-86198-8_31

For described situations, we need to work with the documents in two formats (hard-copy and electronic soft-copy). This type of documents is called hybrid documents [6]. For hybrid documents, the integrity check must work identically for both soft-copy and hard-copy documents. That means that if the document was printed and captured several times without tampering, the document integrity check should label this document as authentic.

One of the first hybrid protection system was presented in [25]. There was proposed to construct the hash digests for the text in electronic and printed documents. The technique based on the use of Optical Character Recognition (OCR) software and a classical cryptographic message authentication code gave a good performance. Later it was shown that the OCR software cannot give us stable results due to Print-and-Scan (P&S) process impact [8,22].

In this paper, we do not want to improve the accuracy and stability of OCR methods, we would like to find some features extracted from character skeletons that can be robust to P&S impact and used for character representation. These features are then matched with a template in order to identify a character. We consider this work as a first step for the construction of text fuzzy hash that can then easily be stored in a high capacity barcode and integrated to documents (or stored in a database) as a document representation.

The rest of the paper is organized as follows. We introduce the existing document authentication methods and the impact of P&S process to hard-copy and printed-and-scanned documents in Sect. 2. The proposed feature extraction method as well as the proposed matching methods are presented in Sect. 3. We show the experimental results in Sect. 4. Several future paths are discussed in Sect. 5. Finally, we conclude in Sect. 6.

2   Challenges of Printed Document Protection

There exist several approaches for hard-copy document authentication. The first one is a forensics approach that aims at identifying the printer and scanner [4] that were used to produce a given hard-copy document and its scanned version. Here, some specific features are extracted from the printed characters according to different techniques as gray-level co-occurrence matrix [13,14], noise energy, contour roughness and average gradient of character edges [19]. These features are then classified using different machine learning methods (LDA, SVM, *etc.*) in order to identify the printer. In the last few years, some forensics methods based on deep learning appear [9,15]. In [15], authors presented human-interpretable extensions of forensics algorithms that can assist to human experts to understand the forensics results. This approach cannot be used for hybrid document authentication as we cannot control the printer and scanner used, and thus, the forensics features cannot ensure the authenticity of a hybrid document.

The second approach aims to add a specific copy-sensitive code to the document that is used to detect unauthorized duplication of the document [17,23]. These solutions take an advantage of the stochastic nature of Print-and-Scan (P&S) process. Nevertheless, these copy sensitive codes can only make the difference between first print and all other re-prints of the document. Thus, this approach cannot be also extended to hybrid document authentication. However, if such code has a high storage capacity as [23], it could be used for document hash storage.

The third approach works with hybrid documents [6]. The contributors of this approach introduce the term of stability in the document processing domain. The main idea of this approach is to separate the document into primary elements as images [5], text [8], layout [7] and tables [1], and to represent these elements by stable features. These stable features are unchanged when the document is printed-and-scanned using different resolutions.

The text integrity check can be done by another approach that consists of the construction of document hash using the specific feature code extracted from each character [21]. The authors show that the proposed solution can resist to affine transformations, JPEG compression and low-level noise, but is not robust to median filtering. Specific features based on character skeleton can also be used for character recognition [12]. The authors reported the recognition results comparable with those obtained by the deep learning approach. In [22], the authors suggest to use the PCA for character feature extraction and a minimal euclidean distance for character recognition. The main problem of this approach is the extraction of correct bounding boxes and stable features as a P&S process impacts to the shape and color of the printed characters. In addition, this machine learning based method cannot be generalized, thus it is necessary to re-train the model for each font type.

The images after P&S process are affected by noises, blur and other changes [20, 29]. Therefore, the P&S communication channel is always characterized by loss of information. The loss could be minimal and imperceptible by the naked eye, but it is significant for authentication test or integrity check.

When the soft-copy document is printed and scanned, some noise is added by the printer and the scanner. Therefore, if a hard-copy document was scanned and re-printed, it suffers from double impact of the P&S process. In general case, the document can be scanned and reprinted several times. Nevertheless, the most realistic situations are: 1) one P&S operation - when the soft-copy was printed and then scanned or captured with a camera by a person before sending to authority center; 2) double P&S - when the hard-copy was scanned and then reprinted by an authority center (or a person). That is why, in this paper, we work with characters printed once (P&S) or printed twice (double P&S).

3   Proposed Method

When a hard-copy document is scanned several times, each time a slightly different document image is obtained [31] due to the optical characteristics of captured devices. The similar problem can be found in biometrics: we know that the enrolled fingerprint has several differences with the stored fingerprint template. In this work, we want to adapt the biometrical features for character feature extraction and matching with a template.

3.1 Pre-processing Operations

The pre-processing steps of fingerprint matching process consists of binarization and thinning (or skeletonization) processes. As a P&S process impacts a lot to the character shape, we need to apply some morphological operations before the binarization step in order to fill the holes (appeared due to the inhomogeneous spread of ink during printing or quantization and compression operations during scanning). In this paper, we do not focus on the search of the best pre-processing operations, we use 1) the opening operation to correct possible errors of P&S process, 2) the classical Otsu's binarization method [16] and 3) the classical thinning method based on medial axis transform [11].

3.2 Feature Extraction

The feature extraction is done from the skeleton image of a character, in digital (during the template construction phase) or printed-and-scanned form.

We analyze pixels of the binary image considering their neighborhood. Two pixel values are possible: 0 for black and 1 for white (the skeleton is represented by white pixels). In order to extract significant feature points, we compute the crossing numbers [30]. For each pixel, the associated crossing number is defined as half of the sum of differences between two adjacent pixel values. Depending on the value of its associated crossing number (CN), five pixel types can be defined, as presented in Fig. 1.

CN = 0	CN = 1	CN = 2	CN = 3	CN = 4
Isolated point	Ending point	Connective point	Bifurcation point	Crossing point

Fig. 1. Five pixel types as a function of their associated crossing number (the centered pixel framed in red is the pixel of interest). (Color figure online)

In biometrics, only ending and bifurcation points are generally considered for minutia extraction from fingerprint. However, in the case of printed characters, it should be interesting to consider isolated and crossing points in addition. Indeed, isolated points should be relevant for characters 'i' and 'j' and crossing points for 'x'.

Furthermore, for some characters, serifs can be present. A serif is a small line or stroke attached to the end of a longer stroke in a character. They can induce the extraction of additional feature points. Indeed, in case of serif, there are a bifurcation point and one or two close ending points. These last extracted features are not significant. Therefore, we propose a smoothed version (S) of the feature extraction step consisting to remove these ending points from the feature point set. In addition, the serifs are not presented in all kinds of fonts, so the smoothed version of features is more adapted for generalization of the matching methods.

In order to perform the smoothing operation, we compute the euclidean distance between a bifurcation point and each extracted ending point. If this distance is lower than the threshold th, the ending point is considered as being part of a serif and is then removed from the feature point set. Note that the value of the threshold th is experimentally fixed and depends on the template database. The practical interest of this smoothing operation is illustrated in Sect. 4.2 and Fig. 4.

3.3 Template Matching

In order to compare a printed-and-scanned character with the digital template, we suggest to test three different matching methods:

- M_1: For each feature point extracted from the printed-and-scanned character, we are looking for the closest (in terms of euclidean distance between coordinates) reference point in the template of digital character. Therefore, two extracted feature points can be associated to the same reference point. We then average the distances computed for each extracted feature point.
- M_2: For each reference point in the template of digital character, we are looking for the closest (in terms of euclidean distance between coordinates) feature point extracted from the printed-and-scanned character. We then average the distances computed for each reference point.
- M_3: Same as M_1, but we only consider the reference points which have the same crossing number type as the extracted feature point to compare. Moreover, we focus on template of digital characters which have approximately the same number of reference points as the number of extracted features from the printed-and-scanned character (equal numbers ±2).

The template of digital character which has the smallest difference score with the printed-and-scanned character allows us to find the associated letter in the alphabet. The performances with each of these methods are discussed in Sect. 4.

4 Experimental Results

In this section, we present the database used and the pre-processing steps done. Then, we show the extracted features for matching and we discuss the results obtained for character images printed-and-scanned once with 300 dpi and 600 dpi resolutions.

4.1 Database Description

In our database[1], we have 10 images per low-case character with Times New Roman font, that gives us in total 260 images per P&S resolution. All images are of size 100×100 pixels with a character centered in the image. For images printed-and-scanned with 300 dpi, we apply the $\times 2$ resize function before centering these characters in the images. We illustrate some images from our database in the Fig. 2. We can notice that there exist small imperfections for characters printed with 300 dpi in comparison with those printed with 600 dpi or digital sample. In addition, the printed characters are in grayscale. Thus, we need to do some pre-processing before the skeleton extraction process.

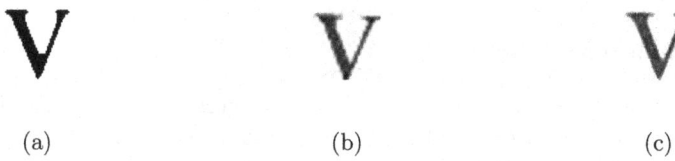

(a)   (b)   (c)

Fig. 2. The letters from our database a) digital sample, b) sample printed and scanned with 300 dpi, c) printed and scanned with 600 dpi.

After several experiments, we have found that the best results for our database are obtained using an open morphological operation with square structural element of size 3×3. After this operation, the character color is more homogeneous and the binarization process works better. We binarize the characters using classical Otsu's binarization. After the binarization, the skeletonization is done using the build-in Matlab function (*bwskel*) that uses the medial axis transform based on thinning algorithm introduced in [11]. This function uses 4-connectivity with 2-D images and gives us sufficient results for character skeleton extraction. Examples of skeletons extracted using this function are illustrated in Fig. 3.

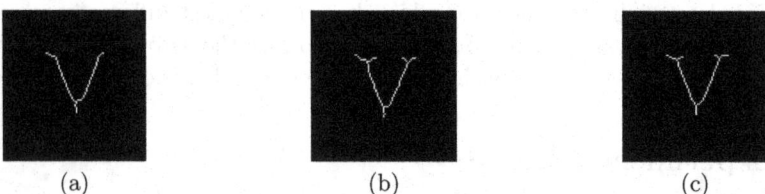

(a)   (b)   (c)

Fig. 3. The skeletons extracted from characters a) digital sample, b) sample printed and scanned with 300 dpi, c) printed and scanned with 600 dpi.

---

[1] The database is available on demand. Contact: iuliia.tkachenko@liris.cnrs.fr.

From Fig. 3, we notice that the skeleton is less noisy in the case of digital sample and a sample printed and scanned with 600 dpi. In addition, several strokes detected in samples Fig. 3b–c are not present in the template (Fig. 3a). Therefore, we have decided to compare the proposed matching methods using non-smoothed and smoothed features.

4.2   Feature Extraction

The comparison of non-smoothed and smoothed features is illustrated in Fig. 4. We present the extracted features from the skeletons of both digital and printed and scanned with 600 dpi versions of the 'v' character. The feature points are displayed in red for a better visualization. Moreover, each point coordinates and type are also indicated. From the skeleton of the digital character, we extract four feature points: one bifurcation point (CN = 3) and three ending points (CN = 1). From the skeleton of the printed-and-scanned character, the number of feature points is bigger. Indeed, there are three bifurcation points and five ending points. However, due to the presence of serifs, some feature points are not relevant for comparison and matching. Using the smoothing operation described in Sect. 3.2 with a threshold $th = 15$ according to our experiments on the template database, they are removed from the feature point set. One can then note that, after this operation, the same number of feature points is obtained for the digital character and the printed-and-scanned version. In addition, we can remark that their coordinates are quite close from each others, which is an important property to ensure a good match.

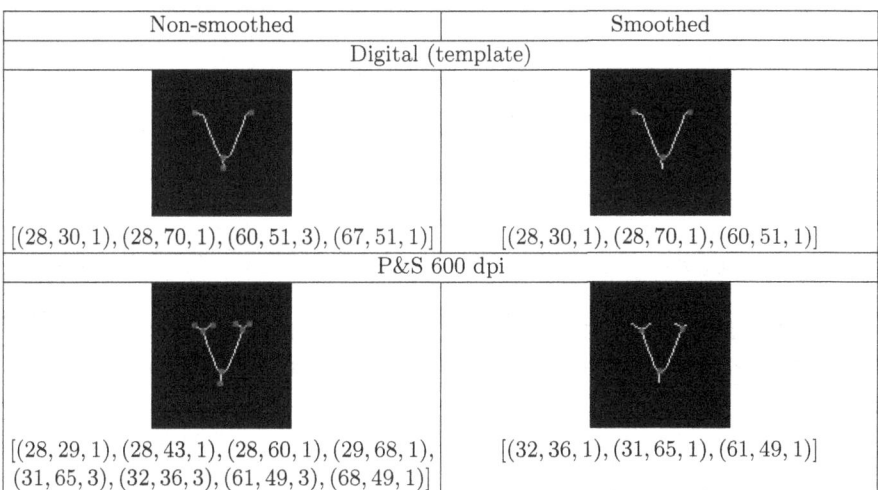

Non-smoothed	Smoothed
Digital (template)	
$[(28, 30, 1), (28, 70, 1), (60, 51, 3), (67, 51, 1)]$	$[(28, 30, 1), (28, 70, 1), (60, 51, 1)]$
P&S 600 dpi	
$[(28, 29, 1), (28, 43, 1), (28, 60, 1), (29, 68, 1),$ $(31, 65, 3), (32, 36, 3), (61, 49, 3), (68, 49, 1)]$	$[(32, 36, 1), (31, 65, 1), (61, 49, 1)]$

Fig. 4. Example of skeleton extraction and crossing number extraction.

4.3   Character Matching

For character matching experiments, we have constructed a database with 26 templates of digital characters. These templates are used during the matching process in order to recognize 260 character images printed-and-scanned with 300 dpi resolution and 260 character images printed-and-scanned with 600 dpi resolution. We do not test the matching methods with digital characters as they always have the same features and, thus, they are always correctly recognized.

Table 1 shows the results obtained for each letter while the characters were printed-and-scanned with 300 dpi resolution. A big number of characters are matched correctly in 100% of cases even if we use non-smoothed features (see rows M_1, M_2, M_3 in Table 1). Nevertheless, for some characters, the use of smoothed features significantly improves the matching results (for example, for letters 'k', 'm', 'v'). The smoothed features improve the results for the characters that have additional strokes in the end points. In general, we can make the conclusion that the matching method M_3 with smoothed features gives us the best matching results. This is proved by the mean values shown in Table 2 (Pre-processing 1). Indeed, we see that the mean correct matching rate for characters printed-and-scanned with 300 dpi resolution, obtained with SM_3 method, is equal to 95%.

Table 1. Character matching rates using different crossing number comparison techniques (P&S 300 dpi).

	a	b	c	d	e	f	g	h	i	j	k	l	m
M_1	1	0.9	1	1	0.9	1	1	1	1	0.8	1	0.5	1
M_2	1	0.9	1	1	1	1	1	0.1	0.8	1	0.5	0.8	0.7
M_3	1	0.9	1	1	0.9	1	1	0.4	1	0.8	0.9	0.5	0.5
SM_1	1	0.9	1	1	0.9	1	1	1	0.9	0.8	1	0.5	1
SM_2	1	0.9	1	1	1	1	1	0.1	0.7	1	0.1	0.5	0.7
SM_3	1	0.9	1	1	0.9	1	1	1	0.9	0.8	1	0.5	1
	n	o	p	q	r	s	t	u	v	w	x	y	z
M_1	1	1	1	1	1	0.8	1	1	0.1	1	1	1	0.9
M_2	1	1	1	1	1	0.8	1	1	0.5	0.6	1	0.9	0.9
M_3	0.3	1	1	1	1	1	1	1	0.1	1	1	0.8	0.9
SM_1	1	1	1	1	1	0.8	1	1	1	1	1	0.8	0.9
SM_2	1	1	1	1	1	0.8	0.7	1	0.7	0.9	0.9	0.8	0.7
SM_3	1	1	1	1	1	1	1	1	1	1	1	0.8	0.9

Figure 5 illustrates the results obtained for each letter while the characters were printed-and-scanned with 600 dpi resolution. From these results, the same conclusions can be done: the matching method M_3 with smoothed features (SM_3)

gives us the best matching results. The mean correct matching rate with SM_3 method is equal to 97.31% from Table 2 (Pre-processing 1). We can conclude that the proposed feature extraction and matching methods work better for these images as the resolution and, thus, the image quality are higher. We only need to improve the results for letters 'b', 'e', 'o' and 'z' by adjusting the skeletonization step. These results are very promising and show us that we can extract stable features and construct a text hash.

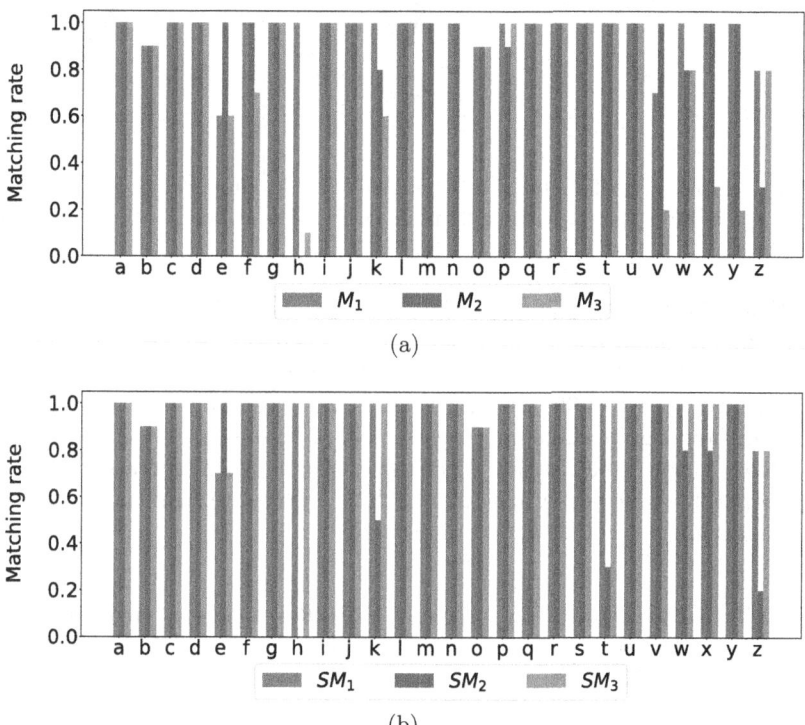

Fig. 5. Character matching rates using different crossing number comparison techniques (P&S 600 dpi): a) non-smoothed methods, b) smoothed methods.

5   Discussion

The results presented in the previous section show that the suggested method can work well for character images printed-and-scanned using different resolutions. Nevertheless, the characters after double P&S have more distortions [22]. Therefore, when we tested our matching methods with images after double P&S process, we have obtained the results presented in Table 2 under the title "Preprocessing 1". We notice that the recognition results for characters after double

P&S process are drastically lower than the results after P&S with 300 dpi and 600 dpi and equal to 74.23%. Analyzing these results, we have visualized the character images after the pre-processing operations (introduced in Sect. 4) in Fig. 6b. From this illustration, we can conclude that suggested pre-processing operations are not adapted to the character images after double P&S process.

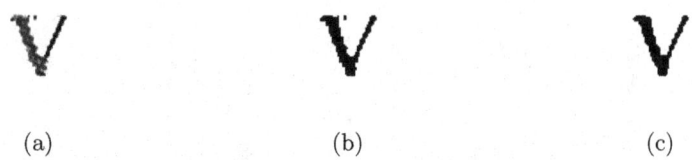

(a)   (b)   (c)

Fig. 6. Examples of a character a) after double P&S, b) after "pre-processing 1" used in Sect. 4.1, c) after "pre-processing 2" presented in this section.

The double P&S process changes the character shape (see Fig. 6a) due to the scanner quantization and compression steps. Therefore, we need to apply specific pre-processing to improve the character shape. In order to fill the holes in the images (see Fig. 6b), we use the 2×2 open-close operation. This pre-processing step significantly improves the character shape and makes the color more homogeneous (see Fig. 6c). After these morphological operations, the thinning operation works better and the matching results are significantly improved (see Table 2 under the title "Pre-processing 2").

Table 2. Percentage of correctly recognized characters using suggested crossing number comparison techniques.

	M_1	M_2	M_3	SM_1	SM_2	SM_3
Pre-processing 1						
P& S 300 dpi	91.92%	86.54%	84.62%	94.23%	82.69%	**95.00%**
P& S 600 dpi	95.77%	90.77%	73.46%	**97.31%**	86.15%	**97.31%**
double P& S 600 dpi	66.92%	65.77%	54.23%	73.08%	66.54%	**74.23%**
Pre-processing 2						
P& S 300 dpi	85.38%	86.15%	83.46%	84.62%	78.08%	**88.08%**
P& S 600 dpi	89.62%	88.85%	90.00%	90.00%	85.38%	**91.15%**
double P& S 600 dpi	78.08%	75.38%	76.54%	80.00%	70.77%	**83.46%**

Table 2 shows that the correctly used pre-processing operations can significantly improve the recognition results: the recognition rate improves up to 83.46%. Nevertheless, the recognition results of characters printed once drop off from

95–97% to 88–91%. Thus, for that moment, we do not find unique set of pre-processing operations for both characters printed once and characters printed twice.

This study helps us to identify several future paths:

- Find the unique pre-processing operations for characters printed once and twice. We can test image noise reduction and image sharpening methods, or even try to do the pre-processing step using a deep learning approach [3].
- Find stable skeleton extraction method that ensures the stable extraction of proposed features. For this, we can, for example, extract the noise resistant digital euclidean connected skeletons introduced in [10].
- Make the text processing by lines in order to avoid the problems with bounding boxes extraction. Indeed, we can extract the same features by lines and then use a fuzzy hash functions for text hashing. This solution can be similar to sketchprint presented in [26].

Finally, we have decided to compare our matching method with PCA based approach from [22]. The results are reported in Table 3, where abbreviations F_1F_1 and F_2F_2 mean that the training and testing were done using the same font (F_1 - Times New Roman, F_2 - Arial) and abbreviation F_1F_2 means that F_1 was used for training and F_2 for testing. We note that the PCA approach gives slightly better results while training and testing sets come from the same font. Nevertheless, the generalization is better using our proposed matching method SM_3: the matching accuracy is 17–20% higher in case of P&S 300 dpi and P&S 600 dpi, and it is 8% higher in case of double P&S with "pre-processing 2".

Table 3. Comparison of proposed SM_3 matching technique with PCA based approach from [22].

	Proposed matching SM_3			PCA approach [22]		
	F_1F_1	F_1F_2	F_2F_2	F_1F_1	F_1F_2	F_2F_2
Pre-processing 1						
P& S 300 dpi	95%	**85.77%**	85.38%	99.62%	66.54%	94.62%
P& S 600 dpi	97.31%	**82.69%**	85%	91.15%	65.00%	97.69%
double P& S 600 dpi	74.23%	70.00%	68.08%	96.54%	**76.15%**	91.15%
Pre-processing 2						
P& S 300 dpi	88.08%	**86.15%**	84.62%	94.23%	66.92%	95.00%
P& S 600 dpi	91.15%	**86.54%**	85.38%	97.31%	66.54%	98.08%
double P& S 600 dpi	83.46%	**83.84%**	78.85%	90.77%	76.15%	90.77%

From the results of Table 3 and from the feature extraction method, we can conclude that the proposed matching methods are more adapted for fuzzy text hash construction than the PCA based method.

6   Conclusions

Nowadays, the use of both soft-copy and hard-copy documents increases significantly. In the same time, due to accessibility of editing tools and printing/capturing devices as well as the improvements of deep learning techniques, the number of document counterfeits increases each year. For fighting against the document counterfeits, it is important to find a novel integrity check systems that work well for both versions of documents (soft- and hard-copies).

In this paper, we have proposed crossing numbers based features for printed character matching. These features were extracted from the character skeletons. We have explored the use of non-smoothed and smoothed features. The smoothed features show us more stable and high accuracy results of 95% and 97.31% both for characters printed with 300 dpi and with 600 dpi, respectively. The additional experiments were done for the characters printed-and-scanned twice. When the pre-processing operations are chosen correctly, the matching accuracy for double P&S characters comes up to 83.46%. In addition, it was shown that the proposed matching methods can be generalized when different fonts are used for template construction and matching.

In the future, we would like to modify the pre-processing operations (noise reduction and skeletonization) in order to improve the matching results and construct compact text hash using the proposed features for printed document integrity check.

References

1. Alhéritière, H., Cloppet, F., Kurtz, C., Ogier, J.M., Vincent, N.: A document straight line based segmentation for complex layout extraction. In: IAPR International Conference on Document Analysis and Recognition (ICDAR), vol. 1, pp. 1126–1131. IEEE (2017)
2. Artaud, C., Sidère, N., Doucet, A., Ogier, J.M., Poulain D'Andecy, V.: Find it! Fraud detection contest report. In: International Conference on Pattern Recognition (ICPR), pp. 13–18. IEEE (2018)
3. Bui, Q.A., Mollard, D., Tabbone, S.: Selecting automatically pre-processing methods to improve OCR performances. In: IAPR International Conference on Document Analysis and Recognition (ICDAR), vol. 1, pp. 169–174. IEEE (2017)
4. Chiang, P.J., et al.: Printer and scanner forensics: models and methods. In: Sencar, H.T., Velastin, S., Nikolaidis, N., Lian, S. (eds.) Intelligent Multimedia Analysis for Security Applications, pp. 145–187. Springer, Heidelberg (2010). https://doi.org/10.1007/978-3-642-11756-5_7
5. Eskenazi, S., Bodin, B., Gomez-Krämer, P., Ogier, J.M.: A perceptual image hashing algorithm for hybrid document security. In: IAPR International Conference on Document Analysis and Recognition (ICDAR), vol. 1, pp. 741–746. IEEE (2017)
6. Eskenazi, S., Gomez-Krämer, P., Ogier, J.-M.: When document security brings new challenges to document analysis. In: Garain, U., Shafait, F. (eds.) IWCF 2012/2014. LNCS, vol. 8915, pp. 104–116. Springer, Cham (2015). https://doi.org/10.1007/978-3-319-20125-2_10
7. Eskenazi, S., Gomez-Krämer, P., Ogier, J.M.: The Delaunay document layout descriptor. In: Symposium on Document Engineering, pp. 167–175. ACM (2015)

8. Eskenazi, S., Gomez-Krämer, P., Ogier, J.M.: A study of the factors influencing OCR stability for hybrid security. In: IAPR International Conference on Document Analysis and Recognition (ICDAR), vol. 9, pp. 3–8. IEEE (2017)

9. Ferreira, A., et al.: Data-driven feature characterization techniques for laser printer attribution. IEEE Trans. Inf. Forensics Secur. **12**(8), 1860–1873 (2017)

10. Leborgne, A., Mille, J., Tougne, L.: Noise-resistant digital Euclidean connected skeleton for graph-based shape matching. J. Visual Commun. Image Representation **31**, 165–176 (2015)

11. Lee, T.C., Kashyap, R.L., Chu, C.N.: Building skeleton models via 3-D medial surface axis thinning algorithms. CVGIP: Graph. Models Image Process. **56**(6), 462–478 (1994)

12. Lipkina, A., Mestetskiy, L.M.: Grapheme approach to recognizing letters based on medial representation. In: International Joint Conference on Computer Vision, Imaging and Computer Graphics Theory and Applications - Volume 4: VISAPP, pp. 351–358 (2019)

13. Mikkilineni, A.K., Khanna, N., Delp, E.J.: Texture based attacks on intrinsic signature based printer identification. In: Media Forensics and Security, vol. 7541, p. 75410T. International Society for Optics and Photonics (2010)

14. Mikkilineni, A.K., Khanna, N., Delp, E.J.: Forensic printer detection using intrinsic signatures. In: Media Watermarking, Security, and Forensics, vol. 7880, p. 78800R. International Society for Optics and Photonics (2011)

15. Navarro, L.C., Navarro, A.K., Rocha, A., Dahab, R.: Connecting the dots: toward accountable machine-learning printer attribution methods. J. Vis. Commun. Image Representat. **53**, 257–272 (2018)

16. Otsu, N.: A threshold selection method from gray-level histograms. IEEE Trans. Syst. Man Cybern. **9**(1), 62–66 (1979)

17. Picard, J.: Digital authentication with copy-detection patterns. In: Electronic Imaging, pp. 176–183. International Society for Optics and Photonics (2004)

18. Roy, P., Bhattacharya, S., Ghosh, S., Pal, U.: STEFANN: scene text editor using font adaptive neural network. In: Conference on Computer Vision and Pattern Recognition (CVPR), pp. 13228–13237. IEEE/CVF (2020)

19. Shang, S., Memon, N., Kong, X.: Detecting documents forged by printing and copying. EURASIP J. Adv. Sig. Process. **2014**(1), 1–13 (2014)

20. Solanki, K., Madhow, U., Manjunath, B.S., Chandrasekaran, S., El-Khalil, I.: Print and scan resilient data hiding in images. IEEE Trans. Inf. Forensics Secur. **1**(4), 464–478 (2006)

21. Tan, L., Sun, X.: Robust text hashing for content-based document authentication. Inf. Technol. J. **10**(8), 1608–1613 (2011)

22. Tkachenko, I., Gomez-Krämer, P.: Robustness of character recognition techniques to double print-and-scan process. In: IAPR International Conference on Document Analysis and Recognition (ICDAR), vol. 09, pp. 27–32 (2017)

23. Tkachenko, I., Puech, W., Destruel, C., Strauss, O., Gaudin, J.M., Guichard, C.: Two-level QR code for private message sharing and document authentication. IEEE Trans. Inf. Forensics Secur. **11**(3), 571–583 (2016)

24. Van Renesse, R.L.: Optical document security. Appl. Opt. **13528**, 5529–34 (1996)

25. Villán, R., Voloshynovskiy, S., Koval, O., Deguillaume, F., Pun, T.: Tamper-proofing of electronic and printed text documents via robust hashing and data-hiding. In: Security, Steganography, and Watermarking of Multimedia Contents, vol. 6505, p. 65051T. International Society for Optics and Photonics (2007)

26. Voloshynovskiy, S., Diephuis, M., Holotyak, T.: Mobile visual object identification: from SIFT-BoF-RANSAC to sketchprint. In: Media Watermarking, Security, and Forensics 2015, vol. 9409, p. 94090Q. International Society for Optics and Photonics (2015)

27. Wu, L., et al.: Editing text in the wild. In: International Conference on Multimedia, pp. 1500–1508. ACM (2019)

28. Yang, Q., Huang, J., Lin, W.: SwapText: image based texts transfer in scenes. In: Conference on Computer Vision and Pattern Recognition (CVPR), pp. 14700–14709. IEEE/CVF (2020)

29. Yu, L., Niu, X., Sun, S.: Print-and-scan model and the watermarking countermeasure. Image Vis. Comput. **23**(9), 807–814 (2005)

30. Zhao, F., Tang, X.: Preprocessing and postprocessing for skeleton-based fingerprint minutiae extraction. Pattern Recogn. **40**(4), 1270–1281 (2007)

31. Zhu, B., Wu, J., Kankanhalli, M.S.: Print signatures for document authentication. In: Conference on Computer and Communications Security, pp. 145–154. ACM (2003)

Writer Characterization from Handwriting on Papyri Using Multi-step Feature Learning

Sidra Nasir$^{(\boxtimes)}$ ⓘ, Imran Siddiqi ⓘ, and Momina Moetesum ⓘ

Department of Computer Science, Bahria University, Islamabad, Pakistan
{imran.siddiqi,momina.buic}@bahria.edu.pk

Abstract. Identification of scribes from historical manuscripts has remained an equally interesting problem for paleographers as well as the pattern classification researchers. Though significant research endeavors have been made to address the writer identification problem in contemporary handwriting, the problem remains challenging when it comes to historical manuscripts primarily due to the degradation of documents over time. This study targets scribe identification from ancient documents using Greek handwriting on the papyri as a case study. The technique relies on segmenting the handwriting from background and extracting keypoints which are likely to carry writer-specific information. Using the handwriting keypoints as centers, small fragments (patches) are extracted from the image and are employed as units of feature extraction and subsequent classification. Decisions from fragments of an image are then combined to produce image-level decisions using a majority vote. Features are learned using a two-step fine-tuning of convolutional neural networks where the models are first tuned on contemporary handwriting images (relatively larger dataset) and later tuned to the small set of writing samples under study. The preliminary findings of the experimental study are promising and establish the potential of the proposed ideas in characterizing writer from a challenging set of writing samples.

Keywords: Scribe identification · FAST keypoints · ConvNets · IAM Dataset · Greek Papyrus

1 Introduction

The recent years have witnessed an increased acceptability of digital solutions by handwriting experts [1] thanks primarily to the success of joint projects between paleographers and computer scientists. Handwriting develops and evolves over years and, in addition to being the most common form of non-verbal communication, it also carries rich information about the individual producing the writing. Furthermore, the geographical, cultural and social attributes are also reflected in the writing style as well as its evolution. These present interesting research

ⓒ Springer Nature Switzerland AG 2021
E. H. Barney Smith and U. Pal (Eds.): ICDAR 2021 Workshops, LNCS 12916, pp. 451–465, 2021.
https://doi.org/10.1007/978-3-030-86198-8_32

avenues for paleographers who are typically interested in identifying the scribe, place and origin of a manuscript or testify the authenticity of a document.

The digitization of ancient documents has increased significantly in the last few decades [2,3] and serves multiple purposes. On one hand, the digitization serves to preserve the rich cultural heritage and make it publicly available while on the other hand, such digitized collections offer a spectrum of challenges to the pattern classification community [4]. Among such digitization projects, the prominent ones include the Madonne [4], International Dunhaung Project (IDP) [5], and NAVIDOMASS (NAVIgation in Document MASSes) [6] etc.

In addition to digitization, such projects also include development of computerized tools for paleographers facilitating tasks like keyword spotting, transcription and retrieval etc. The key idea of such systems is to assist rather than replace the human experts. These semi-automatic tools narrows down the search space to enable the experts to focus on limited set of samples for in-depth and detailed analysis [7]. The SPI (System for Paleographic Inspection) [8] tool for example has been developed to assist paleographers in classifying and identifying scripts and study the morphology of writing strokes. Exploiting the idea that similar strokes are likely to be produced in similar temporal and geographical circumstances, such systems allow paleographers to infer useful information about the origin of a manuscript. Furthermore, the notion of similarity can also be effectively employed to conclude the authenticity of a document and/or identify its scribe. Identification of scribe can also serve to implicitly estimate the date and origin of a manuscript by correlating it with a scribe's active time period [9]. This scribe identification from historical manuscripts also makes the subject of our current study.

As opposed to contemporary documents, identification of writer from historical manuscripts offers a number of challenges mainly due to degradation of documents over time. Typical challenges include removal of noise, segmentation of handwriting (from background) and extraction of writing components for feature extraction and subsequent classification hence hampering the direct application of established writer identification techniques to historical documents. In addition, the writing instruments as well as the writing mediums have also witnessed an evolution over time from stones, to papyrus, parchment, paper etc. and must be taken into account when designing computational features to characterize the writer.

Identification of writers from handwriting typically relies on capturing the writing style which has been validated to be unique for an individual [10]. The writing style can be captured from computational features either at global (page or paragraph) or at local (words, characters or graphemes etc.) levels. At global level for instance, textural features have been widely employed to capture the writing style [11–13] of the writer. Likewise, at relatively smaller scales of observation, low level statistical features are typically computed from parts of handwriting like characters or graphemes [10]. Another well-known and effective technique is to characterize the writer by the probability of producing certain writing patterns i.e. the codebook [14,15] that is similar in many aspects to the bag of words model. With the recent paradigm shift from hand-engineered to machine-learned features, feature learning using (deep) convolutional neural networks has also been investigated for writer identification [16].

This paper investigates the problem of writer characterization from handwritten scripts written on papyrus and extends our previous work presented in [17]. Handwriting images are first pre-processed and key points in the writing are identified. Subsequently, small fragments of handwriting are extracted using the key points as centers. These fragments are then employed to extract features using ConvNets. Since the dataset under study is scarce and the existing pre-trained CNN models are mostly trained of images very different from handwriting, a two-step fine-tuning of CNNs is employed. The networks are first tuned on a large collection of contemporary handwriting images and then tuned on the papyrus images. Writer identification decisions for patches are combined to image level decisions using a majority vote. Experiments are carried out GRK-Papyri [18] dataset and the performance of key points is compared with that of dense sampling.

We organize this paper as follows. Section 2 discusses the recent notable studies on similar problems followed by the introduction of the dataset and the proposed methods in Sect. 3. Section 4 introduces the experimental protocol and the reported results along with accompanying discussion. Finally, we summarize the key findings of our research and conclude the paper in Sect. 5.

2   Related Work

Development of computerized solutions for analysis of ancient documents and handwriting has received a renewed interest of the document analysis and handwriting recognition community in the recent years [19–22]. From the view point of writer characterization, the key challenge is to identify the discriminative features that allow capturing the writer-specific writing style of an individual. Another important design choice is the scale of observation at which the handwriting is analyzed and computational features are extracted that may vary from a complete page to partial words or small fragments in writing.

As discussed previously, there has been an increased tendency to learn features from data (samples under study) rather than designing the features for different classification tasks. The problem of writer identification has also witnessed a similar trend where features learned using ConvNets are known to outperform the conventional hand-crafted features. Hence, our discussion will be more focused on machine learning-based techniques for writer identification. The comprehensive reviews covering writer identification can be found in [23,24].

Among significant recent contributions to identification of writers, He et al. [25] proposed a deep neural network (FragNet) to characterize writer from limited handwriting (single words or small blocks). The proposed network comprises of two pathways, a feature pyramid that maps the input samples to feature maps and the classification pathway to predict the writer identity. The method was evaluated on four different datasets (IAM, CVL, Firemaker & CERUG-ER) and reported promising identification rates. In another study, Kumar et al. [26] proposed a CNN model to identify writers from handwriting in Indic languages and carried out evaluations on word as well as document levels.

Tang and Wu [27] employed CNN with joint Bayesian technique for writer identification. Data augmentation is applied to generate multiple samples for each writer and the CNN is used as a feature extractor. The experimental study was carried out on the CVL datasets and ICDAR2013. The Top-1 identification rates of more than 99% are achieved in different experiments. In another relevant study, pre-trained AlexNet is employed as feature extractor to identify the writer from Japanese handwritten characters [28]. Likewise, the authors in [29] employed ResNet with a semi-supervised learning approach. The unlabeled data was regulated using WLSR (Weighted Label Smoothing Regularization). Words from the CVL data set were used as labelled data, while those from the IAM dataset were used as unlabeled data in the experiment.

Among other studies, Fiel et al. [30] maps the features from the handwritten images to feature vectors using ConvNet and identify writer using the k-nearest neighbor classifier. Christlein et al. [21] study was more focused on unsupervised feature learning using the SIFT descriptors and a residual network. Keglevic et al. [22] proposed the use of a triplet network for learning the similarities between handwriting patches. The network is trained by maximizing the interclass and minimizing the intra-class distances. He et al. [13] proposed a multitask learning method focused on a deep adaptive technique for the identification writers from single word handwritten images. Feature set is enhanced by reusing the features learned on auxiliary tasks. A new adaptive layer was introduced that enhanced the accuracy of the deep adaptive method as compared to simple adaptive and non-adaptive methods. Among other studies exploiting deep neural networks for writer identification, Nguyen et al. [31] present and end-to-end system to characterize writer while Vincent et al. [32] propose to employ the CNN feature maps as local descriptors which are mapped to global descriptors using the Gaussian Mixture Model supervector encoding. In a recent study, Javidi et al. [33] extend the ResNet architecture and combine the handwriting thickness information to improve the identification rates.

From the view point of historical documents, in some cases, the techniques developed for contemporary handwriting have been investigated [34]. Lai et al. [35], for instance, employ pathlet and SIFT features and encode them using a bagged VLAD technique to characterize the writer. In [36] exploit the textural information in handwriting to identify the writer from historical manuscripts. The textural information is captured through a combination of oriented Basic Image Features (oBIFs) at different scales. Classification is carried out using a number of distance metrics which are combined to arrive at a final decision and an accuracy of 77.39% on ICDAR 2017 Historical WI dataset was achieved. In another work, Chammas et al. [37] extract patches from handwriting using the SIFT descriptor and use CNN for feature extraction. The features are encoded using multi-VLAD and an exemplar SVM was employed for classification. Experimental study was carried on the ICDAR2019 HDRC-IR dataset and yielded an accuracy of 97%.

Among other studies targeting historical documents, Cilia et al. [38] proposed transfer learning using pre-trained ConvNets to recognize writers of digitized images from a Bible of the XII century. The work was later extended [39] and evaluated on the medieval manuscripts (Avila Bible). Likewise, writer identification was adapted for historical documents in [40] and the same method was employed to the GRK-Papyri dataset in [18] with FAST keypoints. In another similar study, Studer et al. [41] also investigate the performance of different pre-trained CNNs on different task like character recognition, dating and identification of writing style. In one of our recent works [17], we applied dense sampling (using square windows) on binarized images of handwriting and employed different pre-trained CNNs for feature extraction. Experiments on the GRK-papyri dataset reported a writer identification rate of 54%.

The discussion on different writer identification methods reveals that the domain has been heavily dominated by deep learning-based system in the recent years. The literature is very rich when it comes to contemporary documents, however, analysis of historical manuscripts, calls for further research endeavors. Nonetheless, given the challenges in historical documents, automatic feature learning represents an attractive choice for writer identification as well as other related tasks.

3   Materials and Methods

This section presents in detail the proposed technique for identification writers from papyrus handwriting. We first provide an overview of the dataset used in our study then the details of pre-processing, sampling and classification using pre-trained CNNs. Small fagments are extracted from handwriting and are mapped to feature vectors using a two-step fine-tuned CNN. The key steps of our study is presented in Fig. 1 and each of these steps is elaborated in the following sections.

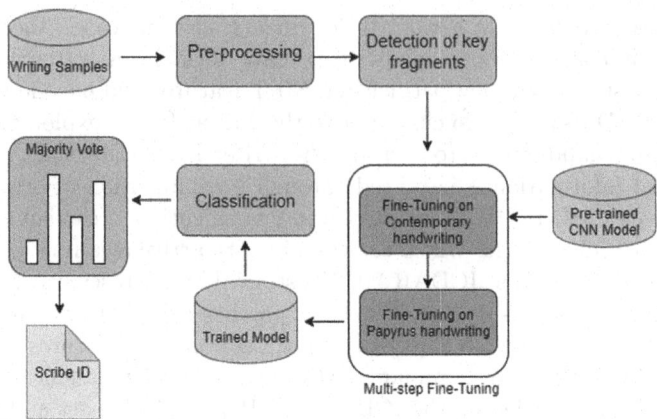

Fig. 1. System overview

3.1 Dataset

We carry out the experimental study of our system on papyri dataset (GRK-Papyri) [18]. The dataset is comprised of images of Greek handwriting on papyri from the 6th century A.D. There are a total of 50 images in the dataset produced by 10 different scribes. In Fig. 2 sample images from the dataset are exhibited. All images are digitized as JPEGs and have varied spatial resolution and DPIs.

Fig. 2. Sample images in the GRK-Papyri dataset [18]

Few of the images are digitized as three channel RGB images while others are in grayscale. Typical degradations in the documents include low contrast, reflection of glass and the varying papyrus fiber in the background. The number of samples per writer also varies as summarized in Fig. 3.

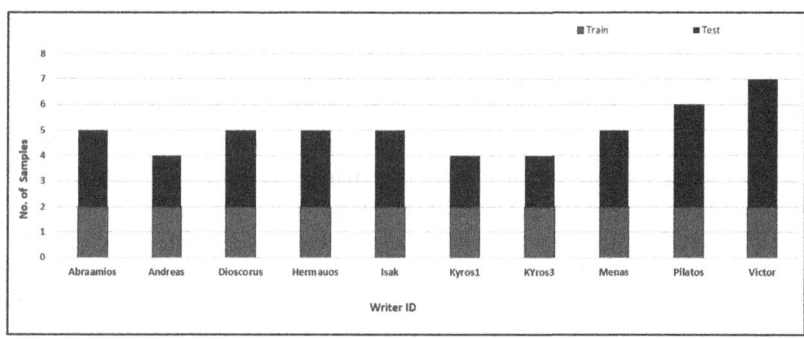

Fig. 3. Distribution of samples in training and test set in the GRK-Papyri dataset

3.2 Pre-processing

The handwriting images under study contain various papyrus fiber backgrounds hence prior to feature extraction, we convert all images to grayscale. Furthermore, in an attempt to ensure that the learned features are dependent on handwriting and not on the image background, we investigate different image preprocessing techniques prior to feature learning. These include segmentation of handwriting from background using adaptive binarization [42], extraction of edges in writing strokes using Canny edge detector, edge detection on adaptively binarized images and segmentation using a deep learning based binarization method - the DeepOtsu [43]. The result from these different types of techniques are illustrated in Fig. 4.

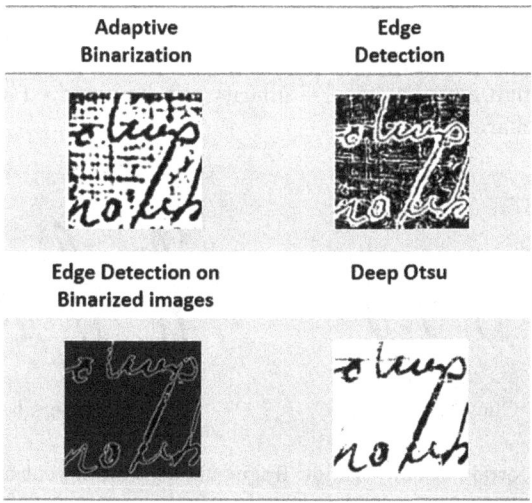

Fig. 4. Images of output from various Image Enhancement Techniques

3.3   Data Preparation

An important choice in extracting writer-specific features from handwriting is the scale of observation. As discussed earlier, page, paragraph, line, word and sub-word levels have been investigated as units of feature extraction. Since we employ fine-tuned ConvNets to extract features, resizing complete images or paragraphs to a small matrix (to match the input layer of the network) would not be very meaningful. Another approach commonly employed in such cases to carry out a dense sampling of handwriting using small windows (patches) which are then employed for feature extraction and subsequent classification [17]. Extracting such patches, though simple, does not ensure that important writer-specific writing strokes are always preserved (Fig. 5). Since the position of windows with respect to handwriting is random, important writing strokes could be separated in multiple windows resulting in loss of important information. Furthermore, a number of studies have established that each writer employs a specific set of writing gestures to produce the strokes and these redundant patterns can be exploited to characterize the writer [14, 15].

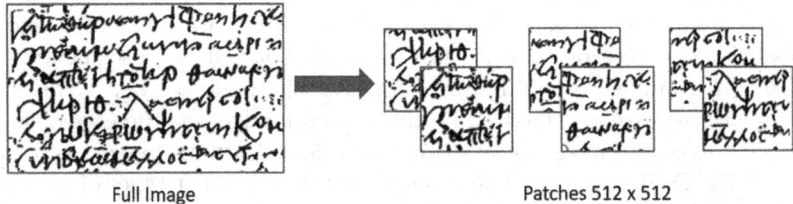

Full Image   **Patches 512 x 512**

Fig. 5. Patches produced using dense sampling with square windows

These patterns can also be shared across different characters as illustrated in Fig. 6 where high morphological similarity can be observed among fragments across different characters.

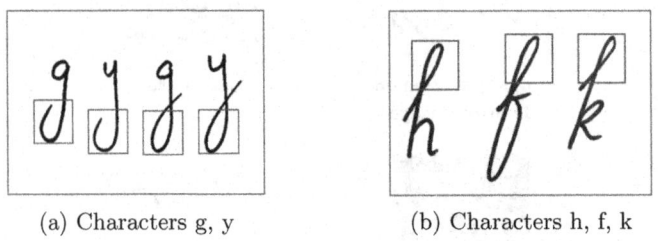

(a) Characters g, y   (b) Characters h, f, k

Fig. 6. Morphologically similar fragments across different characters

To extract the potentially discriminative fragments in handwriting, we apply keypoint detection on the binarized images of handwriting. Subsequently, using these keypoints as centers, small fragments around them are extracted and are employed as units of feature extraction. Similar ideas have been presented in [30, 37,44,45] and have been effective in characterizing the writer. Keypoints are typically the locations where writing stroke changes direction abruptly. Likewise, intersecting strokes etc. also represent potential keypoints. A number of keypoint detectors have been proposed by the computer vision community; among these, we have chosen to employ the FAST keypoint detector [46]. Fragments of size 50×50 are extracted using FAST keypoints from one of the binarized image in the dataset are illustrated in Fig. 7.

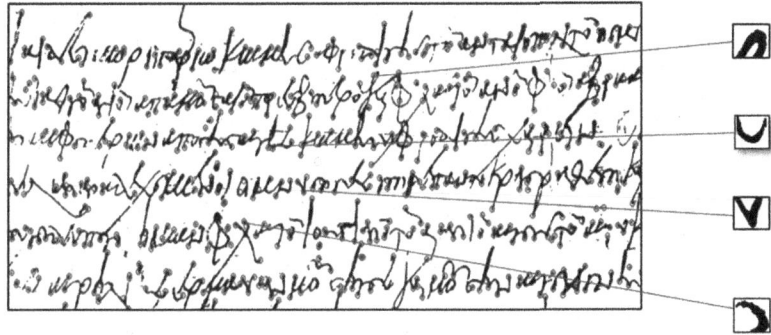

Fig. 7. Fragments extracted from a Binarized (DeepOtsu) image using FAST keypoints

3.4 Two-Step Fine Tuning

Once writing fragments are extracted, we proceed to extraction of features using ConvNets. For many classification tasks, it is common to borrow the architecture and weights of a network trained on large dataset(s) and adapt it for the problem under study. The convolutional-base of such pre-trained models can be used as feature extractor and these features can then be employed with any of the classifiers. It is also common to fine-tune a model by continuing back propagation on all or a subset of layers and replacing the classification layer to match the requirements of the problem at hand.

An important consideration in adapting CNNs for feature extraction and classification is the similarity of source and target datasets. It is important to mention that most of available models are trained on the ImageNet [47] dataset which is quite different from handwriting images under study. To address this point, we employ a multi-step fine-tuning approach. We first fine-tune the standard models trained on ImageNet dataset on writing samples in the IAM dataset [48].

The IAM dataset contains more than 1500 writing samples from 650 different writers. Although these are contemporary documents and do not suffer from the degradations encountered in case of historical documents, nonetheless, since the images contain handwriting, the features learned on these images are likely to be more effective in characterizing writer from the papyrus handwriting. The effectiveness of such multi-step fine-tuning has also been established in a number of other studies [17,49]. Once a network is fine-tuned on writing samples in the IAM dataset, we subsequently tune it to the relatively smaller set of writing samples in the dataset. The softmax layer of each investigated model is adapted to match the 10 classes (10 scribes).

4   Experiments and Results

This section presents the details of the experiments and the realized results along with the accompanying discussion. The dataset (GRK-Papyri) is organized to support the task of writer identification in two different experimental protocols, leave-one-out and a split of train (20 images) and test (30 images) sets. Since our technique relies on feature learning, a leave-one-out protocol is not very feasible as it will require fine-tuning the models for each of the 50 runs. Hence, we use the part of dataset that is distributed into training and test sets.

The pre-trained models investigated include three well-known architectures namely VGG16 [50], InceptionV3 [51] and ResNet [52]. All models are first fine-tuned on writing samples in the IAM database and subsequently on the Greek handwriting images.

In the first experiment, we compare the performance of the employed pre-processing techniques (using Inception v3). The identification rates are reported at fragment (patch) level as well as at image level by combining the decisions using a majority voting. It can be seen from Fig. 8 that DeepOtsu outperforms other techniques reporting identification rates of 32% and 57% at patch and document levels respectively. Consequently, for the subsequent evaluations, we employ DeepOtsu as the pre-processing technique.

In the next series of experiments, we first compute writer identification by directly fine-tuning the standard pre-trained models on the Greek writings followed by a two-step fine-tuning. The results of these experiments are presented in Table 1. It can be seen that in all cases, two-step fine-tuning results in a notable improvement in identification rates. The highest reported identification rate is 64% with ResNet. Taking into account the small dataset and the associated challenges, the reported performance seems to be very promising.

For comparison purposes, we outline the performance of a couple of other studies who employ the same dataset for evaluation purposes. One of the well-known works on this problem is presented in [18] where authors employ the NLNBNN with FAST key points. Identification rates of 30.0% and 26.6% are reported in this study using leave-one-out and train/test set respectively. Nasir et al. [17] employ a similar two-step fine-tuning with DeepOtsu as the binarization method. However, rather than extracting writing patches using keypoints, authors use a dense sampling to extract rectangular windows from handwriting images and report an identification rate of 54% with ResNet. Using the same

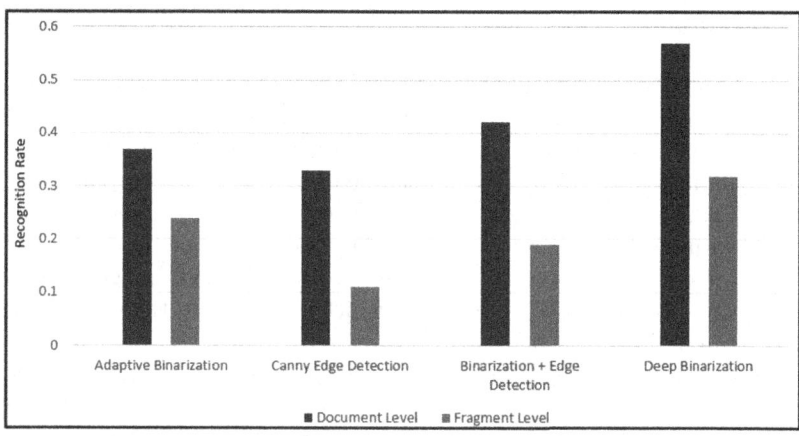

Fig. 8. Identification rates from different pre-processing techniques using Two-step fine-tuning of Inceptionv3

Table 1. Single and two step fine tuning performance on different pre-trained ConvNets (Fragments).

Networks	Fine-Tuning Scheme	Fragment Level	Image Level
VGG16 [50]	ImageNet→Papyri	0.16	0.32
	ImageNet→IAM→Papyri	0.19	0.41
InceptionV3 [51]	ImageNet→Papyri	0.25	0.45
	ImageNet→IAM→Papyri	0.32	0.57
ResNet-50 [52]	ImageNet→Papyri	0.35	0.58
	ImageNet→IAM→Papyri	**0.38**	**0.61**
ResNet-34 [52]	ImageNet→Papyri	0.33	0.60
	ImageNet→IAM→Papyri	**0.40**	**0.645**

experimental protocol, we report an identification rate of 64.5% representing an enhancement of 10%. It also signifies the importance of choice of handwriting unit that is employed to extract features. Extracting patches around keypoints is likely to preserve important writer-specific patterns which might be lost in a random dense sampling using windows (Table 2).

Table 2. Comparison with the state of the art

Study	Technique	Identification Rate
Mohammed et al. [18]	NLNBNN with FAST	30%
Nasir et al. [17]	Dense sampling+two-step fine-tuning	54%
Proposed	FAST keypoints+two-step fine-tuning	**64%**

5   Conclusion

This paper addressed the problem of identification of scribes from historical manuscripts and more specifically Greek handwriting on papyrus as a case study. Handwriting is segmented from the noisy documents and keypoints in the writing are identified using FAST keypoint detector. Using the keypoints as centers, small patches around these points are extracted from handwriting and are employed as units for feature extraction. Patches are mapped to features using a two-step fine-tuning of well-known ConvNet architectures. Image level identification rates of up to 64% are reported which, considering the challenging set of images, are indeed promising.

In our further study on this problem, we aim to handle images where the writer information is not available in the ground truth. In such cases, a similarity measure can be designed that allows assessing how 'similar' two writing styles are hence implicitly providing information on scribe and date of origin etc. Pre-processing in such documents is a critical step and can be further investigated to allow more effective feature learning and eventually leading to enhanced performance.

Acknowledgments. Authors would like to thank Dr. Isabelle Marthot-Santaniello from University of Basel, Switzerland for making the dataset available.

References

1. Hamid, A., Bibi, M., Siddiqi, I., Moetesum, M.: Historical manuscript dating using textural measures. In: 2018 International Conference on Frontiers of Information Technology (FIT), pp. 235–240. IEEE (2018)
2. Baird, H.S., Govindaraju, V., Lopresti, D.P.: Document analysis systems for digital libraries: challenges and opportunities. In: Marinai, S., Dengel, A.R. (eds.) DAS 2004. LNCS, vol. 3163, pp. 1–16. Springer, Heidelberg (2004). https://doi.org/10.1007/978-3-540-28640-0_1
3. Le Bourgeois, F., Trinh, E., Allier, B., Eglin, V., Emptoz, H.: Document images analysis solutions for digital libraries. In: First International Workshop on Document Image Analysis for Libraries, 2004, Proceedings, pp. 2–24. IEEE (2004)
4. Sankar, K.P., Ambati, V., Pratha, L., Jawahar, C.V.: Digitizing a Million books: challenges for document analysis. In: Bunke, H., Spitz, A.L. (eds.) DAS 2006. LNCS, vol. 3872, pp. 425–436. Springer, Heidelberg (2006). https://doi.org/10.1007/11669487_38
5. Klemme, A.: International dunhuang project: the silk road online. Reference Reviews (2014)
6. Jouili, S., Coustaty, M., Tabbone, S., Ogier, J.-M.: NAVIDOMASS: structural-based approaches towards handling historical documents. In: 2010 20th International Conference on Pattern Recognition, pp. 946–949 (2010)
7. Hamid, A., Bibi, M., Moetesum, M., Siddiqi, I.: Deep learning based approach for historical manuscript dating. In: 2019 International Conference on Document Analysis and Recognition (ICDAR), pp. 967–972 (2019)
8. Aiolli, F., Ciula, A.: A case study on the system for paleographic inspections (SPI): challenges and new developments. Comput. Intell. Bioeng. **196**, 53–66 (2009)

9. He, S., Samara, P., Burgers, J., Schomaker, L.: Image-based historical manuscript dating using contour and stroke fragments. Pattern Recogn. **58**, 159–171 (2016)
10. Srihari, S.N., Cha, S.-H., Arora, H., Lee, S.: Individuality of handwriting. J. Forensic Sci. **47**(4), 1–17 (2002)
11. Said, H.E., Tan, T.N., Baker, K.D.: Personal identification based on handwriting. Pattern Recogn. **33**(1), 149–160 (2000)
12. He, Z., You, X., Tang, Y.Y.: Writer identification using global wavelet-based features. Neurocomputing **71**(10–12), 1832–1841 (2008)
13. He, S., Schomaker, L.: Deep adaptive learning for writer identification based on single handwritten word images. Pattern Recogn. **88**, 64–74 (2019)
14. Bulacu, M., Schomaker, L.: Text-independent writer identification and verification using textural and allographic features. IEEE Trans. Pattern Anal. Mach. Intell. **29**(4), 701–717 (2007)
15. Siddiqi, I., Vincent, N.: Text independent writer recognition using redundant writing patterns with contour-based orientation and curvature features. Pattern Recogn. **43**(11), 3853–3865 (2010)
16. Xing, L., Qiao, Y.: DeepWriter: a multi-stream deep CNN for text-independent writer identification. In: 2016 15th International Conference on Frontiers in Handwriting Recognition (ICFHR), pp. 584–589. IEEE (2016)
17. Nasir, S., Siddiqi, I.: Learning features for writer identification from handwriting on Papyri. In: Djeddi, C., Kessentini, Y., Siddiqi, I., Jmaiel, M. (eds.) MedPRAI 2020. CCIS, vol. 1322, pp. 229–241. Springer, Cham (2021). https://doi.org/10. 1007/978-3-030-71804-6_17
18. Mohammed, H. Marthot-Santaniello, I., Märgner, V.: GRK-Papyri: a dataset of greek handwriting on papyri for the task of writer identification. In: 2019 International Conference on Document Analysis and Recognition (ICDAR), pp. 726–731 (2019)
19. Rehman, A., Naz, S., Razzak, M.I., Hameed, I.A.: Automatic visual features for writer identification: a deep learning approach. IEEE Access **7**, 17149–17157 (2019)
20. Xing, L., Qiao, Y.: DeepWriter: a multi-stream deep CNN for text-independent writer identification. In: 2016 15th International Conference on Frontiers in Handwriting Recognition (ICFHR), pp. 584–589 (2016)
21. Christlein, V., Gropp, M., Fiel, S., Maier, A.: Unsupervised feature learning for writer identification and writer retrieval. In: 2017 14th IAPR International Conference on Document Analysis and Recognition (ICDAR), vol. 01, pp. 991–997 (2017)
22. Keglevic, M., Fiel, S., Sablatnig, R.: Learning features for writer retrieval and identification using triplet CNNs. In: 2018 16th International Conference on Frontiers in Handwriting Recognition (ICFHR), pp. 211–216 (2018)
23. Awaida, S.M., Mahmoud, S.A.: State of the art in off-line writer identification of handwritten text and survey of writer identification of Arabic text. Educ. Res. Rev. **7**(20), 445–463 (2012)
24. Tan, G.J., Sulong, G., Rahim, M.S.M.: Writer identification: a comparative study across three world major languages. Forensic Sci. Int. **279**, 41–52 (2017)
25. He, S., Schomaker, L.: FragNet: writer identification using deep fragment networks. IEEE Trans. Inf. Forensics Secur. **15**, 3013–3022 (2020)
26. Kumar, B., Kumar, P., Sharma, A.: RWIL: robust writer identification for Indic language. In: 2018 Second International Conference on Intelligent Computing and Control Systems (ICICCS), pp. 695–700 (2018)

27. Tang, Y., Wu, X.: Text-independent writer identification via CNN features and joint Bayesian. In: 2016 15th International Conference on Frontiers in Handwriting Recognition (ICFHR), pp. 566–571, October 2016

28. Nasuno, R., Arai, S.: Writer identification for offline Japanese handwritten character using convolutional neural network. In: Proceedings of the 5th IIAE (Institute of Industrial Applications Engineers) International Conference on Intelligent Systems and Image Processing, pp. 94–97 (2017)

29. Chen, S., Wang, Y., Lin, C.-T., Ding, W., Cao, Z.: Semi-supervised feature learning for improving writer identification. Inf. Sci. **482**, 156–170 (2019)

30. Fiel, S., Sablatnig, R.: Writer identification and retrieval using a convolutional neural network. In: Azzopardi, G., Petkov, N. (eds.) CAIP 2015. LNCS, vol. 9257, pp. 26–37. Springer, Cham (2015). https://doi.org/10.1007/978-3-319-23117-4_3

31. Nguyen, H.T., Nguyen, C.T., Ino, T., Indurkhya, B., Nakagawa, M.: Text-independent writer identification using convolutional neural network. Pattern Recogn. Lett. **121**, 104–112 (2019)

32. Christlein, V., Bernecker, D., Maier, A., Angelopoulou, E.: Offline writer identification using convolutional neural network activation features. In: Gall, J., Gehler, P., Leibe, B. (eds.) GCPR 2015. LNCS, vol. 9358, pp. 540–552. Springer, Cham (2015). https://doi.org/10.1007/978-3-319-24947-6_45

33. Javidi, M., Jampour, M.: A deep learning framework for text-independent writer identification. Eng. Appl. Artif. Intell. **95**, 103912 (2020)

34. Schomaker, L., Franke, K., Bulacu, M.: Using codebooks of fragmented connected-component contours in forensic and historic writer identification. Pattern Recogn. Lett. **28**(6), 719–727 (2007)

35. Lai, S., Zhu, Y., Jin, L.: Encoding pathlet and SIFT features with bagged VLAD for historical writer identification. IEEE Trans. Inf. Forensics Secur. **15**, 3553–3566 (2020)

36. Abdeljalil, G., Djeddi, C., Siddiqi, I., Al-Maadeed, S.: Writer identification on historical documents using oriented basic image features. In: 2018 16th International Conference on Frontiers in Handwriting Recognition (ICFHR), pp. 369–373 (2018)

37. Chammas, M., Makhoul, A., Demerjian, J.: Writer identification for historical handwritten documents using a single feature extraction method. In: 19th International Conference on Machine Learning and Applications (ICMLA 2020) (2020)

38. Cilia, N.D., et al.: A two-step system based on deep transfer learning for writer identification in medieval books. In: Vento, M., Percannella, G. (eds.) CAIP 2019. LNCS, vol. 11679, pp. 305–316. Springer, Cham (2019). https://doi.org/10.1007/978-3-030-29891-3_27

39. Cilia, N., De Stefano, C., Fontanella, F., Marrocco, C., Molinara, M., Di Freca, A.S.: An end-to-end deep learning system for medieval writer identification. Pattern Recogn. Lett. **129**, 137–143 (2020)

40. Mohammed, H., Märgner, V., Stiehl, H.S.: Writer identification for historical manuscripts: analysis and optimisation of a classifier as an easy-to-use tool for scholars from the humanities. In: 2018 16th International Conference on Frontiers in Handwriting Recognition (ICFHR), pp. 534–539 (2018)

41. Studer, L., et al.: A comprehensive study of ImageNet pre-training for historical document image analysis. arXiv preprint arXiv:1905.09113 (2019)

42. Sauvola, J., Pietikäinen, M.: Adaptive document image binarization. Pattern Recogn. **33**(2), 225–236 (2000)

43. He, S., Schomaker, L.: DeepOtsu: document enhancement and binarization using iterative deep learning. Pattern Recogn. **91**, 379–390 (2019)

44. Fiel, S., Hollaus, F., Gau, M., Sablatnig, R.: Writer identification on historical Glagolitic documents. In: Document Recognition and Retrieval XXI, vol. 9021, p. 902102. International Society for Optics and Photonics (2014)
45. Bennour, A., Djeddi, C., Gattal, A., Siddiqi, I., Mekhaznia, T.: Handwriting based writer recognition using implicit shape codebook. Forensic Sci. Int. **301**, 91–100 (2019)
46. Rosten, E., Drummond, T.: Machine learning for high-speed corner detection. In: Leonardis, A., Bischof, H., Pinz, A. (eds.) ECCV 2006. LNCS, vol. 3951, pp. 430–443. Springer, Heidelberg (2006). https://doi.org/10.1007/11744023_34
47. Deng, J., Dong, W., Socher, R., Li, L.-J., Li, K., Fei-Fei, L.: ImageNet: a large-scale hierarchical image database. In: 2009 IEEE Conference on Computer Vision and Pattern Recognition, pp. 248–255. IEEE (2009)
48. Marti, U.-V., Bunke, H.: The IAM-database: an English sentence database for offline handwriting recognition. Int. J. Doc. Anal. Recogn. **5**(1), 39–46 (2002)
49. Gazda, M., Hireš, M., Drotár, P.: Multiple-fine-tuned convolutional neural networks for Parkinson's disease diagnosis from offline handwriting. IEEE Trans. Syst. Man Cybern. Syst. (2021)
50. Simonyan, K., Zisserman, A.: Very deep convolutional networks for large-scale image recognition. arXiv preprint arXiv:1409.1556 (2014)
51. Szegedy, C., Vanhoucke, V., Ioffe, S., Shlens, J., Wojna, Z.: Rethinking the inception architecture for computer vision. In: Proceedings of the IEEE Conference on Computer Vision and Pattern Recognition, pp. 2818–2826 (2016)
52. Targ, S., Almeida, D., Lyman, K.: Resnet in resnet: Generalizing residual architectures. arXiv preprint arXiv:1603.08029 (2016)

Robust Hashing for Character Authentication and Retrieval Using Deep Features and Iterative Quantization

Musab Al-Ghadi$^{(\boxtimes)}$ ⓘ, Théo Azzouza, Petra Gomez-Krämer ⓘ,
Jean-Christophe Burie ⓘ, and Mickaël Coustaty ⓘ

L3i, La Rochelle University, Avenue Michel Crépeau,
17042 La Rochelle Cedex1, France
{musab.alghadi,petra.gomez,jean-christophe.burie,
mickael.coustaty}@univ-lr.fr

Abstract. This paper proposes a hashing approach for character authentication and retrieval based on the combination of a convolutional neural network (CNN) and the iterative quantization (ITQ) algorithm. This hashing approach is made up of two steps: feature extraction and hash construction. The feature extraction step involves the reduction of high-dimensional data into low-dimensional discriminative features by applying a CNN model. While, the hash construction step quantizes continuous real valued features into discrete binary codes by applying ITQ. These two steps are combined together in this work to achieve two objectives: (i) a hash should have a good anti-collision (discriminative) capability for distinct characters. (ii) a hash should also be quite robust to the common image content-preserving operations. Experiments were conducted in order to analyze and identify the most proper parameters to achieve higher authentication and retrieval performances. The experimental results are performed on two public character datasets including MNIST and Font-Char74K. The results show that the proposed approach builds hashes quite discriminative for distinct characters, and is also quite robust to the common image content-preserving operations.

Keywords: Hashing · CNN · Iterative quantization · Character authentication and retrieval

1 Introduction

Nowadays, a large part of the documents that we are using in our daily life exist in hybrid forms (i.e. native digital, digitized, printed/scanned). And with the widespread use of advanced editing tools, document content can be easily forged and tampered. This challenges lead to the emerging need for developing efficient document authentication and retrieval systems. Image hashing is widely applied in image authentication and image retrieval [1,2]. Generally, hashing scheme maps the high-dimensional data from its original space (such as images,

ⓒ Springer Nature Switzerland AG 2021
E. H. Barney Smith and U. Pal (Eds.): ICDAR 2021 Workshops, LNCS 12916, pp. 466–481, 2021.
https://doi.org/10.1007/978-3-030-86198-8_33

videos, documents) into low-dimensional compact codes [3]. And the hash for an image is constructed based on two basic steps: image feature extraction and hash construction. Hashing approaches can be classified into two main groups: cryptographic hashing and perceptual hashing. Cryptographic hashing is very sensitive to any change in the input data, it maps the input data into a hash code without attention to the semantic features (the most relevant features to the human perception) of the input data. Whereas, the recent perceptual hashing (learning based hashing) has less sensitivity to any alteration in the perceptual (visual) features of the input data. Cryptographic hashing achieves randomization (the hash is unpredictable and secure against malicious attacks) and fragility (discrimination; the hash for any image perceptually distinct to another is different), in addition to these two properties the perceptual hashing achieves robustness (the hashes for two perceptually similar images are the same) and compactness (the size of the hashing value should be much smaller than that of the original space). These desired properties of perceptual hashing make it more preferable than cryptographic hashing for achieving content-based authentication and retrieval of hybrid documents. Indeed, perceptual hashing can be classified as traditional hashing or learning based hashing. Traditional hashing approaches which are based on hand-crafted features to build the hash have not advanced much in terms of authentication and retrieval performances in recent years due to their limitations to fully capture the semantics of images [1,4]. Recently, learning based hashing was proposed instead of traditional hashing to overcome their drawbacks [1]. Learning based hashing approaches can be classified afterwards into non-deep learning and deep learning hashing. The difference between them is the nature of features that are used to build a hash. The non-deep learning approach extracts the features from the raw or transformed data, while deep learning hashing employs deep features that are extracted via CNN to build a hash. Both approaches aim to build a hash that have discriminative capability for visually distinct images [1], and fulfill four properties: robustness, compactness, randomization, and fragility.

As the character is the basic unit for the text and the text is the most important content of the documents, most content-based hashing methods were applied on characters. A skeleton-based hashing approach for document authentication is proposed in [5]. This approach used the properties of character skeleton including the conjunctions of skeleton lines and their directions to build hash vectors and then to match between the genuine character and the character in control. The method is robust to low-level geometric transformations and noises, but not to the median filtering. In [6], a text hashing approach is proposed based on random sampling of the text into components (characters or words). Each component, obtained by a segmentation, is sampled into rectangles of random sizes and positions. Afterwards, each rectangle is represented by the mean of its pixels values. The vectors of means are then quantified and randomized to produce the final hash of each component. Recently, a character classification based hashing approach is proposed in [7]. It used the principal component analysis (PCA) in the case of a double print-and-scan process to build the character hash. The hash generation in this work is too sensitive even with light content-preserving operations.

The motivation of this paper is to provide a new approach for authenticating and retrieving the entirety of the content of a text document through a robust hashing, in order to fight against fraud and falsification. Concretely, this objective is based on two functional themes: document authentication and document retrieval. In this paper, we work on the basic unit of the text, which is the character, and we propose a robust hashing approach for character authentication and retrieval by taking the advantages of a CNN into account. The CNN is used here to extract highly discriminative features from the character and then to build a hash which has a good anti-collision (discriminative) capability for distinct characters and is also quite robust to the common image content-preserving operations. Two influences on hash anti-collision and robustness are studied: (i) the influence of the different typefaces and fonts, where the characters have been printed/written in both digital and handwritten forms (i.e. same character may have different shapes if it is written in two fonts or by two hands), (ii) the influence of the common character content-preserving operations.

Along the text of this paper, the distinct characters are defined as the characters which are distinct visually like: ('A',' a'), ('K', 'M'), ('n', 'N'); except the same characters which are typed using different fonts, in this case we consider them as similar characters like: ('K', '$\mathscr{K}$'), ('m', 'm').

The contributions of this paper can be summarized as follows:

- To the best of our knowledge, the proposed approach is the first attempt to design a robust hashing approach for character authentication and retrieval w.r.t. different fonts and handwriting.
- The proposed approach uses a CNN model and the ITQ algorithm to construct a hash in a supervised way.
- The proposed approach fulfills the four desired properties of a learning based hashing scheme (robustness, compactness, randomization, and fragility).
- The results show that the proposed approach builds hashes quite discriminative for distinct characters, and is also quite robust to the common image content-preserving operations.
- Finally, the experimental results on two public datasets (MNIST and Font-Char74K) achieve state-of-the-art performance in the character authentication and retrieval applications.

The rest of this paper is organized as follows: the literature review is presented in Sect. 2. Section 3 presents the proposed approach. Section 4 is dedicated to the experimental results and the comparisons with previous studies. The paper ends with a conclusion in Sect. 5.

2   Literature Review

Basically, any learning based hashing approach includes three elements: the hash function, the similarity measures, and the loss function for optimization objectives [8].

A hash function maps an input item x into a binary code y (i.e. $y = h(x) \in \{0,1\}$) aiming to retrieve the true nearest neighbor search result for a query q. In learning based hashing approaches the input item x of the hash function is a discriminative (semantic) feature that is extracted from a CNN.

The similarity in the hash code space d_{ij}^h between two hash codes (y_i and y_j) is expressed by the Hamming distance or Euclidean distance. The Hamming distance between two hash codes, each in length L, is formulated as $d_{ij}^h = \Sigma_{l=1}^L \delta[y_{il} \neq y_{jl}]$, then the similarity is $s_{ij}^h = L - d_{ij}^h$. The Euclidean distance is formulated as $d_{ij}^h = ||y_i - y_j||_2$.

A loss function is defined to preserve the similarity between the input space and the hash code space. This is achieved by minimizing the gap between the nearest hash codes in the hash code space and the true search result in the input space. The loss function can be applied in three forms: pairwise similarity, multiwise similarity, and quantization-based similarity [8].

2.1   Non-deep Learning Based Hashing

Most of non-deep learning approaches learn the extracted features of the image to build a hash that have discriminative capability for visually distinct images, and robust to different content-preserving operations.

Locality-Sensitive Hashing (LSH) [9] is a data-independent hashing; no learning is involved and it quantizes different data dimensions into a same bit length regardless to its importance. Spectral Hashing (SH) [10] assigns more bits to more relevant directions (which has more information). Anchor Graph Hashing (AGH) [11] and Spherical Hashing (SpH) [12] represent the input data as a graph, and the data on the same dimension are quantized to similar hash codes. Principal Component Analysis Hashing (PCAH) [13] performs PCA on feature vectors X, then uses the top m eigenvectors of the matrix XX^T as columns of the projections matrix $W \in \mathbb{R}^{d \times m}$ to generate a hash code of m bits. The PCAH is formulated as: $h(x) = sgn(W^T x)$. PCAH uses the same number of bits for different projected dimensions, which leads to low similarity preservation. This challenge is solved in Isotropic Hashing (IsoH) [14] by balancing the variance of the input data and making variances equal along different data dimensions. Iterative Quantization (ITQ) [15] seeks to maximize the bit variance when transforming given vectors into a binary representation by minimizing the quantization error through finding the optimal rotation matrix. K-means Hashing (KMH) [16] performs k-means clustering for the input space and then quantizes the centroids based on the Euclidean distance. An input item is associated to the approximation nearest neighbor (centroid) via the Hamming distance.

2.2   Deep Learning Based Hashing

The learning based hashing allows to represent the data in a binary space from the extracted deep features via CNN. The Deep Hashing (DH) and Supervised Deep Hashing (SDH) [17] approaches are based on a CNN network to extract deep features that are highly discriminative and often get better performance search. Indeed, these approaches learn the binary codes by maintaining three constraints along the deep network layers. These constraints aim to minimize the quantization loss, balance the bit distribution in binary codes and allow bit independency. The Unsupervised Triplet Hashing (UTH) [18] feds three forms of images (the anchor images, the anchor rotated images and the random images) into a triplet-CNN network to maximize the similarity preservation by extracting more discriminative image features. Furthermore, UTH incorporated a latent layer in order to optimize the discriminative features for fast image retrieval. Similarity-Adaptive Deep Hashing (SADH) [19] is an unsupervised hashing approach that manipulates three modules in order to achieve high similarity-preserving hash codes. The first module consists of a training deep hash model that helps to generate discriminative features for the processed images. The second module updates the similarity graph matrix, which is then input to the third module to optimize the hash codes. Deep Quantization (Deep-Quan) [20] is a deep unsupervised hashing approach that manipulated deep features via product quantization to minimize the quantization error. Hash Generative Adversarial Network (HashGAN) [21] is an unsupervised deep hashing approach, composed of three networks: a generator, a discriminator and an encoder. The discriminator and encoder network are incorporated together to train an unlabeled dataset efficiently. A novel loss function is proposed in HashGAN to balance the bit distribution and to allow bit independency. Recently, Sparse Graph based Self-supervised Hashing (SGSH) [22] was proposed to solve the performance degradation problem in traditional graph based hashing approaches such as AGH [11]. The SGSH approach adopts a self-supervised reconstruction to enhance the retrieval performance, and allows more scalability by conserving the sparse neighborhood relationship rather than the fully connected neighborhood relationship.

In general, most of the mentioned approaches in this section have a limitation regarding to establish a good trade-off in terms of building a hash robust to the common content-preserving operations and have sensitivity to non-authorized image content operations.

3   Proposed Model

The proposed hashing approach for efficient character authentication and retrieval is made up of two steps: feature extraction and hash construction. The feature extraction step involves the reduction of high-dimensional data into low-dimensional discriminative features via a CNN model. While, the hash extraction step quantizes continuous real-valued features into discrete binary codes via the ITQ algorithm. The general structure of the proposed approach is shown in Fig. 1.

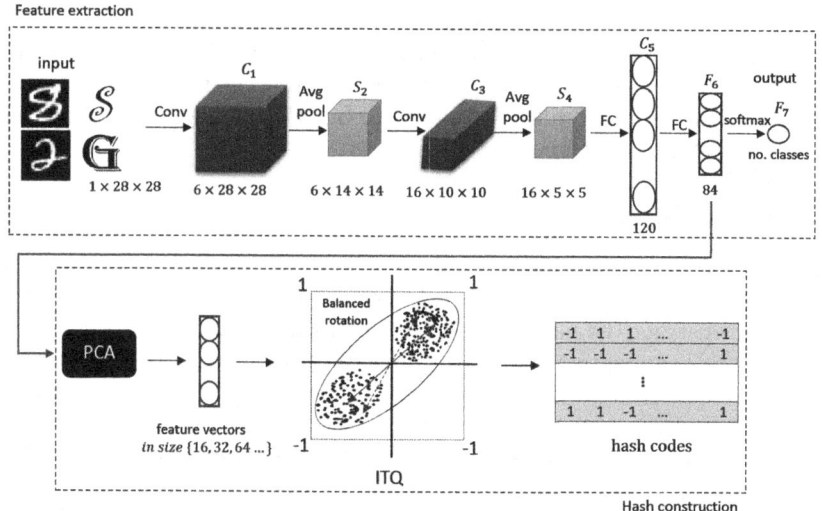

Fig. 1. Structure of the proposed hashing approach.

3.1 Feature Extraction

As CNNs are excellent feature extractors and are used efficiently in down-scaling tasks to extract characteristics relevant enough to represent the images and to distinguish them correctly. The proposed approach starts by applying a CNN architecture to extract the discriminative features from the processed character in order to maximize the similarity preservation between the input space and the hash code space. The LeNet-5 architecture, which was proposed by Yann Lecun et al. [23], is used in the design of the proposed approach. The first block in Fig. 1 illustrates the LeNet-5 architecture in the feature extraction step. The LeNet-5 architecture consists of two convolutional (Conv) layers and two average pooling (Avg pool) layers, followed by a flattening convolutional layer, two fully-connected (FC) layers and ends with a softmax classifier. The configuration of the LeNet-5 architecture is illustrated in Table 1.

Table 1. Configuration of LeNet-5 network.

Layer	Feature maps	Filter	Stride	Padding/Pooling	Activation function
C_1	$6 \times 28 \times 28$	5×5	1×1	Pad 2	Hyperbolic tangent (tanh)
S_2	$6 \times 14 \times 14$	2×2	2×2	Pool 2×2	Sigmoid
C_3	$16 \times 10 \times 10$	5×5	1×1	Pad 0	Hyperbolic tangent (tanh)
S_4	$16 \times 5 \times 5$	2×2	2×2	Pool 2×2	Sigmoid
C_5	$120 \times 1 \times 1$	5×5	1×1	Pad 0	Hyperbolic tangent (tanh)
F_6	84				Hyperbolic tangent (tanh)
F_7	10 (output)				Softmax

The proposed approach selects the obtained 84 values from the LeNet-5 model as discriminative features for each character to construct the hash. It is worth noting that the LeNet-5 architecture was chosen to prove the feasibility and the relevance of the proposed hashing approach, but this step could be replaced by any other CNN architecture from the literature.

3.2   Hash Construction

This step works to map the extracted features of the character into hash codes with least quantization error via the ITQ algorithm. The ITQ algorithm [15] is applied to construct the hash codes based on a random orthogonal matrix that minimizes the quantization error. ITQ starts by applying the PCA algorithm on the centered data for reducing the d dimensions of data. The top m eigenvectors of the data covariance matrix $X^T X$ are kept to obtain the projection matrix W. The PCA algorithm is used in ITQ to obtain another representation of the features in a space dimension with maximum variance. The features become independent of each other and more discriminative. Hence, the projected data is formulated as $V = WX$. Afterwards, ITQ alternates two stages in order to find a random orthogonal matrix R that minimizes the quantization error and to learn a binary code matrix $B \in \{-1, 1\}$. The first stage involves setting R and updating B. Here, R is fixed and is initialized randomly. Then, B is updated by assigning each data dimension to the nearest binary hypercube $\in \{-1, 1\}$. Updating B in this stage is formulated as $B = sgn(VR)$. The second stage involves setting B and updating R, this is achieved by computing the Singular Value Decomposition (SVD) of $B^T V$ as $SVD(B^T V) = U\Sigma A^T$ and set $R = U^T A$. Generally, the iterative quantization is formulated as follows to find the optimal solution for the function F:

$$F(B, R) = \|B - VR\|_F^2$$

3.3   Character Authentication and Retrieval

In order to verify or retrieve the character based on a hash, the Hamming distance (hd) between the hash codes of the query character y_c from the test set and the top return character y_t from the gallery set is calculated. The distance hd between y_c and y_t, each in length L, is formulated according to Eq. 1:

$$hd = hd(y_c, y_t) = \sum_{l=1}^{L} \delta[y_{cl} \neq y_{tl}]. \tag{1}$$

If hd is less than the pre-defined threshold λ, then the two characters are considered as similar (identical/not-tampered); otherwise, the two characters are considered as distinct (falsified/tampered).

4   Experimental Results

This section details the experimental results of the proposed approach.

4.1   Character Datasets and Performance Metrics

To evaluate the performance of the proposed hashing approach, the MNIST [24] and Font-Char74K [25] datasets were selected. The MNIST dataset consists of 70,000; 28×28 gray-scale images in 10 classes of handwritten digits from '0' to '9'. Each image is represented by a 784-dimensional vector with real-valued pixels as elements. Specifically, 1,000 images (100 images per class) are sampled randomly to form the test set (i.e. the queries) and the remaining 69,000 images are taken as the training set and also as the gallery set. The Font-Char74K dataset consists of 62,992; 128×128 gray-scale images in 62 classes of digits from '0' to '9' and English letters from 'a' to 'z' and 'A' to 'Z'. All characters in the Font-Char74K dataset were synthesized in 254 different fonts in 4 styles (normal, bold, italic, and bold+italic). From the Font-Char74K dataset specifically, 6,200 images (100 images per class) are sampled randomly to form the test set (i.e. the queries) and the remaining 56,792 images are taken as the training set and also as the gallery set. These proportions for training and testing sets have been selected to make our performance comparison consistent with some recent related work (DeepQuan [20]; HashGAN [21]; SGSH [22]). Some example images from MNIST and Font-Char74K datasets are displayed in Fig. 2.

Fig. 2. Examples of test image.

The implementation of the proposed approach has been carried out in Python using the PyTorch library. The implementation starts by training the LeNet-5 model using the training set, the batch size = 128, the learning rate = 0.001 and the cross-entropy as loss function. The 84 values, that are obtained from the fully connected layer (F_6) in the LeNet-5 model, are used as a feature vector for each digit/character in the MNIST/Font-Char74K datasets. PCA reduces the feature vectors length according to the preferable length of hash codes. ITQ iterates 50 times in order to find the optimal orthogonal matrix. This factor is reported in [15].

We evaluate the character authentication and retrieval performances based on four standard evaluation metrics: mean Average Precision (mAP), mAP curves w.r.t. a different number of bits as hash code length (mAP w.r.t. number-of-bits), mAP curves w.r.t. a different number of return samples (mAP@top-return-samples), and the true positive rate – false positive rate curves (TPR-FPR).

The mAP@k refers to the mean average precision @ k. This metric is adopted for the most of the related work. The AP@k ratio is computed according to Eq. 2:

$$AP@k = \frac{1}{GTP} \sum_{k}^{n} P@k \times rel@k \tag{2}$$

where GTP refers to the total number of ground truth positives, P@k refers to the precision@k and rel@k is a relevance function which equals 1 if the return image at rank k is relevant and which equals 0 otherwise.

The receiver operating characteristics (ROC) curve is used to evaluate the authentication performance as well as the retrieval output quality. The TPR and FPR of the ROC curve indicate the robustness and the discriminative capability of the character hash, respectively. The TPR and FPR ratios, as reported in [26], are calculated according to Eq. 3:

$$TPR(\lambda) = \frac{n_1(hd < \lambda)}{N_1}, \quad FPR(\lambda) = \frac{n_2(hd < \lambda)}{N_2} \tag{3}$$

where n_1 is the amount of identical pairs of characters classified as similar characters, n_2 is the amount of distinct pairs of characters classified into similar characters, N_1 and N_2 correspond to the total number of identical pairs and distinct pairs of characters, respectively. λ is the authentication threshold to consider pairs of characters as identical characters or distinct characters.

4.2   Parameter Determination

The authentication threshold λ has a direct influence on the performance of the proposed approach and it needs to be determined. To do so, the Hamming distance hd between the hash codes of each query character and the top return character from the gallery set is calculated. Figure 3 (a) and (b) show respectively the hd distributions for hashes of identical characters (visually identical) and for hashes of different characters (visually distinct) in the MNIST dataset for the ITQ-only [15] approach and the proposed approach. Figure 4 (a) and (b) show respectively the hd distributions for hashes of identical characters and for hashes of different characters of the Font-Char74K dataset for the ITQ-only [15] approach and the proposed approach.

From Fig. 3 it can be seen that the proposed approach is capable to define λ in comparison to ITQ-only [15], where almost all Hamming distances for identical characters are less than 2 while almost all Hamming distances between distinct characters are greater than or equal 1. Hence, we can define $\lambda = 1$ as an authentication threshold for the MNIST dataset.

On the other hand, from Fig. 4 it can be seen that with $\lambda = 0$ the proposed approach capable to discriminate more identical characters in comparison to

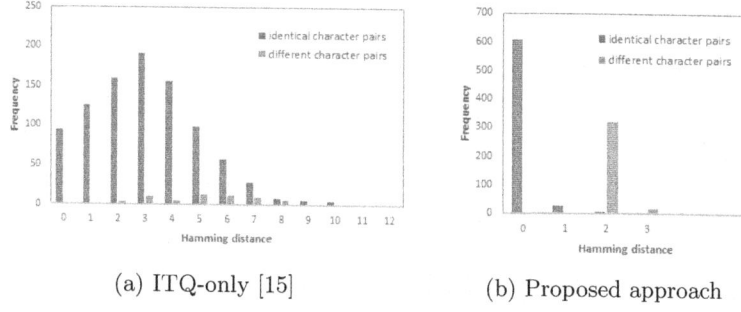

(a) ITQ-only [15]   (b) Proposed approach

Fig. 3. Distribution of the Hamming distances of MNIST dataset.

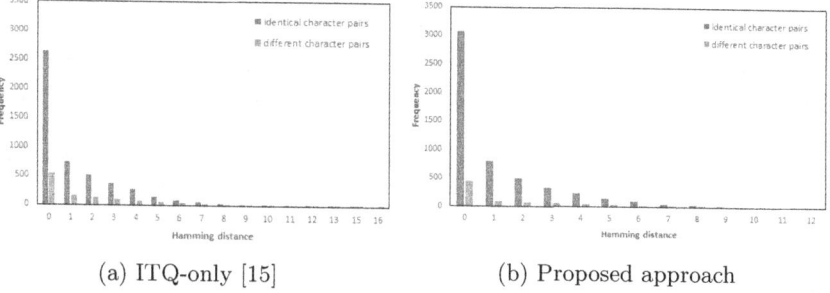

(a) ITQ-only [15]   (b) Proposed approach

Fig. 4. Distribution of the Hamming distances of Font-Char74K dataset.

ITQ-only [15]. Indeed, for the proposed approach the Hamming distance converge for identical characters and diverges for distinct characters. Hence, we can define $\lambda = 0$ as an authentication threshold for the Font-Char74K dataset.

4.3   Performance Test

Figure 5 shows the authentication and retrieval performances of the proposed approach and ITQ-only [15] through the mAP curves w.r.t. the number-of-bits and mAP curves w.r.t. the top-return-samples on the MNIST and Font-Char74K datasets. We can find that the proposed approach outperforms the ITQ-only [15] approach by large margins on the MNIST and Font-Char74K datasets w.r.t the two performance metrics. From Fig. 5 (a) it can be seen that the proposed approach brings an increase of at least 14.7% and 21.1% in mAP for the different numbers of bits on the MNIST and Font-Char74K datasets comparing with ITQ-only [15], respectively. Specifically, the mAP w.r.t. number-of-bits is increased as the length of hash code increases. The mAP@top-return-characters with 64 bits in Fig. 5 (b) show that the proposed approach brings an enhancement in mAP results on MNIST and Font-Char74K datasets comparing with ITQ-only [15].

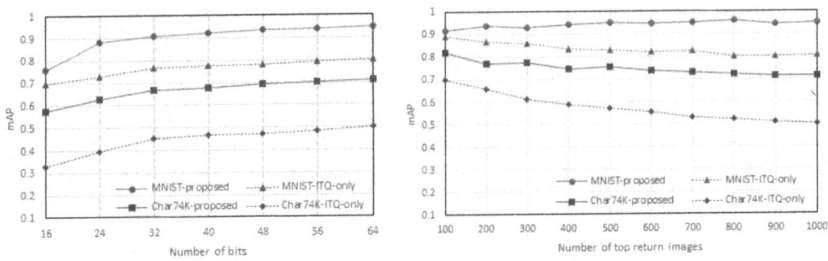

(a) mAP w.r.t. the number-of-bits   (b) mAP@top-return-characters at 64bits

Fig. 5. mAP w.r.t. the number of bits and the number of top return characters of proposed approach and ITQ-only [15] on MNIST and Font-Char74K datasets .

Moreover, Fig. 6 shows the authentication and retrieval performances in *TPR-FPR* curves @64bits w.r.t. the top-return-character and with a varying authentication threshold λ. The presented *TPR* and *FPR* curves demonstrate the robustness and discriminative capability of the hashes of the proposed approach over ITQ-only [15].

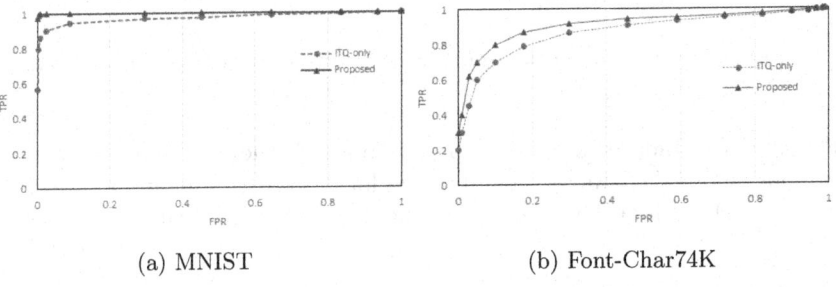

(a) MNIST   (b) Font-Char74K

Fig. 6. FPR–TRP curves of the proposed approach and ITQ-only [15] on MNIST and Font-Char74K dataset @64bits w.r.t. the top return character .

To evaluate the robustness of the proposed hashing approach to the common content-preserving operations. Table 2 presents mAP@1000 results against eleven kinds of corruptions that are applied to the characters of MNIST and Font-Char74K. In the same time from Table 2 we can hold a comparison between the achieved mAP@1000 of the proposed hashing approach and the ITQ-only [15] approach. The obtained mAP@1000 results are calculated with 64-bit hash codes (64-bit hash codes are selected here because they present the highest performance in Fig. 5). For this target, half of the query characters (50 characters out of 100 characters per each class for the MNIST and Font-Char74K datasets) are exposed to one of eleven kinds of corruptions (including JPEG compression, median filtering, salt&pepper noise, and finally Gaussian noise) with different factors in each experiment. Accordingly, the total size of the test images are shuffled randomly to allow random selections of the query character in each test. The mAP@1000 results in Table 2 show that the proposed approach presents more stability by about 16% and 21% for the MNIST and Font-Char74K datasets comparing with ITQ-only [15], respectively. And, these results show that the robustness of a hash of the proposed approach remains appealing even with corrupted samples.

To visualize the authentication and retrieval process in table 2; the top 10 retrieved characters of digit '3' are presented in Fig. 7 with the ITQ-only [15] approach and the proposed approach. The visualized results demonstrate the robustness and discriminative capability of the generated hash via the proposed approach over ITQ-only [15]. Red-boxes indicate retrieved characters, which are false positives.

Table 2. mAP@1000 of the proposed approach and ITQ-only [15] against some kinds of corruptions on MNIST and Font-Char74K.

Distortion	Factor	MNIST		Font-Char74K	
		ITQ-only [15]	Proposed	ITQ-only [15]	Proposed
		mAP@1000 (64 bits)	mAP@1000 (64 bits)	mAP@1000 (64 bits)	mAP@1000 (64 bits)
JPEG compression	90	38.41	53.47	39.07	60.57
	80	37.95	54.24	39.39	60.10
	70	38.62	52.42	39.15	59.59
Median filtering	Window(3 × 3)	39.49	53.93	38.07	60.65
	Window(5 × 5)	39.41	54.07	38.80	61.25
Salt and Pepper	Noise density = 0.01	37.08	55.63	38.70	58.64
	Noise density = 0.05	40.49	55.32	39.76	60.02
	Noise density = 0.10	38.16	54.41	38.56	59.42
Gaussian noise	$\sigma = 0.1$	37.60	52.83	38.83	61.46
	$\sigma = 0.5$	37.54	56.20	39.04	60.70
	$\sigma = 1$	38.93	54.27	38.95	61.35

4.4   Performance Comparison

A comparison study between the proposed approach and some of the state-of-the-art approaches on the MNIST dataset is presented in Table 3. To the best of our knowledge, these works only show results on the MNIST dataset as a character dataset.

Table 3 shows that the mAP results of the proposed approach outperform all mentioned approaches except the HashGAN [21] approach. The improvement in mAP results is achieved with a large margin, especially with long-bit hash codes (with 64 bits). By considering the results of mAP@1000, the 16-bit, the 32-bit and the 64-bit experiments show that the proposed approach outperforms the second best approach ITQ-only [15] by about 6%, 14%, and 14% respectively. And by considering the results of mAP@All, the 16-bit, the 32-bit and the 64-bit experiments show that the proposed approach outperforms the second best approach SGSH [22] by about 14%, 21% and 21% respectively. On the other

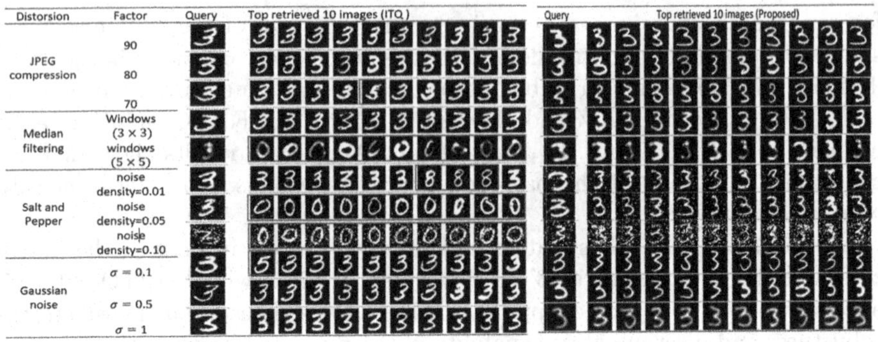

Fig. 7. Visualization of the top 10 retrieved images for ITQ-only [15] and the proposed approach with 64-bit hash codes.

Table 3. mAP@1000 and mAP@All results of the proposed approach and other state-of-the-art approaches on MNIST dataset, with 16, 32 and 64-bits hash codes. '−' indicates the value has not been reported.

Method	mAP@1000			mAP@All		
	16 bits	32 bits	64 bits	16 bits	32 bits	64 bits
LSH [9]	42.10	50.45	66.23	20.88	25.83	31.71
SH [10]	52.97	65.45	65.45	25.81	30.77	24.10
AGH [11]	−	−	−	39.92	33.39	28.64
SpH [12]	59.72	64.37	67.60	26.64	25.72	34.75
PCAH [13]	60.98	64.47	63.31	27.33	24.85	31.71
ITQ-only [15]	70.06	76.86	80.23	41.18	43.82	45.37
KMH [16]	59.12	70.32	67.62	32.12	33.29	35.78
DH [17]	−	−	−	43.14	44.97	46.74
SDH [17]	−	−	−	46.75	51.01	52.50
UTH [18]	43.15	46.58	49.88	−	−	−
SADH [19]	−	−	−	46.22	43.03	41.00
DeepQuan [20]	−	−	−	60.30	55.50	52.54
HashGAN [21]	94.31	95.48	96.39	91.13	92.70	93.93
SGSH [22]	−	−	−	62.66	65.25	69.11
Proposed	76.04	90.79	94.95	76.99	86.88	90.28

hand, Table 3 shows that the mAP results of HashGAN [21] outperform the mAP results of the proposed approach. The superiority of HashGAN [21] was explained in the literature [22,27] as due to the application of the reconstructive loss function in the encoder module. However, the HashGAN [21] outperforms the proposed approach by about 3% in 64-bit experiments for both mAP@1000 and mAP@All. Nevertheless, the results in Table 3; especially with 64-bit hash codes, ensure the robustness and discriminative capability of the generated hash for distinct characters in the proposed approach.

5   Conclusion

A robust hashing approach for character authentication and retrieval has been proposed in this paper. With respect to different fonts and handwriting, a CNN model has used to extract highly discriminative features of characters in order to build a hash. The generated hashes have good discriminative capability for distinct characters and also quite robust to the common image content-preserving operations. The authentication and retrieval performances were evaluated on two public character datasets and are compared with related approaches. Compared to the related state of art, our approach provides on the overall a better performance. Our future work will focus on exploring the proposed approach to achieve authentication and retrieval on words, sentences, or text-lines.

Acknowledgements. This work is financed by the Nouvelle-Aquitaine Region (SVP-IoT project, reference 2017-1R50108-00013407), the ANR CHIST-ERA SPIRIT project (reference ANR-16-CHR2-0004), the ANR LabCom IDEAS (reference ANR-18-LCV3-0008) and the FUI IDECYS+ project.

References

1. Ng, W., Li, J., Tian, X., Wang, H., Kwong, S., Wallace, J.: Multi-level supervised hashing with deep features for efficient image retrieval. Neurocomputing **399**, 171–182 (2020). https://doi.org/10.1016/j.neucom.2020.02.046
2. Wang, Y., Song, J., Zhoua, K., Liua, Y.: Image alignment based perceptual image hash for content authentication. Sig. Process. Image Commun. **80**, 115642 (2020). https://doi.org/10.1016/j.image.2019.115642
3. Chaidaroon, S., Park, D.H., Chang, Y., Fang, Y.: node2hash: graph aware deep semantic text hashing. Inf. Process. Manage. **57**(6), 102143 (2020). https://doi.org/10.1016/j.ipm.2019.102143
4. Du, L., Ho, A., Cong, R.: Perceptual hashing for image authentication: a survey. Sig. Process. Image Commun. **81**, 115713 (2020). https://doi.org/10.1016/j.image.2019.115713
5. Tan, L., Sun, X.: Robust text hashing for content-based document authentication. Inf. Technol. J. **10**(8), 1608–1613 (2011). https://doi.org/10.3923/itj.2011.1608.1613
6. Villán, R., Voloshynovskiy, S., Koval, O.J., Deguillaume, F., Pun, T.K.: Tamper-proofing of electronic and printed text documents via robust hashing and data-hiding. In: Proceedings of SPIE - The International Society for Optical Engineering, San Jose, CA, United States, vol. 6505, pp. 1–12 (2007). https://doi.org/10.1117/12.704097
7. Tkachenko, I., Gomez-Krämer, P.: Robustness of character recognition techniques to double print-and-scan process. In: Proceedings of the 14th IAPR International Conference on Document Analysis and Recognition (ICDAR), Kyoto, Japan, pp. 27–32 (2017). https://doi.org/10.1109/ICDAR.2017.392

8. Wang, J., Zhang, T., Song, J., Sebe, N., Shen, H.T.: A survey on learning to hash. IEEE Trans. Pattern Anal. Mach. Intell. **40**(4), 769–790 (2018). https://doi.org/ 10.1109/TPAMI.2017.2699960

9. Andoni, A., Indyk, P.: Near-optimal hashing algorithms for approximate nearest neighbor in high dimensions. In: Proceedings of 47th Annual IEEE Symposium on Foundations of Computer Science (FOCS), Berkeley, CA, pp. 459–468 (2006). https://doi.org/10.1109/FOCS.2006.49

10. Weiss, Y., Torralba, A., Fergus, R.: Spectral hashing. In: Proceedings of the 21st ACM International Conference on Neural Information Processing Systems (NIPS), Vancouver, British Columbia, Canada, pp. 1753–1760 (2008)

11. Liu, W., Wang, J., Kumar, S., Chang, S-F.: Hashing with graphs. In: Proceedings of the 28th International Conference on Machine Learning (ICML), Bellevue, WA, USA, pp. 1–8 (2011)

12. Heo, J., Lee, Y., He, J., Chang, S., Yoon, S.: Spherical hashing. In: Proceedings of IEEE Conference on Computer Vision and Pattern Recognition (CVPR), Providence, RI, USA, pp. 2957–2964 (2012). https://doi.org/10.1109/CVPR.2012. 6248024

13. Wang, J., Kumar, S., Chang, S.-F.: Semi-supervised hashing for large-scale search. IEEE Trans. Pattern Anal. Mach. Intell. **34**(12), 2393–2406 (2012). https://doi. org/10.1109/TPAMI.2012.48

14. Kong, W., Wu-Jun Li, W.J.: Isotropic hashing. In: Proceedings of the 25th ACM International Conference on Neural Information Processing Systems, 57 Morehouse Lane, Red Hook, NY, United States, pp. 11646–1654 (2012)

15. Gong, Y., Lazebnik, S., Gordo, A., Perronnin, F.: Iterative quantization: a procrustean approach to learning binary codes for large-scale image retrieval. IEEE Trans. Pattern Anal. Mach. Intell. **35**(12), 2916–2929 (2013). https://doi.org/10. 1109/TPAMI.2012.193

16. He, K., Wen, F., Sun, J.: K-means hashing: an affinity-preserving quantization method for learning binary compact codes. In: Proceedings of the IEEE Conference on Computer Vision and Pattern Recognition (CVPR), Portland, OR, USA, pp. 2938–2945 (2013). https://doi.org/10.1109/CVPR.2013.378

17. Liong, V.E., Lu, J., Wang, G., Moulin, P., Zhou, J.: Deep hashing for compact binary codes learning. In: Proceedings of IEEE Conference on Computer Vision and Pattern Recognition (CVPR), Boston, MA, pp. 2475–2483 (2015). https:// doi.org/10.1109/CVPR.2015.7298862

18. Huang, S., Xiong, Y., Zhang, Y., Wang, J.: Unsupervised triplet hashing for fast image retrieval. In: Proceedings of the on Thematic Workshops of ACM Multimedia, California, USA, pp. 84–92 (2017). https://doi.org/10.1145/3126686.3126773

19. Shen, F., Xu, Y., Liu, L., Yang, Y., Huang, Z., Shen, H.T.: Unsupervised deep hashing with similarity-adaptive and discrete optimization. IEEE Trans. Pattern Anal. Mach. Intell. **40**(12), 3034–3044 (2018). https://doi.org/10.1109/TPAMI. 2018.2789887

20. Chen, J., Cheung, W.K., Wang, A.: Learning deep unsupervised binary codes for image retrieval. In: Proceedings of the 27th International Joint Conference on Artificial Intelligence (IJCAI), Stockholm, pp. 613–619 (2018). https://doi.org/ 10.24963/ijcai.2018/85

21. Dizaji, K.G., Zheng, F., Nourabadi, N.S., Yang, Y., Deng, C., Huang, H.: Unsupervised deep generative adversarial hashing network. In: Proceedings of IEEE Conference on Computer Vision and Pattern Recognition (CVPR), Salt Lake City, UT, pp. 3664–3673 (2018). https://doi.org/10.1109/CVPR.2018.00386

22. Wang, W., Zhang, H., Zhang, Z., Liu, L., Shao, L.: Sparse graph based self-supervised hashing for scalable image retrieval. Inf. Sci. **547**, 622–640 (2021). https://doi.org/10.1016/j.ins.2020.08.092

23. Lecun, Y., Bottou, L., Bengio, Y., Haffner, P.: Gradient-based learning applied to document recognition. Proc. IEEE **86**(11), 2278–2324 (1998). https://doi.org/10.1109/5.726791

24. The MNIST database. http://yann.lecun.com/exdb/mnist/. Accessed 30 Jan 2021

25. Campos, T.D., Babu, B.R., Varma, M.: Character recognition in natural images. In: Proceedings of the 4th International Conference on Computer Vision Theory and Applications (VISIGRAPP), vol. 2, pp. 273–280 (2009). https://doi.org/10.5220/0001770102730280

26. Ouyang, J., Coatrieux, G., Shu, H.: Robust hashing for image authentication using quaternion discrete Fourier transform and log-polar transform. Digit. Sig. Process. **41**, 98–109 (2015). https://doi.org/10.1016/j.dsp.2015.03.006

27. Wang, W., Shen, Y., Zhang, H., Liu, L.: Semantic-rebased cross-modal hashing for scalable unsupervised text-visual retrieval. Inf. Process. Manage. **57**(6), 102374 (2020). https://doi.org/10.1016/j.ipm.2020.102374

Correction to: Accurate Graphic Symbol Detection in Ancient Document Digital Reproductions

Zahra Ziran, Eleonora Bernasconi, Antonella Ghignoli,
Francesco Leotta, and Massimo Mecella

Correction to:
Chapter "Accurate Graphic Symbol Detection in Ancient Document Digital Reproductions" in: E. H. Barney Smith and U. Pal (Eds.): *Document Analysis and Recognition – ICDAR 2021 Workshops*, LNCS 12916, https://doi.org/10.1007/978-3-030-86198-8_12

The updated version of this chapter can be found at
https://doi.org/10.1007/978-3-030-86198-8_12

© The Author(s) 2021
E. H. Barney Smith and U. Pal (Eds.): ICDAR 2021 Workshops, LNCS 12916, p. C1, 2021.
https://doi.org/10.1007/978-3-030-86198-8_34

Author Index

Printed in the United States
by Baker & Taylor Publisher Services